Reviews in Fluorescence

Editors
Dr Chris D. Geddes, Professor
Joseph R. Lakowicz

More information about this series at http://www.springer.com/series/6946

Chris D. Geddes
Editor

Reviews in Fluorescence 2016

Springer

Editor
Chris D. Geddes
Institute of Fluorescence
University of Maryland Baltimore County
Baltimore, MD, USA

ISSN 1573-8086
Reviews in Fluorescence
ISBN 978-3-319-48259-0 ISBN 978-3-319-48260-6 (eBook)
DOI 10.1007/978-3-319-48260-6

Library of Congress Control Number: 2017930597

Printed on acid-free paper

This Springer imprint is published by Springer Nature
The registered company is Springer International Publishing AG
The registered company address is: Gewerbestrasse 11, 6330 Cham, Switzerland

Preface

This is the ninth volume in the very popular fluorescence series, *Reviews in Fluorescence*, by Springer (http://www.springer.com/series/6946). To date, eight volumes have been both published and well received by the scientific community, the very first volume 12 years ago in 2004. Since that time, we have seen the continued growth of fluorescence techniques, as well as recognition for two fluorescence-based Nobel Prizes. In addition, the *Journal of Fluorescence*, http://link.springer.com/journal/10895, continues to be the major repository for fluorescence-based peer-reviewed publications, also a Springer journal, celebrating 25 years of publishing excellence this year.

In this 2016 volume, we are pleased again with the broad and timely fluorescence content from contributors around the world. We subsequently thank the authors for their very timely and exciting contributions again this year. We hope you all will find this volume as useful as the past volumes.

In closing, I would like to thank both Tanja Koppejan and Meran Owen at Springer for their help in compiling this volume and with the broader series.

http://theinstituteoffluorescence.com/

http://www.chrisgeddes.com/

Baltimore, MD, USA
September 12, 2016

Chris D. Geddes

Contents

Chapter 1
Fluorescence Made Easier: Fluorescence Techniques for the Novice. Episode 1: The Basics

Richard B. Thompson and Suzanne Scarlata

Keywords Fluorescence • Monochromator • Spectra • Excitation spectrum • Emission spectrum • Quantum yield • Golden Rules

First Things First This is one of a series of short articles designed to help those new to fluorescence do successful experiments with these methods. Our particular goal is to help people who may have little or no background in the field (like from the life sciences) avoid pitfalls and artifacts, and generate results they can trust, and maybe even publish. We assume you have some kind of a science background with an undergraduate and maybe a graduate degree; we expect you know some basic chemistry and physics. We will mostly avoid theory and detailed mathematical treatments as much as we can, because not everybody thinks they're as much fun as we do. There's a glossary at the end with definitions of words spelled out in *italics.* J.R. Lakowicz's text *Principles of Fluorescence Spectroscopy* and David Jameson's *Introduction to Fluorescence* both cover all this in greater depth and are recommended.

1.1 Basic Concepts, with a Minimum of Annoying Physics: The Golden Rules

Everyone is acquainted with the phenomenon of *absorbance*, where atoms or molecules actually absorb the energy of light: a clear glass of Merlot wine appears red in sunlight because molecules in the wine absorb blue and green light, leaving the red to pass through the wine to our eye. In the case of the wine, the energy of the blue and green light (*photons*) absorbed is mostly lost in the wine as heat. If we take a fluorescent dye like Rhodamine B, dissolve it in water, and shine a green

R.B. Thompson (✉) • S. Scarlata
BioChemistry and Molecular Biology, University of Maryland, 108 N. Greene Street, Baltimore, MD 21201, USA
e-mail: RThompson@som.umaryland.edu

C.D. Geddes (ed.), *Reviews in Fluorescence 2016*, Reviews in Fluorescence,
DOI 10.1007/978-3-319-48260-6_1

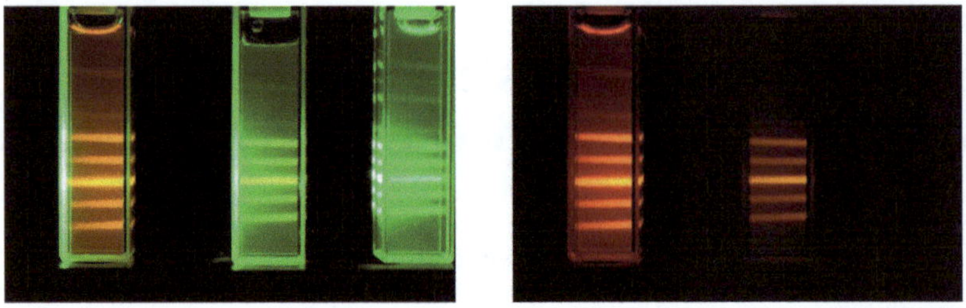

Fig. 1.1 *Left panel*: Samples with Rhodamine B (*left* and *middle samples*) and none (*right sample*) illuminated with *green* laser light and photographed with no filter. *Right panel*: Same sample and conditions, but photographed through an *orange* filter which blocks the *green* laser light

light on it from a laser, it glows orange (Fig. 1.1). The orange glow is *fluorescence*: what is happening is that the Rhodamine in the water is absorbing energy from the green photons and subsequently re-emitting it as orange light. The really key thing here is the color (*wavelength*) of the emitted fluorescence (orange, peaking about 570 nanometers) is different than that of the green light at 532 nm being absorbed (sometimes called the excitation). You can verify that the color of the exciting light is different from that of the fluorescence by putting an orange filter between the fluorescing sample and your eye: it blocks > 99 % of the green exciting light but lets the orange fluorescence pass through to your eye. This is illustrated in Fig. 1.1, which shows photographs of Rhodamine B samples excited with green light, with and without an orange filter.

The fact that the light absorbed by the fluorescent Rhodamine molecules is re-emitted as a different color is the main reason fluorescence is so useful. In the left panel of Fig. 1.1 you can easily see the orange fluorescence from Rhodamine B in the sample on the left, and just barely in the sample in the middle (which has 1/10th the Rhodamine); the sample on the right has no Rhodamine in it. When an orange filter that blocks the green light is placed between the samples and the camera (Fig. 1.1, right panel), you can now easily see the orange fluorescence in the middle sample, and the sample without Rhodamine essentially vanishes because it doesn't emit any light the camera can see. Because we can block the (strong) exciting light so effectively we can see very weak fluorescence, and thus even very low concentrations of fluorophores, much lower than we can detect by absorbance. In fact, fluorescence is so sensitive that it is now fairly common to detect the fluorescence from single molecules, which would seem to be about as much sensitivity as anybody would need.

This brings us to the first of the **Golden Rules**, for which there are very few exceptions:

1.1.1 Golden Rule 1: The Fluorescence Is Always at Longer Wavelengths than the Excitation

All this really says is that the molecule generally loses some energy between when it absorbs the excitation and reemits it as fluorescence.

To talk about the other two Golden Rules and what a *spectrum* is, we have to use the *Jablonski Diagram*, (Fig. 1.2) which depicts the changes in energy of the molecule or atom when it absorbs and emits photons. In case you slept through most of quantum mechanics, we'll keep this brief. The vertical axis is energy, so bigger distances vertically means a bigger energy change. The horizontal groups of lines represent electronic energy levels of the molecule: if the molecule absorbs a photon it moves from the *ground state* (S_0) to an *excited state* (S_1 or S_2, referring to the 1st and 2nd excited singlet states). A molecule or atom can absorb a photon if the photon's energy matches the difference in energy between S_0 and S_1 or S_2; raising the molecule to S_2 (purple arrow in Fig. 1.2) requires more energy (bigger jump) than to S_1 (blue arrow), therefore shorter wavelength photons since wavelength is inversely proportional to energy.

A molecule in S_1 can return to S_0 by emitting a photon: this is fluorescence (green arrows) and the molecule is termed a *fluorophore*. It can also lose the energy (by different routes) and return to S_0 without emitting a photon—so no emission is observed. How rapid these rate(s) of loss are compared with the rate of emission determines how efficient the fluorophore is: if the loss rate is fast compared to emission you see little fluorescence. The loss rate(s) depend on both the fluorophore and its surroundings. Sometimes the loss process is built into the molecular structure (and is thus very efficient); and sometimes another molecule interacts with the excited fluorophore to "steal" its energy, which is called quenching.

The finer horizontal lines labeled 0, 1, and 2 in the S_0 and S_1 refer to "vibronic substates", basically vibrational energy levels of the molecule within each electronic state that change the electronic state energy slightly. This group of substates means

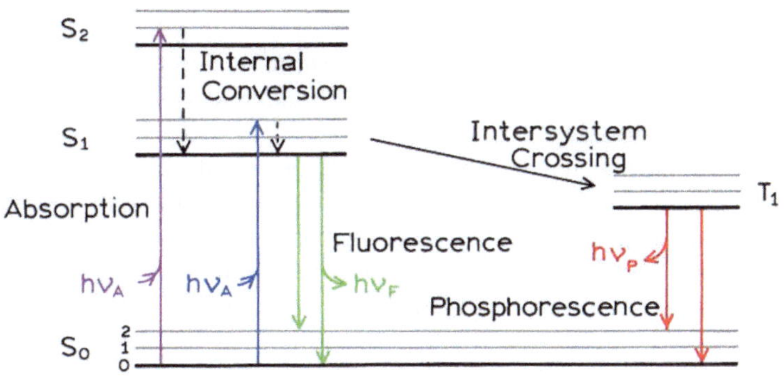

Fig. 1.2 Jablonski Diagram (from Lakowicz, *Principles of Fluorescence Spectroscopy, 3rd Ed.*; used with permission of copyright holder)

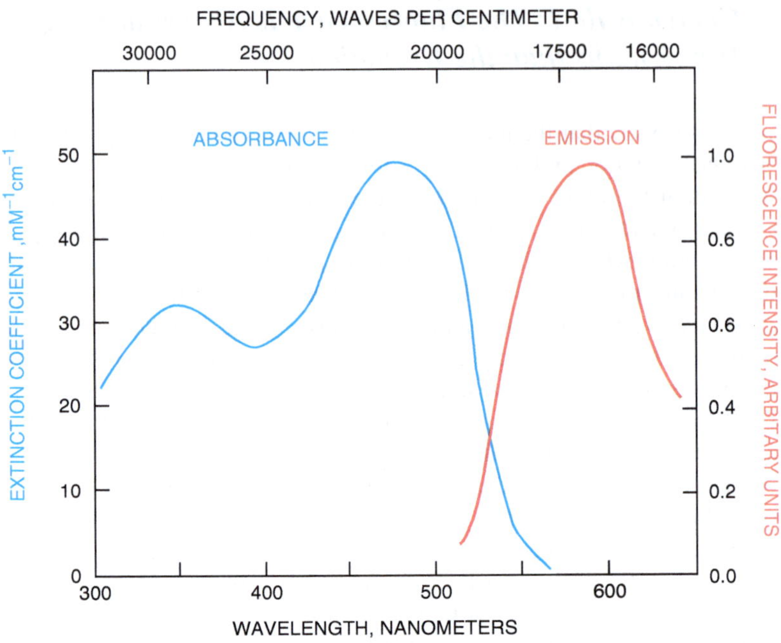

Fig. 1.3 Absorbance and emission spectra of a fluorophore. The absorbance spectrum, expressed in units of extinction coefficient, is in *blue*, whereas the emission spectrum is in *red* and is expressed in arbitrary units. The top scale expresses the frequency of the light in terms of waves per cm, often abbreviated as cm^{-1}; note the frequency scale is inverted (increases as you go to the *left*) compared with wavelength, and is non-linear

photons of a range of slightly different energies (wavelengths) can be absorbed and emitted between different electronic states (green fluorescence arrows). This range of wavelengths means the molecule will absorb and emit light over a continuous band of wavelengths; the relative efficiency with which photons of each particular wavelength are absorbed is the *absorbance spectrum* and is a property of the molecule depending on its structure. The corresponding distribution of emission over different wavelengths is the fluorescence *emission spectrum*. Such spectra are typically broad, with bands ranging over 50 nanometers width at half the peak absorbance or emission (Fig. 1.3)

You might think that if you excited to S_2, you might observe emission corresponding to both $S_2 \rightarrow S_0$ and $S_2 \rightarrow S_1$; in fact, these emissions are hardly ever seen due to something called *internal conversion*, whereby the molecule in S_2 or S_3 rapidly loses energy without emission and ends up in S_1. Once it's in S_1, it can emit fluorescence in the ordinary way and relax down to S_0. This leads to the second Golden Rule:

1.1.2 *Golden Rule 2: The Shape of the Emission Spectrum Does Not Change as the Excitation Wavelength Is Varied*

Basically, even if you excite with higher energy photons (shorter wavelength) into S_2 or S_3, you only see emission from $S_1 \rightarrow S_0$. If the fluorophore happens to be an atom or monatomic ion (like Ar^+), there are no vibrations so you only get the discrete electronic energy levels and not the range of substates. This actually is pretty handy, since you always get the same color emission no matter what color you excite with – the only thing is, the intensity of fluorescence you get out is proportional to the amount of light the fluorophore absorbs: if you excite at a wavelength the fluorophore absorbs poorly, the emission will also be weak.

1.1.2.1 Quantum Yield

Quantum yield (or quantum efficiency) is a fancy term that expresses how efficient a fluorophore is at fluorescing. Basically, it is the number of fluorescence photons emitted per 100 photons absorbed, so it is a percentage. Thus a "good" fluorophore (like fluorescein or a rhodamine) might exhibit a quantum yield of more than 50 % under some conditions, whereas a poor fluorophore (like a nucleotide) might have a quantum yield of less than 0.01 %: e.g., less than 1 absorbed photon in 10,000 is reemitted as fluorescence, and so nucleotides are nearly nonfluorescent.

1.1.2.2 Measuring Fluorescence Intensities and Spectra

At this point, before we get to the third Golden Rule we need to consider how fluorescence intensities and spectra are actually measured. We presume you have at least a nodding acquaintance with absorbance spectra and how they are measured: basically you shine a light of a given wavelength through a solution of a colored compound and see how much the light intensity is reduced by the solution compared with the solvent alone, without the colored solute. To collect a spectrum, you vary the wavelength of the light and measure the attenuation as a function of wavelength.

For fluorescence, the story is a little different: whereas for an absorbance measurement you just need "something" to select the wavelength at which you want to measure the absorbance, for fluorescence, you need to select two wavelengths: the wavelength of excitation and the wavelength of emission, and they are different. The "somethings" can be different optical devices: most often the something is an optical filter, or a monochromator. Monochromators are used in many fluorometers (or in the old days, spectrophotofluorometers), whereas filters are usually used in fluorescence microscopes, plate readers, and simple fluorometers. Filters pass a

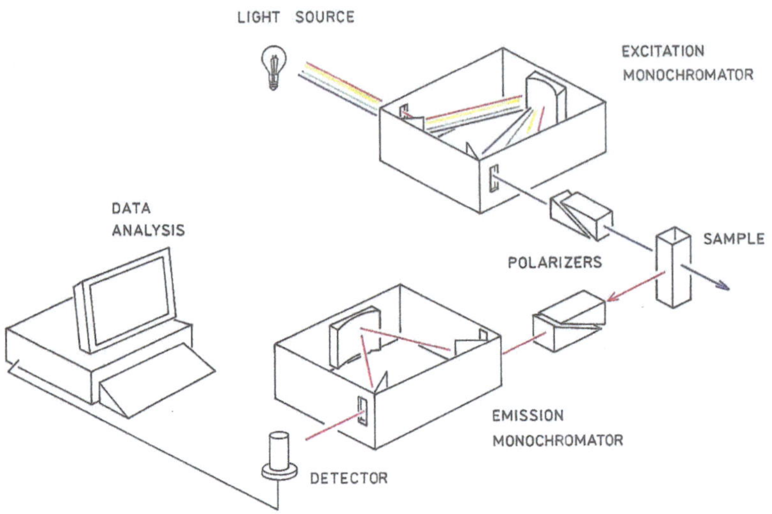

Fig. 1.4 Configuration of the fluorometer

fixed band of wavelengths and are cheaper and simpler, whereas a monochromator can be "tuned" to pass light whose peak wavelength and bandwidth can be varied over a wide range.

The configuration of the fluorometer is shown in Fig. 1.4. The excitation monochromator selects a wavelength band which the fluorophore will absorb (blue in the figure), and it passes light from the light source to the sample through the polarizer (wedge-like things – we'll come to them later) and excites the fluorescence. Light emitted from the sample (red) passes through the emission monochromator to the detector, usually a photomultiplier tube, charge-coupled device (CCD), or avalanche photodiode. While the absorbance measurement is made with the light source, sample, and detector all in a line, in the fluorometer the light path from the light source to the sample is at right angles to the path the emission (**red**) takes from the sample to the detector:

This right angle configuration is deliberately chosen to minimize scattered exciting light getting into the detector and being mistaken for fluorescence. Under most circumstances the fluorescence is emitted from the sample in all directions, but at right angles scattering is weakest, and interferes least. Since the physical property that makes fluorescence very sensitive is the difference between the exciting and emitted light, a key feature of the monochromator (or filter) is its ability to pass the desired wavelengths of light while blocking the undesirable ones. Left out of the picture are lenses used to collect and focus the light. Obviously various light sources can be used, including incandescent and arc lamps, light emitting diodes, and lasers; lasers are desirable for many purposes because they are monochromatic, intense, and their highly collimated light needn't be focused.

The fluorometer in Fig. 1.4 enables us to take two kinds of spectra. The first is the *emission spectrum* (red in Fig. 1.2), where we would keep the excitation wavelength constant, and vary the emission wavelength. Less common is the *excitation spectrum*, where the emission wavelength is kept fixed and the excitation wavelength is varied over a broad range. This brings us to the third of the Golden Rules:

1.1.3 Golden Rule 3: Ordinarily, the Excitation Spectrum Is the Same Shape as the Absorption Spectrum

This seems straightforward: in order to see the emission, the fluorophore must be excited at a wavelength it will absorb, and it will emit most strongly when excited with the wavelengths it absorbs best. There are lots of good reasons to take emission spectra (mainly to avoid or identify artifacts, which we will discuss), but one big reason to take excitation spectra: to see if the excitation spectrum of the emitting species is the same as its absorbance spectrum. Usually it is, but there is an important and useful exception when it is not: *Förster resonance energy transfer* or *FRET* (about which more later).

Summary This article summarizes the basics of the fluorescence phenomenon, spectra, the Jablonski Diagram, and the Golden Rules.

Glossary

Absorbance, Absorbance Spectrum Classically, a molecule will absorb light energy through a resonant process when the plane orientation and frequency of the oscillating electric field of incident light are close to or match those of a characteristic transition dipole of the molecule, such as S_0 to S_1. The efficiency of the absorbance is expressed as a molar extinction coefficient ε, specified at a particular wavelength λ. The absorbance spectrum is the variation of the extinction coefficient with wavelength.

Emission, Emission Spectrum The emission spectrum expresses how the intensity of fluorescence varies with wavelength: it is measured by exciting at a fixed wavelength that the fluorophore absorbs, and measuring the intensity as the wavelength transmitted by the emission monochromator is scanned over some range.

Excitation, Excitation Spectrum The excitation spectrum is the converse of the emission spectrum: one varies the excitation wavelength and measures intensity at a fixed emission wavelength. Since the molecule must absorb to fluoresce, the 3rd Golden Rule states the excitation spectrum is usually congruent to the absorbance spectrum (but the units differ).

Excited State A molecule or atom enters an excited state when it absorbs energy somehow, such that an electron (usually a valence electron) jumps from a lower to a higher energy orbital. For typical fluorescence experiments with UV to IR wavelengths of absorbance and emission, the energy differences between ground (*q.v.*) and excited states are large enough that at mild temperatures (c. 300 K) the excited states are not populated thermally.

Fluorescence Phenomenon whereby a molecule or atom in an excited state after absorbing a photon loses that energy by emitting a photon of longer wavelength.

Fluorophore Generic term for an atom, ion, molecule, or portion of a molecule that is capable of fluorescing.

FRET, Förster resonance energy transfer Resonant, non-photon mediated energy transfer from an excited molecule (usually a fluorophore) to a second molecule whose absorbance roughly matches the first molecule's emission wavelength; the efficiency of the transfer depends on proximity, wavelength overlap, relative orientation of the molecules, and other factors.

Ground State Lowest energy state of a molecule; often but not always, the ground state is a singlet.

Internal Conversion Rapid (picosecond) relaxation from upper level excited states to the lowest excited state of the same multiplicity (e.g., singlet or triplet).

Jablonski Diagram Diagram that schematically depicts energy levels and transitions between them; see Fig. 1.1.

Monochromator An optical device that takes a collimated beam of light as input and emits a beam with only a small portion of the incoming wavelengths; usually the center wavelength can be varied over several hundred nanometers, and the bandwidth can often be varied as well.

Photon A quantum mechanical "packet" of light energy.

Spectrum The distribution of light energy over a range of wavelengths.

Wavelength Classically, the distance peak-to-peak of electromagnetic waves; sometimes denoted as λ. For green light, the wavelength is approximately 500 nanometers, or 0.5 micrometers.

Chapter 2
Fluorescence Made Easier: Fluorescence Techniques for the Novice. Episode 2: Using Fluorescence Anisotropy or Polarization to View Intermolecular Associations

Suzanne Scarlata and Richard B. Thompson

Keywords Fluorescence anisotropy • Fluorescence polarization • Fluorescence lifetime • Rotational motion • Hydrodynamic volume • Molecular associations

2.1 What's Behind Fluorescence Anisotropy?

Fluorescence anisotropy (or polarization, see below) measures how easily a fluorophore can rotate. Many factors can contribute to this rotation. First and most importantly to the experimentalist, fluorescence anisotropy depends on the size of the fluorophore, or the hydrodynamic volume (V). The larger the molecule, the slower it rotates. This size dependence makes fluorescence anisotropy an attractive method to view molecular associations. Sometimes there is confusion about the exact nature of the rotational volume. While this is pretty obvious for an isolated and freely rotating probe, most biochemists measure the anisotropy of a probe attached to a protein or the side chain rotations of tyrosines or tryptophans within a protein. Because the rotations of small probes are much faster than the rotations of the whole protein, then during the fluorescence lifetime of the fluorophore (usually a few nanoseconds) both the fast and slow rotations can be observed and even delineated if desired (we won't discuss this here). To view molecular associations by fluorescence anisotropy, the different rotational contributions need not be known, but instead the change in volume (usually the slower component) is important. Because of this dependence on hydrodynamic volume, fluorescence anisotropy is very good at accurately detecting and quantifying such events as the binding of proteins to membranes, to DNA, to each other, or other supramolecular aggregates.

S. Scarlata • R.B. Thompson (✉)
BioChemistry and Molecular Biology, University of Maryland, 108 N. Greene Street, Baltimore, MD 21201, USA
e-mail: RThompson@som.umaryland.edu

C.D. Geddes (ed.), *Reviews in Fluorescence 2016*, Reviews in Fluorescence,
DOI 10.1007/978-3-319-48260-6_2

2.2 To Understand Fluorescence Anisotropy or Polarization, You Need to Know One Simple Equation

The Perrin-Weber equation (Eq. 2.1) directly relates the ratio of the measured anisotropy (A, or sometimes r in the literature, just to confuse matters) and its value in the absence of rotation (A_o) to rotational motion of the fluorescent species:

$$A_o/A = 1 + RT\tau/\eta V \tag{2.1}$$

where R is the gas constant, T is the temperature (Kelvin), τ is the lifetime, and η is the viscosity. Values for A_o can often be found in the literature, or can be determined by measuring the anisotropy of the fluorescent species under circumstances where it doesn't move, such as placing it in a very viscous solvent and lowering the temperature (i.e. 80 % glycerol, −40 °C). However, since the goal of most anisotropy experiments is to determine changes in association, then the value of A_o does not need to be known. For many fluorophores (tryptophan is an important exception—see below!), A_0 is close to 0.4 when excited at the peak of the longest wavelength absorption band; exciting at much shorter wavelengths can dramatically lower the anisotropy. Note that anisotropy is usually the same throughout the fluorophore's emission band.

As mentioned, the hydrodynamic volume is the key parameter in viewing molecular associations. The equation above assumes a sphere, and unless the change in shape of the species can compensate for the change in size, then a change in anisotropy with association can be detected.

An important parameter in these measurements is the fluorescence lifetime (τ). Fortunately, most probes spend a few nanoseconds in the excited state allowing us to view a significant amount of rotation. If the fluorescence lifetime is too short, then only fast motions will be seen and changes in anisotropy won't be sensitive to protein associations. This may happen if the fluorophore is close to a quenching group. It is also possible, but not all that probable, that the fluorophore becomes quenched during the association making the changes in anisotropy difficult to see. Therefore, if you expect a change in anisotropy but do not detect one, check to see whether this could be due to a large reduction in the intensity. Of course, if the lifetime becomes greater with association, a smaller change in anisotropy will be detected which will reduce the resolution of the measurement. You can think of the fluorophore as having a race between emitting and rotating: if it rotates more slowly than it emits (big volume or short lifetime), the anisotropy will be high, and vice versa.

The other two parameters in the Perrin-Weber are pretty self-explanatory. The temperature is important since more rotation will be seen with higher temperature. The viscosity (η) is important since higher viscosities impede motion. Movement of a species from a high to a low viscosity environment might compensate for any increases in size, and changes in anisotropy would not be seen. Usually, molecular associations tend to move a fluorophore from aqueous to a protein environment and so this typically isn't a problem.

2.3 To Measure Fluorescence Anisotropy or Polarization, You Need to Think About Light

Light propagates in a given direction (x) with its electronic (E) and magnetic (M) vectors perpendicular to the propagation vectors (Fig. 2.1). Usually, the electronic and magnetic vectors take on many orientations in this perpendicular plane. Certain films or crystals will pass light with electronic vectors in the same direction causing the light to be polarized (Fig. 2.1). If you now use this polarized light to excite a fluorophore, then only the probes whose excitation vectors are closely aligned to the excitation vector (actually, within $\cos^2$ of the angle between them) will be excited. Because the alignment between the exciting light and the excitation dipole may not be exact, and because the electronic clouds of the fluorophore deform in the excited state, the emitted light will be somewhat depolarized even in the absence of rotation. For this reason, A_o is specific for a specific probe. Rotational motion will further depolarize the light which can be related to hydrodynamic parameters as described above.

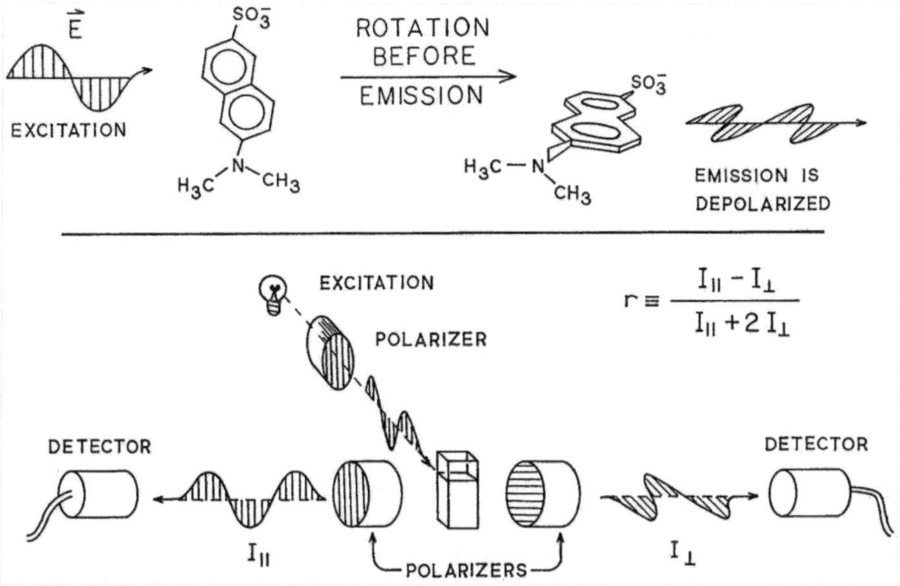

Fig. 2.1 Principle of anisotropy measurement. The upper panel shows that rotational motion of the fluorophore (dimethylaminonaphthalene sulfonate) depolarizes its emission. The *bottom panel* shows a fluorescent sample in a square cuvette being excited with vertically polarized light, with fluorescence emission that is also vertically polarized (e.g., $I_{\parallel}$) passing through the polarizer and being detected on the *left*, and fluorescence that is horizontally polarized ($I_{\perp}$) passing through the polarizer and being detected on the *right*. Filters, lenses, and monochromators are omitted for clarity

Fluorescence anisotropy or polarization is experimentally determined by exciting the sample with vertically polarized light and measuring the fluorescence intensity in the vertical (v) or horizontal (h) directions.

$$A = (I_v - I_h) / (I_v + 2I_h) \quad p = (I_v - I_h) / (I_v + I_h) \quad A = 1/ (1/p - 1/3) \qquad (2.2)$$

where the total intensity $T = I_v + 2I_h$

The relations above show how the anisotropy and polarization are measured and how they are related to each other. As can be seen, the two parameters are nearly the same; polarization measures the differences between the vertically and horizontally polarized fluorescence over the sum of the two, while anisotropy measures the differences between the vertically and horizontally polarized fluorescence over the total intensity. This denominator encompasses emission in all three dimensions, and as a result, the anisotropy is additive, meaning that if you have several fluorophores with different anisotropies, then the total observed anisotropy will be the sum of the individual values and their fractional contributions. Because of this additivity, anisotropy is computationally easier than polarization for measurements. Unfortunately, a lot of the literature (especially for clinical chemistry and high throughput screening) still uses polarization.

To measure the anisotropy, you need to excite your samples with vertically polarized light and measure the emission in the parallel and perpendicular directions. This means that you need at least 2 polarizers – one for the exciting light and one for the emitting light. While most fluorometers have one detector situated at 90° of the exciting light ('L' format), others have 2 detectors on either side of the sample compartment, ('T' format) and so you will need a polarizer for each side [4] (see Fig. 2.1). Polarizers can usually be purchased from whatever company furnished the fluorometer. If you expect to have a large change in anisotropy and don't require high resolution, then you can use film polarizers such as the ones found in many sunglasses, but for most biological systems, optically pure, crystalline polarizers (e.g., Glan-Thompson calcite polarizers) are best and totally worth it. It is important to keep in mind that most laser systems produce linearly polarized light. The orientation of the laser light can be easily checked by placing a perfect scatterer (I use a very, very, very small amount of coffee creamer suspended in water) in the sample compartment, and measuring the emission at the same wavelength as the exciting light (be VERY careful to reduce the amount of light getting to the detector by closing the slits, the iris, adding neutral density filters, etc., to avoid trashing your detector). If the laser light is vertically polarized, the resulting anisotropy will be ~1.0 (e.g., $I_\perp \approx 0$) and can be used directly without a focusing lens. If the light is horizontally polarized then its orientation can be changed by bouncing the beam off a couple of mirrors arranged as a periscope mirrors. Someone told me that you can simply turn the laser on its side, but unless it's a small diode laser, that may be a bad (or very bad) idea. . Fiber optic cables will depolarize light and so a polarizer needs to be placed at the end of the cable.

Although anisotropy measurements appear very simple, there is one small complication; a typical monochromator does not pass vertical and horizontal light identically. This problem can be easily remedied by determining the 'g' or grating factor. The 'g' factor can be thought of as a normalization factor to correct the polarization, and is simply measured by exciting the sample with horizontally polarized light and measuring the emission at vertical and horizontal directions (Eq. 2.3). The 'g' factor only needs to be measured once for a specific wavelength in a specific instrument, but it is a good idea to measure it at the beginning of each experiment.

$$A = \left(I_{vv} - g^{*} I_{vh}\right) / \left(I_{vv} + 2g^{*} I_{vh}\right) \quad \text{where } g = I_{vh} / I_{hh} \tag{2.3}$$

where the first letter of the subscript (v or h) denotes the orientation of the excitation polarizer while the second denotes the orientation of the emission polarizer.

Technical Comments It is a good idea to test your system before starting to take measurements. A small molecule fluorophore in water should give an anisotropy close to 0 while a perfect scatterer should give you an anisotropy of 1.0. As mentioned above, perfect scatterers can be easily prepared by immersing a small amount of milk into an aqueous solution or non-dairy coffee creamer if you are lactose intolerant. Fluorescent anisotropy standards have been published [2].

If you are using a typical white light source, keep in mind that you will lose more than half of your intensity as the light passes through the polarizers and so you may have to work at higher sample concentrations or at wider slit widths. The main technical issue for anisotropy measurements is the contribution of scattered light. Vertically polarized light will be scattered in the vertical direction (i.e., singly scattered light does not change orientation of the electronic vector), and if this light bleeds through the monochomators, the resulting anisotropy will be artificially high. A little scattering goes a long way: if only 2 % of the light your detector sees is scattered (polarized) excitation and you're trying (for example) to measure an anisotropy of 0.04, what you will actually observe is an anisotropy of 0.06 (0.04 + (2 % x 1.0)), which is a 50 % error. Therefore, it is important to look carefully at your cuvette, make sure the contents are well mixed and that there isn't any lint or other particles floating around (hold it up to the light – it helps). Also keep in mind that pipetting a sample that has been sitting on ice, into a cuvette at room temperature can result in tiny little bubbles appearing as the sample warms. Light scattering from these bubbles can bleed through and result in a high observed anisotropy and so you should gently vortex or tap the cuvette to get rid of the bubbles. Many experiments involve highly scattering samples such as whole cells, organelles, chromatin or membranes. For these studies, you need to prepare a control cuvette that contains the identical concentrations of the scatterers but without the fluorophore. The background from the control must be subtracted from the sample intensity at each configuration (i.e. I_{vv}, I_{vh}, I_{hv}, I_{hh}). Many times scattered exciting light can be excluded from your detector by use of cutoff filters that block the excitation and transmit the fluorescence.

How Do You Know if Your Measurement Is Valid? The initial way to assess if your measured values are valid is to check if they are within the theoretical limits – if all the fluorophores in the cuvette are aligned with their excitation and emission dipoles perfectly parallel with the polarization of the exciting light, A= (1–0)/(1+0) = 1, and perpendicular alignment, A = (0–1)/ (0+2) = −0.5. Of course, this never happens in solution where the fluorophores are expected to have a random distribution, and in this case, the limit for motionless fluorophores will be 0.4 to −0.2. Additionally, because the excitation dipoles are rarely collinear with the exciting light and because the molecule may deform when its electrons are in the excited state, the range for anisotropy values is expected to be even narrower. All of this is in the absence of motion. Therefore, you should be extremely suspicious if you obtain an anisotropy much above 0.3 for a labeled protein in solution, and you can be sure there's a problem if it's above 0.4. For small fluorescent probes in solution, this number should usually be close to 0.

Keep in mind that anisotropy is a ratio and should be independent of concentration and so if you dilute or concentrate your sample (assuming you are not monitoring homo-oligomerization) the value shouldn't change. If the anisotropy increases with dilution under conditions where you do not expect changes in intermolecular association, it is possible that you are increasing the relative amount of scattering due to the reduced fluorescence signals.

Measuring the Anisotropy of Tryptophan in Proteins Lots of people like to measure the anisotropy of tryptophan in proteins to observe conformational changes or association reactions—after all, the tryptophan is already there, so it's like a free label. However, there are some annoying things to bear in mind. First, the A_0 of tryptophan tops out at 0.32 at wavelengths above 305 nm [3] where the tryptophan absorbance is weak, whereas the A_0 at 280 nm (the usual peak of absorbance everybody excites at) is only 0.17. Since the A_0 changes so dramatically with wavelength, a good compromise is 300 nm: the absorbance is good enough that you'll have a decent signal, but the A_0 is high enough to give easily measurable changes. The other challenge is that tryptophan in proteins usually has a short lifetime (3 nanoseconds or less) so that a typical protein in water at room temperature doesn't rotate very far in that time: for instance, a modest size protein like carbonic anhydrase (30,000 Daltons molecular weight) doesn't rotate very far (16 ° on average) in 3 nsec, so its anisotropy free in solution is nearly as high as its A_0 when it's immobilized, and tryptophan anisotropy measurements will show little increase if the carbonic anhydrase binds to itself or something much larger. To study proteins and DNA associating with one another it's often helpful to label the macromolecule with longer lifetime fluorescent labels [1]; you can predict the change in anisotropy using the Perrin-Weber equation (Eq. 2.1).

References

1. Terpetschnig E, Szmacinski H et al (1995) Fluorescence polarization immunoassay of a high molecular weight antigen based on a long lifetime Ru-ligand complex. Anal Biochem 227:140–147
2. Thompson RB, Gryczynski I et al (2002) Fluorescence polarization standards for high-throughput screening and imaging. Biotechniques 32(1):34–42
3. Valeur B, Weber G (1977) Resolution of the fluorescence excitation spectrum of indole into 1La and 1Lb excitation bands. Photochem Photobiol 25(5):441–444
4. Weber G (1956) Photoelectric method for the measurement of the polarization of the fluorescence of solutions. J Opt Soc Am 46(11):962–970

Chapter 3
Time-Gated FRET Detection for Multiplexed Biosensing

Jurriaan M. Zwier and Niko Hildebrandt

Keywords Fluorescence • Terbium • Europium • Immunoassays • Imaging • Spectroscopy

3.1 Introduction

The very high sensitivity of photoluminescence (PL) detection has made this technology to an irreplaceable tool for biosensing. Many different fluorophores (as the term luminophore is less common, we use the term fluorophore for all photoluminescent compounds), ranging from organic dyes (including dark quenchers), over light-harvesting complexes, fluorescent proteins, and metal complexes, to nanoparticles, have been successfully used for studying biological systems from the ensemble down to a single molecule [1–4]. Driven by the myriad of biological interactions that define biological function and happen on a nanometric scale, Förster resonance energy transfer (FRET) plays a major role in PL-based biosensing [5–13]. PL is characterized by energy or wavelength (PL spectrum), intensity, excited-state lifetime, quantum yield, and polarization [14–17], and, in particular, intensity and lifetime are frequently used in FRET to statically and dynamically quantify concentrations and distances in biological systems. In addition to sensitivity, another highly important aspect of biosensing is multiplexing capability, i.e. the possibility to measure (in parallel or sequentially) several different biological molecules and/or their interactions from a single sample. The combination of spectral, intensity, and lifetime information of PL can be used to perform multiplexed FRET biosensing with different types of fluorophores [2–4, 10, 18–19]. Within one fluorophore, the probability of photon emission decays exponentially over time. In the case of different configurations and different fluorophores such decays can also become

J.M. Zwier (✉)
R&D, Cisbio Bioassays, Parc Marcel Boiteux, BP 84175, 30200 Codolet, France
e-mail: jzwier@cisbio.com

N. Hildebrandt (✉)
NanoBioPhotonics, Institute for Integrative Biology of the Cell, Université Paris-Sud, Université Paris-Saclay, CNRS, CEA, 91405 Orsay Cedex, France
e-mail: niko.hildebrandt@u-psud.fr; http://www.nanofret.com

C.D. Geddes (ed.), *Reviews in Fluorescence 2016*, Reviews in Fluorescence,
DOI 10.1007/978-3-319-48260-6_3

multiexponential and, in the best case, each decay time (or lifetime) can identify a specific configuration and/or fluorophore. Most fluorophores have lifetimes in the nanosecond range but some can have lifetimes that reach micro or even milliseconds. The most prominent ones are lanthanides, which have been used as ions, in complexes with antenna ligands, or in nanoparticles for various biosensing applications [12, 19–27]. Their extremely long PL lifetimes can lead to intensity decays over several milliseconds after pulsed excitation and this long PL decay can be used for time-gated (TG) intensity detection. This approach uses a temporal detection window that opens at a specific time point after the excitation pulse. If the excitation pulse is considered as time zero, the PL intensity is measured in a specific time interval (or time gate or window) $\Delta t = x$ starting at $t = 0 + y$ and ending at $t = y + x$. The big advantage for biosensing and FRET is that all PL signal that occurs within $0 < t < y$, which can be, *e.g.*, autofluorescence of the biological sample or direct excitation of a FRET acceptor, is completely suppressed and only the long-lifetime signal is detected. In other words, TG PL detection allows for an efficient background suppression. Another important feature of long PL lifetime FRET donors is their capability of transferring their FRET-quenched PL decay time to short-lifetime acceptor fluorophores. This means that the decay time of the donor in the presence of the acceptor (which is usually used for FRET analysis) equals the decay time of the acceptor in the presence of the donor because the intrinsic lifetime of the acceptor is negligible (orders of magnitude shorter). Therefore, acceptor fluorophores of different PL colors can be combined with the same lanthanide donor and multiplexed detection of the different FRET-sensitized acceptors can be performed. This chapter reviews the theory and applications of TG FRET (and TG PL) for multiplexed biosensing using spectroscopy and microscopy. Representative examples from the recent literature, which used lanthanides as donors and dyes, fluorescent proteins, or quantum dots as acceptors for the development of assays for nucleic acids, proteins, and enzyme kinetics, the design of molecular logic gates, or the investigation of extracellular protein-protein interactions and intracellular detection, will be discussed to show the versatility and wide applicability of this powerful optical biosensing technology.

3.2 Theory of Time-Gated (TG) FRET

3.2.1 Time-Gated vs. Time-Resolved Luminescence

Photon emission due to a transition from a higher to a lower energy level follows an exponential probability over time, which is usually expressed in a time-dependent exponential PL intensity decay:

$$I(t) = A \sum_i \alpha_i \exp\left(-\frac{t}{\tau_i}\right) \tag{3.1}$$

In this equation A is the intensity at $t = 0$ and α_i (with $\sum \alpha_i = 1$) are the pre-exponential factors that give the amplitude fractions of each decay time τ_i. This decay can be mono-exponential (in the case of a single fluorophore with a single conformation) or multi-exponential (in the case of different fluorophores or a single fluorophore with different configurations that can lead to additional de-excitation pathways). Knowledge of the decay time or decay times can provide important information about the configuration of the fluorophore because these decay times are representative for different de-excitation pathways of the excited state. The fractions of each decay time represent the concentration fractions of each fluorophore configuration. Because FRET opens an additional de-excitation pathway to an excited donor fluorophore (FRET quenching), decay time analysis can be used to analyze FRET. Figure 3.1 shows several single-exponential decay curves with decay times ranging from $\tau = 20$ ns to $\tau = 5$ ms. The intensities (heights) of the curves can also vary (different values of A in Eq. 3.1). Integration of the decay curve over the entire decay results in the steady-state PL intensity. An accurate determination of decay times requires acquisition of the entire decay curve with sufficient intensity (photons) and can therefore be a time-consuming procedure. In time-correlated single photon counting (TCSPS) the arrival time of the

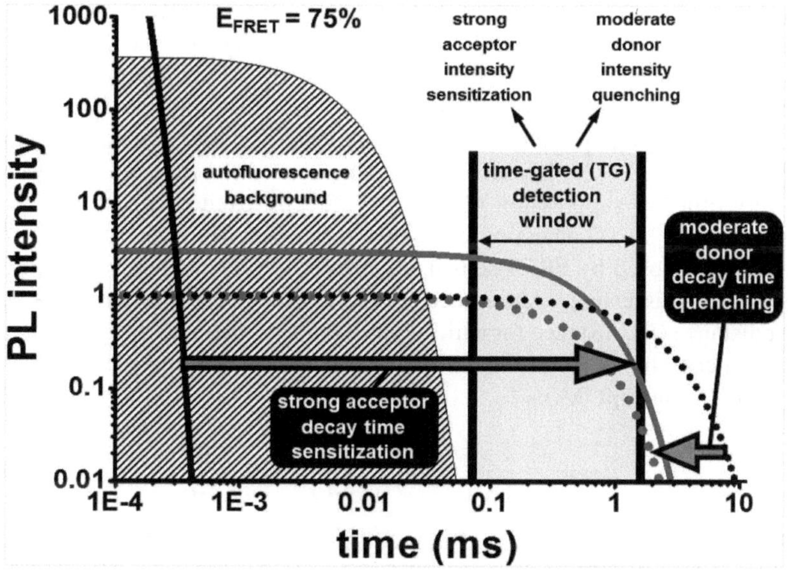

Fig. 3.1 The PL decay curve of a long-lived fluorophore (black dotted line, $\tau = 2$ ms) can be efficiently distinguished from those of a short-lived fluorophore (black solid line, $\tau = 20$ ns) and autofluorescence background (filled curve in the background, $\tau = 5$ μs) by using a TG detection window that opens with a delay that is longer than the short-lived PL decays. In the case of FRET from the long-lived donor to the short-lived acceptor, these fluorophores get quenched and sensitized (indicated by the large arrows), respectively. The resulting PL decay curves (gray) both have the same PL decay time ($\tau = 0.5$ ms in the case of 75 % FRET efficiency) and can both be detected in the TG detection window (Modified from Ref. [20])

first photon on the detector after sample excitation is measured, and then the decay curve (photon counts over time) is build up excitation pulse by excitation pulse (or photon by photon). This technique is only efficient for short-lived (ns) fluorophores, for which excitation sources with MHz repetition rates can be used. For long-lived (μs-ms) fluorophores all photons within certain temporal windows that are shifted over the entire time of the PL decay (boxcar method) are counted. This method is often called multi-channel scaling (MSC). If no decay time information but still a time-dependent intensity is required, then a single time-window with a certain delay after the excitation pulse can be used to count all photons within this time range. The principle of this TG detection is shown in Fig. 3.1. As simply all photons are counted within the TG detection window no matter of their exact arrival time and only within a single window, this TG intensity detection is significantly faster than time-resolved (decay time) detection. However, it does not contain any (or very limited) decay time information.

3.2.2 TG FRET (Förster Resonance Energy Transfer)

FRET has been described in detail in many text books [5, 8, 10–11, 14, 16–17], and here, we focus on the most important aspects for TG FRET. TG FRET uses donors with much longer excited-state lifetimes than the acceptors. Alternatively, abbreviations such as TR(Time Resolved)-FRET or L(Lanthanide)RET are used in the literature. Usually, lanthanide-based donors with millisecond lifetimes and dye or quantum dot acceptors with nanosecond lifetimes are applied for TG FRET biosensing [19–20, 24, 26, 28–29]. This large difference in lifetimes allows for a very efficient suppression of PL from directly excited acceptors or sample autofluorescence by TG detection, as shown in Fig. 3.1. Moreover, the TG acceptor PL is entirely caused by FRET-sensitization from the donor. The FRET efficiency E_{FRET} (Eq. 3.2) is usually determined by the donor-acceptor distance (r) and the Förster distance (R_0, distance for which FRET is 50 % efficient), or by PL intensities (I) and/or decay times (τ) of the donor alone (subscript D) and in the presence of the acceptor (subscript DA).

$$E_{FRET} = \frac{R_0^6}{R_0^6 + r^6} = 1 - \frac{I_{DA}}{I_D} = 1 - \frac{\tau_{DA}}{\tau_D}, \tag{3.2}$$

In TG FRET the large difference of excited-state lifetimes and the background-free detection of acceptor allow for two additional ways to determine E_{FRET}. First, the decay time of the donor in the presence of the acceptor (τ_{DA}) equals the decay time of the acceptor in the presence of the donor (τ_{AD}) and Eq. 3.2 can be changed to:

$$E_{FRET} = 1 - \frac{\tau_{DA}}{\tau_D} = 1 - \frac{\tau_{AD}}{\tau_D} \tag{3.3}$$

Because the FRET-sensitized acceptor adapts the long PL decay time of the FRET-quenched donor, it can be easily distinguished from the short-lived PL signals

(Fig. 3.1). Second, TG detection permits determination of E_{FRET} by a single measurement of the FRET-pair, without additional fluorescence measurements of the donor and the acceptor alone, by using the integrated PL spectra (area under the spectra) of FRET-quenched donor (F_{DA}) and FRET-sensitized QD acceptor (F_{AD}) and the PL quantum yield of the acceptor (Φ_A) [20, 30–31]:

$$E_{FRET} = \frac{F_{AD}/\Phi_A}{F_{AD}/\Phi_A + F_{DA}} \tag{3.4}$$

3.2.3 *Multiplexed TG FRET*

In addition to their extremely long excited-state lifetimes, a particularly important aspect of lanthanide emitters is their PL spectrum with several, often well-separated, emission bands. Terbium (Tb) is an ideal example for such multiband PL emission spectra. As shown in Fig. 3.2, Tb^{3+} has four relatively strong PL bands around 495 nm, 545 nm, 575 nm, and 620 nm, and some weaker PL bands between ca. 640 and 700 nm. These emission bands can overlap with the absorption spectra of several different acceptors (Fig. 3.2a left for five different organic dyes and right for five different quantum dots), which allows for multiplexed FRET from one type of Tb complex to all of these acceptors. The very strong absorption of quantum dots over a wide spectral range can result in Förster distances of up to 11 nm [19], which is beyond the usual FRET distance range of *ca.* 1–10 nm [14, 17].

Because FRET from one Tb donor to different acceptors cannot be easily distinguished by measuring Tb PL quenching, the acceptors have to be selected such that their sensitized PL, which is then a long-lived PL (cf. Eq. 3.3), emits in a spectral range that is in between or beyond the Tb PL bands (Fig. 3.2b left for five different organic dyes and right for five different quantum dots). Spectral distinction can be done by measuring the full TG spectra or by using bandpass filters (as shown in Fig. 3.2b), which is usually simpler and faster. Examples of multiplexed biosensing using TG FRET from a single Tb complex to several different acceptors will be presented in Sect. 3.4.3.

3.3 Advantages of TG FRET

3.3.1 *Single Analyte Detection*

Fluorescent technologies have become the method of choice to detect biological interactions [32]. The power of TG FRET methods over other fluorescent technologies are due to several factors [24, 33]. The first obvious parameter is the introduction of a time delay of ca. 50–150 μs, which reduces the background

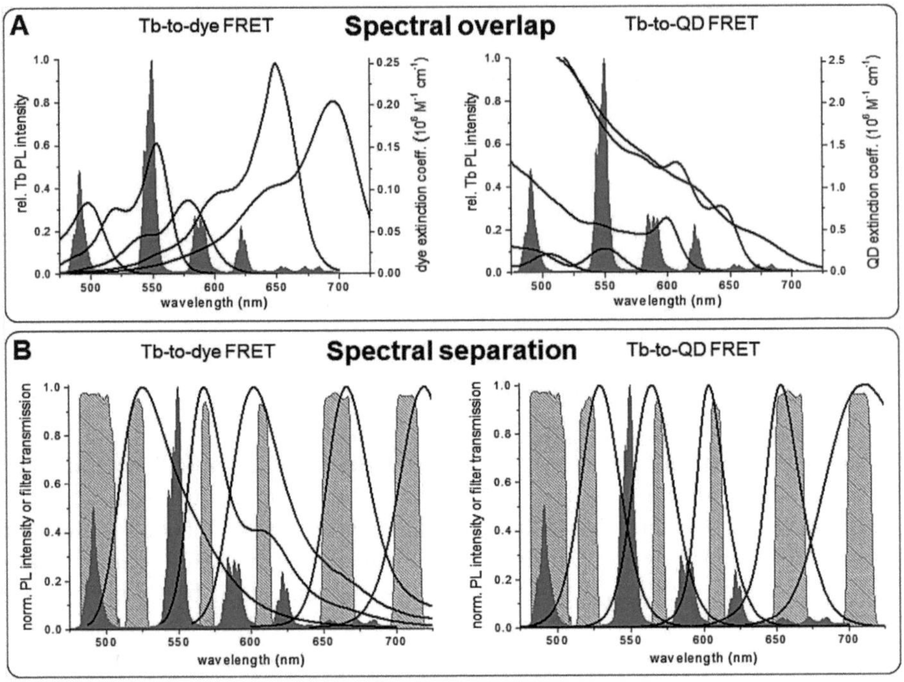

Fig. 3.2 Spectral overlap of donor PL and acceptor absorption (**a**) and PL emission spectra (**b**) of multiplexed TG-FRET donor-acceptor pairs that use a common donor (Tb) and different dye or quantum dot (QD) acceptors. (**a**): The PL spectrum of Tb overlaps with several different dyes (left) or QDs (right), which allows for FRET from Tb to all the different acceptors. The broader overlap and the significantly higher extinction coefficients of QDs allow for larger Förster distances when QDs are used as acceptors. (**b**): Careful selection of dye or QD PL bands that fit in between or beyond the Tb PL bands opens the possibility of separately measuring all fluorophores in different detection channels using optical bandpass filters (filled spectra in the background). The significantly narrower PL bands of QDs (compared to dyes) allow for a much more efficient spectral separation (Modified from Ref. [19])

emission of non-interacting fluorophores or biological materials and of scattering of excitation light. The gate time can vary depending on the instrumentation and the kind of assay, but is generally around 400–500 μs. The particular emission spectrum of lanthanides is very advantageous as well. Because of the spectral windows present in their spectra, TG FRET can be more easily detected in these regions, which is in contrast with the significant cross-talk often experienced in regular FRET. Furthermore Förster radii between lanthanide complexes and the bright fluorophores are elevated (4–11 nm) and in accord with biological interactions. Another advantage is the possibility for ratiometric detection, which consists of delayed acceptor emission output divided by the donor intensity. Although now used intensively in non-TG FRET as well, its main purpose in TG FRET is the correction for the perturbation and absorption caused by media such as serum/blood, tissue or highly absorbing screening compounds and to correct for interwell differences in multititer plates. Related to this, TG-FRET assays can be done on rather inexpensive instrumentation and is therefore attainable for laboratories with limited financial resources. Microtiter plates can be measured at high speed with TG-FRET and thus is the method of choice for screening campaigns in high or ultra-high throughput done in the pharmaceutical industry. Methods such as HTRF® and TRACE® are homogeneous, which therefore avoids tenuous washing steps necessary in ELISA formats and kinetics studies can be performed.

3.3.2 *Multiplexed Detection*

The above advantages hold for multiplexed TG-FRET detection as well. In practice, TG-FRET is most convenient using Tb^{3+} complexes because of its emission spectrum (*vide infra*). In duplex mode, the most convenient combination is using a dye emitting around 520 nm (i.e., fluorescein-like) and one emitting around 665 nm (i.e., cyanine5-like). In triplex or multiplex mode, adequate correction procedures to account for spectral crosstalk are necessary to properly detect the interaction of interest [34]. Measuring in TG-FRET multiplex mode will reduce the operational costs because fewer experiments are necessary to obtain the results needed. This will prove especially worthwhile on 'valuable' samples, where only little is available, such as blood, tissue, or organ samples.

3.4 Spectroscopy-Based TG FRET Biosensing

This section presents an overview of TG FRET biosensing using spectroscopy. After a short introduction to the necessary equipment, single analyte and multiple analyte detection will be exemplified by some recent commercial and academic applications using lanthanide donors. For more examples and explanations of this technique the interested reader is referred to recent review articles [18–21, 24, 29].

3.4.1 *Equipment and Detection Formats*

The equipment for TG FRET experiments is only a little more complicated than the standard fluorescence equipment, which is mainly related to the necessity of pulsed excitation because TG detection counts the photons in a specific time-window after a short excitation pulse (cf. Sect. 3.2). Moreover, lanthanide complexes require UV excitation and therefore nitrogen lasers (337.1 nm) or Xe flash lamps (the excitation wavelength range needs to be selected, *e.g.*, by bandpass filters, from a broad emission spectrum) are often used as excitation sources. As TG detection is most often based on very long PL decay times (in the time scale of tens of microseconds to a few milliseconds) the relatively long afterglow of flash lamps is in most cases not problematic. Another possibility is continuous excitation with fast off-switching, for which the moment of switching to the off-state is used as time zero (equivalent to the time of the excitation pulse). Unless optical shutters are used, detection requires time-resolved detectors, such as photomultiplier tubes (PMTs). These detectors must be interconnected (*e.g.*, via photon counting modules) to the excitation sources, in order to link the time settings of excitation pulse and detection window. Time-resolved fluorescence spectrometers can use the MCS mode to perform TG detection. Fluorescence plate readers very often have so-called TR-FRET or TR-fluorescence modes, which are in fact not time-resolved but simply TG (detection of PL intensity within a certain time window after excitation). In the case of FRET it is usually advantageous to measure the TG PL intensities of both the quenched donor and the sensitized acceptor, which may require two detectors (at least for simultaneous detection) and the appropriate dichroic mirrors and bandpass filters for spectral separation.

3.4.2 Single Analyte Biosensing Applications

TG FRET with lanthanide donors has been successfully used in commercial biosensing applications for many years. One of the most frequently used lanthanide complexes is Europium-tris(bipyridine) (Eu-TBP) [24, 33, 35–36]. This europium cryptate is used in many TG FRET bioassays such as in HTRF® (homogeneous time-resolved fluorescence) [24, 36–37], and TRACE® (time-resolved amplified cryptate emission) [38]. The most successfully applied terbium complex (also used in HTRF and TRACE), which is based on 2-hydroxyisophthalamide ligands [39], is Lumi4®-Tb from Lumiphore [40]. Both HTRF and TRACE are homogeneous TG FRET assay methods that are based on FRET from the lanthanide donor to an organic dye acceptor, which is most often cross-linked allophycocyanin (XL665) [41] or d2 [24, 36]. TRACE is based on sandwich FRET immunoassays, whereas HTRF uses sandwich, competitive [24, 36], ligand binding (Tag-lite®) [42], or conformational [43] TG-FRET assays. The principle of TG FRET assays is exemplified by homogeneous sandwich immunoassays in Fig. 3.3. Donors and acceptors are conjugated to two different (usually primary) antibodies (AB1 and AB2) against the antigen of interest, which leads to AB1-antigen-AB2 immunocomplexes in the presence of the correct antigen. Only for the sandwich complexes (Fig. 3.3b) donor and acceptor are in close enough distance to engage in FRET and the FRET-sensitized acceptor emission will display the same long PL decay time as the FRET-quenched donor. The PL intensity in the TG detection window is directly proportional to the antigen concentration, which allows for a quantitative detection of the antigen. Moreover, the simultaneous detection of both TG acceptor and donor PL intensities can be used for ratiometric measurements, which provide higher accuracy (lower coefficients of variation) even in challenging biological media,

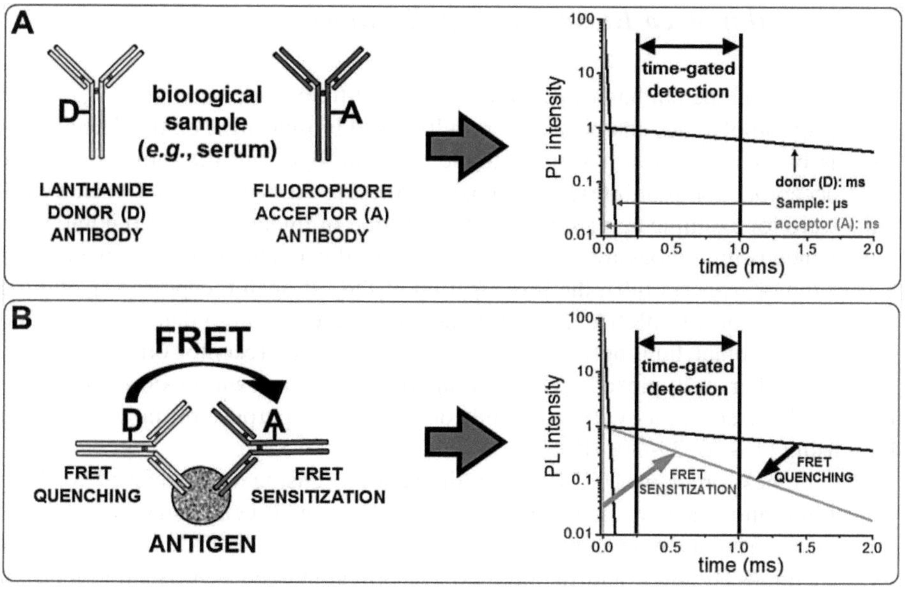

Fig. 3.3 Principle of homogeneous TG-FRET sandwich immunoassays. Assay kits contain two primary antibodies against different epitopes of the antigen (biomarker). These antibodies are conjugated with a lanthanide donor and a fluorophore acceptor. If there is no antigen present in the sample (usually serum or plasma), only the donor can be measured in the TG detection window (**a**). If an antigen is present, both antibodies will bind it and FRET can occur, which leads to donor-quenching and acceptor-sensitization, both of which can be measured in the TG detection window (**b**). Usually a ratiometric measurement of TG acceptor intensity and TG donor intensity is used for antigen quantification

such as serum or plasma. For competition assays, such as for the detection of cAMP or Inositol-monophosphate [24, 36, 44], either the donor or the acceptor is conjugated to an antibody, whereas the other FRET partner is labeled to the antigen itself, which leads to a competition between conjugated and free antigen in the sample. Therefore, the FRET-sensitized acceptor PL is inversely proportional to the antigen concentration. PL from the sample components (i.e., directly excited acceptors as well as autofluorescence from the biological medium, both occurring in the nanosecond time range) are not measured in TG detection and a very sensitive, nearly background-free antigen quantification becomes possible. Due to the homogeneous liquid-phase format, the assays require only short reaction times without the necessity of any washing and separation steps.

Other homogeneous TG FRET assay methods are LANCE® (lanthanide chelate excitation) [45], which uses europium chelates based on dichlorotriazinyl-activated pyridine or terpyridine-based tetra-acetates as donors [46], or LanthaScreen® [47], which uses europium or terbium complexes with TTHA-cs124 ligands [48]. Both technologies have a limited assay portfolio compared to HTRF®.

3.4.3 Multiplexed Biosensing Applications

Conjugation of one lanthanide donor and several different acceptors (cf. Fig. 3.2) to different biomolecules can be used for multiplexed biosensing, i.e., the detection of various biomarkers from a single sample. Organic dyes are the most commonly used acceptors for multiplexed biosensing [19, 21, 34, 49–55], and such FRET pairs were for example applied to the TG detection of ligand-receptor binding [55], homogeneous immunoassays against prostate specific antigen [54] and several lung cancer tumor markers [49], the investigation of the estrogen receptor [52], and the quantification of microRNA [34]. Protein-protein interaction to detect soluble and aggregated mutant huntingtin protein [56] and $GABA_B$ receptor-oligomerization [57] is another multiplexed TG-FRET approach. Also quantum dots have been used as FRET acceptors with lanthanide donors for the purpose of multiplexed TG biosensing [18–20, 29–30, 58–65].

The highest order of TG FRET multiplexed biosensing was achieved for lung cancer immunoassays, with the aim of distinguishing small-cell from non-small-cell lung cancer from a single sample [49]. The homogeneous TG FRET immunoassay used a Lumi4-Tb donor and the organic dyes OregonGreen 488, AlexaFluor 555, AlexaFluor 568, Cy5, and AlexaFluor 700 as acceptors, and these FRET pairs provided Förster distances between 4.4 and 6.0 nm. Besides spectral crosstalk, which could be efficiently corrected, the immunoassay (Fig. 3.4) needed to be performed under biologically very challenging conditions, because among the 15 different biological components (five tumor markers and 10 antibodies) in the 50 μL serum samples the five tumor markers had to be quantified at very low concentrations with low picomolar detection limits. The possibility of detecting concentrations below the clinical cut-off values showed the high sensitivity of these multiplexed TG FRET biosensors and their applicability to *in vitro* diagnostics.

3.5 Microscopy-Based TG FRET Biosensing

Here we present the recent development in TG fluorescence and TG FRET microscopy using lanthanide based systems. TG imaging is still under development and one of its promises is imaging of biological events in native tissue or animals, where fluorescence of native tissue is an issue and biochemical engineering with fluorescent proteins is not possible. The requirement for proper instrumentation will be briefly discussed including some recent developments. TG microscopy is discussed in more detail and TG FRET microscopy and its use in multiplexing, which has developed strongly over the past years, will be featured in the last subsection.

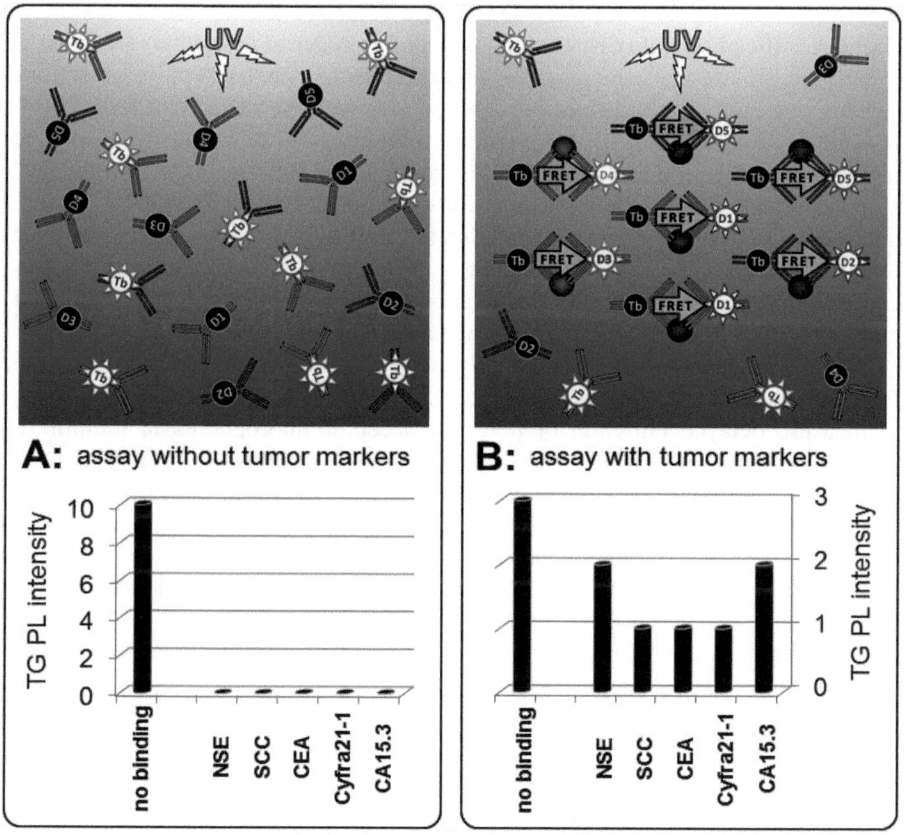

Fig. 3.4 Principle of a multiplexed homogeneous TG-FRET immunoassay for lung cancer diagnostics. Each of the five tumor markers NSE, SCC, CEA, Cyfra21–1 and CA15.3 is recognized by a specific Tb-antibody/dye-antibody pair. In the absence of tumor markers (**a**), there is no antibody-antigen binding and only TG Tb PL is measured. In the presence of tumor markers (**b**) antigen-specific FRET complexes are formed and the energy is transferred from Tb to the antigen-specific dyes. The FRET-sensitized TG dye PL intensity depends on the antigen concentration, which is used for multiplexed tumor marker quantification

3.5.1 Equipment and Detection Formats

Lanthanide-based TG microscopy needs specific requirements of the fluorescence microscope set-up [66–67]. Most instruments use an adapted inverted wide field fluorescence microscope using a pulsed excitation source, which needs to operate in the UV region due to the absorption properties of the lanthanide complexes. These can be flash lamps operating with mechanical choppers but better results are obtained by using either laser or LED excitation operating at low frequency (generally between 20–1000 Hz). Due to the extended lifetime of the lanthanide complexes the photon flux is relatively low and in order to detect the rare emission events intensified CCD cameras or high level PMTs are needed. A pulse generator to synchronize excitation and detection is required as well. Because of the long lifetime of lanthanide emission it is also possible to monitor TG emission by shifting the pinhole in two photon scanning confocal microscopy [68–69]. Using bright Eu^{3+} complexes and 355 nm excitation emission can be easily monitored [70] in confocal microscopy and, using a phase modulation technology, super resolution (*ca.* 60 nm) imaging of lanthanide complexes can be achieved [71]. A combination of scanning confocal microscopy and TCSPC led to another time gated technology of lanthanide emission called Photon Arrival Time Imaging (PArTI) [72].

TG FRET microscopy is potentially a very powerful technology due to the line-like emission spectra of lanthanides. This property gives the opportunity to measure delayed emission of potential acceptor fluorophores in multiplexed mode within the spectral windows if proper emission filters are used. Due to the low spectral crosstalk of lanthanide emission TG FRET can easily be identified in intensity imaging, while FRET microscopy using fluorescent proteins needs extensive analyses [8]. Quantification and detection in TG FRET microscopy has recently been thoroughly described by the Miller group [73–74], and its application will further be detailed in Sect. 3.5.3.

3.5.2 TG Luminescence Biosensing

Discussion of TG FRET cannot do without a brief introduction in the recent history of TG PL imaging [21–22, 25–26, 66–67, 70, 74], which has resided in the limelight of more 'spectacular' advances in fluorescence microscopy. Most notably, the development of monomeric fluorescent proteins and super-resolution-based microscopy has resulted in a plethora of new powerful imaging technologies. TG microscopy based on long lived excited states of lanthanide complexes has continued to evolve and its initial promises are now appearing due to the development of bright and kinetically stable Eu^{3+} and Tb^{3+} complexes and lanthanide-based nanoparticles. Recent reports on using time gating in STED [75] or other long-lifetime excited states have appeared as well [76].

In the 1980s it was already recognized [77] that the use of time-gating using lanthanide complexes could enhance the detection limit compared to regular fluorescent dyes and the first microscopy systems adapted to time-gating appeared in the early 1990s. The group of Tanke [78] were the first to demonstrate the use of TG imaging by imaging human lymphocites with antibody-labeled Y_2O_2S:Eu nanoparticles. Shortly after, Seveus et al. [79] showed TG images of malignant mucosa of the human colon using an Eu^{3+} chelate, which demonstrated that suppression of autofluorescence by time-gating is very beneficial for the quality of tissue analyses. This was further demonstrated by the work of Marriot et al. [80], who used a more stable chelate (Eu^{3+}–SBMC). In this paper, luminescence lifetime imaging of this compound on a glass slide was shown as well. Vereb et al. [81] demonstrated the possibility to do TG lifetime imaging and spectral imaging on Eu-DTPA-cs124 crystals.

Since these days, lanthanide complexes have improved in brightness and kinetic and thermodynamic stability [21, 24–26, 70, 82], allowing them to be used as luminescent probes in (TG) luminescence microscopy. A second technological platform is based on improved optical and biological properties of nanoparticles such as lanthanide dots or upconversion nanocrystals [83–85].

Improved dyes were first demonstrated by incorporating them in nanoparticles [86–87], but where soon used for sensors or as organelle stains within the cell [25, 70, 88–98]. Many of these lanthanide systems are bright enough to be detected by regular fluorescence microscopy and time-gating was not always applied. Multi-photon excitation of lanthanide complexes has been shown to work [99–101], and was later applied [102–104] as well. As the advantages of TG microscopy will be most beneficial for the analyses of biological specimen, it is of prime importance to have complexes available that can be easily bioconjugated to bioactive molecules, such as antibodies or peptides, and have limited non-specific interaction to cells. Lehn's Eu cryptates [35] are very robust systems [24] and are after antibody bioconjugation suitable to detect GABA-B membrane proteins using TG microscopy [105]. Fernandez-Moreira et al. [106] demonstrated that by labeling Eu^{3+} and Tb^{3+} complexes to secondary antibodies one can detect the estrogen receptor (ER) and the human epidermal growth factor (HER2) on breast cancer tissue sample in TG multiplexed-mode (see Fig. 3.5).

The most used lanthanide compound in TG microscopy these days is Lumi4®-Tb, which has near optimal emission properties for a Tb^{3+} complex with an absorption maximum at *ca.* 339 nm (extending to about 365 nm), an overall quantum yield of more than 50 % and a PL lifetime of *ca.* 2.6 ms in water [39]. It is watersoluble, easy to bioconjugate, and has limited non-specific interactions with cells, plastics or glass due to its (near) neutral charge [107–108]. Recently, a novel series of compounds has been developed based on nonadentate triazacyclononane compounds [98]. They combine the chromophores proposed by Maury et al.

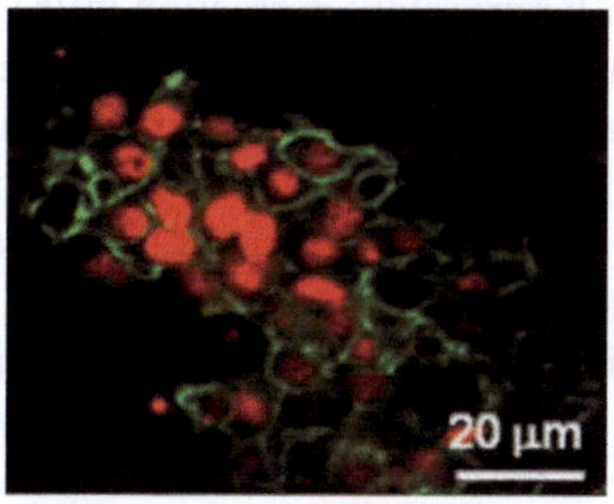

Fig. 3.5 On-chip immunohistochemical detection of HER2/neu and ER in a breast cancer tissue sample. TG image, HER2/neu detected by green emitting TbB5 and ER stained with red-emitting EuB4 (Reproduced with permission from Ref. [106]. Copyright 2010 The Royal Society of Chemistry)

[104, 109] and a triazacyclononane moiety [110] to produce very bright and kinetically stable complexes. By changing the electron donating moiety on the outer phenyl-ring, the absorption maximum can be shifted in order to be better adapted for microscopy purposes [70, 111–112]. The water-solubility of these compounds was further increased using either sulfonates [113–114], carboxylic acid [111, 113–114], or sulfobetaine moieties [115]. Bioconjugation was achieved by either attaching vectors on the outer rim [111] or on the triazacyclonane (TACN) moiety [113], which, in combination with the sulfonate groups, avoids non-specific labeling to cells due a reduction of hydrophobic interactions of the aryl-alkynyl groups. Using benzylguanine functionalization these compound could be covalently labeled to SNAP-tagged [116] cholecystokinin receptors expressed in HEK293 cells which were imaged in TG mode [113]. Terbium [112, 117], ytterbium [103, 118], and samarium [118] complexes were exploited as well, mostly using two-photon excitation to image these often very bright complexes. Multiplexed imaging without biological relevance of Tb^{3+} and Eu^{3+} TACN complexes was possible too [117].

Cytometry multiplexing using fluorescence has been achieved with up to 17 different colors [119]. TG microscopy for cytometry purposes has recently seen considerable progress as well, mainly due to the efforts of Jin, Piper, and coworkers [84, 120–125]. In 2009 they described the development of the incorporation of an Eu^{3+} chelate absorbing in the visible into monodisperse 50 nm silicananoparticles [124]. Using these streptavidin labeled nanoparticles they were able to detect with TG microscopy the intestinal parasite *Giardia lamblia* using anti-Giardia antibody and a biotinylated secondary antibody in concentrated water samples eliminating autofluorescence. Similar results were reported using streptavidin-bovine serum albumin-labeled chelate [121]. Improving their setup resulted in easy detection of *Giardia lamblia* in waste water or *Cryptosporidium parvum* oocyst in fruit juice concentrate [125]. Using a streptavidin labeled chelate or streptavidin-conjugated nanoparticles incorporating this chelate it was possible to easily detect low expressing CD34 cells to CD34 negative cells using a biotinylated antiCD34 antibody [120]. Multiplexed luminescence detection was achieved as well using a multicolor TG microscope. By labeling specific antibodies with a Eu^{3+} or a Tb^{3+} chelate *Giardia lamblia* and *Cryptosporidium parvum* could be detected in different channels simultaneously after crosstalk correction [126].

Scanning a complete microscope slide using time gating can take significant time due to the reduced number of photons detected per unit time compared to regular fluorescence microscopy. By introducing [123] TG-orthogonal scanning automated microscopy (TG-OSAM) the precise location of luminescent events can be detected much faster (up to 15 times). This technology was then further developed into the time-domain by engineering of 40 nm $NaYF_4$:Yb, Tm nanocrystals [84, 122]. By enhancing the doping ratio of thulium from 0.2 to 4 mol% compared to ytterbium (20 mol%) they were able to tune the lifetime of 8 different composites from 669 μs to 49 μs with only a small bandwidth. This technology opens optical multiplexing applications in life sciences, medicine, and data security. An example of an application of these 'τ-dots' in lifetime-encoded document security is shown in Fig. 3.6.

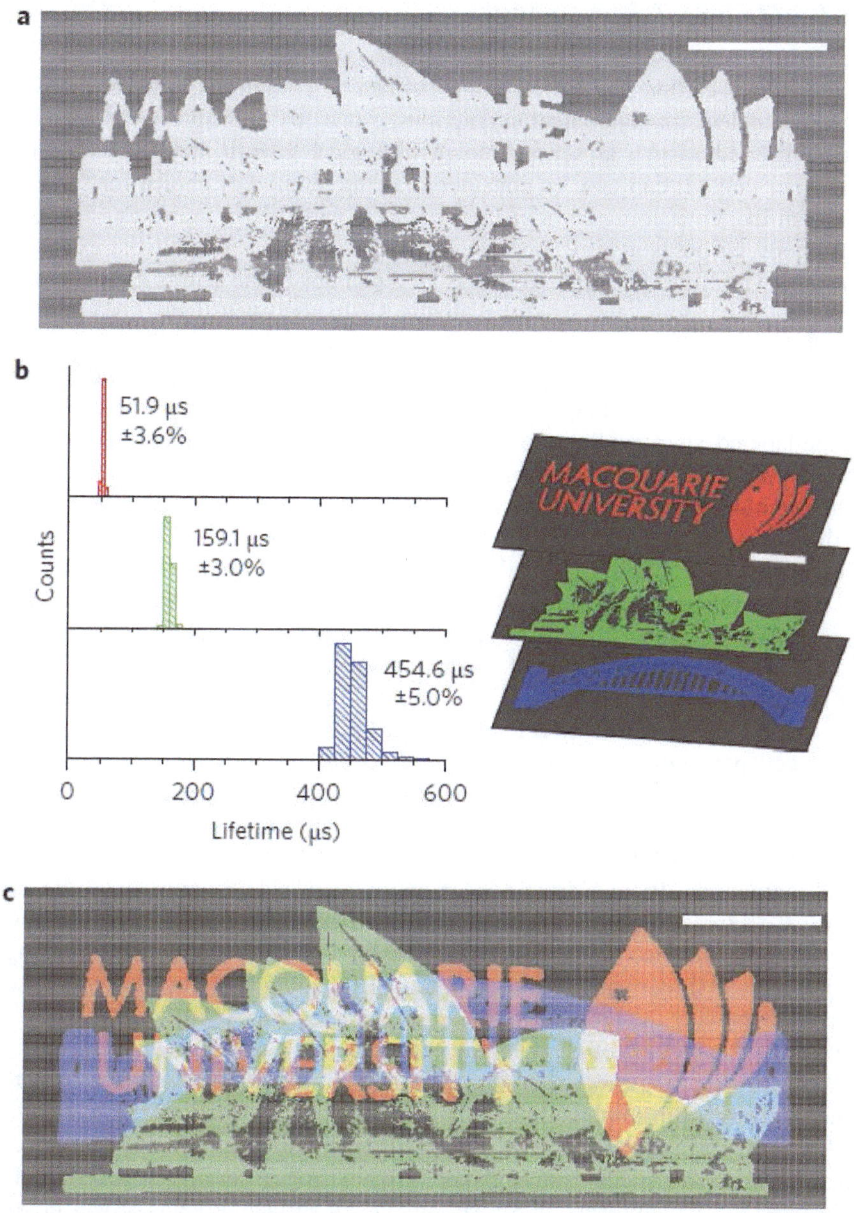

Fig. 3.6 Demonstration of lifetime-encoded document security and photonic data storage. **a–c**, Three overlapping patterns are printed with different Tm τ-dots: ([Yb]:[Tm]) 20:4 for the 'Macquarie University' logo, 20:1 for the Sydney Opera House image, and 20:0.5 for the Sydney Harbour Bridge image. Intensity-based luminescence imaging only gives a complex picture (**a**), but time-resolved scanning separates the patterns based on the lifetime components of every pixel (**b**), so that genuine multiplexeding information, contained in the same overlapping space of the document, can be decoded (**c**; pseudocolor is used to indicate the luminescence lifetime for each pixel). Scale bars (all images), 5 mm (Reproduced with permission from Ref. [84]. Copyright 2014 Macmillan Publishers Limited (Nature Photonics))

3.5.3 TG FRET Biosensing

As mentioned earlier, TG FRET using lanthanide complexes is a very powerful technique in detecting biological interactions, especially in the presence of optically disturbing media, such as serum, blood, biological growth media, colored compounds, or cells. It can now be considered as the method of choice for screening. The atom-like emission spectra of lanthanides is particularly interesting for FRET-microscopy purposes as well, because the FRET signal can be almost exclusively measured with little spectral crosstalk from the donor if one adapts its acceptor dye to the optical windows available for each dye. Up to now exclusively Tb^{3+} and Eu^{3+} complexes have been used for TG FRET microscopy although FRET microscopy (without time gating) using upconverting nanoparticles has been shown to work as well. An early observation concerning TG FRET microscopy was when Tb^{3+} chelate-labeled streptavidin was mixed with sufficiently high concentrations of biotinylated quantum dots. Aggregates where formed that showed delayed emission in the quantum dot emissive region [127]. The first cellular applications were guided by the development of Lumi4-Tb. In 2010, Rajapakse et al. reported [128] the recognition of the antibiotic trimethoprim (TMP)-labeled Lumi4-Tb to *Escherichia coli* dihydrofolate reductase (eDHFR) fused to green fluorescent protein (GFP) expressed in the cytosol after membrane permeabilization. A delayed GFP emission lifetime of 0.78 ms was reported indicating a FRET process between Lumi4-Tb and GFP. Addition of non-labeled TMP removed the TG GFP emission indicating specific binding between eDHFR and TMP. Following up on this study was the use of cell penetrating peptides to internalize TMP-Lumi4-Tb derivatives [129]. Expressing the nuclear protein histone-2B to a red fluorescent protein (TagRFP-T) linked to eDHFR in MDCKII cells revealed that upon adding TMP-Lumi4-Tb derivative a TG-FRET signal was observed indicating the successful delivery of these compounds. Protein-protein interactions were also monitored by the same group. Co-expressing GFP-FKBP12 and FRB-eDHFR in MDCKII cells were exposed to rapamycin, which induces dimerization of FKBP12 and FRB. Adding of a Lumi4-Tb-TMP derivative resulted in TG FRET images of the delayed emission from GFP showing the presence of the dimers in the cytosol [74].

FRET is the method of choice to detect biological interactions and is especially powerful in detecting protein-protein interactions. Previously, we have seen that, using TG-FRET in combination with suicide enzymes such as the SNAP-tag® and/or CLIP-tag®, it is possible to detect oligomerization of GPCR in model cells [43, 57, 108, 130] and in native tissue [107]. Using a benzylguanine-Lumi4-Tb derivative [24], which reacts on SNAP-tagged cell surface proteins, it was possible to visualize Lumi4-Tb labeled GPCR at the cell surface with TG microscopy and to follow the internalization of the VPAC1-SNAP protein after stimulation with its natural agonist VIP [24]. The bright water soluble Eu^{3+} TACN complexes mentioned previously were also conjugated to the benzylguanine moiety and covalently attached to the cholecystokinin-2-SNAP receptor [113]. Using TG FRET microscopy the interaction of a fluorescently labeled agonist to the receptor could be monitored. Biotinylated Eu-TACN complexes were shown to internalize through macropinocytosis and label lysosomes of NIH-3 T3 cells [70, 111]. After adding Alexa633-labeled streptavidin, TG-FRET imaging at the lysosomes was possible, including spectral imaging and lifetime imaging of the whole FOV. Faklaris et al. showed [131] that using such suicide systems labeled with Lumi4-Tb it is possible to monitor TG-FRET on living cells between Lumi4-Tb and a green and a red fluorophore and two different quantum dots in four different emission channels. Expressing different cells with either CLIP-Vasopressin-2, Halo-Dopamine-2 or SNAP-Dopamine-2 receptors and adding their respective Lumi4-Tb and fluorophore substrates, the homo-oligomerization could be monitored in separate channels. Interestingly, using this technology it was possible to monitor that V2 or V1a oligomers can internalize. Moreover, when both V2 and V1a were co-expressed, stimulation with their agonist AVP led to preferential V2 oligomer internalization. If V2 was blocked with a V2 specific antagonist, the V1a-V2 internalized (cf. Fig. 3.7) after stimulation with agonist. Using Lumi4-Tb-NAPS (a D2 receptor antagonist) [107] and NAPS-red, D2 receptor oligomerization could be monitored using TG FRET microscopy as well.

Another bio-orthogonal strategy using SNAP, CLIP, and eDHFR targeting cell penetrating Lumi4-Tb substrates was used by Zou et al. [133]. They showed that simultaneous labeling of intracellularly expressed proteins such as Histone2B by these substrates is possible and can be monitored by multiplexed TG FRET imaging using green (GFP) and red (mCherry) fluorescent protein constructs.

Detection of cell surface receptor expression and dimerization using antibody labeling and TG FRET has some history [24, 134–137]. Linden et al. showed its applicability in TG FRET microscopy as well [138], being able to detect the interaction between two antibodies labeled with Lumi4-Tb and either FITC or AlexaFluor dyes on E-cadherin expressing MCF-7 cells. Recently, the same group [139] showed that by anti- or nanobody labeling of quantum dots and Lumi4-Tb, it was possible to detect EGFR expression on A431 cells using TG FRET microscopy. Interestingly, they could also demonstrate microinjection of Lumi4-Tb-quantum dot and Lumi4-Tb-quantum dot-dye nanoassemblies to the cytosol and intracellular Tb to quantum dot FRET and Tb to quantum dot to dye FRET relays [19, 140]. Endocytosis of nanoassemblies containing up to 40 cell penetrating peptides and 40 Lumi4-Tb per quantum dot and TG Tb to quantum dot FRET in endosomes or lysosomes was observed as well. These FRET nanoassemblies showed enhanced intramolecular TG FRET and constructs based on these principles might be interesting to use in future (multiplexed) biosensor applications.

In contrast to the upconversion ‘τ-dots’ for cytometry purposes (*vide supra*), Lu et al. also developed downconversion τ-dots [122] by doping porous polystyrene beads with the Eu^{3+} complex of thenoyltrifluoroacetonate and various concentrations of a cationic coumarin, which reduced the PL decay time of the Eu complex due to FRET. Five different τ-dots were obtained with lifetime variations between 1.9 and 5.7 %, sufficiently low to discriminate the particles. Four different pathogen DNAs, i.e., human immunodeficiency virus (HIV), Ebola virus (EV), hepatitis

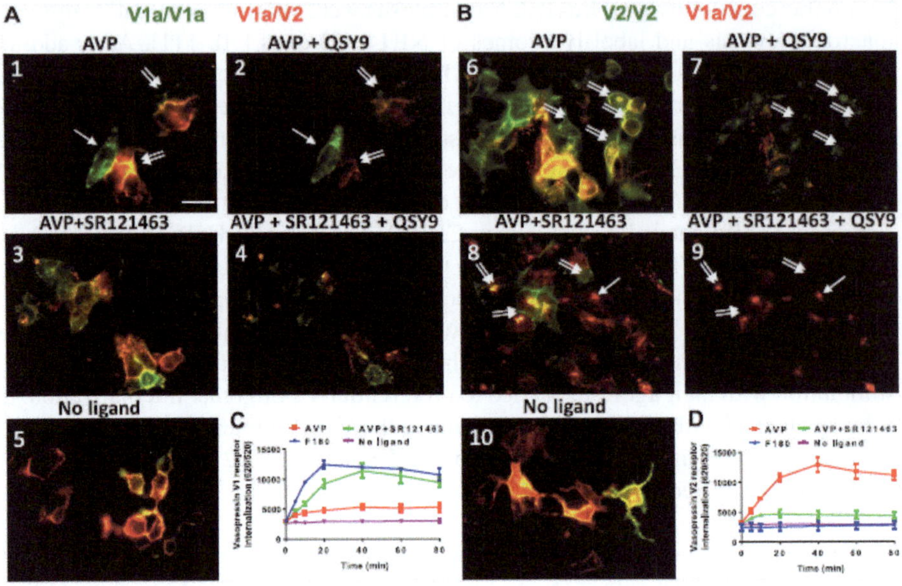

Fig. 3.7 Vasopressin V1a and V2 oligomer internalization when receptors are coexpressed. HEK 293 cells expressing vasopressin SNAP-V1a and CLIP-V2 receptors are labeled with (A) SNAP-Lumi4-Tb, SNAP-green, and CLIP-red leading to a green FRET for V1a oligomers and red FRET for V1a/V2 hetero-oligomers, or with (B) CLIP-Lumi4-Tb, CLIP-green, and SNAP-red leading to a green FRET for V2 oligomers and red FRET for V1a/V2 hetero-oligomers. As indicated on the figure, cells were incubated in the presence of vasopressin (25 nM) (images 1, 2, 6, and 7) or vasopressin (25 nM) and SR121463 (250 nM) (images 3, 4, 8, and 9) or without any ligand (images 5–10). Images 2, 4, 7, and 9 were acquired after addition of QSY9 (10 mM). Quantification of vasopressin V1a (C) and V2 (D) receptor internalization on cells plated in 96-well plates. [132] Receptors, labeled with SNAP-Lumi4-Tb and CLIP-Lumi4-Tb, respectively, were incubated in the presence of an excess of free acceptor (24 mM). Ratio of the emissions 620:520 was plotted as a function of time. Single or double arrows point to cells expressing only 1 or 2 receptors, respectively. Scale bar, 25 μm (Reproduced with permission from Ref. [131]. Copyright 2015 FASEB)

B virus (HBV) and human papillomavirus 16 (HPV-16), could be detected. The complementary strands for the target DNAs were linked to four types of these microspheres and using the TR-OSAM approach [84] multiplexed detection of these four microspheres could be decoded.

Although approaches for (multiplexed) FRET imaging of lanthanide based upconverting nanoparticles have been suggested [141], such particles have to our knowledge not been used for TG applications. Obviously, excitation in the NIR already reduces the background for in vivo applications considerably, but in combination with time gating more reduction of background might be achieved in the near future.

3.6 Conclusions and Future Perspectives

Progress in detection technologies and development of brighter and better lanthanide complexes in combination with an almost infinite choice of acceptor molecules and nanoparticles have advanced TG FRET from a very specialized and rarely used technique to a stand-alone technology that has proven to be very useful for many biosensing applications both for single and multiple analyte detection. In particular the spectral and temporal photophysical advantages of lanthanide complexes to perform homogeneous (no laborious separation and washing procedures), rapid, simple, and still highly sensitive FRET assays are responsible for the ever growing interest in TG FRET. In spectroscopic applications, often on fluorescence plate readers, TG FRET has reached a technical maturity that has proven to complement or even outperform other fluorescence or non-fluorescence based technologies. Recently, also TG FRET microscopy has witnessed a renaissance and especially in this field we can expect many contributions and new developments concerning multiplexed and sensitive biosensing in the near future. Applications such as nucleic acid and protein assays, study of enzyme kinetics, the design of molecular logic gates, or the investigation of extracellular protein-protein interactions and intracellular detection can largely profit from TG FRET and the recent progress toward versatility and wide applicability of this powerful optical biosensing technology promises interesting research studies and the development of commercial applications. We hope that this chapter has provided an interesting and informative overview of TG FRET technology and its application and we are confident that the coming years will witness many new and innovative TG FRET applications for multiplexed biosensing.

Acknowledgements NH thanks the Institut Universitaire de France.

References

1. Haugland RP (2005) *The handbook. a guide to fluorescent probes and labeling technologies*, 10 edn. Invitrogen, San Diego
2. Hötzer B, Medintz IL, Hildebrandt N (2012) Fluorescence in nanobiotechnology: sophisticated fluorophores for novel applications. Small 8(15):2297–2326
3. Sapsford KE, Berti L, Medintz IL (2006) Materials for fluorescence resonance energy tansfer analysis: beyond traditional donor-acceptor combinations. Angew Chem Int Ed 45(28): 4562–4589
4. Sapsford KE, Wildt B, Mariani A, Yeatts AB, Medintz IL (2014) Materials for FRET analysis: beyond traditonal dye-dye combinations. In: Medintz IL, Hildebrandt N (eds) FRET – Förster Resonance Energy Transfer. From theory to applications. Wiley-VCH, Singapore
5. Andrews DL, Demidov AA (1999) Resonance energy transfer. Wiley, Chichester
6. Clegg RM (1996) Fluorescence resonance energy transfer. In: Wang XF, Herman B (eds) Fluorescence imaging spectroscopy and microscopy. Wiley, New York, pp 179–251

7. Clegg RM (2009) Förster resonance energy transfer - FRET what is it, why do it, and how it's done. In: Gadella TWJ (ed) Laboratory techniques in biochemistry and molecular biology. Academic, Burlington, pp 1–57
8. Gadella TWJ. (2009); FRET and FLIM techniques. Elsevier, Amsterdam/London p xxiii, 534 p
9. Jares-Erijman EA, Jovin TM (2003) FRET imaging. Nat Biotechnol 21(11):1387–1395
10. Medintz IL, Hildebrandt N (2014) FRET - Förster Resonance Energy Transfer. From theory to applications. Wiley-VCH, Weinheim, p. 105
11. Periasamy A, Day RN (2005) Molecular imaging : FRET microscopy and spectroscopy. Oxford University Press, New York; Oxford. p xv, 312 p
12. Selvin PR (1996) Lanthanide-based resonance energy transfer. IEEE J Sel Top Quant Electron 2(4):1077–1087
13. Vogel, S. S.; Thaler, C.; Koushik, S. V. (2006) Fanciful FRET.. Sci. STKE *2006* (331), re2
14. Lakowicz J R (2006) Principles of fluorescence spectroscopy, 3rd ed. Springer, New York. p xxvi, 954 p
15. Turro NJ, Ramamurthy V, Scaiano JC (2009) Principles of molecular photochemistry: an introduction. University Science Books, Sausalito
16. Turro N J, Ramamurthy V, Scaiano J C (2010) Modern molecular photochemistry of organic molecules. University Science Books, Sausalito, Calif. p xxxiii, 1084 p
17. Valeur B, Berberan-Santos MN (2013) Molecular fluorescence: principles and applications, 2nd edn. WIley-VCH Verlag & Co. KGaA, Weinheim
18. Algar WR, Kim H, Medintz IL, Hildebrandt N (2014) Emerging non-traditional Förster resonance energy transfer configurations with semiconductor quantum dots: Investigations and applications. Coord Chem Rev 263:65–85
19. Hildebrandt N, Wegner KD, Algar WR (2014) Luminescent terbium complexes: superior Förster resonance energy transfer donors for flexible and sensitive multiplexed biosensing. Coord Chem Rev 273:125–138
20. Cardoso Dos Santos M, Hildebrandt N (2016) Recent developments in lanthanide-to-quantum dot FRET using time-gated fluorescence detection and photon upconversion. TrAC Trends Anal Chem 84:60–71
21. Sy M, Nonat A, Hildebrandt N, Charbonniere LJ (2016) Lanthanide-based luminescence biolabelling. Chem Commun 52(29):5080–5095
22. Bünzli JCG (2010) Lanthanide luminescence for biomedical analyses and imaging. Chem Rev 110(5):2729–2755
23. Eliseeva SV, Bünzli JCG (2010) Lanthanide luminescence for functional materials and bio-sciences. Chem Soc Rev 39(1):189–227
24. Zwier JM, Bazin H, Lamarque L, Mathis G (2014) Luminescent lanthanide cryptates: from the bench to the bedside. Inorg Chem 53(4):1854–1866
25. Montgomery CP, Murray BS, New EJ, Pal R, Parker D (2009) Cell-penetrating metal complex optical probes: targeted and responsive systems based on lanthanide luminescence. Acc Chem Res 42(7):925–937
26. New EJ, Parker D, Smith DG, Walton JW (2010) Development of responsive lanthanide probes for cellular applications. Curr Opin Chem Biol 14:238–246
27. Scholler P, Zwier JM, Trinquet E, Rondard P, Pin J-P, Prézeau L, Kniazeff J (2013) Time-resolved Förster resonance energy transfer-based technologies to investigate G protein-coupled receptor machinery: high-throughput screening and future development. Prog Mol Biol Transl Sci 113:275–312
28. Charbonnière LJ, Hildebrandt N (2008) Lanthanide complexes and quantum dots: a bright wedding for resonance energy transfer. Eur J Inorg Chem 21:3241–3251
29. Geißler D, Linden S, Liermann K, Wegner KD, Charbonnière LJ, Hildebrandt N (2014) Lanthanides and quantum dots as Förster resonance energy transfer agents for diagnostics and cellular imaging. Inorg Chem 53(4):1824–1838
30. Geißler D, Hildebrandt N (2011) Lanthanide complexes in FRET applications. Curr Inorg Chem 1(1):17–35

31. Selvin PR (2002) Principles and biophysical applications of lanthanide-based probes. Annu Rev Biophys Biomol Struct 31:275–302
32. Trinquet E, Mathis G (2006) Fluorescence technologies for the investigation of chemical libraries. Mol BioSyst 2(8):380–387
33. Mathis G (1995) Probing molecular interactions with homogeneous techniques based on rare earth cryptates and fluorescence energy transfer. Clin Chem 41:1391–1397
34. Jin ZW, Geißler D, Qiu X, Wegner KD, Hildebrandt N (2015) A rapid, amplification-free, and sensitive diagnostic assay for single-step multiplexed fluorescence detection of microRNA. Angewandte Chemie-International Edition 54(34):10024–10029
35. Alpha B, Lehn JM, Mathis G (1987) Energy-transfer luminescence of europium(III) and terbium(III) cryptates of macrobicyclic polypyridine ligands. Angewandte Chemie-International Edition 26(3):266–267
36. Degorce F, Card A, Soh S, Trinquet E, Knapik GP, Xie B (2009) HTRF: a technology tailored for drug discovery – a review of theoretical aspects and recent applications. Curr Inorg Chem 3:22–32
37. Cisbio Bioassays HTRF® Technology (2016) http://www.cisbio.com/drug-discovery/htrf-technology
38. Cezanne/BRAHMS/ThermoFisher Scientific: TRACE-Technology (2016) http://www.brahms-instruments.com/
39. Xu J, Corneillie TM, Moore EG, Law G-L, Butlin NG, Raymond KN (2011) Octadentate cages of Tb(III) 2-hydroxyisophthalamides: a new standard for luminescent lanthanide labels. J Am Chem Soc 133(49):19900–19910
40. Lumiphore (2016) http://www.lumiphore.com
41. Trinquet E, Maurin F, Preaudat M, Mathis G (2001) Allophycocyanin 1 as a near infrared fluorescent tracer: isolation, characterization, chemical modification and use in a homogeneous fluorescence energy transfer system. Anal Biochem 296:126–135
42. Zwier JM, Roux T, Cottet M, Durroux T, Douzon S, Bdioui S, Gregor N, Bourrier E, Oueslati N, Nicolas L, Tinel N, Boisseau C, Yverneau P, Charrier-Savournin F, Fink M, Trinquet E (2010) A fluorescent ligand-binding alternative using Tag-lite® technology. J Biomol Screen 15(10):1248–1259
43. Doumazane E, Scholler P, Fabre L, Zwier JM, Trinquet E, Pin J-P, Rondard P (2013) Illuminating the activation mechanisms and allosteric properties of metabotropic glutamate receptors. Proc Natl Acad Sci 110(15):E1416–E1425
44. Trinquet E, Fink M, Bazin H, Grillet F, Maurin F, Bourrier E, Ansanay H, Leroy C, Michaud A, Durroux T, Maurel D, Malhaire F, Goudet C, Pin J-P, Naval M, Hernout O, Chrétien F, Chapleur Y, Mathis G (2006) d-myo-Inositol 1-phosphate as a surrogate of d-myo-inositol 1,4,5-tris phosphate to monitor G protein-coupled receptor activation. Anal Biochem 358(1):126–135
45. Perkin Elmer LANCE Technology (2016) http://www.perkinelmer.com/category/lance-tr-fret
46. Hemmilä I, Webb S (1997) Time-resolved fluorometry: an overview of the labels and core technologies for drug screening applications. Drug Discov Today 2(9):373–381
47. ThermoFisher Scientific LanthaScreen Technology (2016) http://www.thermofisher.com/us/en/home/industrial/pharma-biopharma/drug-discovery-development/target-and-lead-identification-and-validation/kinasebiology/kinase-activity-assays/lanthascreen-tr-fret-toolbox.html
48. Ge P, Selvin PR (2008) New 9-or 10-dentate luminescent lanthanide chelates. Bioconjug Chem 19(5):1105–1111
49. Geißler D, Stufler S, Löhmannsröben H-G, Hildebrandt N (2013) Six-color time-resolved förster resonance energy transfer for ultrasensitive multiplexed biosensing. J Am Chem Soc 135(3):1102–1109
50. Hilal T, Puetter V, Otto C, Parczyk K, Bader B (2010) A dual estrogen receptor TR-FRET assay for simultaneous measurement of steroid site binding and coactivator recruitment. J Biomol Screen 15(3):268–278

51. Jeyakumar M, Katzenellenbogen JA (2009) A dual-acceptor time-resolved Förster resonance energy transfer assay for simultaneous determination of thyroid hormone regulation of corepressor and coactivator binding to the thyroid hormone receptor: mimicking the cellular context of thyroid hormone action. Anal Biochem 386(1):73–78
52. Kim SH, Gunther JR, Katzenellenbogen JA (2010) Monitoring a coordinated exchange process in a four-component biological interaction system: development of a time-resolved terbium-based one-donor/three-acceptor multicolor FRET system. J Am Chem Soc 132(13):4685–4692
53. Kokko T, Kokko L, Soukka T (2009) Terbium(III) chelate as an efficient donor for multiple-wavelength fluorescent acceptors. J Fluoresc 19(1):159–164
54. Kokko T, Liljenback T, Peltola MT, Kokko L, Soukka T (2008) Homogeneous dual-parameter assay for prostate-specific antigen based on fluorescence resonance energy transfer. Anal Chem 80(24):9763–9768
55. Kupcho KR, Stafslien DK, DeRosier T, Hallis TM, Ozers MS, Vogel KW (2007) Simultaneous monitoring of discrete binding events using dual-acceptor terbium-based LRET. J Am Chem Soc 129(44):13372
56. Baldo B, Paganetti P, Grueninger S, Marcellin D, Kaltenbach LS, Lo DC, Semmelroth M, Zivanovic A, Abramowski D, Smith D, Lotz GP, Bates GP, Weiss A (2012) TR-FRET-based duplex immunoassay reveals an inverse correlation of soluble and aggregated mutant huntingtin in Huntington's Disease. Chem Biol 19(2):264–275
57. Comps-Agrar L, Kniazeff J, Brock C, Trinquet E, Pin J-P (2012) Stability of GABAB receptor oligomers revealed by dual TR-FRET and drug-induced cell surface targeting. FASEB J 26(8):3430–3439
58. Bhuckory S, Lefebvre O, Qiu X, Wegner KD, Hildebrandt N (2016) Evaluating quantum dot performance in homogeneous FRET immunoassays for prostate specific antigen. Sensors 16(2):197
59. Geißler D, Charbonniere LJ, Ziessel RF, Butlin NG, Löhmannsröben H-G, Hildebrandt N (2010) Quantum dot biosensors for ultrasensitive multiplexed diagnostics. Angew Chem Int Edn 49(8):1396–1401
60. Geißler D, Hildebrandt N (2016) Recent developments in Förster resonance energy transfer (FRET) diagnostics using quantum dots. Anal Bioanal Chem 408(17):4475–4483
61. Morgner F, Stufler S, Geißler D, Medintz IL, Algar WR, Susumu K, Stewart MH, Blanco-Canosa JB, Dawson PE, Hildebrandt N (2011) Terbium to quantum dot FRET bioconjugates for clinical diagnostics: influence of human plasma on optical and assembly properties. Sensors 11(10):9667–9684
62. Qiu X, Hildebrandt N (2015) Rapid and multiplexed microRNA diagnostic assay using quantum dot-based forster resonance energy transfer. ACS Nano 9(8):8449–8457
63. Wegner KD, Jin ZW, Linden S, Jennings TL, Hildebrandt N (2013) Quantum-dot-based förster resonance energy transfer immunoassay for sensitive clinical diagnostics of low-volume serum samples. ACS Nano 7(8):7411–7419
64. Wegner KD, Lanh PT, Jennings T, Oh E, Jain V, Fairclough SM, Smith JM, Giovanelli E, Lequeux N, Pons T, Hildebrandt N (2013) Influence of luminescence quantum yield, surface coating, and functionalization of quantum dots on the sensitivity of time-resolved FRET bioassays. ACS Appl Mater Interfaces 5:2881–2892
65. Wegner KD, Linden S, Jin ZW, Jennings TL, el Khoulati R, van Bergen en Henegouwen PMP, Hildebrandt N (2014) Nanobodies and nanocrystals: highly sensitive quantum dot-based homogeneous FRET immunoassay for serum-Based EGFR Detection. Small 10(4):734–740
66. Connally R, Jin D, Piper J (2006) High intensity solid-state UV source for time-gated luminescence microscopy. Cytometry Part A 69A(9):1020–1027
67. Gahlaut N, Miller LW (2010) Time-resolved microscopy for imaging lanthanide luminescence in living cells. Cytometry Part A 77A(12):1113–1125

68. Grichine A, Haefele A, Pascal S, Duperray A, Michel R, Andraud C, Maury O (2014) Millisecond lifetime imaging with a europium complex using a commercial confocal microscope under one or two-photon excitation. Chem Sci 5(9):3475–3485
69. Ramshesh VK, Lemasters JJ (2008) Pinhole shifting lifetime imaging microscopy. J Biomed Opt 13(6):064001–064010
70. Butler SJ, Delbianco M, Lamarque L, McMahon BK, Neil ER, Pal R, Parker D, Walton JW, Zwier JM (2015) EuroTracker® dyes: design, synthesis, structure and photophysical properties of very bright europium complexes and their use in bioassays and cellular optical imaging. Dalton Trans 44(11):4791–4803
71. Pal R (2015) Phase modulation nanoscopy: a simple approach to enhanced optical resolution. Faraday Discuss 177:507–515
72. Liao Z, Tropiano M, Faulkner S, Vosch T, Sorensen TJ (2015) Time-resolved confocal microscopy using lanthanide centred near-IR emission. RSC Adv 5(86):70282–70286
73. Rajendran M, Miller LW (2015) Evaluating the performance of time-gated live-cell microscopy with lanthanide Probes. Biophys J 109(2):240–248
74. Rajendran M, Yapici E, Miller LW (2014) Lanthanide-based imaging of protein–protein interactions in live cells. Inorg Chem 53(4):1839–1853
75. Vicidomini G, Moneron G, Han KY, Westphal V, Ta H, Reuss M, Engelhardt J, Eggeling C, Hell SW (2011) Sharper low-power STED nanoscopy by time gating. Nat Methods 8(7):571–573
76. Vaasa A, Ligi K, Mohandessi S, Enkvist E, Uri A, Miller LW (2012) Time-gated luminescence microscopy with responsive nonmetal probes for mapping activity of protein kinases in living cells. Chem Commun 48(68):8595–8597
77. Soini E, Lövgren T, Reimer CB (1987) Time-resolved fluorescence of lanthanide probes and applications in biotechnology. CRC Crit Rev Anal Chem 18(2):105–154
78. Beverloo HB, van Schadewijk A, Bonnet J, van der Geest R, Runia R, Verwoerd NP, Vrolijk J, Ploem JS, Tanke HJ (1992) Preparation and microscopic visualization of multicolor luminescent immunophosphors. Cytometry 13(6):561–570
79. Seveus LVM, Syrjänen S, Sandberg M, Kuusisto A, Harju R, Salo J, Hemmilä I, Kojola H, E. S (1992) Time-resolved fluorescence imaging of europium chelate label in immunohistochemistry and in situ hybridization. Cytometry 13(4):329–338
80. Marriott G, Heidecker M, Diamandis EP, Yan-Marriott Y (1994) Time-resolved delayed luminescence image microscopy using an europium ion chelate complex. Biophys J 67(3):957–965
81. Vereb G, Jares-Erijman E, Selvin PR, Jovin TM (1998) Temporally and spectrally resolved imaging microscopy of lanthanide chelates. Biophys J 74(5):2210–2222
82. Moore EG, Samuel APS, Raymond KN (2009) From antenna to assay: lessons learned in lanthanide luminescence. Acc Chem Res 42(4):542–552
83. Härmä H, Soukka T, Lövgren T (2001) Europium nanoparticles and time resolved fluorescence for ultrasensitive detection of prostate specific antigen. Clin Chem 47:561–568
84. Lu Y, Zhao J, Zhang R, Liu Y, Liu D, Goldys EM, Yang X, Xi P, Sunna A, Lu J, Shi Y, Leif RC, Huo Y, Shen J, Piper JA, Robinson JP, Jin D (2014) Tunable lifetime multiplexing using luminescent nanocrystals. Nat Photonics 8(1):32–36
85. Wang F, Deng R, Wang J, Wang Q, Han Y, Zhu H, Chen X, Liu X (2011) Tuning upconversion through energy migration in core–shell nanoparticles. Nat Mater 10(12):968–973
86. Beeby A, Botchway SW, Clarkson IM, Faulkner S, Parker AW, Parker D, Williams JAG (2000) Luminescence imaging microscopy and lifetime mapping using kinetically stable lanthanide(III) complexes. Journal of Photochemistry and Photobiology B: Biology 57:83–89
87. Charbonnière L, Ziessel R, Guardigli M, Roda A, Sabbatini N, Cesario M (2001) Lanthanide tags for time-resolved luminescence microscopy displaying improved stability and optical properties. J Am Chem Soc 123(10):2436–2437

88. Hanaoka K, Kikuchi K, Kobayashi S, Nagano T (2007) Time-resolved long-lived luminescence imaging method employing luminescent lanthanide probes with a new microscopy system. J Am Chem Soc 129(44):13502–13509
89. Liu M, Ye Z, Wang G, Yuan J (2012) Development of a novel europium(III) complex-based luminescence probe for time-resolved luminescence imaging of the nitric oxide production in neuron cells. Talanta 99:951–958
90. Liu M, Ye Z, Xin C, Yuan J (2013) Development of a ratiometric time-resolved luminescence sensor for pH based on lanthanide complexes. Anal Chim Acta 761:149–156
91. Montgomery CP, New EJ, Palsson LO, Parker D, Batsanov AS, Lamarque L (2009) Emissive and Cell-Permeable 3-Pyridyl- and 3-Pyrazolyl-4-azaxanthone lanthanide complexes and their behaviour in cellulo. Helvetica Chimica Acta 92(11):2186–2213
92. Pandya S, Yu J, Parker D (2006) Engineering emissive europium and terbium complexes for molecular imaging and sensing. Dalton Trans 23:2757–2766
93. Poole RA, C. M, EJ N, Congreve A, Parker D, Botta M (2007) Identification of emissive lanthanide complexes suitable for cellular imaging that resist quenching by endogenous antioxidants. Org Biomol Chem 5(13):2055–2062
94. Smith DG, McMahon BK, Pal R, Parker D (2012) Live cell imaging of lysosomal pH changes with pH responsive ratiometric lanthanide probes. Chem Commun 48(68):8520–8522
95. Song B, Ye Z, Yang Y, Ma H, Zheng X, Jin D, Yuan J (2015) Background-free in-vivo imaging of Vitamin C using time-gateable responsive probe. Sci Rep 5:14194
96. Song B, Wang G, Tan M, Yuan J (2006) A europium(III) complex as an efficient singlet oxygen luminescence probe. J Am Chem Soc 128(41):13442–13450
97. Sun J, Song B, Ye Z, Yuan J (2015) Mitochondria targetable time-gated luminescence probe for singlet oxygen based on a β-Diketonate–Europium complex. Inorg Chem 54(24):11660–11668
98. Walton JW, Bourdolle A, Butler SJ, Soulie M, Delbianco M, McMahon BK, Pal R, Puschmann H, Zwier JM, Lamarque L, Maury O, Andraud C, Parker D (2013) Very bright europium complexes that stain cellular mitochondria. Chem Commun 49(16):1600–1602
99. D'Aléo A, Pompidor G, Elena B, Vicat J, Baldeck PL, Toupet L, Kahn R, Andraud C, Maury O (2007) Two-photon microscopy and spectroscopy of lanthanide bioprobes. ChemPhysChem 8(14):2125–2132
100. Piszczek G, Maliwall BP, Gryczynskil I, Dattelbaum J, Lakowicz JR (2001) Multiphoton Ligand-Enhanced Excitation of Lanthanides. J Fluoresc 11(2):101–107
101. Werts MHV, Nerambourg N, Pelegry D, Le Grand Y, Blanchard-Desce M (2005) Action cross sections of two-photon excited luminescence of some Eu(III) and Tb(III) complexes. Photochem Photobiol Sci 4(7):531–538
102. Bourdolle A, D'Aléo A, Philippot C, Baldeck PL, Guyot Y, Dubois F, Ibanez A, Andraud C, Brasselet S, Maury O (2016) NIR-to-NIR two-photon scanning laser microscopy imaging of single nanoparticles doped by YbIII complexes. ChemPhysChem 17(1):128–135
103. D'Aléo A, Bourdolle A, Brustlein S, Fauquier T, Grichine A, Duperray A, Baldeck PL, Andraud C, Brasselet S, Maury O (2012) Ytterbium-based bioprobes for near-infrared two-photon scanning laser microscopy imaging. Angew Chem 124(27):6726–6729
104. Picot A, D'Aléo A, Baldeck PL, Grichine A, Duperray A, Andraud C, Maury O (2008) Long-lived two-photon excited luminescence of water-soluble europium complex: applications in biological imaging using two-photon scanning microscopy. J Am Chem Soc 130(5):1532–1533
105. Ghose S, Trinquet E, Laget M, Bazin H, Mathis G (2008) Rare earth cryptates for the investigation of molecular interactions in vitro and in living cells. *J Alloys Compd 451*(1–2):35–37
106. Fernandez-Moreira V, Song B, Sivagnanam V, Chauvin AS, Vandevyver CDB, Gijs M, Hemmilä I, Lehr HA, Bünzli JCG (2010) Bioconjugated lanthanide luminescent helicates as multilabels for lab-on-a-chip detection of cancer biomarkers. Analyst 135:42–52

107. Albizu L, Cottet M, Kralikova M, Stoev S, Seyer R, Brabet I, Roux T, Bazin H, Bourrier E, Lamarque L, Breton C, Rives M-L, Newman A, Javitch J, Trinquet E, Manning M, Pin J-P, Mouillac B, Durroux T (2010) Time-resolved FRET between GPCR ligands reveals oligomers in native tissues. Nat Chem Biol 6(8):587–594
108. Doumazane E, Scholler P, Zwier JM, Trinquet E, Rondard P, Pin J-P (2011) A new approach to analyze cell surface protein complexes reveals specific heterodimeric metabotropic glutamate receptors. FASEB J 25(1):66–77
109. Bourdolle A, Allali M, Mulatier J-C, Le Guennic B, Zwier JM, Baldeck PL, Bünzli J-CG, Andraud C, Lamarque L, Maury O (2011) Modulating the photophysical properties of azamacrocyclic europium complexes with charge-transfer antenna chromophores. Inorg Chem:4987–4999
110. Takalo H, Hemmilä I, Sutela T, Latva M (1996) Synthesis and luminescence of novel EuIII complexing agents and labels with 4-(Phenylethynyl)pyridine subunits. Helv Chim Acta 79(3):789–802
111. Butler SJ, Lamarque L, Pal R, Parker D (2014) EuroTracker dyes: highly emissive europium complexes as alternative organelle stains for live cell imaging. Chem Sci:1750–1756
112. Soulié M, Latzko F, Bourrier E, Placide V, Butler SJ, Pal R, Walton JW, Baldeck PL, Le Guennic B, Andraud C, Zwier JM, Lamarque L, Parker D, Maury O (2014) Comparative analysis of conjugated alkynyl chromophore-triazacyclononane ligands for sensitized emission of europium and terbium. Chem Eur J 20:8636–8646
113. Delbianco M, Sadovnikova V, Bourrier E, Mathis G, Lamarque L, Zwier JM, Parker D (2014) Bright, highly water-Soluble Triazacyclononane europium complexes to detect ligand binding with time-resolved FRET microscopy. Angew Chem Int Ed 53:10718–10722
114. Starck M, Pal R, Parker D (2016) Structural control of cell permeability with highly emissive europium(III) complexes permits different microscopy applications. Chem Eur J 22(2): 570–580
115. Placide V, Pitrat D, Grichine A, Duperray A, Andraud C, Maury O (2014) Design and synthesis of europium luminescent bio-probes featuring sulfobetaine moieties. Tetrahedron Lett 55(7):1357–1361
116. Keppler A, Gendreizig S, Gronemeyer T, Pick H, H V, Johnsson K (2003) A general method for the covalent labeling of fusion proteins with small molecules in vivo. Nat Biotechnol 21:86–89
117. Placide V, Bui AT, Grichine A, Duperray A, Pitrat D, Andraud C, Maury O (2015) Two-photon multiplexing bio-imaging using a combination of Eu- and Tb-bioprobes. Dalton Trans 44(11):4918–4924
118. Bui AT, Grichine A, Brasselet S, Duperray A, Andraud C, Maury O (2015) Unexpected efficiency of a luminescent samarium(iii) complex for combined visible and near-infrared biphotonic microscopy. Chem Eur J 21(49):17757–17761
119. Perfetto SP, Chattopadhyay PK, Roederer M (2004) Seventeen-colour flow cytometry: unravelling the immune system. Nat Rev Immunol 4(8):648–655
120. Lu J, Martin J, Lu Y, Zhao J, Yuan J, Ostrowski M, Paulsen I, Piper JA, Jin D (2012) Resolving low-expression cell surface antigens by time-gated orthogonal scanning automated microscopy. Anal Chem 84(22):9674–9678
121. Lu Y, Jin D, Leif RC, Deng W, Piper JA, Yuan J, Duan Y, Huo Y (2011) Automated detection of rare-event pathogens through time-gated luminescence scanning microscopy. Cytometry 79A(5):349–355
122. Lu Y, Lu J, Zhao J, Cusido J, Raymo FM, Yuan J, Yang S, Leif RC, Huo Y, Piper JA, Paul Robinson J, Goldys EM, Jin D (2014) On-the-fly decoding luminescence lifetimes in the microsecond region for lanthanide-encoded suspension arrays. Nat Commun 5:3741
123. Lu Y, Xi P, Piper JA, Huo Y, Jin D (2012) Time-gated orthogonal scanning automated microscopy (osam) for high-speed cell detection and analysis. Sci Rep 2:837
124. Wu J, Ye Z, Wang G, Jin D, Yuan J, Guan Y, Piper J (2009) Visible-light-sensitized highly luminescent europium nanoparticles: preparation and application for time-gated luminescence bioimaging. J Mater Chem 19(9):1258–1264

125. Jin D, Piper JA (2011) Time-gated luminescence microscopy allowing direct visual inspection of lanthanide-stained microorganisms in background-free condition. Anal Chem 83(6): 2294–2300
126. Zhang L, Zheng X, Deng W, Lu Y, Lechevallier S, Ye Z, Goldys EM, Dawes JM, Piper JA, Yuan J, Verelst M, Jin D (2014) Practical implementation characterization and applications of a multi-colour time-gated luminescence microscope. Sci Rep 4:6597
127. Charbonnière LJ, Hildebrandt N, Ziessel RF, Löhmannsröben H-G (2006) Lanthanides to quantum dots resonance energy transfer in time-resolved fluoro-immunoassays and luminescence microscopy. J Am Chem Soc 128(39):12800–12809
128. Rajapakse HE, Gahlaut N, Mohandessi S, Yu D, Turner JR, Miller LW (2010) Time-resolved luminescence resonance energy transfer imaging of protein-protein interactions in living cells. Proc Natl Acad Sci 107:13582–13587
129. Mohandessi S, Rajendran M, Magda D, Miller LW (2012) Cell-penetrating peptides as delivery vehicles for a protein-targeted terbium complex. Chem Eur J 18(35):10825–10829
130. Maurel D, Comps-Agrar L, Brock C, Rives M-L, Bourrier E, Ayoub MA, Bazin H, Tinel N, Durroux T, Prézeau L, Trinquet E, Pin J-P (2008) Cell-surface protein-protein interaction analyses with time-resolved FRET and snap-tag technologies: application to GPCR oligomerization. Nat Methods 5:561–567
131. Faklaris O, Cottet M, Falco A, Villier B, Laget M, Zwier JM, Trinquet E, Mouillac B, Pin J-P, Durroux T (2015) Multicolor time-resolved Förster resonance energy transfer microscopy reveals the impact of GPCR oligomerization on internalization processes. FASEB J 29(6):2235–2246
132. Levoye A, Zwier JM, Jaracz-Ros A, Klipfel L, Cottet M, Maurel D, Bdioui S, Balabanian K, Prézeau L, Trinquet E, Durroux T, Bachelerie F (2015) A broad g protein-coupled receptor internalization assay that combines snap-tag labeling, diffusion-enhanced resonance energy transfer, and a highly emissive terbium cryptate. Front Endocrinol 6:167
133. Zou X, Rajendran M, Magda D, Miller LW (2015) Cytoplasmic delivery and selective, Multicomponent labeling with oligoarginine-linked protein tags. Bioconjug Chem 26(3): 460–465
134. Gaborit N, Larbouret C, Vallaghe J, Peyrusson F, Bascoul-Mollevi C, Crapez E, Azria D, Chardes T, Poul M-A, Mathis G, Bazin H, Pélegrin A (2011) Time resolved fluorescence resonance energy transfer (TR-FRET) to analyze the disruption of EGFR/HER2 dimers: a new method to evaluate the efficiency of targeted therapy using monoclonal antibodies. J Biol Chem 286:11337–11345
135. Ho-Pun-Cheung A, Bazin H, Gaborit N, Larbouret C, Garnero P, Assenat E, Castan F, Bascoul-Mollevi C, Ramos J, Ychou M, Pélegrin A, Mathis G, Lopez-Crapez E (2012) Quantification of HER expression and dimerization in patients tumor samples using time-resolved Förster resonance energy transfer. PLoS ONE 7(7):e37065 EP
136. Maurel D, Kniazeff J, Mathis G, Trinquet E, Pin J-P, Ansanay H (2004) Cell surface detection of membrane protein interaction with homogeneous time-resolved fluorescence resonance energy transfer technology. Anal Biochem 329(2):253–262
137. Rondard P, Huang S, Monnier C, Tu H, Blanchard B, Oueslati N, Malhaire F, Li Y, Trinquet E, Labesse G, Pin J-P, Liu J (2008) Functioning of the dimeric GABAB receptor extracellular domain revealed by glycan wedge scanning. EMBO J 27:131–144
138. Linden S, Singh MK, Wegner KD, Regairaz M, Dautry F, Treussart F, Hildebrandt N (2015) Terbium-based time-gated Forster resonance energy transfer imaging for evaluating protein-protein interactions on cell membranes. Dalton Trans 44(11):4994–5003
139. Afsari HS, Cardoso Dos Santos M, Lindén S, Chen T, Qiu X, van Bergen en Henegouwen PMP, Jennings TL, Susumu K, Medintz IL, Hildebrandt N, Miller LW (2016) Time-gated FRET nanoassemblies for rapid and sensitive intra- and extracellular fluorescence imaging. Sci Adv 2(6):e1600265

140. Hildebrandt, N.; Spillmann, C. M.; Algar, W. R.; Pons, T.; Stewart, M. H.; Oh, E.; Susumu, K.; Díaz, S. A.; Delehanty, J. B.; Medintz, I. L.,2016 Energy Transfer with Semiconductor Quantum Dot Bioconjugates: A Versatile Platform for Biosensing, Energy Harvesting, and Other Developing Applications. Chem Rev. doi:10.1021/acs.chemrev.6b00030
141. Cheng L, Yang K, Shao M, Lee S-T, Liu Z (2011) Multicolor in vivo imaging of upconversion nanoparticles with emissions tuned by luminescence resonance energy transfer. J Phys Chem C 115(6):2686–2692

Chapter 4
Pharmacologically Active Plant Flavonols as Proton Transfer Based Multiparametric Fluorescence Probes Targeting Biomolecules: Perspectives and Prospects

Pradeep K. Sengupta

Keywords Plant flavonols • Excited state intramolecular proton transfer (ESIPT) • Two color fluorescence • Multiparametric probes • Interactions with therapeutically important bio-relevant target molecular systems

4.1 Introduction: Background and Historical Perspectives

4.1.1 Fluorescence as an Exquisitely Sensitive Approach for Exploring Biomolecular Interactions: A Brief Overview

Fluorescence spectroscopy and imaging techniques have developed into powerful tools for non-invasive and non-destructive sensing of interactions of bioactive fluorophores with various biological targets at molecular, cellular, and tissue levels. What distinguishes fluorescence from other methods for probing intermolecular interactions are its exquisitely high sensitivity and high speed of response (typically on 0.1 ns −10 ns timescales). These features allow detection of fluorophores at physiologically relevant concentrations on the one hand, and, on the other, to explore dynamic processes in solution and at macromolecular binding sites. Moreover, fluorescence has the unique capability to perform, in addition to in vitro measurements, ex vivo as well as in vivo explorations to sense target binding [1–5]. Latest in vivo applications have especially benefited from technological breakthroughs based on two-photon excitation methodology, with its inherent advantages of high spatial resolution, low scatter, and high tissue penetration [2, 4].

P.K. Sengupta (✉)
Department of Biophysics, Molecular Biology and Bioinformatics, University of Calcutta, 92 Acharya Prafulla Chandra Road, Kolkata 700009, India
e-mail: pradeepsinp@yahoo.co.in

C.D. Geddes (ed.), *Reviews in Fluorescence 2016*, Reviews in Fluorescence,
DOI 10.1007/978-3-319-48260-6_4

4.1.2 Plant Flavonoids: Natural Distribution, Chemical Classification, Therapeutic Importance, and Intrinsic Fluorescence Behavior

Nature has endowed many bioactive, plant based compounds with exquisitely sensitive intrinsic fluorescence properties which can be fruitfully exploited to probe their interactions with their physiologically relevant intra-cellular targets. Flavonols (3-hydroxyflavones) are one such class of naturally occurring intrinsically fluorescent bioactive compounds of potential therapeutic importance, which have received remarkable attention. [5–7]. Flavonols and other phenolic compounds of the flavonoid group have a common structure of diphenylpropanes, C6-C3-C6 (consisting of two aromatic rings linked through a three carbon bridge) and are produced as secondary metabolites in plants. Based on the degree of unsaturation and oxidation of the central three carbon segments, flavonoids are divided into different chemical classes (Fig. 4.1). Over 6000 structurally distinct flavonoids have been identified in plants until now. They are ubiquitous in higher plants (occuring in flowers, fruits, leaves and other parts of the plant body) and are abundant in common plant based food and beverages such as onion, broccoli, citrus fruits, berries, apple, soy products, tea and red wine.These phytochemicals serve as pigments/copigments in flowers and fruits (and thus attract pollinating vectors), act as photo-screens in leaves (thereby protecting photosynthetic organelles against harmful solar UV B radiation), and fulfill many other useful functions which are essential for the survival of plants in the natural environment [7–10]. Although flavonols are most widely known for their remarkable fluorescence behaviors, intrinsic fluorescence has also been reported and fruitfully exploited for some compounds belonging to other chemical classes of flavonoids including flavones, isoflavones, and anthocyanidins [11–13].

The therapeutically beneficial role of dietary flavonoids was first recognized 80 years back by István Rusznyák and Albert Szent-Györgyi (Nobel laureate in Physiology or Medicine, 1937) [14]. Recent years have witnessed a remarkable renascence of interest on flavonoids as therapeutic and prophylactic drugs. This has been marked by an explosive growth of research on various bioactive flavonoids which are effective against a wide range of free radical mediated and other human diseases including cancers, atherosclerosis, ischemia, cardiovascular diseases, diabetes, allergies, AIDS, and neuronal degeneration. The high potency and low cytotoxicity of such compounds make them potentially viable alternatives to conventional therapeutic drugs [7–10]. In this connection, the question of their physiological targets and the mode of interaction with such targets loom large and constitute an important focus of current pharmacological research on flavonoids [7].

On a different scenario, from the spectroscopic perspective, flavonols (3-hydroxyflavones (3HFs), which comprise the most widely occurring chemical class among natural flavonoids, have been the focus of enormous attention for their exquisitely sensitive and remarkable fluorescence properties. Figure 4.2 shows the chemical structure of 3-hydroxyflavone (3 HF, a model flavonol forming the

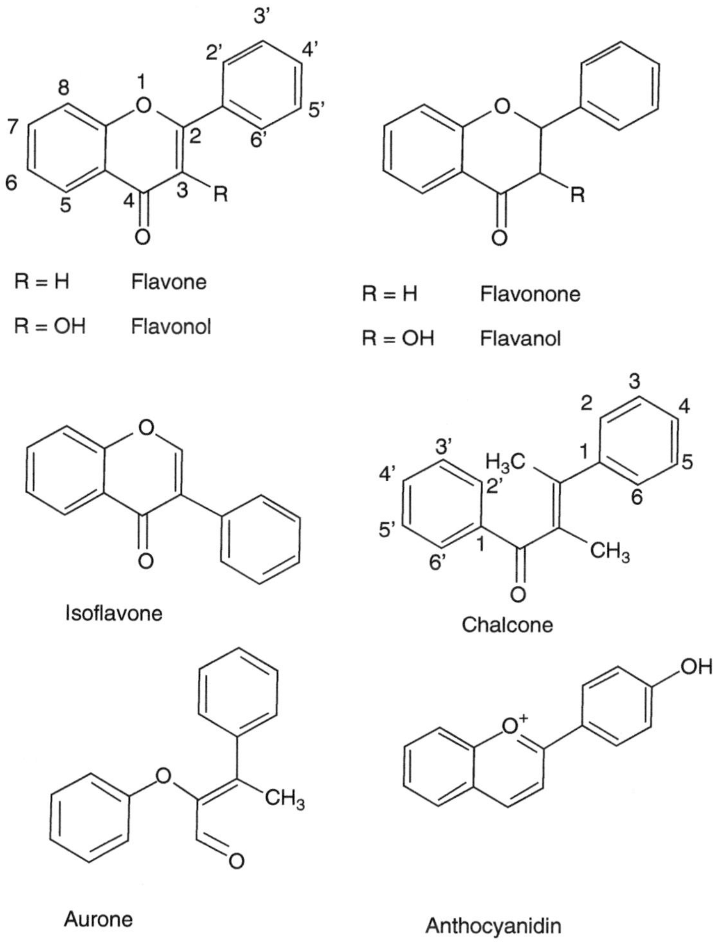

Fig. 4.1 Chemical classification of flavonoids

backbone of natural flavonols) and some typical polyhydroxy substituted naturally occurring plant flavonols we picked up for our sudies. From a photophysical context, interest on these molecules centers around the fact that they undergo a photoinduced excited state intramolecular proton transfer (ESIPT) reaction (via the internal hydrogen bond linking the C=O and 3-OH groups) [15–17]. This reaction is ultrafast [18, 19] and leads to the transformation of the initially excited normal (N*) form to the tautomer (T*) form [15–20]. Consequently, flavonols, in general, exhibit 'two color' fluorescence in the blue-violet and yellow-green regions, which originate from the normal and ESIPT phototautomer forms of the molecule respectively. Apart from the two color fluorescence from the neutral species of flavonols, in specific situations distinct fluorescence emissions have been noted from the anionic species of various flavonols. Such anion formation occurs due to

Fig. 4.2 Chemical structures indicating –OH substitution patterns for typical flavonols with representative fluorescence behaviors chosen for our studies

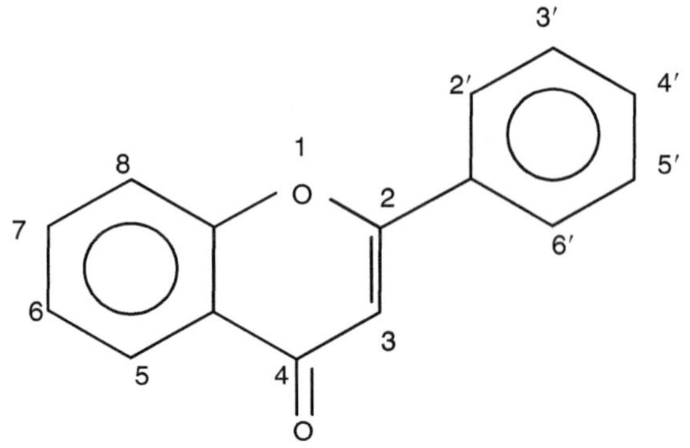

3-OH	3-hydroxyflavone
3, 3′, 4′, 5, 7-OH	Quercetin
3, 3′, 4′, 7-OH	Fisetin
3, 3′, 4′, 5′, 7-OH	Robinetin

ground state deprotonation by proton abstracting groups which are often present at the flavonol binding sites of receptors (e.g. proteins). Since the anion absorption and fluorescence peaks are usually well separated from those of the neutral molecules, anion fluorescence has distinct spectroscopic signatures and can be selectively excited and detected by fluorescence spectroscopy, thereby extending the scope of flavonols as multi -wavelength fluorescence probes [6, 21, 22]. The two color fluorescence (generated by excited state intramolecular proton transfer (ESIPT)) as well as anion fluorescence (generated by ground state proton transfer (GSPT)) show high sensitivity to environmental parameters. This has opened up opportunities for non-invasive sensing of the interactions of various therapeutically important flavonols (at physiologically relevant (in micromolar range) concentrations), with potential bio-relevant targets (related to their therapeutic actions). Representative targets chosen for studies reported by us and other research groups with overlapping research interests encompass proteins, duplex as well as quadruplex DNA, biomembranes, membrane mimetic models , nano-vehicles for drug delivery, and more recently, protein based ternary systems. [11, 21–51]. In this review, I shall present perspectives, based on research findings from our laboratories during the past 15 years, illustrating the potential usefulness of flavonols as their own fluorescent 'reporters' for noninvasive sensing of their interactions with representative biologically relevant targets. The emphasis of this article will be on pharmacologically active naturally occurring, plant flavonols with intrinsic 'two color' fluorescence. The discussion will also include relevant findings relating to the model flavonol 3HF, which constitutes the backbone, and chromophoric moiety of flavonols of therapeutic importance, and moreover, has been found to possess significant antioxidant activities of potential therapeutic utility in biological environments.

4.1.3 Excited–State Intramolecular Proton Transfer (ESIPT) and 'Two Color' Fluorescence of Flavonols: Historical Perspective and Sensitivity to Environment

The aim of this section is to present a brief overview on salient features of ESIPT and 'two color' fluorescence of flavonols, together with the basis for their remarkable environmental sensitivity, which has inspired the extensive use of such molecules as a new generation of fluorescence probes. In this connection it is perhaps worthwhile to begin with a brief historical background. The occurrence of photoinduced ESIPT in flavonols (where a proton is transferred from the 3-OH group to the adjacent 4-C=O group across the internal hydrogen bond between these two groups) was discovered in 1979 by Sengupta and Kasha [15], in course of a spectroscopic study on 3HF (the backbone of natural flavonols). This study was undertaken in an attempt to resolve as to why flavonols, with their lowest (S_0-S_1) electronic absorption band in the near-UV region, showed anomalously large Stokes shifted green-yellow fluorescence upon excitation with a UV (365 nm) lamp. Such fluorescence has been traditionally used as a standard diagnostic test for detection of plant flavonols (e.g. kaempferol, quercetin, fisetin, etc.) following their chromatographic separation from raw plant extracts [15, 16]. The recognition that ESIPT takes place in 3HF and its naturally occurring derivatives immediately resolved the spectroscopic anomaly regarding their large Stokes shifted fluorescence, which had remained unexplained until then [16]. Significantly, it also prompted a remarkable outburst of detailed research on various aspects concerning the photophysics of 3HF, explored via steady state spectroscopy and laser kinetic techniques. The choice of 3HF as a model for natural flavonols is advantageous especially because of its good solubility and fairly high fluorescence quantum yield in a wide range of media, thus permitting detailed and systematic spectroscopic investigations on its photophysics to be performed. In nonpolar solvents, for example cyclohexane, the lowest energy (So-S_1) electronic absorption of 3HF occurs due to a π- π^* transition, resulting in a vibrationally resolved band in the near-UV with the principal peak at 340 nm together with a lower energy peak at 355 nm. The corresponding emission consists of a green fluorescence band with the peak at 526 nm, dramatically Stokes shifted with respect to the S_0-S_1 absorption, its other relevant parameters being, quantum yield) = 0.36, and lifetime = 3 ns. In contrast to 3HF, in 3-methoxyflavone (where the 'H' atom in 3HF is replaced by a 'CH_3 group) there is no scope for proton transfer, and the emission now occurs with a normal Stokes shift, with a peak at 360 nm, which lends credence to the proposal of ESIPT in 3HF [20]. A convenient way of representing ESIPT in 3HF and its derivatives is in terms of proton transfer across an asymmetric double well potential involving four energy states as depicted below [15, 17] (Fig. 4.3).

During the years that have followed the initial discovery of ESIPT in 3HF well over three decades ago, flavonols (comprising 3-hydroxyflavone (3HF) and its natural as well as synthetic derivatives) along with various chromone derivatives of related interest , have emerged as one of the best known and most intensively

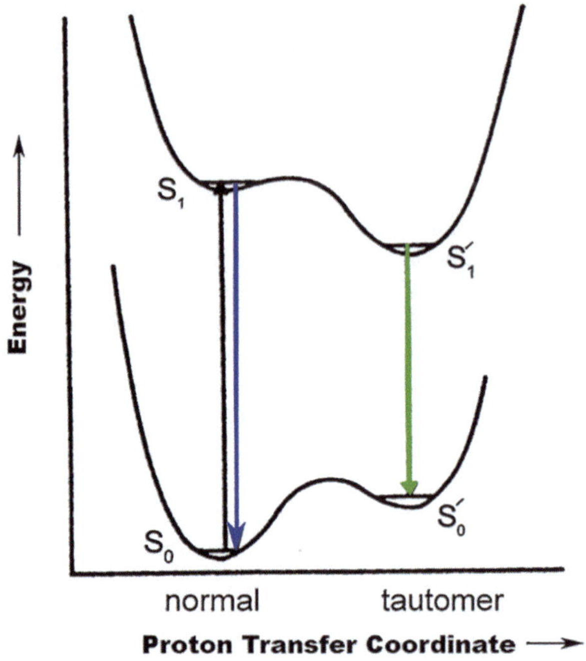

Fig. 4.3 Proton transfer across an asymmetric double well potential involving four energy states. The *arrows* show initial photo-excitation (*black*), normal (non proton transferred) fluorescence (*blue*), and tautomer (generated by ESIPT reaction) fluorescence (*green*) (Refs. [15, 17, 26])

studied group of molecules serving as prototypes for exploring mechanistic aspects of photoinduced ESIPT, the interplay between ESIPT and charge transfer (CT) reactions, and 'two color' fluorescence phenomena [2]. Figure 4.4 shows the steps involved in photoinduced ESIPT in a typical polyhydroxy substituted natural flavonol, fisetin (3,7,3′,4′-OH flavone) which is responsible for its'two color' fluorescence behavior. The photoinduced ESIPT reaction occurs from the 3-OH group to the proximally located C=O group, across the intermolecular hydrogen bond linking the two groups. As a result, the initially excited normal(N*) form is transformed to the ESIPT phototautomer (**T***) form [15–17]. Emission originating from the N* and T* forms results in 'two color' fluorescence, consisting of a blue-violet normal fluorescence with a normal Stokes shift and a yellow-green fluorescence with a dramatically large Stokes shift. Stokes shifts for the ESIPT tautomer fluorescence of flavonols are typically in the range 5000–6000 cm^{-1}. This ESIPT reaction is ultrafast and occurs in sub-picosecond timescales. A time constant of 35 fs has been reported for ESIPT in unsubstituted 3HF in fluid solution in a typical aprotic solvent at room temperature [18]. In Shpol'skii matrix under cryogenic condition the corresponding value has been found to be ca. 40 fs. It is especially noteworthy that the ESIPT reactions in 3 HFs are extraordinarily sensitive to external hydrogen bonding perturbation of the environment on the internal (C(4)=O . . . . . . HO—C(3)) hydrogen bond of the molecule [17]. As a consequence, the relative contributions between the two colors is strongly modulated by the local environments of the fluorophores. Moreover,

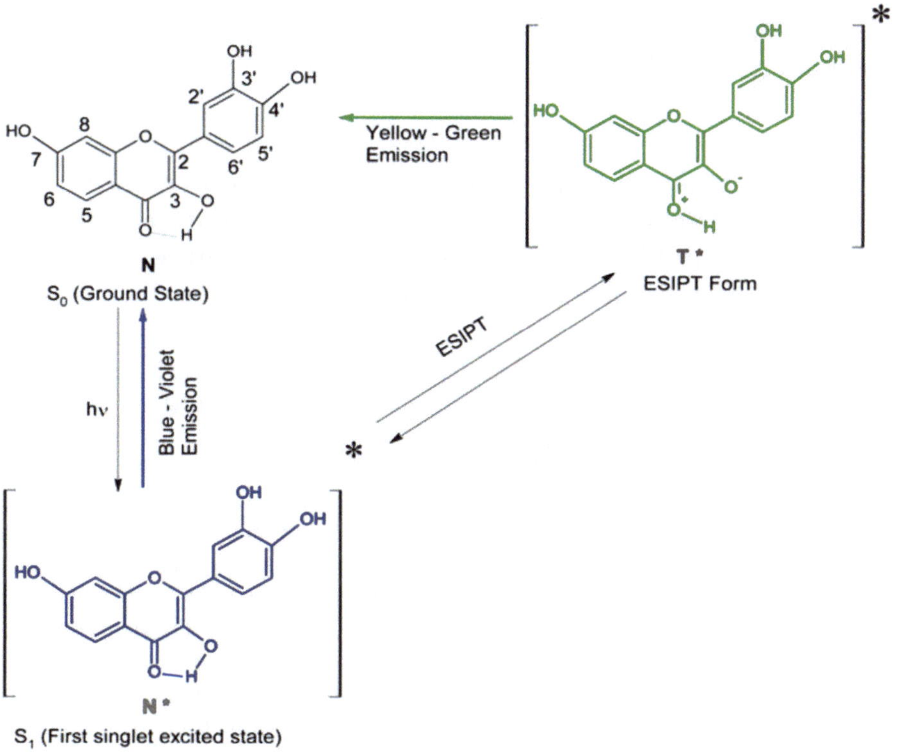

Fig. 4.4 Photoinduced ESIPT in a typical dietary plant flavonol, fisetin, leading to 'two color' fluorescence behavior (Refs. [5, 15, 17])

in flavonol derivatives where, because of the presence of appropriately located substituents on the 3HF chromophore, the N* form shows strong charge transfer (CT) character[43], significant solvent dipolar relaxation effects occur. Thus, the yellow-green ESIPT tautomer fluorescence serves as a 'proton transfer' probe (sensing H-bonding effects), and the blue-violet normal fluorescence serves as a 'polarity probe' (sensing the polarity of the fluorophore environment). This enables multiparametric use of the same fluorophore [2, 22]. A particularly interesting situation occurs in flavonols where 5-OH and 3-OH groups are simultaneously present (e.g. in quercetin (3,5,7,3′,4′-OH flavone)),the most commonly occuring dietary flavonol, where the C(4)=O. . . . . . HO—C(5) hydrogen bond interferes with the C(4)=O. . . . . . HO—C(3) hydrogen bond, which prevents efficient ESIPT and as a result the fluorescence quantum yield is intrinsically low. However, strong fluorescence can be elicited in specific situations e.g. on binding to target proteins with specific structural motifs [23, 38, 44] or in low temperature solvent glass matrix [52] where the internal hydrogen bond involving the 5-OH group is disrupted. These features constitute the basis for the high sensitivity of flavonol emission to the surrounding environment, enabling their applications as exquisitely sensitive fluorescent molecular probes for exploring their interactions with various biological targets in critical detail.

4.2 Applications of ESIPT and 'Two Color' Fluorescence of Therapeutically Active Flavonols with Representative Bio-relevant Targets: Scope of the Present Survey and Fluorescence Methodologies

We exploited the 'intrinsic 'two color' fluorescence properties of selected flavonols to explore their interactions with representative bio-relevant target molecules encompassing proteins (the carrier proteins bovine and human serum albumins (BSA, HSA), and normal human hemoglobin (HbA)), duplex and quadruplex DNA, together with model and natural membranes. Salient findings, exemplifying the use of flavonols as their 'own fluorescence reporters' in relation to such intermolecular interactions, are presented here and discussed in relevant detail. For these studies we chose the model flavonol of synthetic origin, namely 3-hydroxyflavone, along with the naturally occurring, polhydroxy substituted plant flavonols fisetin (3,7, 3′,4′OH-flavone), robinetin (3,7,3′,4′,5′OH- flavones), and quercetin (3,5,7,3′,4′-OH flavones). Steady state as well as time resolved fluorescence, including steady state emission and excitation profiles, steady state anisotropy, time resolved emission and anisotropy decay studies, and use of Förster Resonance Energy transfer (FRET) as a proximity indicator [1, 2], comprise key elements of this research. In the following sections findings based on such studies will be elaborated in appropriate detail. For data obtained from other supporting techniques (such as electronic circular dichroism (ECD), Raman, and theoretical (molecular modeling) which were used to consolidate various aspects of the primarily fluorescence based research which we undertook, the reader is requested to refer to the original articles cited herein. Here we present a brief background on relevant fluorescence parameters.

Fluorescence emission and excitation profiles were obtained using standard procedures [1, 2]. Besides the emission and excitation profile shape, structure, and yield, a parameter that has been frequently used for characterizing two color fluorescence of flavonols is the tautomer/normal emission intensity ratio ($I_{tautomer}/I_{normal}$) which provides a convenient measure of the hydrophobicity of the local environment of the fluorophore. The steady state fluorescence anisotropy (r) values were obtained using the expression,

$$r = \frac{I_{VV} - GI_{VH}}{I_{VV} + 2GI_{VH}}$$

where I_{VV} and I_{VH} are the vertically and horizontally polarized components of probe emission with excitation by vertically polarized light at the respective wavelength and G defines the instrumental correction factor (polarization characteristics of the photometric system) calculated as

$$G = I_{HV}/I_{HH}$$

Each intensity value used in this expression represents the computer-averaged values of several successive measurements.

Time resolved fluorescence decay measurements [1, 2] were carried out with a time correlated single photon counting (TCSPC) spectrometer, using excitation with sub-nanosecond pulses. For some of our recent studies we used 375 nm pulses from a laser diode excitation source (pulse FWHM $\sim$ 120 ps).An emission monochromator was used to block the scattered light and isolate the emissions. Fluorescence intensity decay curves were deconvoluted with the instrument response function and fitted to a multiexponential decay function,

$$F(t) = \sum_i \alpha_i \, exp\,(-t/\tau_i)$$

where *F(t)* represents the fluorescence intensity at time *t,* and α_i and τ_i are the amplitudes and decay times of the individual components in the multiexponential decay profile such that $\sum_i \alpha_i = 1$. The goodness of fit was estimated by using χ^2 values and Durbin- Watson parameter. Average lifetimes $\bar{\tau}$ were calculated from the decay times and preexponential factors using the expression:

$$\bar{\tau} = \frac{\sum_i \alpha_i \tau_i^2}{\sum_i \alpha_i \tau_i}$$

For measurement of fluorescence depolarization kinetics the parallel (I_{VV}) and perpendicular (I_{VH}) components were collected as a function of time in an alternating manner until the difference of fluorescence counts collected reaches $\sim$5000. The time resolved fluorescence anisotropy [1, 2] values *r(t)* were calculated using the expression:

$$r(t) = \frac{I(t)_{VV} - GI(t)_{VH}}{I(t)_{VV} + 2GI(t)_{VH}}$$

where I_{VV}, I_{VH} and G are defined as for the steady state anisotropy measurements. When freely rotating spherically symmetrical molecules are excited with polarized light, anisotropy decays as a function of time according the following equation

$$r(t) = r_0 e^{-t/\tau_{rot}}$$

where r_0 is the limiting anisotropy, which is the anisotropy value just after photo excitation (i.e. at t =0) and τ_{rot} is rotational correlation time, which is a parameter of molecular rotation (i.e. how fast or slow the molecule rotates in the environment).Rotational diffusion depolarizes the fluorescence anisotropy from the initial r_0 value to a final randomized value $r_{\infty.\infty}$. High values of τ_{rot} indicate strong motional constraint in the fluorophore microenvironment , and is therefore useful in diagnosing the binding of small molecule ligands to macromolecular targets [1, 2]

4.3 Protein-Flavonol Interactions

4.3.1 Background

From numerous biochemical and molecular biological investigations there is mounting evidence that various proteins (including enzymes) are important targets for therapeutically active flavonoids in relation to their protective functions [7–10]. Recent years have witnessed a growing trend in the application of fluorescence spectroscopy [5] as an optical spectroscopic tool for exploring protein–flavonoid interactions. In principle, protein-flavonoid interactions can be monitored by means of three types of experiments illustrated in Fig. 4.5 [54]: (a) excitation and detection of flavonoid intrinsic fluorescence (b) excitation and detection of protein (tryptophan) fluorescence, (c) excitation of protein (tryptophan) and detection of flavonoid fluorescence (occurring due to FRET from Trp to flavonoid), (Fig. 4.5). Both steady state and time resolved florescence measurements have proved to be useful for such investigations (5, 11, 21, 39, 40, 41, 54, 57). The three lines of investigations (are applicable, in general, for other types of polyphenols as well, with experimental approaches (A) and (C) being available for intrinsically fluorescent ligands .While

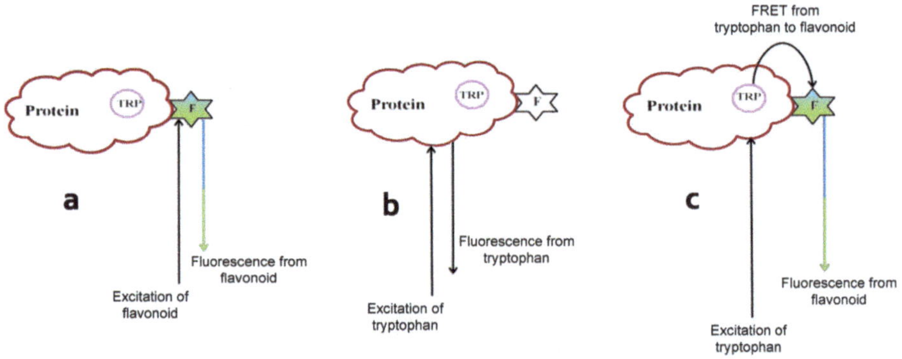

Fig. 4.5 Schematic diagram showing three different approaches for exploring protein-flavonoid interactions via fluorescence spectroscopy: (**a**) Excitation and detection of flavonoid fluorescence, (**b**). excitation and detection of tryptophan (protein) fluorescence, (**c**). excitation of tryptophan and detection of flavonoid fluorescence sensitized by FRET (Adapted with slight modification from Ref. [54])

studies of protein-fluorescent flavonoid interactions can be usually monitored by all three experimental approaches (Fig. 4.5a–c), the methodology based on measuring the quenching of the intrinsic tryptophan fluorescence of the proteins upon their interactions with the flavonoids (Fig. 4.5b) [53–57], has the advantage of applicability to both fluorescent and non-fluorescent ligands. On the other hand, our emphasis on exploiting the intrinsic fluorescence of flavonols has the merit of multiparametric probing opportunities as well as applications to a broader range of biologically relevant receptors (such as DNA, model biomembranes, nano-vehicles for drug delivery etc)., where, unlike proteins , no intrinsic fluorophores are present.

4.3.2 Serum Albumin-Flavonol Interactions

The remarkable 'two color' intrinsic fluorescence of flavonols has been extensively used by us and other researchers to explore protein-flavonol interactions and related studies [5]. Binding to serum albumins (which are the principal "carrier proteins") are important for modulating the bioavailability of flavonoids in the physiological system, and therefore, significant attention has focussed on exploring flavonoid-serum albumin interactions. Xiao et al have shown that the structural differences between different flavonoid classes as well as substituents play a critical role on the affinities of the flavonoids to serum albumins. In particular, the binding affinity increases with hydroxylation of the B ring, decreases with glycosylation, and is not significantly affected by the presence of the C2=C3 double bond [57]. With this scenario in mind,, we explored interactions of various intrinsically fluorescent flavonoids with serum albumins (bovine serum albumin, BSA, and human serum albumin, HSA), which we chose as model proteins for assessing the usefulness of the intrinsic fluorescence of flavonols in exploring protein-flavonol interactions Large Stokes shifted emissions, and 'two color' fluorescence phenomena with distinctive fluorescence signatures are obtained for the flavonols 3-hydroxyflavone, fisetin, quercetin and robinetin in serum albumins, (HSA), with dramatic changes in the emission parameters signifying their binding to the protein(s).

For initial studies we examined the steady state fluorescence emission and excitation spectra of 3-HF in presence of BSA [21]. With increasing BSA concentration the tautomer (green) emission intensity is dramatically enhanced and the emission corresponds to that for polar aprotic solvents [15, 17, 20]. The excitation spectra (shown in Fig. 4.6) were examined in presence of BSA and in reference solvent media used for comparison. Interestingly, in the presence of BSA, the excitation spectrum consists of a vibrationally resolved band with λ_{ex}^{max} = 345 nm, together with a prominent vibrational shoulder at 363 nm. Such features correspond to the spectroscopic signature of 3HF in a predominantly aprotic environment, and is in sharp contrast with the smooth (i.e. lacking vibrational structure) excitation band typically observed in polar protic environments [15, 20]. Furthermore, a band with $\lambda_{ex}^{max} \sim 280$ nm (corresponding to protein (tryptophan) absorption [1]) appears in the excitation profile, indicating Förster-type fluorescence

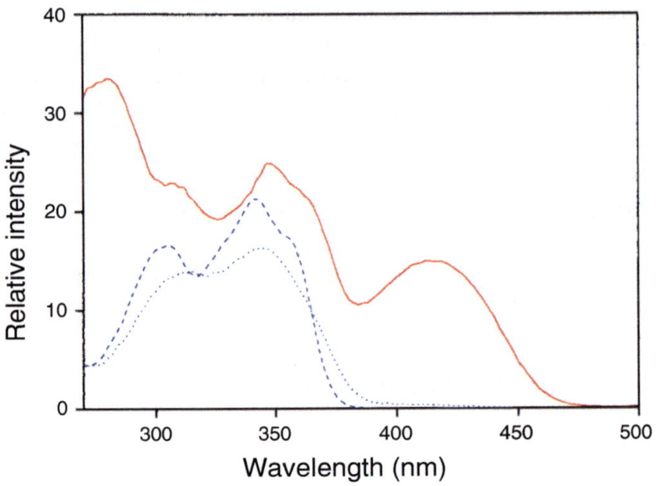

Fig. 4.6 Fluorescence excitation spectra of 3-hydroxyflavone in bovine serum albumin (BSA) (*unbroken curve*), ethyl acetate (*dashed curve*), and aqueous buffer (*dotted cureve*) for the neutral species. $\lambda_{em} = 535$ nm. [3-HF] = 5 μM) (Adapted in modified form, from Ref. [21])

resonance energy transfer (FRET) between the tryptophan (s) (donor) present in BSA and the 3-HF molecule (acceptor), which suggests proximity of the 3-HF binding site to the tryptophan residue(s).

Typical emission spectra of fisetin (3, 7, 3′, 4′-OH flavone, a dietary flavonol abundant in fruits, a particularly rich source being strawberries) in presence of different concentrations of HSA show that the addition of the protein (HSA) induces remarkable changes in the 'two color' emission profile , the tautomer emission becoming predominant as the protein concentration increases [35]. This signifies a hydrophobic microenvironment of the binding site [2, 15, 17]. In the absorption spectrum (not shown, see Ref. [37]), a distinct shoulder appears at ~430 nm which is attributed to the formation of an anionic species of fisetin in HSA [6]. Upon selective excitation of the anionic species (using $\lambda_{ex} = 430$ nm), the emission occurs with $\lambda_{em} \sim 500$ nm, with the emission intensity increasing with increasing protein concentration; the ESIPT tautomer fluorescence now appears as a shoulder at ~ 550 nm [35] . Existence of such a third emission band (apart from the normal–tautomer dual emission) has also been reported for other bioactive flavonols of related interest [21, 22]. The theoretically calculated OH-bond dissociation enthalpies (BDE) of the fisetin molecule [36] suggest that abstraction of a proton from the 3-OH group of the flavonol and the resulting formation of an anion is most probable due to its lower BDE value in comparison to the 3′-OH, 4′-OH and 7-OH groups. The emission spectral data have been interpreted in terms of two distinct binding sites of fisetin in the protein matrix, from one of which the dual (normal – tautomer) fluorescence occurs, the other one being responsible for the anionic form emission band. The latter site is presumably close to an amino acid having a charged side chain, specific interaction with which is presumably responsible for proton abstraction and the consequent formation of the anionic species of fisetin [35]. Interestingly, evidence for two binding sites in HSA was also found in our earlier work on the interaction of quercetin with HSA, from steady state fluorescence data [24]. Fluorescence anisotropy (*r*) is a sensitive monitor for fluorophore binding to motionally constrained regions of proteins [1, 2]. Fluorescence anisotropy (*r*) measurements for the ESIPT tautomer emission and the anionic species of fisetin show high values of '*r*' ($r \approx 0.28$ for the tautomeric species and $r \approx 0.36$ for the anionic species of fisetin at [HSA] = 30 μM). Such high values of r indicate that fisetin molecules are 'locked' in motionally constrained sites of HSA. FRET serves as a reliable indicator for assessing the proximity between donor and acceptor sites in a macromolecule [1, 2]. The fluorescence excitation profiles reveal that in addition to the band(s) corresponding to the absorption of fisetin (including the shoulder at ~ 430 nm, due to the anionic species), a band with $\lambda_{ex} \sim 280$ nm occurs. This is indicative of the occurrence of FRET from the single trp-214 moiety of HSA to the protein bound fisetin. The role of FRET is also evident from the emission spectrum of HSA in presence of fisetin: Besides the emission band of the protein (due to trp) at ~ 340 nm, a band with $\lambda_{em} \sim 550$ nm together with a shoulder at ~ 500 nm are present [35] (Fig. 4.7), which can be attributed to emission from the ESIPT tautomer and the anionic species of fisetin [6, 35], respectively. From the quenching of the donor (trp-214) fluorescence, the overall efficiency of FRET for the donor (trp-214)/acceptor (fisetin) pair has been estimated to be 38 %. From the FRET data we inferred that both the binding sites of fisetin are located in close proximity to trp-214 situated in the inter-domain (between IIA and IIIA domains) cleft region of HSA [58]. Time resolved fluorescence studies indicated that the decay parameters (and consequently the average lifetime) of flavonoid fluorescence are significantly affected by the protein environment. In particular, the average lifetime $\langle\tau\rangle$ of the tautomer fluorescence changes from 0.824 ns in aqueous buffer to 2.55 ns in presence of HSA [35]. Such increased lifetime may be attributed to a large decrease in non-radiative decay rates upon binding of the fisetin molecule to the protein matrix which may be due to its binding to a rigid pocket in the HSA matrix.

The interaction of quercetin (3,5,7,3′,4′-OH flavone, the most abundant dietary flavonol present in onion, broccoli, apple etc.) with HSA exemplifies a particularly interesting situation where protein binding 'activates' prominent fluorescence in a fluorophore which is weakly fluorescent intrinsically. With increased protein concentrations, gradual induction of prominent 'two color' fluorescence behavior of quercetin is observed (Fig. 4.8a). These observations can be explained in terms of interference with the internal H-bonds of quercetin, i.e., with C(4)=O......HO—C(5) (which facilitates non-radiative deactivation [52]) and C(4)=O......HO—C(3) (which permits the ESIPT process [2, 15–17]) at the binding site in HSA. Furthermore, like 3HF and fisetin occurrence of FRET (tryptophan → quercetin) is evident from the excitation profile (Fig. 4.8b). Since HSA contains a single Trp residue (at position 214) located in the inter-domain cleft region of the protein (known from X-ray crystallographic studies [58]), it is evident that the binding site of the flavonoid must be located in this region, in close proximity

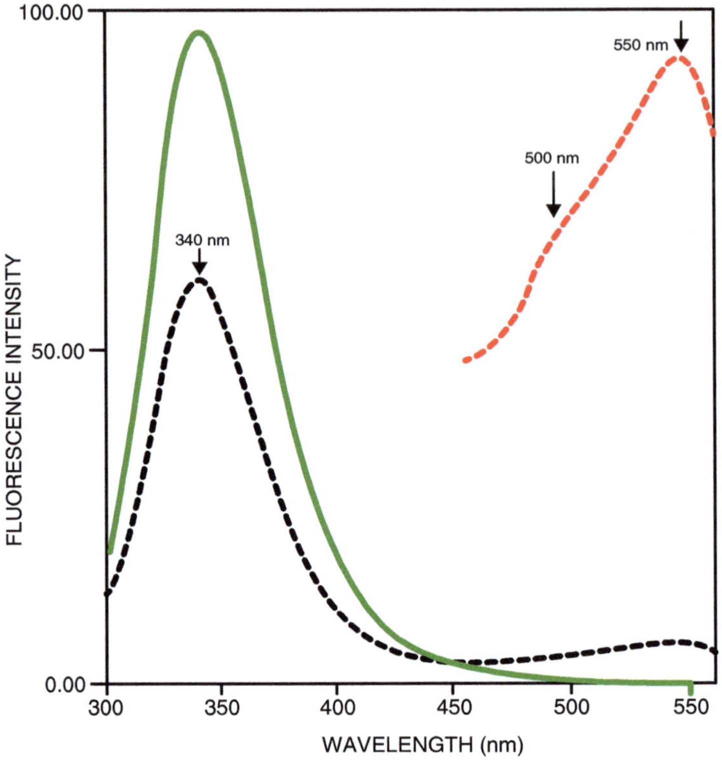

Fig. 4.7 Evidence for FRET (trp→fisetin) from emission studies: Fluorescence emission spectra were recorded for HSA (30 μM) (__) both in the absence (____) and presence (------) of fisetin (15μM), with λ_{ex}= 280 nm (corresponding to the trp absorption maximum). The protein fluorescence peak (due to trp) occurs at 340 nm. The 550 nm peak, and 500 nm shoulder (indicated by vertical arrows in the 16X magnified spectrum) signify FRET from trp to different species of fisetin (Adapted with some modification from Ref. [37])

to trp-214. Furthermore, high fluorescence anisotropy values are obtained for the HSA bound flavonoid ($r \sim 0.18$ at [HSA] = 30 μM) indicating binding to a motionally constrained site of the protein matrix. Gutzeit and co-workers have shown the potential utility of such fluorescence (in quercetin and other medicinally important flavonols) elicited by protein binding to detect specific target proteins in drosophila(fruit fly) follicles [38] and in the nuclei of human leukemia cells [44]. They have emphasized the general utility of this approach for quercetin and other related fluorogenic flavonoids of medicinal importance. Apart from such qualitative understanding on protein-flavonoid interactions, estimates of binding constants are obtained by monitoring variations in fluorescence intensity and/or anisotropy parameters. Recently, Wybranowski and Kruszewski [45] estimated the binding constant for quercetin-HSA interaction at the physiologically relevant temperature (37°C) and found it to be $2.38 +- 0.1X10^5 M^{-1}$.

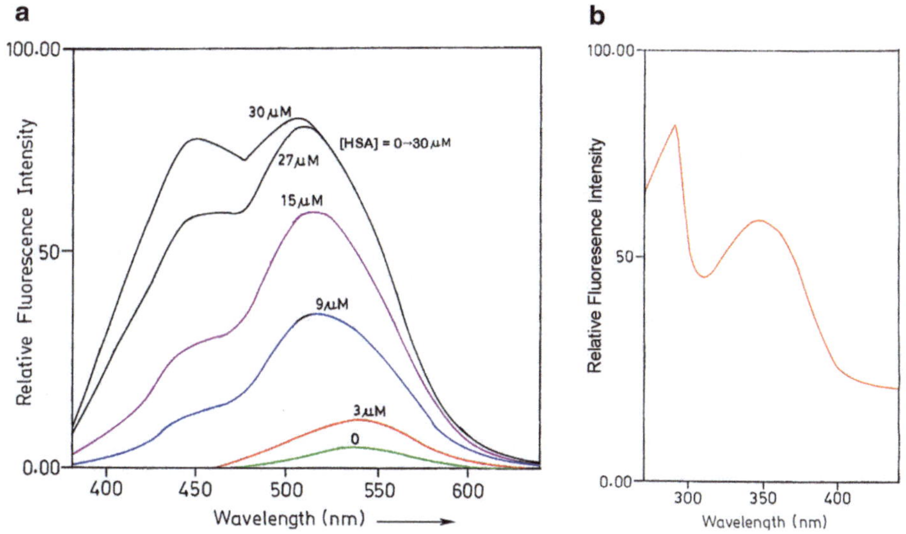

Fig. 4.8 **(a)** Fluorescence emission spectra of quercetin in the presence of various concentrations of human serum albumin (HSA, 0 → 30 μM) (λ_{ex} = 370 nm) (Adapted with modification from Ref. [23]) **(b)** Excitation spectrum of quercetin in the presence of. 18 μM human serum albumin (HSA), Note: Peaks due to both quercetin and protein (trp) appear, signifying FRET from trp to quercetin (Adapted with slight modification from Ref. [23])

Recently, we performed a detailed spectroscopic, combined with theoretical (molecular docking) study on the bioactive plant flavonol robinetin (3,7,3′,4′,5′-OH flavone), which is closely related in its chemical structure, to fisetin [22]. Multi-parametric fluorescence sensing, exploiting the intrinsic 'two color' fluorescence of the flavonol (comprising ESIPT) and CT) emissions) reveals that binding to HSA significantly affects the emission and excitation profiles, with strongly blue shifted (~29 nm) normal fluorescence, and remarkable increase in the ESIPT fluorescence intensity, anisotropy (*r*) and lifetime (τ). The $I_{tautomer}/I_{normal}$ ratio increases from 1.72 to 4.36. The corresponding increase in the steady state anisotropy (r) of the tautomer emission is nearly ~10 times (changing from 0.032 to i0.31), and that of the average lifetime > 20 times (changing from ca. 0.11 ns, to 2.62 ns). All these data are strongly indicative of hydrophobic binding site (s) in the protein (HSA) matrix. Time resolved fluorescence anisotropy decay studies show dramatic (~170 times) increase in the rotational correlation time (τ_{rot}), reflecting highly enhanced restrictions in motion of robinetin in the protein matrix. Furthermore, prominent induced circular dichroism (ICD) bands appear, which indicates that the chiral environment of HSA strongly perturbs the electronic transitions of the intrinsically achiral robinetin molecule. Molecular docking calculations suggest that robinetin binds in subdomain IIA of HSA [58], where specific interactions with basic residues promote ground state proton abstraction and stabilize an anionic species, which is consistent with spectroscopic observations.

4.3.3 Hemoglobin-Flavonol Interaction

The interactions of fisetin and robinetin with normal human hemoglobin (HbA), (which like HSA, is a carrier protein), have been explored via both steady state and time resolved fluorescence emission spectroscopy. The emission spectrum of fisetin [28] consists of characteristic overlapping 'two color' dual fluorescence bands of flavonols consisting of the tautomer and normal (non proton transferred) emissions (with λ_{em}^{max} ca. 495 nm and 540 nm respectively) of comparable intensities. However, binding to HbA does not cause any change in the absorption maximum indicating that no new species of fisetin is generated. Similar spectral behavior was also noted for the structurally related flavonol, robinetin [29]. The fluorescence anisotropy (r) values of the tautomer emission of these flavonols increases significantly upon binding to HbA ($r = 0.04$ in aqueous buffer while in presence of HbA, $r = 0.14$ for fisetin and 0.12 for robinetin), which suggests that these flavonols are firmly bound to the protein matrix. The presence of multiple emissive species of fisetin in HbA has been directly tested by fluorescence anisotropy. For only one emissive species the fluorescence anisotropy should not change over the fluorescence excitation and emission bands belonging to a particular electronic transition [1, 2]. For fisetin no significant change occurs in the fluorescence anisotropy across the excitation or emission band which clearly indicates that the emission bands arise from essentially the same ground state species. (i.e. the neutral form of the molecule) [28].

4.4 DNA-Flavonol Interactions

4.4.1 Background

DNA being the carrier of genetic information , it has been the traditional choice for studies on interactions with various therapeutic drugs., in view of the potential ability to interfere with transcription and DNA replication with implications related to pathogenesis and tumorigenesis. However, since most drugs are non-fluorescent, intrinsic fluorescence based studies have been rather limited,. Besides double stranded DNA with canonical Watson-Crick structure, much recent interest has focused on quadruplex DNA, as the target for small molecule ligands capable of stabilizing quadruplex (QD) structure, and acting as novel cancer therapeutics [25, 59]. Keeping these in view, exploratory studies have been recently performed on the interactions of flavonols and related molecules possessing intrinsic fluorescence, with duplex DNA, and quadruplex DNA and other non-canonical DNA structures [25, 35, 41, 42, 60]. Our contributions in this venture will be briefly highlighted here.

4.4.2 Duplex DNA-Flavonol Interactions

Addition of double stranded(ds) calf thymus (CT) DNA induces a dramatic increase in the fluorescence emission intensity of fisetin accompanied by remarkable changes in its emission profile[35]. Dual fluorescence bands with emission maxima at 480 nm and 524 nm appear with the $I_{tautomer}/I_{normal}$ ratio reaching a value of 1.545 at the highest DNA concentration used (Fig. 4.9a). Thus a predominantly hydrophobic binding site of the flavonol in the DNA matrix is clearly evident. A competitive binding experiment using Ethidium Bromide (EtBr, a classic DNA intercalator)was also performed.Interestingly, the addition of EtBr caused a dramatic reversal of the emission profile which now resembled the situation before addition of DNA (Fig. 4.9b). When the same sample solution (containing fisetin +DNA+EtBr) was selectively excited at 480 nm (where EtBr absorbs maximally, but fisetin has no absorption) a strong orange fluorescence appears with emission maximum at 480 nm, which is the characteristic fluorescence of EtBr bound to ds DNA. This shows that EtBr displaces fisetin from its binding site in DNA, implicating an intercalative mechanism for fisetin binding. A large increase in the anisotropy was also found, indicating a motionally constrained binding site for fisetin in the DNA matrix [35]. Jana et al. have reported, that in contrast to fisetin and

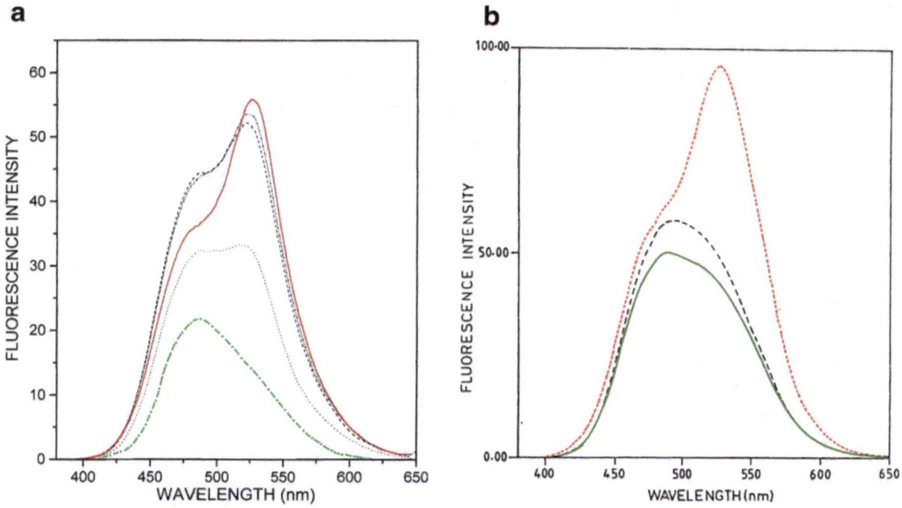

Fig. 4.9 **(a)**. Fluorescence emission spectra (λ_{ex} = 370 nm) of fisetin (12 μM) at different concentrations of double stranded DNA (calf thymus, CT) DNA (_._._) 0 μM, (. . . . . . .)5 μM, (—) 15 μM, (_.._.._) 100 μM, (__) 200 μM (Adapted with modification from Ref. [37]). **(b)** Results of competitive binding studies with the classic intercalator Ethidium bromide (EtBr): Fluorescence emission spectra of fisetin (15 μM; λ_{ex} = 370 nm) (- - - - - -) in absence of DNA, (. . . ..) in presence of DNA (200 μM), and (___) after addition of EtBr (15 μM) to the fisetin-DNA mixture (Adapted with modification from Ref. [37])

related polyhydroxyflavone, the mono-hydroxyflavone 3HF shows a groove binding mechanism [41], indicating a vital role of OH substituents on the flavonol moiety in modulating the binding mode.

4.4.3 Quadruplex DNA-Flavonol Interaction

Studies on quadruplex(QD) DNA-flavonol interactions have attracted much recent prominence. The principal motivation for such work is a quest for small molecules which can serve as QD DNA ligands, with prospective applications as novel anti-cancer drugs [25, 42, 59]. Our recent studies on the interaction of fisetin with the quadruplex DNA $d(T_2AG_4)_4$ (which closely resembles the human telomeric DNA sequence) exemplifies extensive applications of fluorescence, together with other spectroscopic and biophysical techniques to this new direction of research [25]. Salient findings are presented here. From Fig. 4.10 it can be seen that addition of QD DNA induces drastic changes in the emission behavior of fisetin. In aqueous medium, the fluorescence spectrum of fisetin exhibits strong overlap between the normal and tautomer emission bands. With the addition of the $d(T_2AG_4)_4$, dual fluorescence behavior is observed, with the ESPT tautomer emission becoming predominant at high concentrations of DNA, signifying an aprotic binding site [25]. This inference is further corroborated by the spectral characteristics of the excitation profile (monitored for the ESPT fluorescence) which reveals a weak but clearly perceptible vibrational shoulder typical of a predominantly aprotic environment. Moreover, the normal emission (which possesses a strong charge transfer character

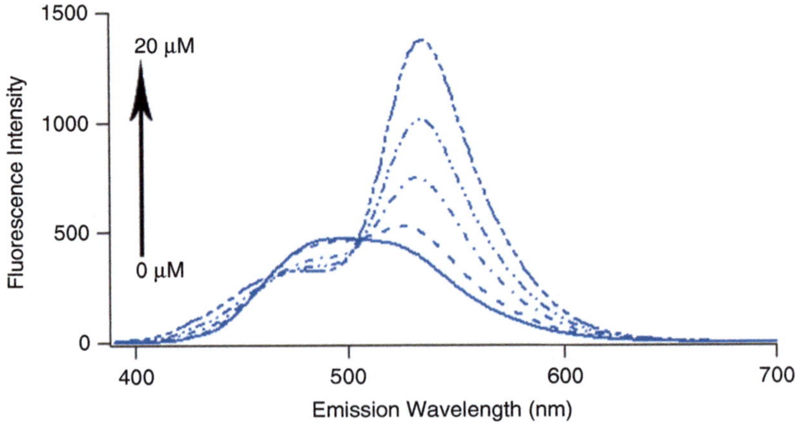

Fig. 4.10 Fluorescence Emission spectra of fisetin (15 μM) in the presence of various concentrations of quadruplex ($d(T_2AG_4)_4$) DNA. $\lambda_{ex} = 370$ nm. Note: Large increase in the tautomer fluorescence intensity, and strong blue shift of the normal emission peak can be seen at high DNA concentrations (Adapted with modification from Ref. [25])

and is therefore susceptible to solvent dipolar relaxation induced spectral shifts) undergoes a substantial blue shift of 14 nm, with $\lambda_{em}^{max} \sim 484$ nm (in the absence of DNA)moving to 470 nm at 20 μM DNA concentration. This provides another line of evidence for the hydrophobic nature of the binding site. These observations nicely exemplifies the use of both tautomer and normal fluorescence of flavonols as independent monitors of the fluorophore micro-environment. Time resolved fluorescence intensity decay studies show that in absence of DNA , fisetin ESPT tautomer species decay fits to a double exponential function with an average lifetime ($\bar{\tau}$) of 0.72 ns which is in agreement with a previous literature data. Upon inclusion into DNA, there is a dramatic change in the decay profile with three discrete decay components now appearing. Compared to free state in aqueous system, $\bar{\tau}$ increases significantly with increasing DNA concentration, reaching a value of 5.21 ns at 27 μM DNA. From these and other relevant data we picture that fisetin is bound at the diagonal loop region of the quadruplex, where it faces a relatively hydrophobic environment where non-radiative decay processes are reduced. This is also reflected in the higher emission yields noted in the steady state emission spectra.

Further insights were obtained from time resolved anisotropy decay studies on both the tautomer and normal fluorescence. Figure 4.11 shows a typical anisotropy decay profile for the ESPT tautomer (in presence of QD DNA). In methanol (chosen as a reference solvent) and DNA matrix (27 μM), the anisotropy decay of fisetin tautomer is found to be single exponential with rotational time constant (τ_{rot}) 0.21 ns and 16.79 ns respectively. The increase in the rotational time constants from MeOH to DNA is attributed to the increase in the restrictions in the motion of the fisetin molecules in the DNA matrix as well as the conformational swelling due to the overall motion of the fisetin-DNA complex. Thus, fisetin is firmly bound in a motionally constrained environment in the QD DNA matrix which further reinforces its role as a QD DNA ligand ,and likely prospect as an anti cancer drug with low systemic toxicity.

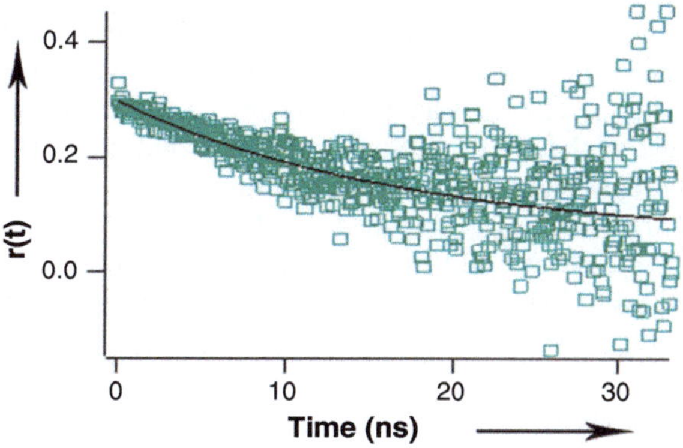

Fig. 4.11 Time resolved fluorescence anisotropy (r(t)) decay for fisetin ESPT tautomer fluorescence in quadruplex DNA($d(T_2AG_4)_4$) matrix. The *solid line* shows the single exponential fit curve (Adapted with modification from Ref. [25])

4.5 Membrane-Flavonol Interactions

4.5.1 Background

Flavonols and other related polyphenols are antioxidants with powerful inhibitory activities against lipid peroxidation in biomembranes [26]. Therefore it is of considerable importance to obtain relevant details about their binding sites, partition coefficients and related aspects in relation to their binding to membranes. This section is devoted to a brief account of studies of flavonols binding to representative model and natural membrane systems, namely model phospholipid membranes (liposomes) and red blood cell (RBC) ghost membranes [26, 33, 36].

4.5.2 Flavonols in Model Membranes (Liposomes)

Liposomes, are small artificial vesicles consisting of one or more phospholipid bilayers, which serve as excellent models simulating the basic physico-chemical features of natural biomembranes. Liposomes prepared from a natural source (egg yolk phosphatidyl choline (EYPC)) as well as those of artificial origin, based on saturated phospholipids, namely dipalmitoyl phosphatidylcholine (DPPC) and dimyristoylphosphatidylcholine(DMPC), figured in our research on membrane-flavonol interactions [26, 29, 31, 36].

Upon gradual addition of lipid (EYPC), the emission profiles of 3HF, fisetin, and robinetin show significant changes with substantial increase in the emission intensity of the tautomer fluorescence band [26, 29, 31]. The remarkably enhanced tautomer fluorescence (which is known to be exquisitely sensitive to external H-bonding perturbation effects) tends to indicate that the flavonol molecules are incorporated in relatively hydrophobic fatty acyl chain regions of the liposomes, where the chromone moiety (which is the part of the molecule mainly relevant to the ESIPT process) is shielded from the water molecules. As already mentioned, the ratio of the intensities of the tautomer to normal emission bands.

$I_{tautomer}/I_{normal}$ is an useful parameter for monitoring the enhancement in relative yield of the ESIPT tautomer emission and provides a convenient indicator of the hydrophobicity of the microenvironment of flavonols [9, 10]. The saturating values of the ratio $I_{tautomer}/I_{normal}$ for 3-HF is $\sim$ 12 whereas those for the polyhydroxy substituted derivatives, fisetin and robinetin, are $\sim$ 9 and $\sim$ 8 respectively. The relatively lower value of this parameter for fisetin and robinetin suggest that their average environment in liposomes are comparatively less hydrophobic than thar for 3-HF. This is consistent with the presence of multiple OH groups in fisetin and robinetin, which might be expected to promote, at least in part, exposure to the polar regions of the membranes.

The partition coefficient (K_p) is estimated from the slope and *1/F* intercept of the double-reciprocal linear plot of *1/F* against *1/[EYPC]*, according to the equation,

$$\frac{1}{F} = \frac{55.6}{K_p.F_{\max}} \times \frac{1}{[Lipid]} + \frac{1}{F_{\max}}$$

where F is the intensity of the tautomer fluorescence for a particular lipid concentration, and F_{max} is maximum fluorescence resulting from total probe incorporation into the membrane. K_p was found to be 4.2×10^4 for 3-HF, 6.2×10^5 for fisetin and 8.65×10^4 for robinetin (at 25°C) [26].

We also performed fluorescence polarization studies and estimated the anisotropy (*r*) parameter, which is a good indicator of the rigidity of the microenvironment of fluorophores, with zero or very low values of *r* in fluid solution where the fluorophore can freely rotate, and high *r* values signifying restricted motional freedom [1, 2]. Progressive increase in anisotropy of the tautomer emission is observed with increasing lipid concentration which is consistent with the picture that more and more fluorophore molecules are occurring in liposome bound state. Temperature dependence of r is illustrated for fisetin ESPT fluorescence for two different phospholipids, EYPC and DPPC (Fig. 4.12). Significantly, the *r* vs temperature plots show characteristic sigmoidal shape, revealing the thermotropic phase transition of the phospholipids from gel to liquid crystalline states. Phase transition temperatures (T_m, 42°C for DPPC and −12°C for EYPC) estimated from the midpoints of the sigmoidal shaped curves, are consistent with existing

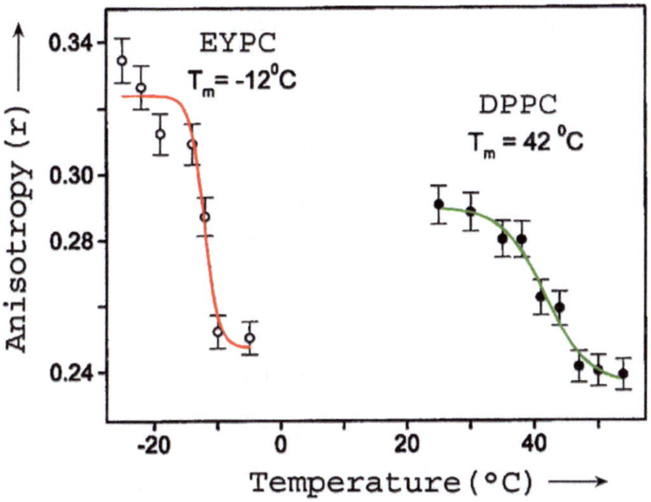

Fig. 4.12 Plot showing the variation of fluorescence anisotropy (*r*) of the ESPT tautomer fluorescence of fisetin in EYPC (*open circles*) and DPPC (*solid circles*) liposomes as a function of temperature. $\lambda_{ex} = 370$ nm, $\lambda_{em} = 535$ nm. The gel→ liquid crystalline state phase transitions of the phospholipids can be clearly discerned (Adapted with modification from Refs. [26, 36])

literature data based on different physical methods [36]. The high values for the anisotropy (r) observed (e.g. 'r'= 0.31 at −14°C in EYPC liposomes), and the fact that this parameter is proving to be a sensitive monitor of the thermotropic phase transitions of the phospholipids, are consistent with the picture that the fisetin molecules are localized in motionally constrained sites in the hydrophobic bilayer. Time resolved fluorescence lifetime measurements show biexponential fluorescence decay kinetics for both 3HF and poly-hydroxy flavonols. Such decay behavior suggests heterogeneity in microenvironments of the fluorophores in the membrane matrix. Moreover, in liposomes the individual components of the fluorescence decay are found to be significantly higher from the corresponding values in water. For example, when robinetin, is incorporated in EYPC liposomes, the average fluorescence lifetime increases by ~50 % for the normal and ~200 % for the tautomer emission, relative to the lifetimes in water. These observations are attributed to decreased water exposure in the membrane lipid bilayer with consequent decrease in non-radiative deactivation rates [29].

4.5.3 *Flavonols in RBC Ghost Membranes*

The fluorescence emission spectra of 3-HF and fisetin incorporated in RBC ghost membranes (along with reference spectra in aqueous buffer) [26, 33] are shown in Fig 4.13a, b respectively. In aqueous buffer (in the absence of membranes) the emission spectrum of 3-HF shows well resolved dual fluorescence bands (consisting of a blue -violet normal fluorescence with λ_{em}^{max} = 408 nm and a green tautomer fluorescence with λ_{em}^{max} = 508 nm). (The weak band at shorter wavelengths is due to Raman scattering from the aqueous solvent medium). Upon binding to

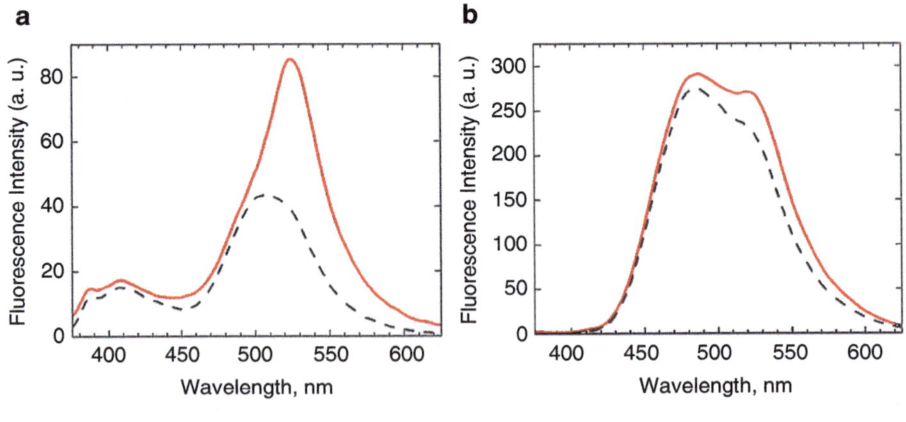

Fig. 4.13 Fluorescence emission spectra of (**a**) 3-HF and (**b**) fisetin in RBC ghost membranes (containing 25 μg protein/ml) (—) and in aqueous buffer (. .), λ_{ex} =340 nm for 3-HF and 360 nm for fisetin. [Flavonol] = 20 μM (Adapted from Ref. [33])

RBC membranes the emission profile shows dramatic changes with significant enhancement in the emission intensity of the green fluorescence band, which shows λ_{em}^{max} at 524 nm, which corresponds to the value in typical hydrocarbon solvents [15, 20]. Thus it is evident that in RBC membranes 3-HF predominantly occurs in environments of low polarity, presumably in the lipid dominated region of the membrane, where external H-bonding perturbation is minimal, which facilitates an efficient ESIPT process with high quantum yield of the tautomer (green) fluorescence band. While dual fluorescence behavior is observed for both 3-HF and fisetin, the spectra are much less resolved in the latter case, where strong overlap occurs between the ESIPT tautomer and normal (non-proton transferred) emissions, the latter possessing notable charge transfer character [43]. Interestingly, in contrast to 3-HF, where the tautomer emission is strongly enhanced upon incorporation in RBC ghost membranes, only a slight increase in intensity occurs in case of fisetin. Hence, we infer that unlike 3-HF, fisetin molecules are localized in relatively polar regions of the RBC membranes, presumably at the lipid-water interface. Thus the differences observed in the protective effects of the various flavonoids towards lipid peroxidation and hypotonic hemolysis (data not shown, see Ref. [35] for details) can be at least partly attributed to the difference in the degree of the penetrations of the flavonoid molecules in intact erythrocytes and ghost membranes.

4.6 Concluding Remarks

The spectroscopic research described in this article exemplify novel uses of flavonols as their own 'reporters' (by exploiting their exquisitely sensitive intrinsic fluorescence properties) for probing and characterizing their interactions with various biomolecular systems which are relevant as targets in relation to the therapeutic actions of these phytochemicals. We can envision that expanding applications of this promising new approach would open the door to new avenues for the 'screening' and 'design' of the most suitable derivatives from among numerous available structural variants of this new generation of therapeutic drugs . Latest developments in ex-vivo and in vivo applications, involving detection of the 'two color' fluorescence of flavonols (via use of two photon excitation technology, and detected through spectroscopic as well as imaging techniques) bound to targets in living cells and live animal models signify a new direction of progress. This should pave the way to expanded use of the intrinsic fluorescence properties of therapeutically potent flavonols and related compounds in medical research, leading to a fruitful transition from the 'lab bench to the clinic'.

Acknowledgements The author gratefully acknowledges all his students, post doctoral associates, and collaborators whose dedicated inputs over the years led to the development and successful growth and progress of the research highlighted in this article. He is thankful to the University Grants Commission (UGC), India for award of an Emeritus Fellowship, and to the Department of Biophysics, Molecular Biology and Bioinformatics, University of Calcutta for facilities and invaluable support during preparation of this manuscript.

References

1. Lakowicz JR (2006) Principles of fluorescence spectroscopy, 3rd edn. Springer, New York Inc.
2. Demchenko AP Introduction to fluorescence Sensing, 2nd ed. Springer 2015
3. Demchenko AP, Klymchenko AS, Pivovarenko VG, Ercelen S (2002) In: Krayenhof R, Visser AJWG, HC G (eds) Fluorescence spectroscopy, imaging and probes-new tools in chemical, physical and life sciences, Springer series on fluorescence methods and applications, vol 2. Springer, Heidelberg, Germany
4. Krasieva TB, Ehren J, O'Sullivan T, Tromberg BJ, Maher P (2015) Cell and brain tissue imaging of the flavonoid fisetin using label-free two-photon microscopy. Neurochem Int 89:243–248
5. Chaudhuri S, Sengupta B, Taylor J, Pahari B, Sengupta PK (2013) Interactions of dietary flavonoids with proteins: insights from fluorescence spectroscopy and other related biophysical studies. Curr Drug Metab 14:491–503 and references cited therein
6. Protti S, Mezzetti A (2012) Any color you like. Excited state and ground state proton transfer in flavonols and applications. In: Albini A (ed) Photochemistry, Vol, vol 40, pp 295–322
7. Sengupta PK, Banerjee A, Sengupta B 2006 Exploring the interactions of therapeutically active plant flavonoids with biological targets: Insights from fluorescence spectroscopy. In: Satoshi K (ed) Photoelectrochemistry and photobiology in the environment, energy & fuel, Research Signpost. Chap.5, 20–226, and references cited therein
8. Andersen OM, Markham KR (eds) (2006) Flavonoiids: chemistry, biochemistry and applications. CRC Press, Boca Raton
9. Rice-Evans CA, Diplock AT (1993) Current status of antioxidant therapy. Free Radic Biol Med 15:77–96
10. Havsteen BH (2002) The biochemistry and medical significance of the flavonoids. Pharmacol Therapeut 96:67–202
11. Pahari B, Sengupta B, Chakraborty S, Thomas B, McGowan D, Sengupta PK (2013) Contrasting binding of fisetin and daidzein in gamma-cyclodextrin nanocavity. J Photochem Photobiol B:Biology 118:33–41
12. Banerjee A, Basu K, Sengupta PK (2008) Interaction of 7-hydroxyflavone with serum albumin: a spectroscopic study. J Photochem Photobiol B Biol 90:33–40
13. Chaudhuri S, Pahari BP, Sengupta PK (2009) Ground- and excited-state proton transfer and antioxidant activity of 7-hydroxyflavone in model membranes: absorption and fluorescence spectroscopic studies. Biophys Chem 139:29–36
14. Rusznyák S, Szent-Görgyi A, Vitamin P (1936) flavonols as vitamins. Nature 138:27–27
15. Sengupta PK, Kasha M (1979) Excited state proton-transfer spectroscopy of 3-hydroxyflavone and quercetin. Chem Phys Lett 68:382–385
16. Demchenko AP, Heldt J, Waluk J, Chou P-T, Sengupta PK, Brizhnik L, del Valle JC (2014) From photochemistry and flowers to spectroscopy and music: the unique and diverse achievements of Michael Kasha. Angew Chem 53:14316–14324
17. Kasha M (1986) Proton-transfer spectroscopy: perturbation of the tautomerization potential. J Chem Soc Faraday Trans 2 82:2379–2392
18. Ameer-Beg S, Ormson SM, Brown RG, Matousek P, Towrie M, Nibbering ETJ, Foggi P, Neuwahl FVR (2001) Ultrafast Measurements of Excited State Intramolecular Proton Transfer (ESIPT) in room temperature solutions of 3-hydroxyflavone and derivatives. J Phys Chem A 105:3709–3718
19. Bader AN, Ariese F, Gooijer C (2002) Proton transfer in 3-Hydroxyflavone studied by high-resolution 10 K laser-excited Shpol'skii spectroscopy. J Phys Chem A 106:2844–2849
20. Demchenko AP, Tang K-C, Chou P-T (2013) Excited-state proton coupled charge transfer modulated by molecular structure and media polarization. Chem Soc Rev 42:1379–1408
21. Guharay J, Sengupta B, Sengupta PK (2001) Protein–flavonol interaction: fluorescence spectroscopic study. Proteins: structure, function. Genetics 43:75–81 and references cited therein

22. Pahari B, Chaudhuri S, Chakraborty S, Sengupta PK (2015) Ground and excited state proton transfer of the bioactive plant flavonol robinetin in a protein environment. J Phys Chem B 119:2533–2545
23. Sengupta B, Sengupta PK (2002) The interaction of quercetin with human serum albumin: a fluorescence spectroscopic study. Biochem Biophys Res Commun 299:400–403
24. Sengupta B, Sengupta PK (2003) Binding of quercetin with human serum albumin: a critical spectroscopic study. Biopolymers 72:427–434
25. Sengupta B, Pahari B, Blackmon L, Sengupta PK (2013) Prospect of bioflavoinoid fisetin as a quadruplex DNA ligand: a biophysical approach. PLoS One 8(6):e65383. doi:10.1371/journal.pone.0065383
26. Pahari B, Chakraborty S, Chaudhuri S, Sengupta B, Sengupta PK (2012) Binding and antioxidant properties of therapeutically important plant flavonoids in biomembranes: Insights from spectroscopic and quantum chemical studies. Chem Phys Lipids 165:488–496
27. Pahari B, Chakraborty S, Sengupta PK (2011) Encapsulation of 3-hydroxyflavone in gamma-cyclodextrin nanocavities: excited state proton transfer fluorescence and molecular docking studies. J Mol Struct 1006:483–488
28. Chaudhuri S, Chakraborty S, Sengupta PK (2011) Probing the interactions of hemoglobin with antioxidant flavonoids via fluorescence spectroscopy and molecular modeling studies. Biophys Chem 154(2011):26–34
29. Chaudhuri S, Pahari BP, Sengupta B, Sengupta PK (2010) Binding of the bioflavonoid robinetin with model membranes and hemoglobin: Inhibition of lipid peroxidation and protein glycosylation. J Photochem Photobiol B:Biology 98:12–19
30. Sengupta PK, Chaudhuri S (2010) "Interactions of therapeutically active plant flavonols with biological targets: insights from fluorescence spectroscopic studies" (Invited article). J Indian Chem Soc 2010(87):213–220
31. Chaudhuri S, Basu K, Sengupta B, Banerjee A, Sengupta PK (2008) Ground and excited state proton transfer and antioxidant activity of 3-hydroxyflavone in egg yolk phosphatidylcholine liposomes: absorption and fluorescence spectroscopic studies. Luminescence 23:1397–1403
32. Banerjee A, Basu K, Sengupta PK (2007) Effect of beta-cyclodextrin nanocavity confinement on the photophysics of robinetin. J Photochem Photobiol B Biol 89:88–97
33. Chaudhuri S, Banerjee A, Basu K, Sengupta B, Sengupta PK (2007) Interaction of flavonoids with red blood cell membrane lipids and proteins: antioxidant and antihemolytic effects. Int J Biol Macromol 2007(41):42–48
34. Banerjee A, Sengupta PK (2006) Encapsulation of 3-hydroxyflavone and fisetin in beta-cyclodextrins: Excited state proton transfer fluorescence and molecular mechanics studies. Chem Phys Lett 424:379–386
35. Sengupta B, Banerjee A, Sengupta PK (2005) Interactions of the plant flavonoid fisetin with macromolecular targets: insights from fluorescence spectroscopic studies. J Photochem Photobiol B Biol 80:79–86
36. Sengupta B, Banerjee A, Sengupta PK (2004) Investigations on the binding and antioxidant properties of the plant flavonoid fisetin in model biomembranes. FEBS Lett 570(2004):77–81
37. Sengupta B, Banerjee A, Sengupta PK (2005) Interactions of the plant flavonoid fisetin with macromolecular targets: insights from fluorescence spectroscopic studies. J Photochem Photobiol B Biol 80:79–86
38. Gutzeit HO, Henker Y, Kind B, Franz A (2004) Specific interactions of quercetin and other flavonoids with target proteins are revealed by elicited fluorescence. Biochem Biophys Res Commun 318:490–495
39. Mohapatra M, Mishra AK (2011) Photophysical behavior of fisetin in dimyristoylphosphatidylcholine liposome membrane. J Phys Chem B 115:9962–9970
40. Shyamala T, Mishra AK (2004) Ground- and excited-state proton transfer reaction of 3-hydroxyflavone in dimyristoylphosphatidylcholine liposome membrane. Photochem Photobiol 80:309–315

41. Jana B, Senapati S, Ghosh D, Bose D, Chattopadhyay N (2012) Spectroscopic exploration of mode of binding of ctDNA with 3-hydroxyflavone: a contrast to the mode of binding with flavonoids having additional hydroxyl groups. J Phys Chem B 116:639–645
42. Sengupta B, MR S, DE D Jr, Harris K, RM W, Ward D, D'Asia G, Hampton C (2015) Excited state proton transfer of natural flavonoids and their chromophores in duplex and tetraplex DNAs. J Phys Chem B 119:2546–2556
43. Sytnik A, Gormin D, Kasha M (1994) Interplay between excited-state intramolecular proton transfer and charge transfer in flavonols and their use as protein-binding-site fluorescence probes. Proc Natl Acad Sci U S A 91:11968–11972
44. Böhl M, Zupalla C, Takalov SV, Hoflack B, Gutzeit HO (2005) Identification of actin as quercetin-binding protein: an approach to identify target molecules for specific ligands. Anal Biochem 34:295–299
45. Wybranowski T, Kruszewski S (2014) Optical spectroscopy study of the interaction between quercetin and human serum albumin. Acta Phys Pol A 125:A-57–A-60
46. Mishra B, Barik A, Priyadarshini KI, Mohan H (2005) Fluorescence spectroscopic studies on binding of a flavonoid antioxidant quercetin to serum albumins. J Chem Sci 117:641–647
47. Böhl M, Tietze S, Sokoll A, Madathil S, Pfennig F, Apostolakis J, Fahmy K, Gutzeit HO (2007) Flavonoids affect actin functions in cytoplasm and nucleus. Biophys J 93:2767–2780
48. Voicescu M, Ionescu S, Nistor C (2017) Spectroscopic study of 3-hydroxyflavone-protein interactions in lipidic bi-layers immobilized on silver nanoparticles. Spectrochim Acta A 170:1–8 (in press)
49. Voicescu M, Ionescu S (2015) 3-hydroxyflavone-Bovine serum albumin interaction in dextran medium. J Serb Chem Soc 80:517–528
50. Voicescu M, Bandula R (2015) 3,6-dihydroxyflavone/ Bovine serum albumin interaction in cyclodextrin medium: absorption and emission monitoring. Spectrochim Acta A 138:628–636
51. Latruffe N, Menzel M, Delmas D, Buchet R, Lancon A (2014) Compared binding properties between resveratrol and other polyphenols to plasmatic albumin: consequences for the health protecting effect of dietary plant microcomponents. Molecules 19:17066–17077
52. Falkovskaia E, Sengupta PK, Kasha M (1998) Photophysical induction of dual fluorescence of quercetin and related hydroxyflavones upon intermolecular H-bonding to solvent matrix. Chem Phys Lett 297:109–114
53. Rolinski OJ, Martin A, Birch DJS (2008) Human serum albumin-flavonoid interactions monitored by means of tryptophan kinetics. Ann N Y Acad Sci 1130:314–319
54. Rolinski OJ, Martin A, Birch JSD (2007) Human serum albumin and quercetin interactions monitored by time-resolved fluorescence: evidence for enhanced discrete rotamer conformations. J Biomed Opt 12(3):034013 -1–034013-7
55. Dufour C, Dangles O (2005) Flavonoid –serum albumin complexation: determination of binding constants and binding sites by fluorescence spectroscopy. Biochim Biophys Acta 172:164–173
56. Singha Roy A, Dinda AK, Dasgupta S (2012) Study of the interaction between fisetin and human serum albumin: a biophysical approach. Protein Pept Lett 19:604–615
57. Xiao J, Kai G (2012) A review of dietary polyphenol-plasma protein interactions: characterization, influence on the bioactivity, and structure-affinity relationship. Crit Rev Food Sci Nutr 52:85–101
58. Carter DC, He XM (1992) Atomic structure and chemistry of human serum albumin. Nature 358:209–215
59. S N, Balasubramanian S (2006) Quadruplex Nucleic acids. RSC, Cambridge, U.K.
60. Wang Y, Y H, Wu T, Zhou X, Shao Y (2015) Trigegered excited-state intramolecular proton transfer fluorescence for selective triplex DNA recognition. Anal Chem 87(23):11620–11624

Chapter 5
Luminescent Nanoparticles for Chemical Sensing and Imaging

Michael Schäferling and Ute Resch-Genger

Keywords Luminescent nanomaterials • Fluorescence probe • Fluorescence label • Chemical sensor • Photon upconversion

5.1 Introduction

The implementation of fluorescent methods is of outstanding importance in the field of optical chemical sensor technology and biosciences. Their bioanalytical applications are manifold including fluorescence microscopy, fluorescence *in situ* hybridization, DNA sequencing, fluorescence-activated cell sorting, immunoassays, analysis of DNA and protein microarrays, and quantitative PCR, just to name a few examples. Particularly, fluorescence microscopy is a valuable method in the versatile field of biomedical imaging methods which nowadays utilizes different fluorescence parameters like emission wavelength/color and lifetime for the discrimination between different targets. Sectional images are available with confocal microscopes. Tissue, cells or single cellular compartments can be stained and visualized with fluorescent dyes and biomolecules can be selectively labeled with fluorescent dyes to monitor biomolecular interactions inside cells or at membrane bound receptors.

On the other hand, fluorophores can act as indicator (or "molecular probe") to visualize intrinsically colorless and non-fluorescent ionic and neutral analytes such as pH, oxygen (pO_2), metal ions, anions, hydrogen peroxide or bioactive small organic molecules such as sugars or nucleotides. Thereby, their photoluminescent properties (fluorescence or phosphorescence intensity, excitation and/or emission wavelength, emission lifetime or anisotropy) respond to the presence of these species in their immediate environment. In general, the use of luminescent probes has the advantage that they can be delivered directly into the sample, and detected in a contactless remote mode. By now, these probes are often encapsulated in different types of nanoparticles (NPs) made from (biodegradable) organic polymers, biopolymers or inorganic materials like silica or bound to their surface. The use

M. Schäferling (✉) • U. Resch-Genger
Division 1.10 Biophotonics, BAM Federal Institute of Materials Research and Testing, Richard-Willstätter-Str. 11, 12489 Berlin, Germany
e-mail: michael.schaeferling@bam.de; ute.resch@bam.de

C.D. Geddes (ed.), *Reviews in Fluorescence 2016*, Reviews in Fluorescence,
DOI 10.1007/978-3-319-48260-6_5

of nanoprobes has several advantages compared to molecular reporters, i.e., single fluorophores, particularly a higher brightness and a higher photostability as well as shielding of the encapsulated fluorophores from interferences leading to unspecific fluorescence quenching. Moreover, this presents a rather simple approach to include analyte-inresponsive, "inert" reference dyes for ratiometric measurements [1]. For this purpose, the surface of NPs can be chemically modified to render them highly cell permeable, e.g., by internalization via endocytosis [2]. Moreover, they can be targeted to cellular compartments or to tumor cells by functionalization e.g. with lipids, peptides or antibodies [3]. It should not be concealed, however, that the use of NPs arises many objections regarding their release to the environment and the subsequent unintended exposure of living organisms including humans [4, 5].

Materials that are frequently used for the preparation of polymeric nanoprobes include silica, polystyrene or derivatives of polyacrylic acid. They can be prepared in diameters lower than 100 nm with high monodispersity. In order to achieve a sensory function, analyte-responsive fluorescent probes are incorporated in or attached to the surface of the NPs. The particles can be prepared by microemulsion polymerization and/or precipitation techniques and stained with organic dyes or metal complexes according to the polarity of the material. The availability of materials covering the range from non-polar (polystyrene) to polar and hydrophilic (polyacrylic acid and its derivatives, silica) provides a high flexibility in terms of encapsulation of probes into materials suited for specific applications. Some polymer materials are highly biocompatible and have already been applied to *in vivo* imaging [6]. Clark et al. referred to such nanoprobes in their pioneering work on intracellular pH and calcium sensing as PEBBLEs (probes encapsulated by biological localized embedding) [7]. Their first polymer-based nanoprobes for intracellular sensing consisted of fluorescein as pH indicator and a reference dye embedded in a polyacrylamide matrix. Further examples of polymeric nanoprobes for the imaging of intracellular oxygen, copper, calcium and hydrogen peroxide are listed in Table 5.1. The co-doping of the polymer particles with internal reference dyes is an important prerequisite for sensor applications, because only this enables reliable quantitative results by ratiometric analysis. Thereby, the analyte-sensitive luminescence of the probe can be referenced against a signal from a reference dye which behaves inert and changes in the fluorescence intensity originating from fluctuations in the excitation light intensity or probe concentration can be properly eliminated [19].

Furthermore, inorganic nanocrystalline materials such as semiconductor quantum dots and rods (QDs and QRs) and lanthanide doped crystal matrices (e.g., upconversion nanoparticles, UCNPs) as well as carbon dots have been used. They either show an intrinsic response to the analyte or can be conjugated to specific indicators to achieve materials with ratiometric sensing properties. Other materials that serve as basis for the preparation of nanoprobes, particularly for applications to intracellular pH sensing, include semiconducting polymers dots, micellar systems, bacteriophages, or metal nanoparticles (gold or silver) [20].

The aim of this contribution is to review the achievements of our working groups in the development of nanoprobes for fluorescent sensing and imaging over the past

Table 5.1 Examples for spherical polymer or silica nano- and microparticles appplied as probes for intracellular sensing and imaging

Analyte	Nanomaterial (size)	Indicator	Reference	Comment	References
pH	PAA (200 nm)	Fluorescein	Sulforhodamine	Dyes encapsulated in polymer	[7]
	PAA copolymer (60 nm)	Fluorescein/OG	RB	Broad dynamic range by two indicators	[8]
	PDMAPA/PAA (38 nm)	Naphthalimid	Sulforhodamine	Covalently linked dyes	[9]
	PS/Phospholipid (500 nm)	OG	Texas Red	Core/shell particle, covalent	[10]
	silica (70 nm)	Fluorescein	TMR	Core/shell particle, covalent	[1]
	silica (42 nm)	Fluorescein	$Ru(bipy)_3$	Core/shell particle,	[11]
Oxygen	MPS (50 nm)	Fluorescein	TMR	Covalently linked dyes	[12]
	PDMA (150–250 nm)	Pt-OEPK	OEP	Dyes encapsulated in polymer	[13]
	ormosil (120 nm)	Pt-OEPK	OEP	Core/shell, dyes encapsulated	[14]
	Polyfluorenes (25 nm)	Pt-OEP	–	FRET system	[15]
Ca^{2+}	PS (500 nm)	Pt-TFPP	PNBE	Dyes encapsulated in polymer	[16]
	PS (2 μm)	Indo-1	2-λ dye	Dye covalently linked to amino-modified PS	[17]
H_2O_2	PAA (200 nm)	Calcium Green	SR 101	Dyes encapsulated in polymer	[7]
	ormosil (100 nm)	2′,7′-dichloro-fluorescin	No referencing	Dyes encapsulated	[18]

MPS mesoporous silica, *OG* Oregon Green 488, *ormosil* organically modified silica, *OEP* octaethylporphyrine, *OEPK* octaethylporphyrine ketone, *PNBE* N-(5-carboxypentyl)-4-piperidino-1,8-naphthalimide butyl ester, *PAA* polyacrylamide, *PDMA* Poly(decyl methacrylate), *PDMAPA* poly(*N*,*N*-dimethyl-3-aminopropyl acrylamide), *PS* polystyrene, *RB* rhodamine B, *SR 101* sulforhodamine 101, *TFPP* 5,10,15,20-tetrakis-(2,3,4,5,6-pentafluorphenyl)-porphyrin, *TMR* tetramethylrhodamine decade. Besides a description of the utilized nanoparticle materials and indicators (sensitive dyes), focusing on polymeric materials and inorganic nanocrystals such as QDs and UCNPs, we will also discuss the following topics:

- General strategies for sensor signal referencing
- Methods for surface functionalization and characterization
- Spectroscopic tools for quantum yield and brightness measurements
- Preparation of sensor films (2D sensors)
- Examples for *in vitro* and in *vivo* imaging

5.2 General Strategies for Sensor Referencing

The detection of fluorescence emission intensity as a measurement parameter in quantitative chemical analysis has some significant drawbacks and limitations. This includes its sensitivity to fluctuations of the excitation light intensity and in the light field (inhomogeneous illumination) as well as inhomogeneities in dye concentration e.g., due to photobleaching or dye accumulation favoured e.g. by non-specific interactions in cell studies. In this respect, it must be considered that the goal of fluorescence sensing studies is the determination of the concentration of a specific analyte and not the determination of the concentration of the fluorophore in the sample. Other factors which may distort signals and hence analyte quantification are scattering of the excitation light, changes in the optical pathlength as well as unspecific background signals or autofluorescence of biological samples.

In fluorescence sensor and imaging technologies, such disadvantages can be elegantly circumvented with intrinsically referenced methods [21]. They eliminate interferences and make measured fluorescence signals independent from probe concentration, thereby leading to an improved robustness and reliability of the sensor output. A particular challenge for the development of internally referenced systems, especially for biological systems such as live cells or in vivo is the prevention of extensive calibration procedures accompanying each experiment. Several referencing methods have been meanwhile established, that were initially developed in fluorescence microscopy, which can be referred to as ratiometric measurements [22]. They include:

- Dual wavelength (2-λ) probes
- Resonance energy transfer systems
- Addition of an inert reference dye
- Time-resolved or fluorescence lifetime detection.

Examples for each of these concepts will be shown below in the applications part. Generally, 2-λ probes are often used for pH sensing or the measurement of certain metal ions like Ca^{2+}. Examples for pH-responsive 2-λ probes are 8-hydroxypyrene-1,3,6-trisulfonic acid (HPTS), 2′,7′-bis(2-carboxyethyl)-5-carboxyfluorescein (BCEFC) and seminaphthorhodafluorescein derivatives. The classical example for metal-ion responsive 2-λ probes are the so-called fura dyes introduced by R. Tsien [23], developed for the determination of Ca^{2+} in bioimaging studies such as the measurement of cytosylic Ca^{2+} concentrations. In these cases, the fact that the acidic (protonated) or metal-complexed forms of 2-λ probes show different excitation and/or emission spectra than the basic (deprotonated) or metal-free form, is exploited for ratiometric measurements.

The utilization of combinations of stimuli-responsive and inert reference dyes has found widespread application in fluorescent sensors such as sensors for pH, oxygen, or metal ions. This approach requires dye combinations which can be excited with the same light source (same excitation wavelength requiring absorption spectra in the same wavelength region) and show emission bands, that can be spectrally discriminated with optical filters. In the case of Förster resonance energy transfer (FRET) systems, the emission of the nanoparticle is mostly used as donor, whereas the analyte-sensitive dye acts as acceptor [20].

The most attractive self-referenced parameter is the decay time of the excited state of a fluorophore. It is directly linked to the compound- and environment-specific fluorescence lifetime. The decay of the fluorescence intensity after excitation with a short pulse of light is in the ideal case of a homogeneous microenvironment (e.g. in solution) monoexponential, given by:

$$I = I_0\, e^{-t/\tau} \tag{5.1}$$

where I_0 is the intensity at $t = 0$ and τ is the fluorescence lifetime, respectively. The latter is the time that is required until the fraction of molecules in the excited state has decreased to 1/e. In the case of dyes in a heterogeneous microenvironment such as a polymeric nanoparticle matrix, or semiconductor nanocrystals with their particle size distribution and complex surface chemistry, however, the photoluminescence decay kinetics are often multiexponential [24, 25]. Unlike intensity, the fluorescence lifetime is barely affected by the concentration of the fluorophores, static quenching effects, photobleaching, and the intensity (i.e., spectral radiance or spectral power density) of the light source. These advantages make the fluorescence lifetime to a widely used parameter in fluorescence sensing and imaging. Generally, methods for lifetime determination can be classified into time-domain and frequency-domain approaches [26].

The latter is based on a phase-modulation technique, in which the sensor is excited with sinusoidally modulated light at a frequency approximately reciprocal to the decay time. The emission of the probe follows the modulation, but with a certain delay. This is measured as a change of the phase angle. Accordingly, τ can be calculated from the phase angle or the modulation ratio. In contrast to dual wavelength methods, lifetime determination can be accomplished applying only one excitation and one emission wavelength.

We favor a straightforward approach termed *Rapid Lifetime Determination* (RLD) for ratiometric fluorescence sensing and imaging [27] where the fluorescence intensity is recorded in two successive time gates, set within defined delays (Δt) after a short pulse of the excitation light (Fig. 5.1).

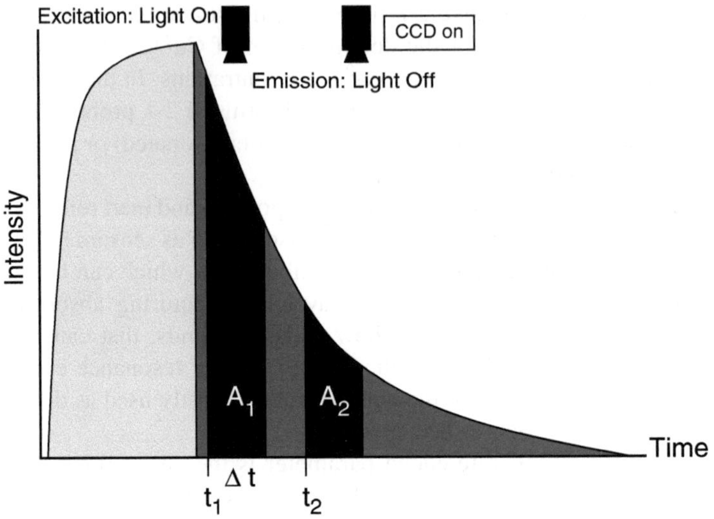

Fig. 5.1 The change of fluorescence lifetime τ can be assessed by the RLD method using time-gated detection. The detector is opened after a short pulse of light at times t_1 and t_2 for an identical period of Δt. The lifetime τ is proportional to the ratio of the integrated photon counts A_1 und A_2

In the case of a monoexponential decay and an identical width/length of the time gates Δt, the lifetime τ can be calculated according to Eq. 5.2, where Δt is the width of the respective integration intervals A_1 and A_2.

$$\tau = \frac{\Delta t}{\ln (A_1/A_2)} \tag{5.2}$$

RLD cannot be applied to calculate absolute lifetimes in the case of multiexponential fluorescence decays, because of its origin in just two images at different delay times. However, the determination of absolute lifetime values is not mandatory for sensor applications. The precision of this method is dependent on the integration times (Δt), which have to be set in a proper relationship with the lifetime ($\Delta t/\tau \approx 2$) and a sufficient number of photon counted [28].

In practice, the images of the two different gates are taken separately in subsequent acquisition cycles. Therefore, the light source and the detector have to be pulsed at a suitable frequency which is in the same magnitude than the reciprocal decay time. The integration of the two sets of images is followed by a subtraction of the corresponding background dark images (detected with the same time gates and frequency, but without illumination). Thus, the significant variable measured is the change of the ratio $R = A_1/A_2$ as a function of the analyte concentration.

We performed a comparative study of different imaging techniques for optical chemical sensors. Thereby, a luminescent film for sensing barometric pressure (via quenching by oxygen) was used as a model system and calibrated by four fluorescence imaging methods including intensity imaging, ratiometric 2-λ imaging and lifetime imaging (according to the RLD method) with a CCD camera, as

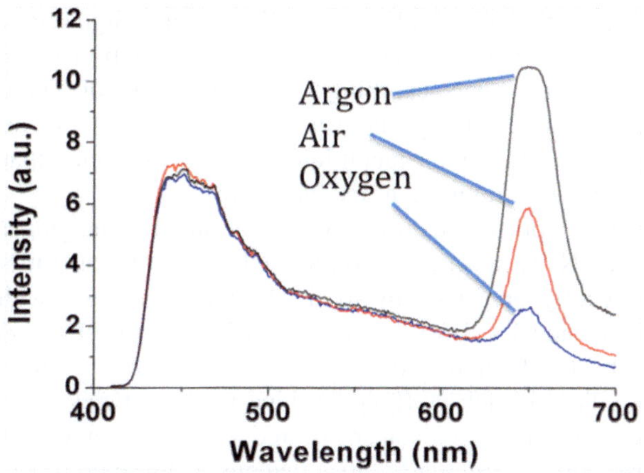

Fig. 5.2 Emission spectrum of the ratiometric oxygen sensor film consisting of DPA (emission $\lambda_{max} = 456$ nm) in PAN microparticles as reference and the oxygen indicator PtTFPP (emission $\lambda_{max} = 650$ nm) incorporated into PS microparticles dispersed in a polyurethane hydrogel. The emission was recorded under various atmospheres (argon, normal air and oxygen). Excitation was done with a 405 nm LED. Reproduced from Ref. [29], Copyright (2013), with permission from Elsevier

well as 2-λ imaging using a customary digital color camera (RGB imaging) [29]. The sensor film used for this study consisted of a 5 μm layer of a polyurethane hydrogel that contained (a) poly(acrylonitrile) (PAN) microparticles (DPA-PAN, size: 0.1–1 μm) dyed with diphenylanthracene (DPA), and (b) polystyrene (PS) microparticles (Pt-PS, size: 0.1–5 μm) stained with the oxygen-sensitive platinum(II) meso-tetrakis(pentafluorophenyl) porphyrin (PtTFPP). In this case, the DPA-PAN microparticles act as reference and the Pt-PS as the oxygen-sensitive indicator. Luminescence intensity and lifetime (around 50 μs in oxygen-free solution at room temperature) of PtTFPP are strongly quenched by oxygen. The emission spectra at different oxygen levels are shown in Fig. 5.2.

This study clearly revealed the superiority of the RLD method compared to the 2-λ read out of the sensor films with respect to standard deviations (from different calibration runs as well as pixel-to-pixel variations of one film), signal to noise ratios, resolution (limit of detection) of pressure determination and suppression of interferences by inhomogeneous illumination.

Fluorophores used as indicators for pH or metal ions are commonly composed of organic dyes which have fluorescence lifetimes of a few nanoseconds. Hence, the determination of lifetime changes in the subnanosecond range is required, which demands CCD cameras with image intensifier and pumped (mode-locked) lasers or laser diodes or light emitting diodes (LEDs) with small pulse widths.

Ratiometric optical chemosensors can be also realized by the combination of an emitter and a colorimetric indicator which undergoes analyte-specific changes in intensity and/or spectral position of its absorption band. Such systems utilize analyte-induced inner filter effects for the modulation of the emission intensity of either a broad band emitter or a fluorophore, which reveals several emission bands like an upconversion phosphor, and hence for fluorescence signalling. A prerequisite for proper function of such systems is the spectral overlap of the emission of the reference dye with at least one absorbing form of the indicator. The great advantage of this strategy is its simplicity and versatility, exploiting the huge number of colorimetric indicators for pH or metal ions which is available, that are also often more robust compared to fluorescent indicator dyes.

A very versatile optical chemosensor for ratiometric imaging and sensing in the near infrared (NIR) spectral region suitable for different ratiometric sensing schemes was recently presented by us in close collaboration with the group of Professor Klimant and Dr. Borisov (Technical University of Graz). The concept was developed by the Klimant/Borisov group and utilizes calcium, strontium, or barium copper silicates such as Egyptian blue microparticles with broad absorption/excitation bands and broad near-infrared emission centered at ca. 920 nm as reference dyes [30]. Particularly advantageous of these commercially available inorganic phosphors is the large spectral shift between their absorption and emission bands ideal for the combination with organic fluorescent indicators, which typically possess rather small Stokes shifts, and the spectral discrimination of both emission bands as well as their high chemical and photochemical stability and high brightness in conjunction with their very low price. This cannot be reached with organic dyes. These phosphors have been utilized for different sensor schemes like ratiometric (2-λ) emission, dual lifetime referencing (DLR), and inner filter effect-modulated excitation or emission. DLR utilizes a reference, here a nanophosphor with a significantly longer luminescence decay time than the analyte-responsive fluorescent indicator, which lifetimes are typically in the microsecond–millisecond and nanosecond time domain, respectively [31, 32]. As exemplarily shown in Fig. 5.3, they can be readily dispersed in a hydrogel, in this case together with a red pH-sensitive aza-Bodipy dye, which emission intensity is switched on upon protonation. As follows from this figure, both fluorophores, which can be excited at 625 nm, reveal spectrally well distinguishable emission spectra in the NIR. The absorption maxima of the protonated and deprotonated forms of the pH indicator are located at 687 and 743 nm, respectively, with the absorption of the protonated form at the excitation wavelength (625 nm) being about three times more efficient. Moreover, Egyptian blue and other copper silicate phosphors are excellent references for the dual lifetime referencing method, since they combine bright long-lived luminescence with inertness to oxygen and can be used for the design of ratiometric inner filter effect sensors.

5.3 Nanoprobe Materials

5.3.1 Polymer Particles

The advantages of the encapsulation of luminescent probes into polymer nano- or microstructures or their covalent binding to polymer matrices for sensing in biological environment have been already outlined. Table 5.1 summarizes some

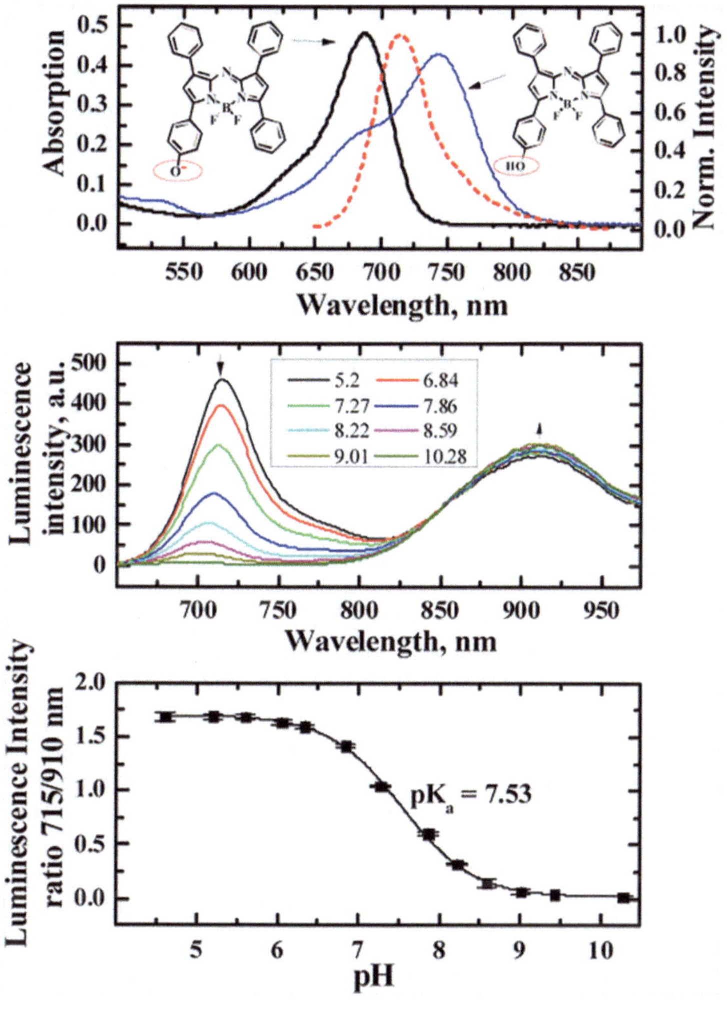

Fig. 5.3 *Top panel*: Absorption (*black*: pH of 5.2; *blue*: pH of 10.3) and emission (*red*) spectrum of a red aza-Bodipy pH indicator in a hydrogel film; *middle panel*: pH dependence of the emission spectra of the pH indicator and micrometer-sized Egyptian blue particles in the hydrogel; *Lower panel*: calibration for the ratio of the luminescence intensities (*squares*) and a sigmoidal fit (*line*). Excitation was at 625 nm. Reprinted from Ref. [30], Copyright (2013) with permission from the American Chemical Society

exemplary milestones in the development of polymer and silica NPs and microspheres for intracellular sensing and imaging of relevant parameters for cellular function including pH, pO_2, Ca^{2+}, and H_2O_2. Some polymer materials such as polystyrene, polyethylene glycol, or polyacrylamide are also suitable for in vivo imaging. For example, the micelle forming block-copolymer poly(ethylene glycol) methyl ether–poly(β-amino ester) was used for pH sensing in tumors [33]. A FRET pair consisting of tetramethyl-rhodamine and a non-emissive black hole quencher was embedded into micelles. This system reveals noticeable fluorescence signal amplification at decreased pH levels, which results from demicellization at the acidic pH environment in tumors. NPs for the imaging of tumor hypoxia were designed on the basis of polyacrylamide nanogels (30 nm diameter) containing the NIR emissive oxygen-sensitive probe Pd-tetraphenyl tetrabenzoporphyrin and Alexa 647 as reference dye [34].

In these examples, two more advantages of polymer, particularly polyelectrolyte nanoparticles emerge: they can accumulate in tumor cells, leading to a high imaging contrast, i.e., a good signal or tumor-to-background ratio. Moreover; they can be easily combined with drug delivery concepts including different types of release strategies initiated by external triggers [35].

A simple approach to the preparation of fluorescent particles presents the swelling of premanufactured polymer nanomaterials and microparticles in the presence of dyes, which must be hydrophobic [36]. This one-step steric incorporation strategy, which is used by many manufacturers of fluorescent polymer particles, is particularly attractive as a broad variety of commercial nm- and μm-sized particles with different surface chemistries are commercially available. The dyes do not have to bear functional groups for a covalent attachment and many NIR-emissive dyes are hydrophobic. Moreover, it can be extended to other hydrophobic compounds.

For the rational design of polymer-based sensors and probes with this approach, we evaluated different swelling procedures exemplarily for the loading of the hydrophobic polarity-probe Nile Red, a neutrally charged oxazine dye, into premanufactured nano- and micrometer sized polystyrene particles (PSP) with different surface chemistries, i.e., non-functionalized, amino-modified and PEG-grafted surfaces. In order to optimize and generalize the swelling conditions, different parameters affecting the encapsulation procedure in PSP were studied. This included different solvent systems varying in volume fractions, thereby covering the solubility of many different organic dyes and metal ion complexes, and the parameters swelling time, dye concentration, particle size and surface modification. The main criterion for the optimization of the dye loading procedures was a high and thermally and photochemically stable fluorescence output of the particles and the colloidal stability of the particles.

Similar studies with dyes representatively chosen from common dye classes differing in chromophore charge and polarity demonstrated the general suitability of such one-step loading procedures for efficient particle staining with neutral, zwitterionic, and charged fluorophores like oxazines, coumarines, squaraines, xanthenes, and cyanines emitting in the visible and near infrared [37]. A suitable tool to estimate the loading efficiency of fluorophores into such polymer particles present dye polarity and log D values as shown in Fig. 5.4. This approach is also suitable for the simultaneous loading with two dyes as required for the preparation of ratiometric sensor particles as shown by us e.g. for a NIR-emissive oxygen nanosensor [38]. Spectroscopically monitored leaking studies with dye-loaded PSP prepared via a swelling procedure in water and ethanol-water mixtures as well as in the buffer containing bovine serum albumin revealed no dye loss. In order to prepare particles with optimum brightness, the composition of the solvent mixture and the dye concentration used for particle swelling must be adapted to dye-specific

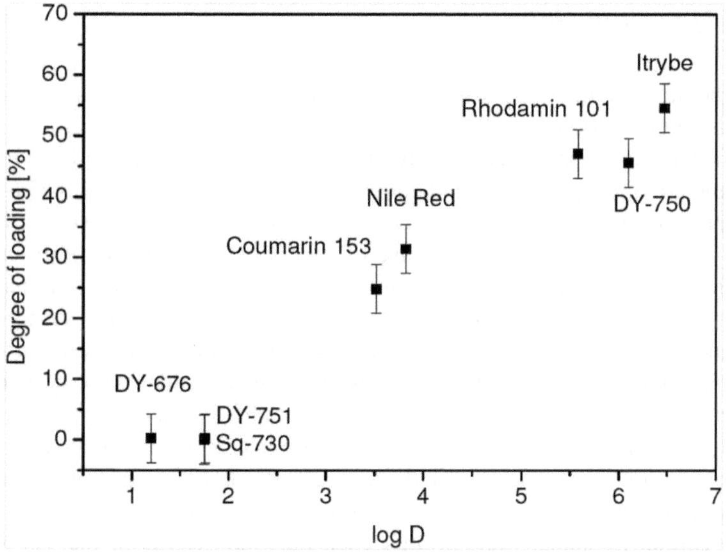

Fig. 5.4 Degree of loadings of common vis and NIR fluorophores in PSP as a function of log D (at pH 5.5) for 1 μm-sized carboxylated PSP. The amount of dye applied was 5×10–8 mol in a 14 % DMF/water staining solution. Reproduced from Ref. [37], Copyright (2012), with permission from Elsevier

properties such as fluorophore solubility and aggregation tendency [37, 39, 40]. In order to facilitate the optimization of dye loading of PSP, we examined different fast and inexpensive spectroscopic methods for the quantification of the fluorophore content of beads loaded or labeled with commercial fluorophores, differing in dye class, charge, and hydrophilicity [41].

Similarly simple preparation strategies relying on premanufactured nanomaterials and noncovalently bound water insoluble fluorophores can be used also for the preparation of fluorescent reporters from another class of inorganic carrier systems, nanoclays like laponites [42, 43]. In this case, charge-related effects have to be considered. The number of dye molecules, which are still highly emissive and not quenched e.g. by dye-dye interactions and by strongly acidic sites is limited in such nanoclays [44–46]. Moreover, the surface functionalization of these nanoclays with biomolecules using silane chemistry seems to be more challenging than the attachment of biomolecules to carboxylated or aminated polymer particles with common bioconjugation strategies [47] using heterobifunctional linkers [38, 48, 49].

Applications of such fluorescent particles range from the use as fluorescent reporters in fluorescence assays and targeted probes and nanosensors in bioimaging studies, here typically after subsequent surface functionalization with different biomolecules, to lifetime multiplexing and barcoding in conjunction with e.g., cellular imaging and sensing [25, 50–53]. Examples for these applications are highlighted in Sect. 5.5.

5.3.2 Quantum Dots

One of the most exciting, yet also controversially debated advances in reporter technology is the development of semiconductor nanocrystals (SCNC), i.e., spherical quantum dots (QDs) and increasingly elongated quantum rods (QRs) with unique optical and chemical properties and complicated surface chemistry [24, 54–57]. The optical properties of these nanocrystals with typical sizes of 2–10 nm (inorganic core or core-shell diameter in the case of core-shell architectures) are controlled by constituent material, particle size and size distribution (dispersity), and surface chemistry. The latter includes the chemical composition, thickness and tightness of the commonly used inorganic surface passivation shell and the chemical nature of the organic ligand shell controlling dispersibility and colloidal stability. Particularly attractive of SCNCs are an absorption that gradually increases toward shorter wavelengths (below the first excitonic absorption band), which allows free selection of the excitation wavelength and thus, straightforward separation of excitation and emission, very high molar absorption coefficients, narrow emission bands of mostly symmetric shape, high fluorescence quantum yields in the vis and NIR, fluorescence lifetimes of a few ten nanoseconds ideal for time-gated emission, and an excellent photostability. The selection of SCNC materials currently used has primarily been driven by the ability to prepare particles with the desired optical properties. The most prominent materials for life science applications are still CdSe and CdTe, although III/V group or ternary semiconductors such as InP and InGaP, which lack cytotoxic cadmium ions, are being increasingly used as possible Cd-free alternatives [24].

In order to achieve high photoluminescence quantum yields close to unity, a high chemical and photochemical stability, and to minimize blinking, surface defects and dangling bonds at the particle surface, favoring nonradiative deactivation, must be properly passivated with tight inorganic shells [58, 59]. The surface passivation by organic ligands, coordinatively bound to surface atoms, is highly significant for high quantum yields and a high stability. In this respect, SCNC-, size-, and ligand-specific ligand desorption/adsorption equilibria must be considered in the case of uptake studies with cells and bioimaging experiments, because ligand desorption leads to a strong reduction in fluorescence quantum yield and loss in colloidal stability as well as possibly to the release of toxic metal ions [60]. Ligand desorption, and thus, SCNC concentration-dependent fluorescence quantum yields, can be prevented by shells made from amphiphilic and cross linked polymers [61], yet on the expenses of an increased particle size of a few ten nanometers, hampering applications for all applications requiring small particles such as FRET assays.

The importance of the ligand shell for the fluorescence properties of SCNC is demonstrated in Fig. 5.5, comparing the optical properties and dilution studies of two TGA-capped CdTe colloids prepared by synthetic routes favoring the formation of a smaller number of nuclei and thus, a thicker shell of Cd-thiolate complexes (sample A) and a larger number of nuclei corresponding to a thin ligand shell (sample B), respectively. Whereas the absorption features of both samples are

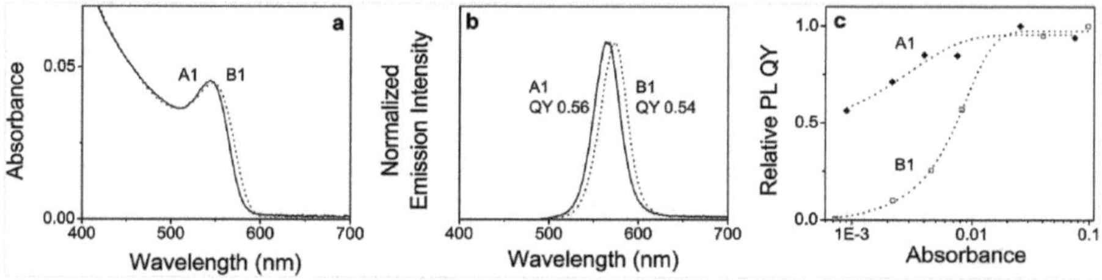

Fig. 5.5 Comparison of the optical properties and dilution studies of two TGA-capped CdTe colloids originating from synthetic routes favoring the formation of small (sample A) and large (sample B) numbers of nuclei and hence large and thin ligand shells. Absorption (panel a) and emission (panel b) spectra and concentration dependent photoluminescence quantum yields (PL QY) with SCNC concentration being provided as absorbance. The *dotted lines* in (**c**) are only a guide to the eye. Reproduced from Ref. [114], with permission from The Royal Society of Chemistry

comparable (Fig. 5.5, panel a) and the corresponding emission spectra (panel b) reveal only small differences, the change of the photoluminescence quantum yield upon dilution (panel c) is much more pronounced for the CdTe colloid with the thin ligand shell compared to that with a large ligand shell. For routine applications, SCNCs are required that can be reproducibly synthesized in a high quality in a large quantity and that do not show concentration-dependent optical properties at typically applied concentrations.

QDs in capped form have been applied as probes for the sensing of temperature [62], pH [63–65], heavy metal ions [66–68] and anions [69, 70].

5.3.3 Photon Upconversion Nanoparticles

Due to its NIR excitation, upconversion luminescence (UCL) is free of an autofluorescence background and thus particularly suited for measurements in complex biological samples such as whole blood [71] as well as for cellular or *in vivo* imaging [72]. Moreover, NIR excitation minimizes the photodamage of tissue and enables a high penetration depth in biological matrices. Photon upconversion occurs in inorganic crystalline materials of low phonon energy like fluorides doped with lanthanide ions such as Yb(III), Er(III), and Tm(III). This multiphotonic process generates higher energy emission from low-energy excitation [73, 74].

Common examples for nm-sized UC materials are hexagonal nanocrystals of $NaYF_4$:Yb^{3+},Er^{3+} , which show two sharp luminescence emission bands in the green ($\sim$ 550 nm) and red ($\sim$ 660 nm) spectral range upon 980 nm excitation. Here, Yb^{3+}acts as sensitizer, which absorbs the excitation light, and Er^{3+}, excited via energy transfer from Yb^{3+}, serves as emitter or so-called activator. Alternatively, Tm^{3+} can be used as activator, resulting in emission bands at in the blue ($\sim$ 475), green ($\sim$ 550 nm), red ($\sim$ 660 nm) and NIR ($\sim$ 800 nm) region. Hexagonal $NaYF_4$:Yb^{3+},Er^{3+} or $NaYF_4$:Yb^{3+},Tm^{3+} particles are usually prepared according to a co-precipitation [75] or a solvothermal [76] synthesis. Efficient UCL systems can be obtained with dopant concentrations of around 20 % for the sensitizer (Yb^{3+}) and 1–3 % for the activator (Er^{3+} or Tm^{3+}). The ratio between sensitizer and activator has a major impact on the peak intensities of the different emission bands. Because of their superior brightness compared to other UC host materials, these materials present the UC nanocrystals mostly used in sensing and bioimaging applications [77, 78].

In order to utilize upconversion nanoparticles (UCNPs) for chemical sensing, it is important to understand and control the influence of the particle environment. This includes the surrounding solvent molecules, pH and ionic strength, temperature, as well as the surface ligands and their effects on UCL intensities, intensity ratios, and lifetimes [79]. As NPs have very large (and size-dependent) surface to volume ratios, processes occuring at the particle/solvent interface become more important and can introduce a size-dependence of the optical properties [80]. Thus, also surface modifications like inactive (undoped) and active (lanthanide ion-doped) protective shells have a strong impact on UCL. Moreover, due to the multiphotonic nature of UC excitation, UCL intensities and intensity ratios are affected by excitation power density [81, 82]. According to our observations, the luminescence of Yb/Er or Yb/Tm doped particles is not prone to oxygen quenching and also changes in pH and ionic strength and the presence of heavy metals have no direct effect on UCL. Depending on the respective surface modification and zeta potential of the particles, these factors can, however, induce UCNP aggregation and precipitation, which is accompanied by changes of the luminescence intensity.

The emission of UCNPs is prone to thermal quenching [83]. However, the most significant drawback for the utilization of UCNPs as reporters, probes, and sensors in biological samples is the strong UCL quenching by water molecules in the first coordination sphere of these NCs. OH vibrations of 3200 cm^{-1} and 3600 cm^{-1} can particularly deactivate the green emitting excited states or intermediate states of this emission efficiently [84]. In case of $NaYF_4$:Yb^{3+},Er^{3+}, vibrations of OH-groups quench the luminescence of $NaYF_4$: Yb^{3+}, Er^{3+} UCNPs by facilitating multiphonon relaxation of $^2H_{11/2}/^4S_{3/2} \rightarrow {}^4F_{9/2}$ and $^4I_{11/2} \rightarrow {}^4I_{13/2}$ transitions of Er^{3+}-ions [85] Both relaxation pathways favor the population of the $^4F_{9/2}$ state (either directly or via $^4I_{13/2} \rightarrow {}^4F_{9/2}$ transition, respectively), from which the red luminescence occurs at the expense of green luminescence [86, 87]. Hence, the observation that smaller particles show a weaker UCL emission particularly in water can be assigned to surface defects, impurities and the impact of high vibrational modes of solvent molecules which facilitate multiphonon relaxation [88, 89].

In cooperation with the group of Prof. Soukka at the University of Turku we studied the mechanisms of water-based quenching of UCNPs via luminescence intensity and decay measurements to reveal the non-radiative deactivation pathways responsible for the rather low upconversion luminescence efficiency in aqueous solutions. Luminescence decay measurements at different excitation and emission wavelengths (UCL and direct excitation of emissive energy states of Er^{3+} and Yb^{3+}) showed that up to 99.9 % of the UCL intensity of both Er^{3+} and $Tm3^{+}$

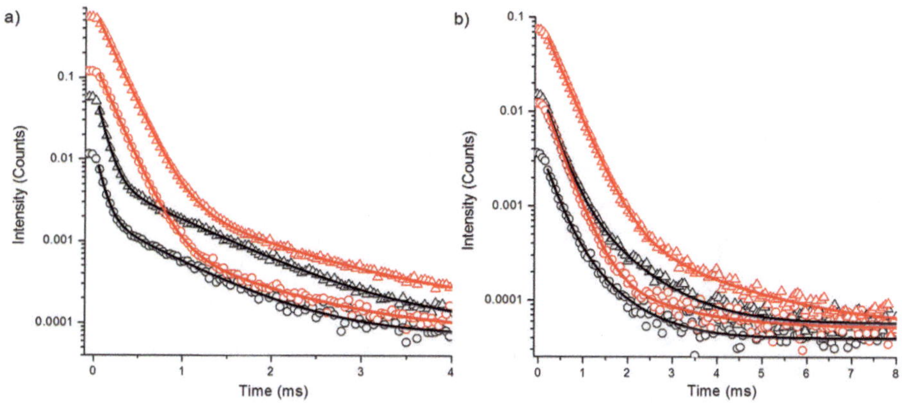

Fig. 5.6 Decay curves of UC emission of bare (*circles*) and silanized (*triangles*) $NaYF_4:Yb^{3+},Er^{3+}$ in H_2O (*black*) and D_2O (*red*) measured (**a**) at 544 nm (*green emission*) and (**b**) at 650 nm (*red emission*) with bi-exponential fits. Excitation was at 980 nm using a pulse width of 20 ms. Reproduced from Ref. [90] with permission from The Royal Society of Chemistry

Table 5.2 Decay times and amplitude fractions of the green and red UC emission of $NaYF_4$: Yb, Er

	Unshelled UCNP, H_2O	Unshelled UCNP, D_2O	SiO_2-UCNP, H_2O	SiO_2-UCNP, D_2O
$^4S_{3/2} \rightarrow {}^4I_{15/2}$ (Green emission)				
τ_1 (μs)	81 ± 1.5 (86.2 %)	177 ± 0.5 (99.1 %)	92 ± 0.5 (91.8 %)	185 ± 0.4 (99.2 %)
τ_2 (μs)	755 ± 10.8 (13.8 %)	1003 ± 55.0 (0.9 %)	843 ± 5.6 (8.2 %)	1070 ± 55 (0.8 %)
$^4F_{9/2} \rightarrow {}^4I_{15/2}$ (Red emission)				
τ_1 (μs)	296 ± 10.8 (85.4 %)	334 ± 2.5 (98.4 %)	300 ± 3.8 (90.7 %)	383 ± 0.9 (99.0 %)
τ_2 (μs)	849 ± 47.3 (14.6 %)	1440 ± 129.3 (1.6 %)	966 ± 24.4 (9.3 %)	1862 ± 72.3 (1.0 %)

doped bare nanophosphors is quenched in water compared to D_2O [90]. The results of the UCL decay measurements are shown in Fig. 5.6 and the decay analysis is summarized in Table 5.2.

Obviously, surface modification of the UCNP with a silica shell provides only little protection against quenching by water as the lifetimes of the green UCL of the SiO_2-coated UCNP in H_2O is similar to that of core-only particles in this solvent and also considerably reduced compared to the lifetimes obtained D_2O. The decay time of the green emitting state $^4S_{3/2} \rightarrow {}^4I_{15/2}$ is more strongly affected by water than that of the red emitting state $^4F_{9/2} \rightarrow {}^4I_{15/2}$. This is attributed to increased multiphonon relaxation of $^4S_{3/2} \rightarrow {}^4F_{9/2}$. When an intermediate energy level is deactivated via coupling to OH-vibrations, the population of the respective emissive states populated via this energy level is also reduced. This leads to a diminution in the intensity of UCL, but should not affect the lifetime of the populated emissive states. Thus, the observed 54 % decrease in the green luminescence lifetime originates most likely only from the increased multiphonon relaxation $^4S_{3/2} \rightarrow {}^4F_{9/2}$ and the deactivation of $^4I_{11/2} \rightarrow {}^4I_{13/2}$ should not affect the lifetime of the green emission. As the decay kinetics of the red emitting state $^4F_{9/2} \rightarrow {}^4I_{15/2}$ is much less affected by water, this cannot account for the observed reduction in UCL intensity.

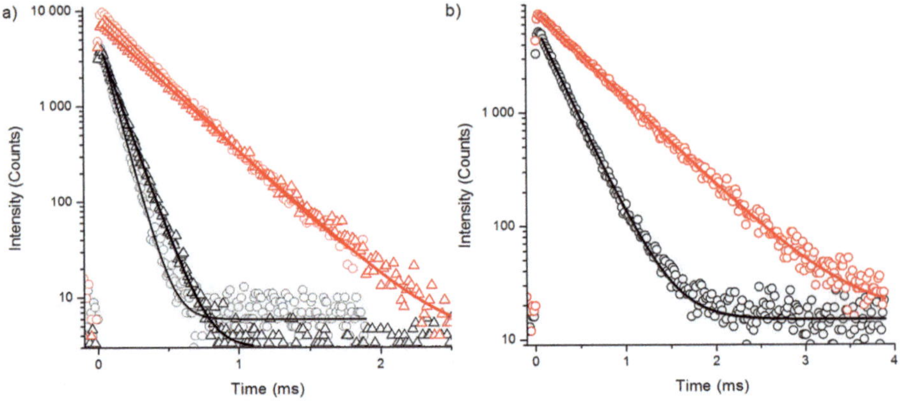

Fig. 5.7 Decay curves of the Yb^{3+} emission at 980 nm excited at 930 nm in H_2O (*black*) and D_2O (*red*) of (**a**) unshelled UCNP (*circles*) and SiO_2-coated UCNP (*triangles*) $NaYF_4$: Yb,Er and (**b**) bare $NaYF_4$:Yb,Tm with mono-exponential fits using 3 µs excitation pulse width. Reproduced from Ref. [90] with permission from The Royal Society of Chemistry

Table 5.3 Luminescence lifetimes and amplitude fractions of $^2F_{5/2}$ to $^2F_{7/2}$ transition (980 nm) of Yb^{3+} in $NaYF_4$:Yb,Er and $NaYF_4$:Yb,Tm emission upon 930 nm excitation

	Bare in H_2O	Bare in D_2O	Silanized in H_2O	Silanized in D_2O
$NaYF_4$: Yb,Er^{3+}				
τ_1 (µs)	79 ± 0.3 (100 %)	288 ± 0.5 (100 %)	106 ± 0.3 (100 %)	319 ± 0.5 (100 %)
$NaYF_4$: Yb,Tm^{3+}				
τ_1 (µs)	256 ± 0.7 (100 %)	556 ± 1.0 (100 %)		

As demonstrated by measurements of the down converted emission of Yb^{3+} at 980 nm following direct excitation of this ion at 930 nm, revealing a shortened in luminescence lifetime by 73 % in water compared to D_2O (see Fig. 5.7 and Table 5.3), the quenching of the Yb^{3+} by water seems to be the main deactivation channel.

These measurements underline that the deactivation of the $^2F_{5/2}$ state of Yb^{3+} is the main source of the strong UCL quenching in water. Possibly, the third harmonic of the OH-stretching vibration can bridge the gap of $^2F_{5/2} \rightarrow {}^2F_{7/2}$ transition ($\sim$10,200 cm^{-1}) causing non-radiative relaxation to the ground state (Fig. 5.8). Similar results were obtained for Yb/Tm doped particles. The quenching of the sensitizer can be expected to have a higher order effect on the luminescence intensity as upconversion photoluminescence is a multiphoton process. Furthermore, it seems that Yb–Yb energy migration is very efficient in these nanocrystals, making the whole 35 nm-sized UCNP susceptible to surface quenching effects. This underlines the need for proper shielding of the Yb^{3+} ion for sensing and imaging applications. An alternative can be undoped $NaYF_4$ shells [91].

Particularly for all applications involving FRET to e.g. surface bound analyte-responsive dyes, such protective shells should be very thin, with a thickness of

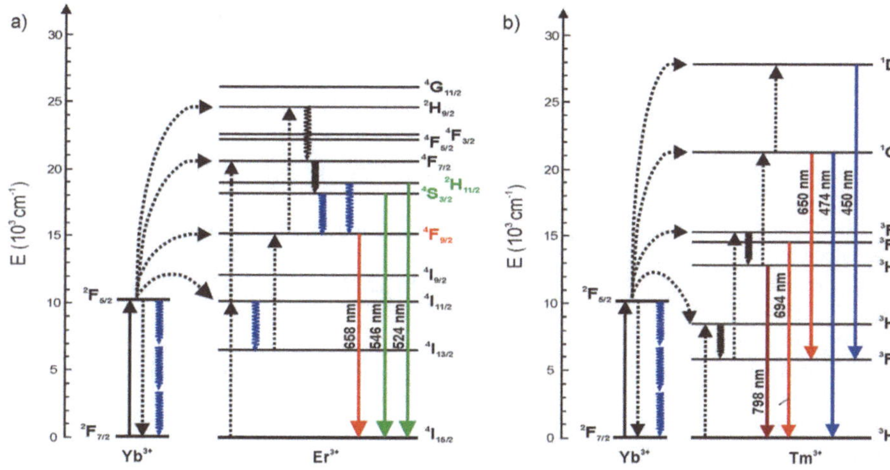

Fig. 5.8 Energy level diagrams of (**a**) $NaYF_4$:Yb,Er and (**b**) $NaYF_4$:Yb,Tm. *Solid*, *dotted* and *wavy arrows* present photon absorption or emission, energy transfer and relaxation processes, respectively. The *blue wavy arrows* denote the increased multiphonon relaxations caused by OH-vibrations. Reproduced from Ref. [90] with permission from The Royal Society of Chemistry

maximum a few nanometers. (Please note here that in connection with UCNPs as donors, the terms luminescence resonance energy transfer (LRET) or UC-RET have been established instead of FRET).Otherwise these FRET processes utilized e.g., for pH or metal ion sensing become too inefficient.

A broad variety of surface-functionalized UCNPs has been meanwhile designed for for chemical sensing and imaging applications. Usually, the analyte-responsive dyes are attached to the surface of the UCNPs and interact with the UCNP via FRET or by a emission-reabsorption process (inner filter effect). There exist meanwhile UCNP-based nanosensors for a large number of target analytes, including pH, NH_3, heavy metal ions (Hg^{2+}, Cu^{2+}, Fe^{3+}), anions (HS^-, CN^-, NO_2^-), ATP, cysteine and glucose, respectively. Table 5.4 summarizes examples of UCNP-based nanosensors that have been used for cellular or in vivo imaging studies.

Functionalized UCNPs can be also applied for chemical sensing in highly absorbing samples and enable e.g., the determination of glucose in serum [98] or of NO_2^- in meat samples [99].

A convenient route to prepare UCL sensors on a solid substrate is the incorporation of UCNPs and stimuli-responsive dyes into polymer materials such as PS or poly(vinyl chloride). Planar sensor films can be easily fabricated from these composite materials which can be subsequently integrated into fiber optic sensors, microtiter plates or reaction vessels. Such sensors have been exploited for analyte measurements in highly colored and scattering sample matrices including the determination of pH in urine [100] or K^+ in plasma or whole blood [101]. We designed a pH sensor film consisting of micrometer-sized UC phosphors and the pH

Table 5.4 Functionalized photon upconversion nanoprobes with chemical sensor function

Sensor system	Mechanism	Analyte	Dynamic range / LOD	Example of use	References
$NaYF_4$:Yb,Er / porphyrin dendrimers	Re-absorption	pH	pH 5.0–8.5	Cell permeable, titration in water	[92]
$NaYF_4$:Yb,Er,Tm / Ru(II)-2.2′-bipyridyl	Re-absorption	Hg^{2+}	0–50 μM/10 nM	Live cells (HeLa)	[93]
$NaYF_4$:Yb,Er,Tm / alkylated Cy7	LRET	$MeHg^{+}$	0–56 μM/4 nM	In vivo imaging with mice	[94]
Mesoporous silica coated NaYF4:Yb, Er / merocyanine	LRET	HS^{-}	0–115 μM / 0.58 μM	Live cells (HeLa)	[95]
Silica coated NaLuF4:Yb,Er,Tm/Ir (III)-complex	LRET	Cysteine	0–900 μM / 28.5 μM	Live cells (KB cells)	[96]
PEG coated $NaYF_4$:Yb,Er / rhodamine B	LRET	Phospho-lipase D	50–800 U/L	Live cells (MCF-10A MDA-MB-231), cell lysate	[97]

indicator Neutral Red (NR) that absorbs the green UC emission, all embedded in a proton permeable polyurethane matrix [102]. In this system, solely the green UC emission is affected by the pH-dependent absorption of the indicator, while the red emission can act as inert reference signal for ratiometric measurements. Figure 5.9 shows the spectral overlap of UCL and the absorption of the indicator and the sensor read out, enabling ratiometric sensing and 2D imaging of pH. As the emission peaks of the particles match the red and green color channels of standard inexpensive digital RGB cameras, which collect the red, green and blue part of the visible light in three independent channels. The pH is calculated by rationing data from the red by the green channel data.

Human serum was selected for the evaluation of the sensor (see Fig. 5.10). The calibration of the sensor film yielded a pK_a of 6.7. Three samples of serum were titrated to pH values of 6.5, 7.0 and 7.5, respectively. Additionally, one non-titrated serum sample was measured. Cut sensor spots were impregnated with the samples. The samples displayed very bright auto-fluorescent background under UV illumination. The second row displays RGB-imaging of the same samples upon 980 nm excitation. It is visible that no detectable auto-fluorescent background is recorded by the camera. Thus, the UC sensor in combination with ratiometric G/R read out offers a pronounced signal to noise ratio. No background signals can be detected even if the sensor is read out in a top-read mode, where the signals were read out through the serum sample. The third row presents the G/R ratios displayed in the identical pseudo-color range as it is used in the calibration row.

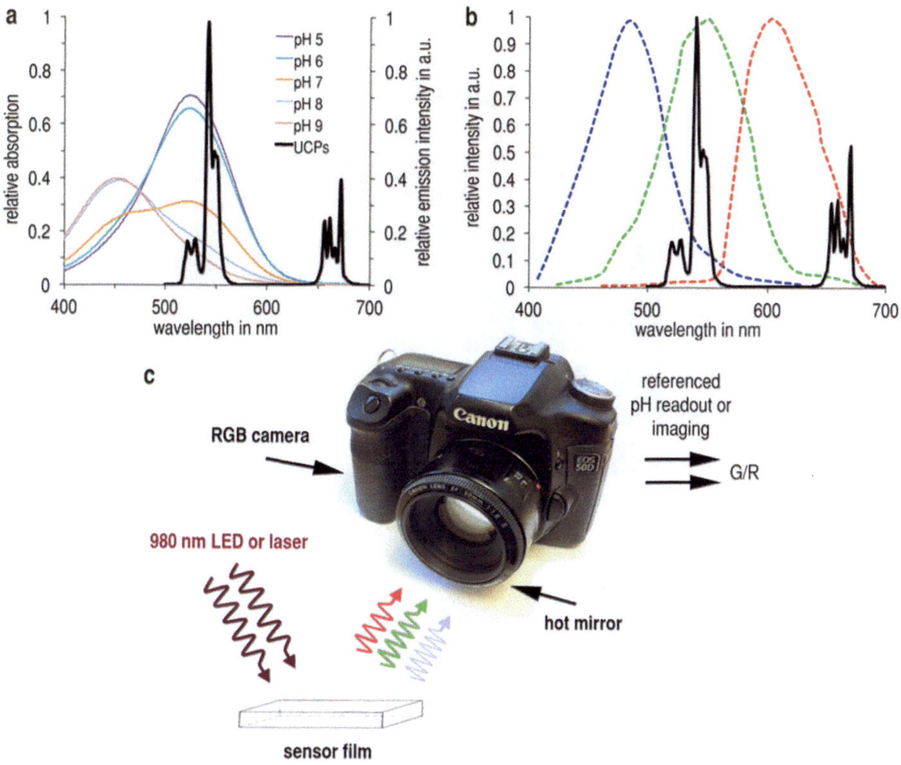

Fig. 5.9 (**a**) Spectral overlap of the *red* absorbance peak of the pH indicator NR and the emission peaks of the UCPs. The *green* emission peaks of the UCPs (*black line*. Excitation: 980 nm) are affected by the absorption of the protonated form of the pH indicator NR, while the *red* emission peaks can act as inert reference signal for ratiometric measurments. (**b**) Spectral match of the *red* and *green* UC emission peaks (*black line*) and the color channels of the digital camera (*dashed red, green and blue lines*). (**c**) Sensing and imaging of the pH sensor film using the RGB technique. A digital camera collects that light that passes through a hot mirror which separates reflected excitation light from the visible wavelengths. Reprinted from Ref. [102], Copyright (2014), with permision from the American Chemical Society

5.4 Surface Functionalization and Characterization

The rational design of nano- and micrometer-sized particles with tailor-made optical properties for biological, diagnostic, and photonic applications requires tools to characterize the application-relevant properties of these particles [103]. This includes particularly size (and shape) and size distribution, charge, crystallinity, surface morphology and surface chemistry, and particle brightness [104–108]. The latter equals the product of the molar absorption/extinction coefficient or absorption cross section (at the excitation wavelength) and the fluorescence quantum yield and determines the signal size from the material side for all fluorescence-based methods [107]. In the case of fluorophore doped or labeled particles, also the number of fluorophores per bead are of interest [39]. As many particle dispersion scatter, except

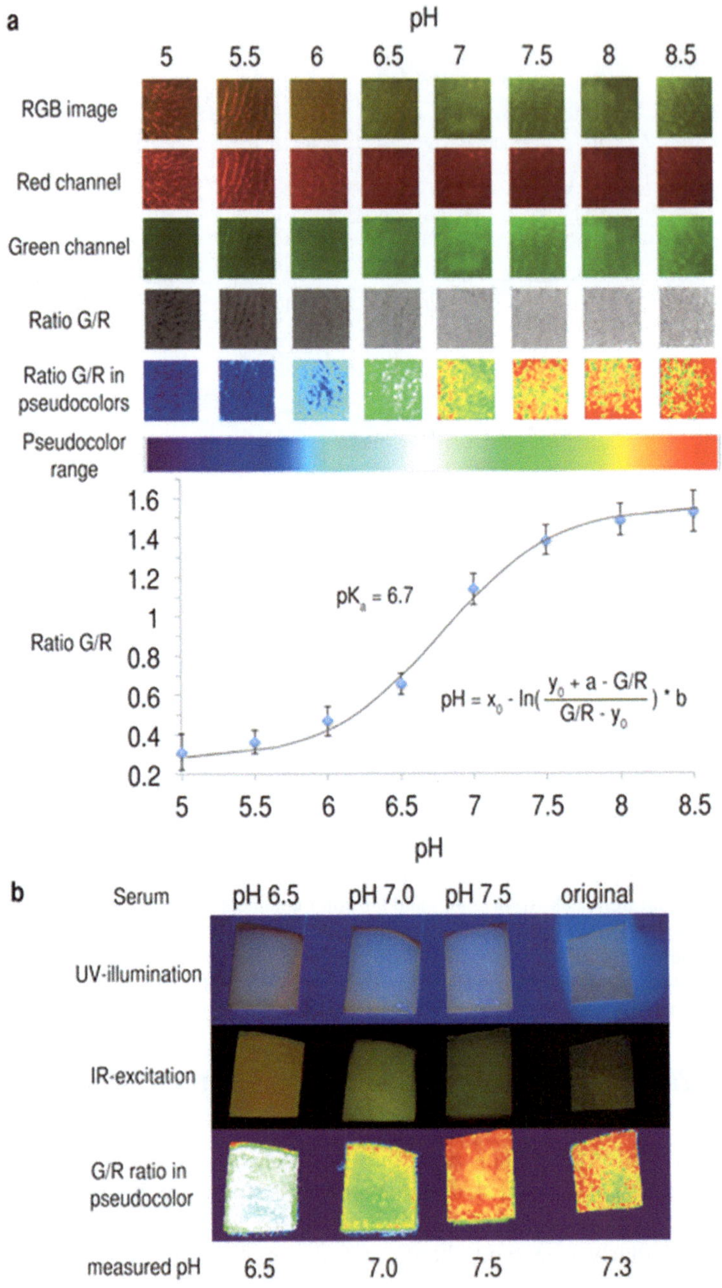

Fig. 5.10 (continued)

for e.g., very small semiconductor nanocrystals, such fluorescence measurements cannot be performed with a simple fluorometer in 0°/90° measurement geometry, yet require an integrating sphere [39, 41, 107].

The characterization of the surface chemistry of functional nanomaterials is of utmost importance for most applications, as functional groups and surface ligands, that can be covalently, coordinatively or electrostatically bound to nanomaterials, significantly determine colloidal stability, subsequent biofunctionalization, and the interaction with biological systems including biodistribution and excretion in *in vivo* studies [47, 103]. This includes the determination of the total and accessible or derivatizable number of functional groups and ligands per particle and, for particles functionalized with biomolecules, also the number of functional biomolecules [103, 109–112]. In this respect, we assessed a broad variety of methods for different nanomaterials and microbeads with special emphasis on polymer particles with covalently bound surface ligands, laponites, and semiconductor nanocrystals and simple optical assays that can be transferred also to other nanomaterials [49, 109, 112–116]. In order to enable the straightforward validation of these strategies, especially multimodal reporters were used which can be read out with different analytical methods relying on different detection principles / principles of signal generation. For dyes containing heteroatoms such as sulfur, nitrogen or fluorine, these methods can be validated e.g. with elemental analysis, ICP-OES (inductively coupled plasma spectroscopy with optical detection) and ICP-MS (inductively coupled plasma spectroscopy with mass spectrometry detection; for sulfur) [41, 112, 115]. In the case of surface-bound fluorophores, also conventional surface-analytical methods like X-Ray photoelectron spectroscopy (XPS) with a limited penetration or information depths of 5–10 nm can be used for method validation [113, 117].

An elegant approach to (absolute) dye quantification can present solid state nuclear magnetic resonance spectroscopy (NMR) as recently demonstrated in a proof-of-concept study for 1000 nm-sized carboxylated polystyrene particles loaded with varying concentrations of the laser dye coumarin 153 containing a CF_3 group and ^{19}F NMR [39]. This approach, which is shown in Fig. 5.11, can be also utilized for surface bound reporters [109, 113].

Due to the special importance of inorganic passivation shells and organic capping ligand for the photoluminescence properties and stability of SCNCs, there is a

Fig. 5.10 (continued) (**a**) pH-RGB-imaging calibration *row* of the sensor including the real color picture, pure splitted signals from the *red* and the *green* channel, and the referenced ratio G/R displayed in grayscale and pseudocolor. The graph at the *bottom* of (**a**) displays the calibration plot of G/R values versus the pH including error bars. The pH values can be calculated from the measured G/R ratio by using the inverse function of the four-parametric sigmoidal fit. (**b**) Four sensor spots were impregnated with real serum samples. The first three serum samples have adjusted pH values and the fourth one is unadjusted. Bright sample and background autofluorescence is visible under UV-illumination (*first row*). IR-excitation and RGB-imaging of the samples is virtually free of background (*second row*). The G/R ratio is displayed in pseudocolors (*bottom row*) and converted into the respective pH values. Reproduced from Ref. [102], Copyright (2014), with permision from the American Chemical Society

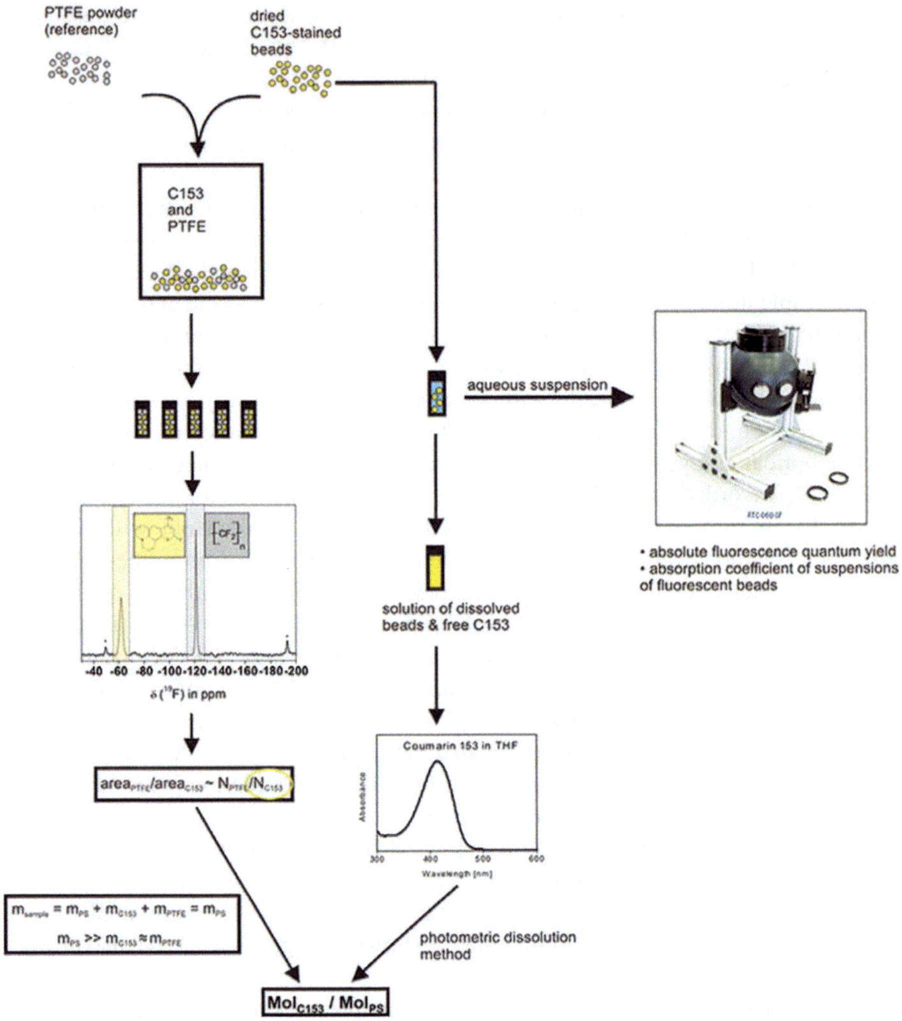

Fig. 5.11 Scheme of the spectroscopic characterization of the C153-stained beads with ^{19}F–NMR (*left*) and optical spectroscopy (*right*) used for the quantification of the dye molecules per bead and the determination of the absolute fluorescence quantum yields and the absorption coefficients of suspensions of the fluorescent beads in suspension. Reprinted from Ref. [39], Copyright (2012), with permission from the American Chemical Society

considerable need for simple analytical and spectroscopic tools to address shell quality and to provide information on shell morphology, tightness, and the number of ligands per SCNC [118, 119]. This has been also addressed e.g., in recent review by [103] for surface-functionalized nanomaterials in general and by [132] for the analysis of the SCNC ligand shell, highlighting the potential of various spectroscopic techniques like NMR, Fourier transform infrared spectroscopy (FTIR), XPS, and optical methods. In this respect, we developed spectroscopic shell quality tests for CdSe/ZnS core shell SCNC [58, 120] and CdTe colloids capped with different thiol ligands [114, 116], which exploit the luminescence quenching of CdSe by phenylthiyl radicals formed photochemically upon illumination of thiophenol and SCNC-dependent luminescence measurements (dilution studies).

Another fluorescence quenching based approach has been recently presented by Weller et al. for the study of the tightness of polymeric ligand shells on CdSe/CdS/ZnS core-shell-shell QDs encapsulated with different amphiphilic poly-(isoprene-block-ethylene oxide) (PI-b-PEO) using the paramagnetic heavy metal ions copper(II) and iron(III) as quencher [119]. A simple and versatile approach for ligand analysis and quantification suitable for process control present simple photometric thiol assays like the Ellman's test [115]. This test has been meanwhile successfully used by us for ligand analysis of SCNCs of varying chemical composition, specifically CdTe and ternary $Cd_xHg_{1-x}Te$ colloids stabilized with mercaptopropionic (MPA) and thioglycolic acid (TGA) and even PEGylated particles and was validated by comparative measurements with ICP-OES [114, 116].

Straightforward tools for the assessment of reporter performance and their comparison provide brightness values and photoluminescence quantum yields [24, 107, 121]. In order to prepare particles with optimum brightness, the dye loading concentration must be optimized, thereby considering dye-specific properties like fluorophore aggregation tendency [37, 39, 40]. In the case of fluorophore doped or labeled particles, also the number of fluorophores per bead are of interest [39]. As many particle dispersion scatter, except for e.g., very small semiconductor nanocrystals, such fluorescence measurements cannot be performed with a simple fluorometer in 0°/90° measurement geometry, yet require an integrating sphere as shown in Fig. 5.11 [39, 41, 107]. The influence of dye concentration on particle brightness derived from transmission, reflection, and absolute fluorescence measurements with a custom designed integrating sphere setup [122] is shown in Fig. 5.12 exemplarily for Nile Red and PSP of different size. Moreover, particularly for solvatochromic dyes like Nile Red, also the sensitivity of the fluorophores located close to the particle surface to polarity effects and hydrogen-bonding induced fluorescence quenching must be taken into account that can result in particle-size dependent spectral shifts and affects the quantum yield of the encapsulated dyes in a particle size dependent manner.

5.5 Application Examples

5.5.1 Lifetime Multiplexing

Thegoaloflifetime multiplexing as a newly emerging multiplexing strategy is to increase the information content from bioassays and imaging studies, which is particularly relevant in the context of current security and health concerns [24]. Lifetime multiplexing, exploiting compound-specific decay kinetics of otherwise

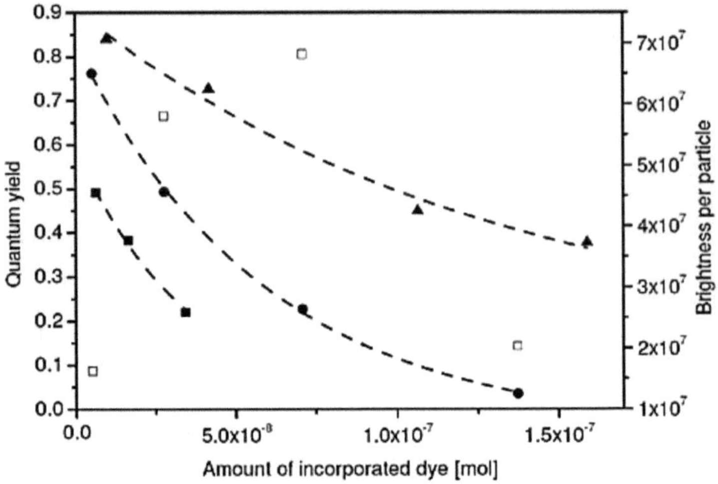

Fig. 5.12 Influence of the amount of incorporated dye on the absolutely measured fluorescence quantum yield of Nile *Red* encapsulated in 25 nm-sized (*full squares*), 100 nm-sized (*full circles*), and 1 μm-sized (*full triangles*) carboxylated PSP and on the brightness per particle (*open squares*) of the 100 nm-sized carboxylated PSP. The brightness equals the product of the fluorescence quantum yield and the molar extinction coefficient or molar absorption cross section. The *lines* present a guide to the eye only. Reprinted from Ref. [122], Copyright (2012), with permission from the American Chemical Society

barely distinguishable chromophores, presents an attractive alternative to spectral multiplexing, which relies on fluorescent labels excitable at the same wavelength, yet sufficiently differing in their emission spectra or color like 2λ-measurements performed for ratiometric sensing studies and can be combined with spectral multiplexing to further enhance the number of codes or analyte-specific variables. Lifetime multiplexing can be done with different types of fluorophores like organic dyes, metal ligand complexes, and semiconductor and upconversion nanocrystals [25, 51, 123–125]. Prerequisites are absorption and emission spectra in a similar wavelength region for excitation at a single excitation wavelength and detection in the same wavelength window and sufficiently different decay kinetics that must not be even monoexponential.

We recently introduced a new strategy based on "pattern-matching" in the lifetime domain for straightforward data evaluation, which was exemplary exploited for the discrimination between organic dyes and quantum dot labels revealing multi-exponential decay kinetics and allowed quantification of these labels [123]. Dye-loaded particles are particularly attractive reporters for lifetime multiplexing. Figure 5.13 presents an examples for the lifetime-based distinction between pairs of fluorescent nanoparticles in cells, here differently sized PSP encoded NIR-emissive Itrybe, a cyanine dye, and Sq730, a squaraine, that can be both excited at 633 nm and emit in the same spectral window of ca. 700–900 nm, yet vary in their fluorescence lifetimes/fluorescence decay kinetics by a factor of about 2. As a first proof-of-concept experiment, the uptake behavior of 3 T3 fibroblast and J774 macrophage

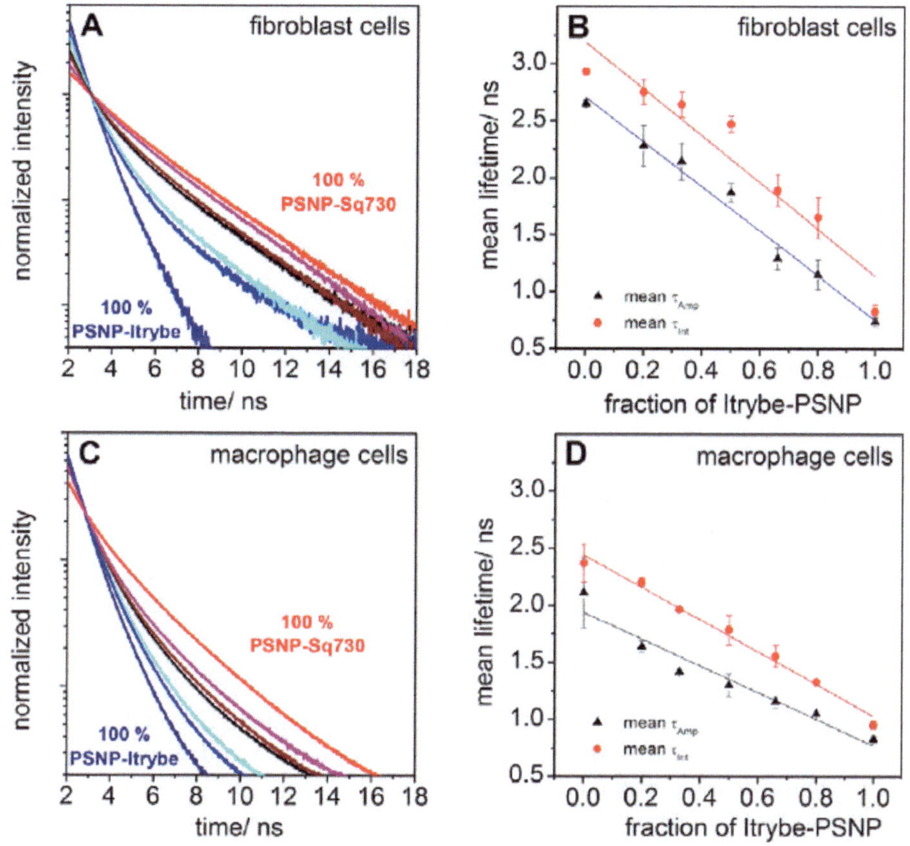

Fig. 5.13 Fluorescence decay curves of 3 T3 fibroblast cells (**a**) and J774 macrophage cells (**c**), exposed to PSP stained with the cyanine dye Itrybe (*dark blue curve*) and to PSP stained with the squaraine dye Sq730 (*red curve*) revealing different (multiexponential) decay kinetics and different mixtures of both components (excitation at 640 nm; detection with a detection filter of 685/70 nm). The resulting average fluorescence lifetimes τ_{Amp} (*triangles*) and τ_{Int} (*circles*) of fibroblast (**b**) and macrophage cells (**d**) were calculated from three lifetimes based on FLIM measurements. Fibroblast cells were incubated with 25 nm-sized PSP, loaded with Itrybe, and 100 nm-sized PSP, stained with Sq730, for one hour, while macrophage cells were exposed to 100 nm-sized PSP, loaded with Itrybe, and to 100 nm-sized PSP, loaded with Sq730. Reprinted from Ref. [123], Copyright (2009), with permission from the American Chemical Society

cells was studied with differently sized dye-loaded PSP, here 25 nm-sized (3 T3 fibroblast cells) and 100 nm-sized (J774 macrophage cells) PSP, stained with Itrybe, and to 100 nm-sized PSP loaded with Sq730 (used for both cell lines) [25, 51]. The fluorescence decay curves obtained from fluorescence lifetime imaging (FLIM) studies with these cells incubated with different mixtures of these dye-loaded PSP are shown in Fig. 5.13 (panels A and C). The corresponding ratio-dependent average lifetimes of the cells are given in panels B and C, underlining the potential of this approach. The potential to differentiate these two cell types based upon particle

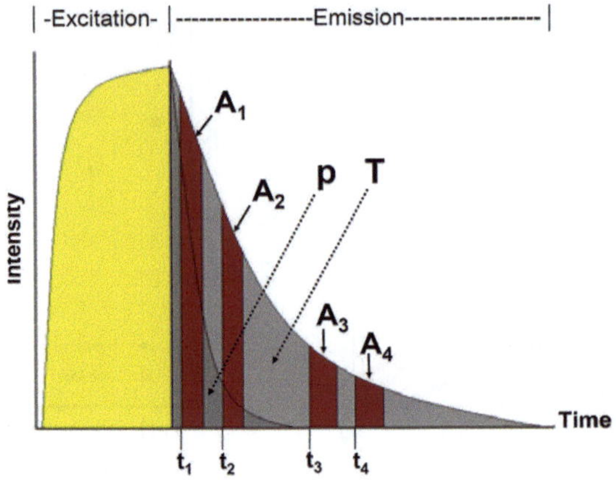

Fig. 5.14 Dual lifetime determination method for signal separation by means of different luminescence decay times. Different luminescence decay curves of two indicators are depicted exemplarily in different *grey* scales. The temperature – dependent lifetime image calculated from A_3 and A_4 can be used to compensate the temperature effect on the lifetime of the pressure sensor determined from A_1 and A_2. The last two gates exclusively detect the luminescence of the temperature indicator whereas the response of the pressure sensor is calculated from the first two gates A_1 and A_2 containing the mixed intensities of both indicators by a data iteration process. Reproduced from Ref. [126], Copyright (2008), with permission from Wiley-VCH

staining and lifetime measurements could be also confirmed in cell co-culture studies [25]. This simple and versatile concept can be extended to all types of fluorophores, thereby expanding the accessible time scale, and can be used, e.g., for the design of labels and targeted probes for fluorescence assays and molecular imaging, cellular imaging studies, and barcoding applications, also in conjunction with spectral and intensity encoding.

Lifetime multiplexing is also a valuable method for the read out of multiple optical sensors. We designed a novel composite material for simultaneous luminescent determination of air pressure and temperature [126]. The dual sensor consists of a fluorinated platinum porphyrin complex (PtTFPP) as an oxygen-sensitive probe, and of a highly temperature-sensitive europium complex. Both are incorporated into different polymer microparticles to control response characteristics and to avoid interferences. Encapsulation of PtTFPP in poly(styrene-*co*-acrylonitrile) (PSAN) results in a broad dynamic range from 0.05 to 2.00 bar for pressure measurements. The europium complex was incorporated into poly(vinyl chloride) to reduce the cross sensitivity towards oxygen. The signals are separated via the different luminescence lifetimes of the indicators (Fig. 5.14). The emission of the temperature-sensitive probe can be monitored by means of time-resolved fluorescence imaging without interferences, because the luminescence lifetime of the temperature indicator is tenfold longer than that of the oxygen indicator. The temperature image can then be used to compensate the cross sensitivity of the pressure indicator towards temperature. In combination with an appropriate time-resolved measurement technique, termed as dual lifetime determination (DLD), this material enables simultaneous imaging of pressure (or oxygen partial pressure) and temperature distributions on surfaces. This approach avoids the need of signal separation by application of different cameras or by use of different optical filters or light sources.

5.5.2 Intracellular Sensors

Photoluminescent methods for intracellular sensing and imaging of pO_2 and pH play an important role in medical research and diagnostics. They are two of the basic parameters characterizing biological samples. The pH impacts the structure and function of all biologically active macromolecules in live cells.These biomolecules have to be sustained in their functional conformation by a thorough control of pH to maintain their various biochemical interactions. Abnormal acidic pH is characteristic of many cellular dysfunctions such as apoptosis, cancer or Alzheimer's disease. A reduced pH between 6 and 4 also occurs in certain cellular compartments such as endosomes or lysosomes. The cytoplasmic pH is an important factor for the success of therapies. For example, the response to therapies and multi drug resistances are related to intracellular acidification. It also regulates the capability of ion transport, and cell proliferation and migration. The same applies for tumor hypoxia, which is the result of inadequate oxygen supply and abnormal tumor metabolism. Hypoxic tumor cells produce a number of factors which can stimulate the growth of new vasculature and promote invasiveness and metastatic spread. Moreover, hypoxic regions of tumors are frequently treatment resistant. Therefore, the development of improved probes for the measurement and imaging of intracellular oxygen supply and pH in real time is of considerable interest in medical research as well as for the evaluation of improved therapies.

We demonstrated a referenced nanoprobe for sensing and imaging of intracellular pH based on an UCL mechanism [127]. Covalent conjugation of the pH indicator pHrodo™ Red to aminosilane coated UCNPs resulted in a pH sensitive nanoprobe with an almost linear response in the range from pH 3.0–7.0. The hexagonal $NaYF_4:Yb^{3+},Er^{3+}$crystals had a diameter of 28×36 nm and were covered with an aminosilane shell of ~ 3 nm thickness. The pH-dependent ratiometric signal is generated by UC-RET from Yb,Er-doped nanocrystals to the fluorogenic indicator dye. The UC-RET mechanism was confirmed by a decrease of the luminescence lifetime of the green UCL emission after conjugation of the pH indicator. A referenced read out can be achieved by measuring the ratio of the green UCL emission (550 nm) and the sensitized red emission of the indicator (590 nm), which rises with decreasing pHs (Fig. 5.15). This probe can be used for ratiometric pH imaging inside living cells and their compartments using a laser scanning luminescence microscope equipped with a 980 nm excitation source. The UCNPs were efficiently internalized by HeLa cells and the green reference signal as well as the red pH-dependent signal could be detected (Fig. 5.16). Both individual particles and particle clusters were visible in the cells. Particle clustering is likely due to

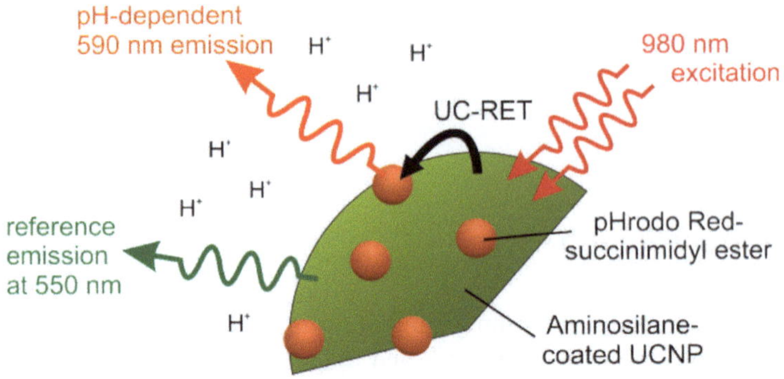

Fig. 5.15 Schematic illustration of a UCL based pH-sensor. The pH-sensitive pHrodo™ Red-succinimidyl ester is conjugated to the aminogroups of the silane-coated UCNP. The UC-RET-sensitized pH-dependent 590 nm emission of the pHrodo™ Red is measured upon 980-nm excitation. The pH-independent 550 nm emission of the UCNP is used as reference signal. Reproduced from Ref. [127] with permission from The Royal Society of Chemistry

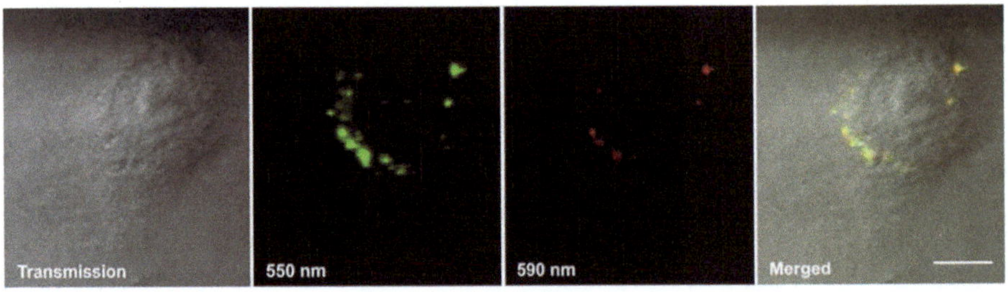

Fig. 5.16 Ratiometric pH imaging with UCNP- pHrodo™ Red in endosomal compartments of living cells. Transmission microscope image of a HeLa cell (*left*) and fluorescence images of the *green* emission (550 nm) and *red* emission (590 nm) of the nanoprobes internalized by the cell. The right image shows the transmission picture superimposed with the *green* and the *red* channel. A *yellow* color indicates a stronger *red* signal, and therefore, an acidic pH. The *white* scale bar is 10 μm. Reproduced from Ref. [127] with permission from The Royal Society of Chemistry

endosomal compartmentalization of the particles. An increased red to green ratio was observed in case of the particle clusters located in the acidic endosomes. Thereby, bright red FRET signals probably originated from late endosomes and less intense spots from early endosomes. No cytotoxic effects of the particles were observed even with the highest nanoprobe concentration of 50 μg/ml. An extensive and proper *in situ* calibration of the probe is owing to enable the determination of quantitative pH values inside cells. We currently evaluate cellular measurements with improved nanoprobes bearing a polyethyleneimine instead of the aminosilane shell. These provide higher indicator loadings and increased cellular uptake. Photostability studies revealed that the indicator dye emission is much more stable if it is sensitized via a UC-RET system then if it is excited directly by the instrumental light source.

5.5.3 *In Vivo Applications*

The ability to identify tumors in an early stage is of enormous importance for effective treatment of cancer. Solid tumors, the treatment of which with radiotherapy or chemotherapy presenting a particular challenge, can be identified based upon their basic characteristics such as hypoxia and acidic pH result from inadequate oxygen supply and abnormal tumor metabolism. For this purpose, we prepared a ratiometric oxygen sensor by doping 100 nm-sized PSP with the oxygen-sensitive near-infrared (NIR)-emissive palladium meso-tetraphenylporphyrin PdTPTBP and an inert reference dye, here Dy-635, both excitable at 635 nm [38]. The working principle of this nanosensor is schematically shown in Fig. 5.17. These nanosensors were subsequently employed for ratiometric dual-wavelength and lifetime-based photoluminescent oxygen sensing. These nanosensors were efficiently taken up by cultured murine alveolar macrophages, yielding a characteristic and reversible change in ratiometric response with decreasing oxygen concentration, which correlates with the cellular hypoxic status. This was verified by analysis of hypoxia inducible factor-1α (HIF-1α) accumulation. Experiments with tumor-bearing mouse and these nanosensors surface-functionalized with the therapeutic monoclonal antibody Herceptin, targeting the cell surface receptor Her2/Neu overexpressed in many cancers revealed a distinctive ratiometric response within the tumor upon hypoxic condition induced by animal sacrifice (Fig. 5.18). These results demonstrate the potential of these referenced NIR nanosensors as optical probes for oncology.

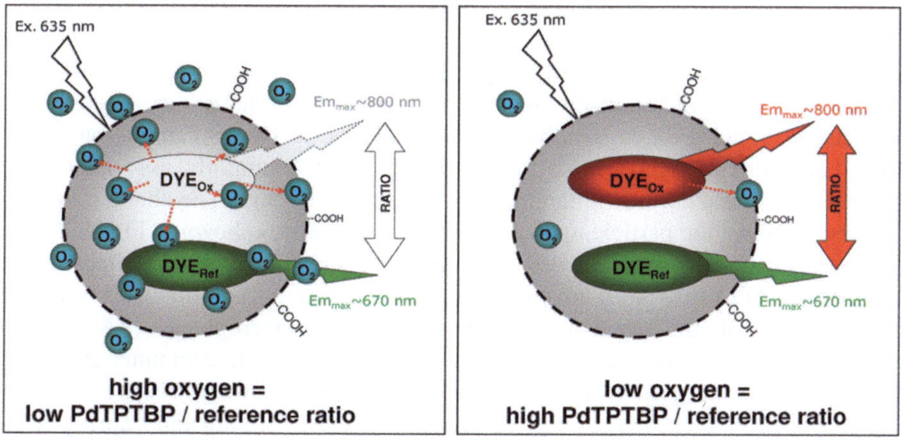

Fig. 5.17 Working principle of a NIR-emissive nanosensor, consisting of 100 nm-sized PSP loaded via a swelling procedure with an oxygen-sensitive near-infrared (NIR)-emissive palladium meso-tetraphenylporphyrin and an inert reference dye, both excitable at 635 nm. *Left*: signaling behavior at high oxygen concentration (air-saturated conditions) and *right* at low oxygen concentration, resulting in a strong increase in the emission of the metalloporphyrin

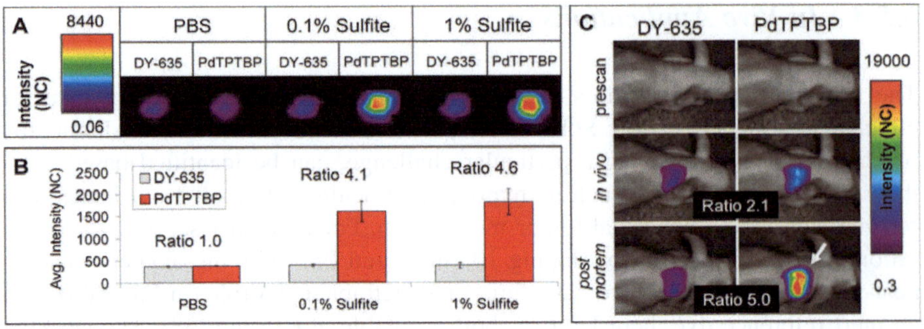

Fig. 5.18 Evaluation of the oxygen nanosensor for in vitro and in vivo imaging of the oxygen concentration in cells and in mice using a typical animal scanner. (**a**) In vitro imaging of the representative luminescence intensities derived from the O_2-sensitive dye PdTPTBP and the inert reference DY-635 in PBS solution in the presence or absence of the oxygen-scavanger Na_2SO_3. (**b**) Summary of the data shown in (**a**). The I_{Pd}/I_{ref} ratios increase in the presence of 0.1 and 1 % sulfite. (C) In vivo and postmortem (~10 min after animal sacrifice) scans of a nude mouse after subcutaneous application of Ox-PS-NPs. An IPd/Iref ratio of 2.1 was measured ca. 20 min after sensor application which increased by a factor of 2.5 upon sacrificing the animal. Reprinted from Ref. [38], Copyright (2011), with permission from the American Chemical Society

As NPs are able of conjunction with different cell types, they can provide interestings tools for cell tracking. This has been explored for stem cell therapy and for cancer research to understand cell migratory pathways and the fate of certain cells. An evolving area of application of fluorescent particles, closely to lifetime multiplexing related, is the staining of macrophages or stem cells by cellular particle uptake followed by imaging of the fate of these fluorescently stained cells. This has been recently demonstrated by us in a proof-of-concept study using 100 nm-sized PS nanoparticles loaded with the NIR fluorescent dye Itrybe (termed Itrybe-NP) for macrophage staining and macrophage tracking by *in vivo* and *ex vivo* fluorescence reflectance imaging of an ovalbumin-based allergic airway inflammation (AAI) model in hairless SKH-1 mice [128]. AAI is most commonly associated with asthma, which is a chronic inflammatory disease of the lungs. The characterization of these NPs and the particle uptake by macrophage cells is shown in Fig. 5.19.

These Itrybe-NPs were inhaled by AAI mice, sensitized with the allergen like ovalbumin (OVA), as well as by healthy mice used as control. This imaging approach exploits the fact that the number of alveolar macrophages (AM), which are part of the immune system and one of the first cell types to encounter allergens, is significantly increased in AAI. Particles inhaled into the lower airways are rapidly phagocytized by these cells, and their number increases significantly during AAI. This imaging approach, the results of which are summarized in Fig. 5.20, may advance preclinical monitoring of AAI in vivo over time and aid the study of the role that macrophages play during lung inflammation. Furthermore, it allows for tracking of inhaled nanoparticles and can hence be utilized for studies of the fate of potential new nanotherapeutics.

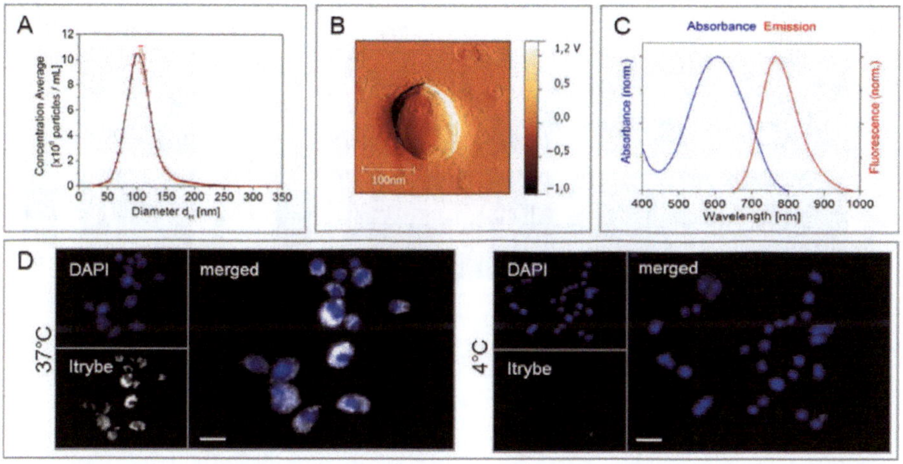

Fig. 5.19 Characterization of the NIR-fluorescent polymeric NPs used for macrophage imagung. (**a**) Size distribution determined via nanoparticle tracking analysis (NTA). (**b**) AFM tapping mode amplitude image of a single NIR-dye loaded NP on silica substrate. (**c**) Absorption and emission spectra (excitation at 600 nm) of the NIR dye used for NP loading. (**d**) In vitro uptake of the NIR-emissive NPs by MH-S macrophages after 24 h incubation at 37 °C (*left*) and uptake inhibition due to the inactivation of phagocytosis by switching the temperature from 37 to 4 °C (*right*). Scale bars in D represent 20 μm. Reprinted from Ref. [128], Copyright (2015), with permission from the American Chemical Society

5.6 Conclusion and Outlook

In this overlook we highlighted our research activities in the field of luminescent nanomaterials for chemical sensing and imaging and the development of tools for the characterization of these systems with special emphasis to surface group analysis and absolute optical measurements. In this respect, different materials for the design of nanoprobes were discussed and compared including dye doped polystyrene particles, quantum dots or upconversion nanocrystals and application-relevant aspects such as strategies for signal referencing, surface functionalization and particle characterization methods were covered. Finally, we demonstrated selected application examples with respect to *in vitro* and *in vivo* imaging of pH or pO_2 as well as new multiplexing strategies like lifetime multiplexing in multiple stained cellular samples and cellular tracking.

Evaluation of the relevant scientific literature reveals still an increasing number of publications per year presenting new nanosensor systems since the group of *Kopelman et al.* introduced their first PEBBLE sensors in 1999. This underlines that the development of appropriate nanoprobes is still a challenging and much-noticed field of research and development. Many systems have been presented in the past years that already get close to the requirements for intracellular or in vivo sensing and imaging such as far red or NIR excitation and emission, high local brightness, photostability, colloidal stability, and cellular uptake, a response in the physiological relevant range and internal referencing.

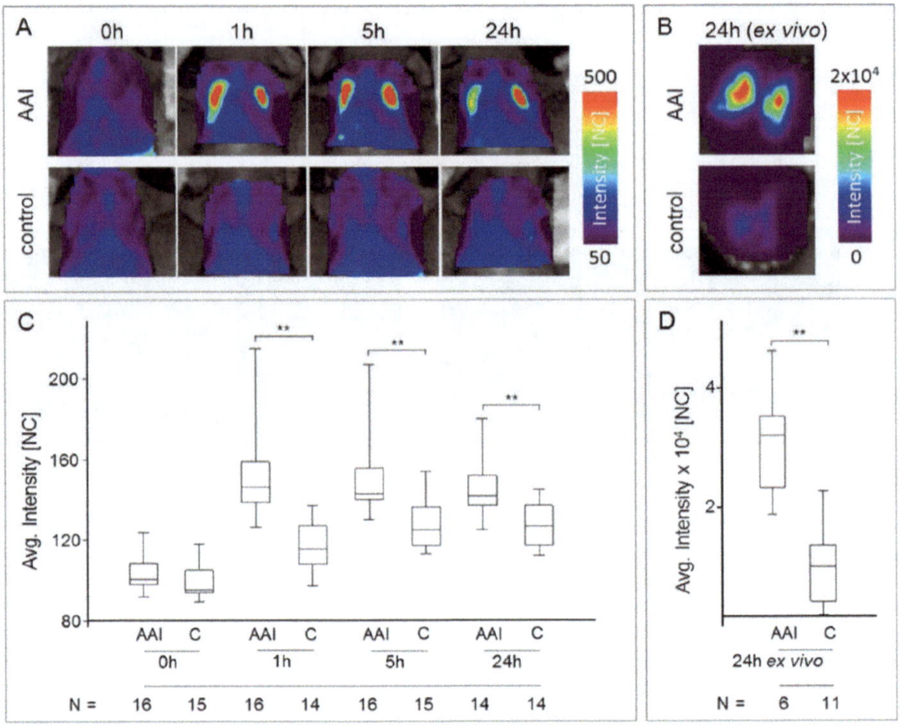

Fig. 5.20 AAI mice demonstrate significantly higher fluorescence intensities over the lungs than controls in vivo (**a**) and ex vivo (**b**). The fluorescence intensity distribution is displayed in normalized counts (NC). On average, AAI mice showed maximum intensities 5 h after i.n. Itrybe-NP application. In vivo (**c**) and ex vivo (**d**) fluorescence imaging results at the indicated time points after i.n. NP application are summarized in box plots of average fluorescence intensities over the lung area. Data were analyzed by Welch t test, and ** represents $P < 0.05$. Reprinted from Ref. [128], Copyright (2015), with permission from the American Chemical Society

However, the applicability of dye doped polymer particles could still be increased by the development of dyes with enhanced photostabilities particularly in the NIR; this concerns the responsive probes as well as the reference dyes. The sensitivity and robustness of the sensor response in biological environment are other factors that await improvements. Many luminescent nanoprobes are based on FRET systems. Small particle sizes (best < 20 nm) coated with very thin, but highly functionalized shells for coupling of biomolecular receptors or indicator dyes as FRET partners, are required to achieve a sufficient FRET efficiency. Thus, synthetic methods yielding particles with highly monodisperse and reproducible sizes are desirable as well as a proper control of the surface chemistry.

The functionalization of the particle surface with proteins, antibodies, peptides or small molecules leads to particles that can be targeted to special cellular compartments (endosomes or lysosomes), organelles such as mitochondrium, endoplasmic reticulum, Golgi apparatus and nucleus, or even to single proteins inside the cell or at the cell membrane. The development of targetable nanoprobes is a demanding task for future improvements of their applicability as it is also a major issue for *in vivo* imaging, e.g. in tumor research. If the particles get accumulated in tumor cells, this leads to a high imaging contrast to healthy surrounding tissue.

Another important advantage of suitable coated nanoprobes is that they can be combined to multifunctional systems, e.g. with drug delivery functions. In the ideal case, an occurrence of pathogenic situations such as an acidic pH or NIR irradiation would trigger the release of drugs. A promising material as drug carrier is mesoporous silica which can be fabricated in form of nanoparticles or of nm-thick shells for other nanoparticle materials. The implementation of magnetic properties, e.g. with Fe_3O_4 clusters, provides also new options for nanoparticle-based theranostics, as these can be navigated or accumulated and separated by means of magnets. Finally, the brightness and performance of sensor dyes and particles can be improved by the use of plasmonic systems. These rely on a fluorescence enhancement by interaction with metal surfaces, e.g. gold and silver nanoclusters [129] and can be realized with a broad variety of luminescent nanomaterials from single fluorophore embedded in polymers over quantum dots [130] to lanthanide doped upconvesion particles [131]. The implementation of luminescence enhancement strategies in combination with NIR excitation/detection is one of the major challenges for the development of bright and sensitive nanoprobes which are applicable to sensing in biological environments and to *in vitro* or *in vivo* imaging.

Acknowledgements The authors gratefully acknowledge the EU CMST COST Action CM1403 "The Euopean Upconversion Network" and the M-era.Net project 2179 "NANOHYPE" for financial support. M.S. thanks the DFG (Deutsche Forschungsgemeinschaft) for a Heisenberg-Fellowship and Prof. Tero Soukka for hosting me as a FiDiPro fellow. U.R. gratefully acknowledges support from DFG (grants RE 1203/17-1 and RE 1203/12-3) and the BMBF (program KMU Nanochance; project NanoGenotox).

References

1. Burns A, Sengupta P, Zedayko T, Baird B, Wiesner U (2006) Core/Shell fluorescent silica nanoparticles for chemical sensing: towards single-particle laboratories. Small 2(6):723–726
2. Shang L, Nienhaus GU (2013) Small fluorescent nanoparticles at the nano–bio interface. Mater Today 16(3):58–66
3. Lee Y-EK, Kopelman R (2009) Optical nanoparticle sensors for quantitative intracellular imaging. Wiley Interdiscip Rev Nanomed Nanobiotechnol 1(1):98–110
4. Nel A, Xia T, Mädler L, Li N (2006) Toxic potential of materials at the nanolevel. Science 311(5761):622–627
5. Yildirimer L, Thanh NTK, Loizidou M, Seifalian AM (2011) Toxicology and clinical potential of nanoparticles. Nano Today 6(6):585–607
6. Koo H et al (2011) Nanoprobes for biomedical imaging in living systems. Nano Today 6(2):204–220
7. Clark HA, Kopelman R, Tjalkens R, Philbert MA (1999) Optical nanosensors for chemical analysis inside single living cells. 2. Sensors for pH and calcium and the intracellular application of PEBBLE sensors. Anal Chem 71(21):4837–4843

8. Søndergaard RV, Henriksen JR, Andresen TL (2014) Design, calibration and application of broad-range optical nanosensors for determining intracellular pH. Nat Protoc 9(12):2841–2858
9. Hornig S et al (2008) Biocompatible fluorescent nanoparticles for pH-sensoring. Soft Matter 4(6):1169–1172
10. Schulz A, Wotschadlo J, Heinze T, Mohr GJ (2010) Fluorescent nanoparticles for ratiometric pH-monitoring in the neutral range. J Mater Chem 20(8):1475–1482
11. Peng J et al (2007) Noninvasive monitoring of intracellular pH change induced by drug stimulation using silica nanoparticle sensors. Anal Bioanal Chem 388(3):645–654
12. Chen Y-P et al (2012) Surface charge effect in intracellular localization of mesoporous silica nanoparticles as probed by fluorescent ratiometric pH imaging. RSC Adv 2(3):968–973
13. Cao Y, Lee Koo Y-E, Kopelman R (2004) Poly(decyl methacrylate)-based fluorescent PEBBLE swarm nanosensors for measuring dissolved oxygen in biosamples. Analyst 129(8):745–750
14. Koo Y-EL et al (2004) Real-time measurements of dissolved oxygen inside live cells by organically modified silicate fluorescent nanosensors. Anal Chem 76(9):2498–2505
15. Wu C, Bull B, Christensen K, McNeill J (2009) Ratiometric single-nanoparticle oxygen sensors for biological imaging. Angew Chem Int Ed 48(15):2741–2745
16. Wang X-d et al (2011) Self-referenced RGB colour imaging of intracellular oxygen. Chem Sci 2(5):901–906
17. Sánchez-Martín RM, Cuttle M, Mittoo S, Bradley M (2006) Microsphere-based real-time calcium sensing. Angew Chem Int Ed 45(33):5472–5474
18. Kim G, Lee Y-EK XH, Philbert MA, Kopelman R (2010) Nanoencapsulation method for high selectivity sensing of hydrogen peroxide inside live cells. Anal Chem 82(6):2165–2169
19. Schäferling M (2012) The art of fluorescence imaging with chemical sensors. Angew Chem Int Ed 51(15):3532–3554
20. Schäferling M (2016) Nanoparticle-based luminescent probes for intracellular sensing and imaging of pH. Wiley Interdiscip Rev Nanomed Nanobiotechnol 8(3):378–413
21. Schäferling M, Duerkop A (2008) Intrinsically referenced fluorimetric sensing and detection schemes: methods, advantages and applications. In: Resch-Genger U (ed) Standardization and quality assurance in fluorescence measurements I: techniques. Springer, Berlin, Heidelberg, pp 373–414
22. Swanson JA (2002) 1 Ratiometric fluorescence microscopy. In: Methods in microbiology, vol 31, Academic, p 1–18
23. Tsien RY (1989) Fluorescent probes of cell signaling. Annu Rev Neurosci 12:227–253
24. Resch-Genger U, Grabolle M, Cavaliere-Jaricot S, Nitschke R, Nann T (2008) Quantum dots versus organic dyes as fluorescent labels. Nat Methods 5(9):763–775
25. Hoffmann K, Behnke T, Drescher D, Kneipp J, Resch-Genger U (2013) Near-infrared-emitting nanoparticles for lifetime-based multiplexed analysis and imaging of living cells. ACS Nano 7(8):6674–6684
26. Bambot SB, Lakowicz JR, Rao G (1995) Potential applications of lifetime-based, phase-modulation fluorimetry in bioprocess and clinical monitoring. Trends Biotechnol 13(3):106–115
27. Woods RJ, Scypinski S, Love LJC (1984) Transient digitizer for the determination of microsecond luminescence lifetimes. Anal Chem 56(8):1395–1400
28. Ballew RM, Demas JN (1989) An error analysis of the rapid lifetime determination method for the evaluation of single exponential decays. Anal Chem 61(1):30–33
29. Meier RJ, Fischer LH, Wolfbeis OS, Schäferling M (2013) Referenced luminescent sensing and imaging with digital color cameras: a comparative study. Sensors Actuators B Chem 177:500–506
30. Borisov SM, Würth C, Resch-Genger U, Klimant I (2013) New life of ancient pigments: application in high-performance optical sensing materials. Anal Chem 85(19):9371–9377
31. Huber C, Klimant I, Krause C, Wolfbeis OS (2001) Dual lifetime referencing as applied to a chloride optical sensor. Anal Chem 73(9):2097–2103

32. Liebsch G, Klimant I, Krause C, Wolfbeis OS (2001) Fluorescent imaging of pH with optical sensors using time domain dual lifetime referencing. Anal Chem 73(17):4354–4363
33. Ko JY et al (2010) pH-sensitive nanoflash for tumoral acidic pH imaging in live animals. Small 6(22):2539–2544
34. Koo Lee Y-E et al (2010) Near infrared luminescent oxygen nanosensors with nanoparticle matrix tailored sensitivity. Anal Chem 82(20):8446–8455
35. Resch-Genger U, Licha K (2011) Probes for optical imaging: new developments. Drug Discov Today 8(2–4):e87–e94
36. Zhu HG, McShane MJ (2005) Loading of hydrophobic materials into polymer particles: implications for fluorescent nanosensors and drug delivery. J Am Chem Soc 127(39):13448–13449
37. Behnke T, Würth C, Laux E-M, Hoffmann K, Resch-Genger U (2012) Simple strategies towards bright polymer particles via one-step staining procedures. Dyes Pigments 94(2):247–257
38. Napp J et al (2011) Targeted luminescent near-infrared polymer-nanoprobes for in vivo imaging of tumor hypoxia. Anal Chem 83(23):9039–9046
39. Huber A, Behnke T, Würth C, Jaeger C, Resch-Genger U (2012) Spectroscopic characterization of Coumarin-stained beads: quantification of the number of fluorophores per particle with solid-state 19F-NMR and measurement of absolute fluorescence quantum yields. Anal Chem 84(8):3654–3661
40. Hennig A, Hatami S, Spieles M, Resch-Genger U (2013) Excitation energy migration and trapping on the surface of fluorescent poly(acrylic acid)-grafted polymer particles. Photochem Photobiol Sci 12(5):729–737
41. Laux E-M, Behnke T, Hoffmann K, Resch-Genger U (2012) Keeping particles brilliant – simple methods for the determination of the dye content of fluorophore-loaded polymeric particles. Anal Methods 4(6):1759–1768
42. Lezhnina MM, Grewe T, Stoehr H, Kynast U (2012) Laponite Blue: Dissolving the Insoluble. Angew Chem Int Ed 51(42):10652–10655
43. Bujdak J (2006) Effect of the layer charge of clay minerals on optical properties of organic dyes. A review. Appl Clay Sci 34:58–73
44. Felbeck T et al (2013) Nile-Red–nanoclay hybrids: red emissive optical probes for use in aqueous dispersion. Langmuir 29(36):11489–11497
45. Felbeck T et al (2015) Multifold fluorescence enhancement in nanoscopic fluorophore-clay hybrids in transparent aqueous media. Chem Eur J 21(20):7582–7587
46. Grabolle M, Starke M, Resch-Genger U (2016) Highly fluorescent dye-nanoclay hybrid materials made from different dye classes. Langmuir 32(14):3506–3513
47. Sapsford KE et al (2013) Functionalizing nanoparticles with biological molecules: developing chemistries that facilitate nanotechnology. Chem Rev 113(3):1904–2074
48. Behnke T et al (2013) Target-specific nanoparticles containing a broad band emissive NIR dye for the sensitive detection and characterization of tumor development. Biomaterials 34(1):160–170
49. Quevedo PD, Behnke T, Resch-Genger U (2016) Streptavidin conjugation and quantification—a method evaluation for nanoparticles. Anal Bioanal Chem 408(15):4133–4149
50. Pauli J et al (2011) Dye-biomolecule conjugates and NIR-fluorescent particles for targeting of disease-related biomarkers. *SPIE Proc* 7910:791014. ISBN: 9780819484475
51. Hoffmann K, Behnke T, Grabolle M, Resch-Genger U (2014) Nanoparticle-encapsulated vis- and NIR-emissive fluorophores with different fluorescence decay kinetics for lifetime multiplexing. Anal Bioanal Chem 406(14):3315–3322
52. Hoffmann K, Behnke T, Drescher D, Kneipp J, & ReschGenger U (2011) Lifetime-based discrimination between spectrally matching vis and NIR emitting particle labes and probes single molecule spectroscopy and imaging IV. *Proc SPIE* 79051F79051-79051F-79059
53. Ashraf S et al (2014) Fluorescence-based ion-sensing with colloidal particles. Curr Opin Pharmacol 18:98–103

54. Petryayeva E, Algar WR, Medintz IL (2013) Quantum dots in bioanalysis: a review of applications across various platforms for fluorescence spectroscopy and imaging. Appl Spectrosc 67(3):215–252
55. Wegner KD, Hildebrandt N (2015) Quantum dots: bright and versatile in vitro and in vivo fluorescence imaging biosensors. Chem Soc Rev 44(14):4792–4834
56. Lesnyak V, Gaponik N, Eychmueller A (2013) Colloidal semiconductor nanocrystals: the aqueous approach. Chem Soc Rev 42(7):2905–2929
57. Palui G, Aldeek F, Wang WT, Mattoussi H (2015) Strategies for interfacing inorganic nanocrystals with biological systems based on polymer-coating. Chem Soc Rev 44(1):193–227
58. Ziegler J, Merkulov A, Grabolle M, Resch-Genger U, Nann T (2007) High-quality ZnS shells for CdSe nanoparticles: rapid microwave synthesis. Langmuir 23(14):7751–7759
59. Greytak AB et al (2012) Alternating layer addition approach to CdSe/CdS core/shell quantum dots with near-unity quantum yield and high on-time fractions. Chem Sci 3(6):2028–2034
60. Grabolle M et al (2009) Determination of the fluorescence quantum yield of quantum dots: suitable procedures and achievable uncertainties. Anal Chem 81(15):6285–6294
61. Ostermann J et al (2015) Tailoring the ligand shell for the control of cellular uptake and optical properties of nanocrystals. Beilstein J Nanotechnol 6:232–242
62. Walker GW et al (2003) Quantum-dot optical temperature probes. Appl Phys Lett 83(17):3555–3557
63. Liu Y-S et al (2007) pH-Sensitive photoluminescence of CdSe/ZnSe/ZnS quantum dots in human ovarian cancer cells. J Phys Chem C 111(7):2872–2878
64. Wang X, Boschetti C, Ruedas-Rama MJ, Tunnacliffe A, Hall EAH (2010) Ratiometric pH-dot ANSors. Analyst 135(7):1585–1591
65. Orte A, Alvarez-Pez JM, Ruedas-Rama MJ (2013) Fluorescence lifetime imaging microscopy for the detection of intracellular pH with quantum dot nanosensors. ACS Nano 7(7):6387–6395
66. Chen Y, Rosenzweig Z (2002) Luminescent CdS quantum dots as selective ion probes. Anal Chem 74(19):5132–5138
67. Li H, Zhang Y, Wang X, Xiong D, Bai Y (2007) Calixarene capped quantum dots as luminescent probes for Hg2+ ions. Mater Lett 61(7):1474–1477
68. Gattas-Asfura KM, Leblanc RM (2003) Peptide-coated CdS quantum dots for the optical detection of copper(ii) and silver(i). Chem Commun 3(21):2684–2685
69. Jin WJ, Fernandez-Arguelles MT, Costa-Fernandez JM, Pereiro R, Sanz-Medel A (2005) Photoactivated luminescent CdSe quantum dots as sensitive cyanide probes in aqueous solutions. Chem Commun 7:883–885
70. Callan JF, Mulrooney RC, Kamila S, McCaughan B (2008) Anion sensing with luminescent quantum dots – A modular approach based on the Photoinduced Electron Transfer (PET) mechanism. J Fluoresc 18(2):527–532
71. Kuningas K et al (2007) Upconversion fluorescence enables homogeneous immunoassay in whole blood. Clin Chem 53(1):145–146
72. Liu Q, Feng W, Yang T, Yi T, Li F (2013) Upconversion luminescence imaging of cells and small animals. Nat Protoc 8(10):2033–2044
73. Auzel F (2004) Upconversion and anti-stokes processes with f and d ions in solids. Chem Rev 104(1):139–174
74. Wang F, Liu X (2009) Recent advances in the chemistry of lanthanide-doped upconversion nanocrystals. Chem Soc Rev 38(4):976–989
75. Zeng JH, Su J, Li ZH, Yan RX, Li YD (2005) Synthesis and upconversion luminescence of hexagonal-phase NaYF4:Yb, Er3+ phosphors of controlled size and morphology. Adv Mater 17(17):2119–2123
76. Chen Z et al (2008) Versatile synthesis strategy for carboxylic acid−functionalized upconverting nanophosphors as biological labels. J Am Chem Soc 130(10):3023–3029
77. Hao S, Chen G, Yang C (2013) Sensing using rare-earth-doped upconversion nanoparticles. Theranostics 3(5):331–345

78. Cao T et al (2010) Water-soluble NaYF4:Yb/Er upconversion nanophosphors: synthesis, characteristics and application in bioimaging. Inorg Chem Commun 13(3):392–394
79. Wilhelm S et al (2015) Water dispersible upconverting nanoparticles: effects of surface modification on their luminescence and colloidal stability. Nanoscale 7(4):1403–1410
80. Wang F, Wang JA, Liu XG (2010) Direct evidence of a surface quenching effect on size-dependent luminescence of upconversion nanoparticles. Angew Chem Int Ed 49(41):7456–7460
81. Gnach A, Bednarkiewicz A (2012) Lanthanide-doped up-converting nanoparticles: merits and challenges. Nano Today 7(6):532–563
82. Liu HC et al (2013) Balancing power density based quantum yield characterization of upconverting nanoparticles for arbitrary excitation intensities. Nanoscale 5(11):4770–4775
83. Vetrone F et al (2010) Temperature sensing using fluorescent nanothermometers. ACS Nano 4(6):3254–3258
84. Dong H, Sun L-D, Yan C-H (2015) Energy transfer in lanthanide upconversion studies for extended optical applications. Chem Soc Rev 44(6):1608–1634
85. Boyer J-C, Manseau M-P, Murray JI, van Veggel FCJM (2010) Surface modification of upconverting NaYF4 nanoparticles with PEG–Phosphate ligands for NIR (800 nm) biolabeling within the biological window. Langmuir 26(2):1157–1164
86. Anderson RB, Smith SJ, May PS, Berry MT (2014) Revisiting the NIR-to-visible upconversion mechanism in beta-NaYF4:Yb3+,Er3+. J Phys Chem Lett 5(1):36–42
87. Berry MT, May PS (2015) Disputed mechanism for NIR-to-red upconversion luminescence in NaYF4:Yb3+,Er3+. J Phys Chem A 119(38):9805–9811
88. Gai SL, Li CX, Yang PP, Lin J (2014) Recent progress in rare earth micro/nanocrystals: soft chemical synthesis, luminescent properties, and biomedical applications. Chem Rev 114(4):2343–2389
89. Liu GK (2015) Advances in the theoretical understanding of photon upconversion in rare-earth activated nanophosphors. Chem Soc Rev 44(6):1635–1652
90. Arppe R et al (2015) Quenching of the upconversion luminescence of NaYF4:Yb3+,Er3+ and NaYF4:Yb3+,Tm3+ nanophosphors by water: the role of the sensitizer Yb3+ in non-radiative relaxation. Nanoscale 7(27):11746–11757
91. Yi G-S, Chow G-M (2007) Water-soluble NaYF4:Yb,Er(Tm)/NaYF4/Polymer Core/Shell/Shell nanoparticles with significant enhancement of upconversion fluorescence. Chem Mater 19(3):341–343
92. Esipova TV et al (2012) Dendritic upconverting nanoparticles enable in vivo multiphoton microscopy with low-power continuous wave sources. Proc Natl Acad Sci 109(51):20826–20831
93. Liu Q, Peng J, Sun L, Li F (2011) High-efficiency upconversion luminescent sensing and bioimaging of Hg(II) by chromophoric ruthenium complex-assembled nanophosphors. ACS Nano 5(10):8040–8048
94. Liu Y et al (2013) A cyanine-modified nanosystem for in vivo upconversion luminescence bioimaging of methylmercury. J Am Chem Soc 135(26):9869–9876
95. Liu S et al (2014) Development of upconversion luminescent probe for ratiometric sensing and bioimaging of hydrogen sulfide. ACS Appl Mater Interfaces 6(14):11013–11017
96. Zhao L et al (2014) Yolk–Shell upconversion nanocomposites for LRET sensing of Cysteine/Homocysteine. ACS Appl Mater Interfaces 6(14):11190–11197
97. Cen Y et al (2014) Phospholipid-modified upconversion nanoprobe for ratiometric fluorescence detection and imaging of Phospholipase D in cell lysate and in living cells. Anal Chem 86(14):7119–7127
98. Zhang C, Yuan Y, Zhang S, Wang Y, Liu Z (2011) Biosensing platform based on fluorescence resonance energy transfer from upconverting nanocrystals to graphene oxide. Angew Chem Int Ed 50(30):6851–6854
99. Han J et al (2014) Upconversion nanoparticles for ratiometric fluorescence detection of nitrite. Analyst 139(12):3032–3038

100. Yan L et al (2014) Biocompatible and flexible graphene oxide/upconversion nanoparticle hybrid film for optical pH sensing. Phys Chem Chem Phys 16(4):1576–1582
101. Xie L, Qin Y, Chen H-Y (2013) Direct fluorescent measurement of blood potassium with polymeric optical sensors based on upconverting nanomaterials. Anal Chem 85(5):2617–2622
102. Meier RJ, Simbürger JMB, Soukka T, Schäferling M (2014) Background-free referenced luminescence sensing and imaging of pH using upconverting phosphors and color camera read-out. Anal Chem 86(11):5535–5540
103. Sapsford KE, Tyner KM, Dair BJ, Deschamps JR, Medintz IL (2011) Analyzing nanomaterial bioconjugates: a review of current and emerging purification and characterization techniques. Anal Chem 83(12):4453–4488
104. Martini M et al (2009) How to measure quantum yields in scattering media: application to the quantum yield measurement of fluorescein molecules encapsulated in sub-100 nm silica particles. J Appl Phys 106(9):094304
105. Sun GR et al (2010) Bright fluorescent nanoparticles for developing potential optical imaging contrast agents. Nanoscale 2(4):548–558
106. Herz E et al (2010) Relative quantum yield measurements of Coumarin encapsulated in core-shell silica nanoparticles. J Fluoresc 20(1):67–72
107. Würth C, Geißler D, Behnke T, Kaiser M, Resch-Genger U (2015) Critical review of the determination of photoluminescence quantum yields of luminescent reporters. Anal Bioanal Chem 407(1):59–78
108. Bonacchi S et al (2011) Luminescent silica nanoparticles: extending the frontiers of brightness. Angew Chem Int Ed 50(18):4056–4066
109. Hennig A et al (2012) Scope and limitations of surface functional group quantification methods: exploratory study with Poly(acrylic acid)-grafted micro- and nanoparticles. J Am Chem Soc 134(19):8268–8276
110. Hennig A et al (2011) Simple colorimetric method for quantification of surface carboxy groups on polymer particles. Anal Chem 83(12):4970–4974
111. Hennig A et al (2011) Quantification of surface functional groups on polymer microspheres by supramolecular host-guest interactions. Chem Commun 47(27):7842–7844
112. Felbeck T, Hoffmann K, Lezhnina MM, Kynast UH, Resch-Genger U (2015) Fluorescent nanoclays: covalent functionalization with amine reactive dyes from different fluorophore classes and surface group quantification. J Phys Chem C 119(23):12978–12987
113. Hennig A et al (2015) En route to traceable reference standards for surface group quantifications by XPS, NMR and fluorescence spectroscopy. Analyst 140(6):1804–1808
114. Leubner S et al (2013) Experimental and theoretical investigations of the ligand structure of water-soluble CdTe nanocrystals. Dalton Trans 42(35):12733–12740
115. Moser M et al (2015) Quantification of PEG-Maleimide ligands and coupling efficiencies on nanoparticles with Ellman's reagent. Anal Chem 87(18):9376–9383
116. Leubner S et al (2014) Influence of the stabilizing ligand on the quality, signal-relevant optical properties, and stability of near-infrared emitting Cd1-xHgxTe nanocrystals. J Mater Chem C 2(25):5011–5018
117. Hoffmann K, Resch-Genger U, Mix R, Friedrich JF (2006) Fluorescence spectroscopic studies on plasma-chemically modified polymer surfaces with fluorophore-labeled functionalities. J Fluoresc 16(3):441–448
118. Zhan N, Palui G, Merkl JP, Mattoussi H (2015) Quantifying the density of surface capping ligands on semiconductor quantum dots. In: Parak WJ, Osinski M, XJ L (eds) Colloidal nanoparticles for biomedical applications X, Proceedings of SPIE, vol 9338
119. Ostermann J et al (2013) Controlling the physical and biological properties of highly fluorescent aqueous quantum dots using block copolymers of different size and shape. ACS Nano 7(10):9156–9167
120. Grabolle M, Ziegler J, Merkulov A, Nann T, Resch-Genger U (2008) Stability and fluorescence quantum yield of CdSe/ZnS quantum dots – Influence of the thickness of the ZnS-shell. *Ann N. Y. Acad Sci* 1130:235–241

121. Hatami S et al (2015) Absolute photoluminescence quantum yields of IR26 and IR-emissive Cd1-xHgxTe and PbS quantum dots – method- and material-inherent challenges. Nanoscale 7(1):133–143
122. Würth C, Pauli J, Lochmann C, Spieles M, Resch-Genger U (2012) Integrating sphere setup for the traceable measurement of absolute photoluminescence quantum yields in the near infrared. Anal Chem 84(3):1345–1352
123. Grabolle M et al (2009) Fluorescence lifetime multiplexing with nanocrystals and organic labels. Anal Chem 81(18):7807–7813
124. Abbasi AZ et al (2011) How colloidal nanoparticles could facilitate multiplexed measurements of different analytes with analyte-sensitive organic fluorophores. ACS Nano 5(1):21–25
125. Lu YQ et al (2014) Tunable lifetime multiplexing using luminescent nanocrystals. Nat Photonics 8(1):33–37
126. Stich MIJ, Nagl S, Wolfbeis OS, Henne U, Schaeferling M (2008) A dual luminescent sensor material for simultaneous imaging of pressure and temperature on surfaces. Adv Funct Mater 18(9):1399–1406
127. Arppe R et al (2014) Photon upconversion sensitized nanoprobes for sensing and imaging of pH. Nanoscale 6(12):6837–6843
128. Markus MA et al (2015) Tracking of inhaled near-infrared fluorescent nanoparticles in lungs of SKH-1 mice with allergic airway inflammation. ACS Nano 9(12):11642–11657
129. Aslan K et al (2005) Metal-enhanced fluorescence: an emerging tool in biotechnology. Curr Opin Biotechnol 16(1):55–62
130. Pompa PP et al (2006) Metal-enhanced fluorescence of colloidal nanocrystals with nanoscale control. Nat Nanotechnol 1(2):126–130
131. Ge W et al (2013) Distance dependence of gold-enhanced upconversion luminescence in Au/SiO 2/Y 2 O 3: Yb 3+, Er 3+ nanoparticles. Theranostics 3(4):282–288
131. Morris-Cohen AJ et al (2013) Chemical, structural, and quantitative analysis of the ligand shells of colloidal quantum dots. Chem Mater 25(8):1155–1165

Chapter 6
Analog Filter Theory Applied to Frequency Domain Fluorescence Spectroscopy

R. Trainham and M. O'Neill

Keywords Fluorescence • Frequency domain • Analog filter • Fluorophore • Laplace transform • Transfer function • Heterodyne • Homodyne

6.1 Introduction

This chapter expands upon an article [1] that we recently published illustrating the parallel between frequency modulated fluorescence and analog filter theory from electronics. All of the material of the original article is included here, and we have added a discussion on non-linear techniques used in fluorescence and how they may be understood in terms of a linear time invariant (LTI) system theory.

Optical detection of fluorophores is often hindered by overlapping emissions from nuisance species emitting into the same spectral band. Frequently the detection limitation is not a problem of the instrumentation, but the result of broad molecular band structure of the fluorophore of interest. For example, compounds containing the uranyl radical [2, 3] emit photons in the green, in five overlapping bands, with a total spectral bandwidth of over 3000 cm^{-1}. In this case, improvement of the spectral resolution of the detector does not improve the measurement, since much of the desired signal would be discarded with the noise. This is a situation where excitation lifetime measurements can be useful to isolate the fluorophore of interest from the nuisance emissions. Emissions from molecules containing the uranyl radical have decay lifetimes of a few hundred of microseconds, whereas most nuisance emitters are much shorter lived, with lifetimes in the hundreds of nanoseconds. The different time regimes of the decays make it possible to isolate the detection of the interesting

R. Trainham (✉) • M. O'Neill
Special Technologies Laboratory, National Security Technologies, LLC, 5520 Ekwill Street, Santa Barbara, CA 93111, USA
e-mail: trainhcp@nv.doe.gov

C.D. Geddes (ed.), *Reviews in Fluorescence 2016*, Reviews in Fluorescence,
DOI 10.1007/978-3-319-48260-6_6

fluorophore, and this can be accomplished by modulating the excitation source and utilizing the techniques of frequency domain spectroscopy for the detection.

The techniques of time domain and frequency domain fluorescence spectroscopy are covered thoroughly in the *Principles of Fluorescence Spectroscopy* by Lakowicz [4]. In that text, however, the frequency domain solutions of the differential equations for fluorescence make use of direct substitution of variables. Although the procedure is correct, the mathematics are tedious and prone to error. The method of direct substitution is also difficult to scale to many coupled differential equations, and it becomes quite complicated for complex driver functions. In other publications, the Fourier transform [5], sine and cosine transforms [6], and the Laplace transform [7] techniques have been applied successfully on the frequency domain fluorescence problem. The multi-lifetime fluorescence problem is mathematically very similar to the filter problem in analog circuit design theory [8, 9], however no publications have appeared to treat the fluorescence problem with this simplified analysis approach. The purpose of the present chapter is to illustrate that fluorescence can be treated mathematically as a discrete component in a network of electronic filters, regardless of the excitation method.

In analog electronics the usual approach for solving coupled differential equations is to take Laplace transforms of the equations and work directly in the frequency domain, where multiplications replace convolutions. This allows the concept of a transfer function to emerge for the response, and the scaling of fluorescence problems to more complicated driving and response functions becomes straight forward. Considerable information about a problem can be extracted from the transfer function itself, without explicit need to apply the inverse transform back into the time domain. The variable s of the Laplace transform is as much an operator as it is a variable; s is a differentiator, and its inverse, s^{-1} is an integrator. This concept is essential when constructing electronic circuits as analogues to differential equations and for using the calculation tools of electronics to solve those equations. The basis of network theory in electronics rests upon these techniques, and the tools of electrical engineering and network analysis can be exploited to solve problems of spectroscopy. In hardware implementations, the signals are usually sampled, so the Z transform and digital filter theory [10] may be a more appropriate approach, however, we restrict ourselves in this chapter to the Laplace transform approach.

Use of analog filter theory allows unification of the optical fluorescence response with that of the system electronics of the driver and detector systems, and a global transfer function can be defined to describe the entire chain. Practical and unavoidable features, such as limited detector bandwidth and resulting phase shifts, or the rounding off of waveforms from driver circuit implementations can be handled elegantly by this approach. In the following sections we discuss the modelling and analysis of fluorescence equations with analog filter theory, and the use of a transfer function to extract the phase and modulation index (*Mod*) without transforming the equations back to the time domain. We present solutions to a few problems involving complicated driving functions, and we present examples of using the circuit tool SPICE [11] to solve fluorescence problems.

6.2 Theory

By defining a transfer function and using the mathematics of analog filter theory the time domain and frequency domain approaches to fluorescence are not fundamentally different. The only difference lies in the appearance of the waveforms of the response functions. The simplest rate equation leading to exponential decay is:

$$\frac{dn}{dt} = -\Gamma n \tag{6.1}$$

where n is the population of the excited state, and Γ is the decay rate. To include the effect of non-radiative de-excitation (at least for first order kinetics), Γ is replaced by a sum of all the transition rates governing the de-population of the excited state. This modification merely changes the value of the lifetime, and does not alter the exponential character of the decay. In order to have multi-exponential decay the excited state needs to be segregated into multiple groups. Effectively, multiple excited states need to be considered. If the detection is not highly discriminating (for example, a band pass filter on a photodiode) then the total measured radiation is the sum of multiple atomic transitions, with each showing its own characteristic exponential decay. Radiative decay from multiple de-excitation paths is thus described by the following equation:

$$n(t) = \sum_i n_i e^{-\Gamma_i t} \tag{6.2}$$

This equation is appropriate for very fast pulsed, or delta function, excitation. However, if the timescale of the driving excitation is comparable to any of the lifetimes, $\tau_i = 1/\Gamma_i$, then the excitation source must be explicitly accounted for in the rate equations. This leads to a convolution of source and response functions. One way of thinking about the problem is to imagine that the excitation function is a sequence of closely spaced fast pulses, where each pulse suddenly adds an excited state population, which then begins to decay away exponentially. This reasoning leads naturally to a limiting sum, which becomes a convolution integral. If we denote the excitation function by $g(t)$, then Eq. 6.2 becomes:

$$n(t) = \sum_i n_i \int e^{-t'/\tau_i} g(t-t')\,\mathrm{d}t' \tag{6.3}$$

The appearance of the convolution integral immediately suggests the use of some kind of transform technique where the convolution theorem can be used to advantage. From analog electronics we recognize Eqs. 6.1, 6.2, and 6.3 as low pass filters.

6.2.1 Single Exponential Decay

In the time domain the rate equations governing the depopulation of the fluorescing excited states are homogeneous first order differential equations, whereas in the frequency domain they are non-homogeneous equations. For both domains, however, the equations can be written as follows:

$$\sum_i \frac{dn_i}{dt} + \Gamma_i n_i = g(t) \tag{6.4}$$

where the n_i are the excited state populations, Γ_i are the decay rates, and $g(t)$ is the driving function. If $g(t)$ is zero, or a delta function, then we have homogeneous equations of the time domain measurement. If $g(t)$ is periodic then we have the non-homogeneous equations of the frequency domain.

The Laplace transform of Eq. 6.4 is

$$\sum_i N_i(s)(s + \Gamma_i) - n_{i_0} = G(s) \tag{6.5}$$

where s is the complex frequency variable, and n_{i_0} represent the initial state populations, which are usually zero. The transfer function for Eq. 6.5 is

$$F_i(s) = \frac{N_i(s)}{G(s)} = \frac{1}{s + \Gamma_i} \tag{6.6}$$

From this we can determine the phase shift and modulation amplitude of the filter response to a periodic driver function g(t). This is accomplished by restricting s to the imaginary axis, and computing the *Mod* of the transfer function for the modulation amplitude, and the ratio of the imaginary and real parts for the tangent of the phase shift. Figure 6.1 shows the theoretical response assuming a single lifetime emission as a function of the drive frequency ω. The phase shift and modulation amplitude evolve on a frequency scale determined by the reciprocal lifetime. The range of frequencies is typically two orders of magnitude centered at the angular frequency of the reciprocal lifetime $\omega = 2\pi/\tau$.

As an example of a Laplace transform solution of the decay problem, the transform Eq. 6.5 of the differential equation Eq. 6.4 with $G(s) = 0$ for just a single i is

$$N(s) = \frac{n_0}{s + \Gamma} \tag{6.7}$$

To convert this back into the time domain we use the Bromwich integral [12] for the inverse Laplace transform

$$n(t) = \frac{n_0}{2\pi i} \int_{\gamma - i\infty}^{\gamma + i\infty} \frac{e^{st}}{s + \Gamma} \, \mathrm{d}s \tag{6.8}$$

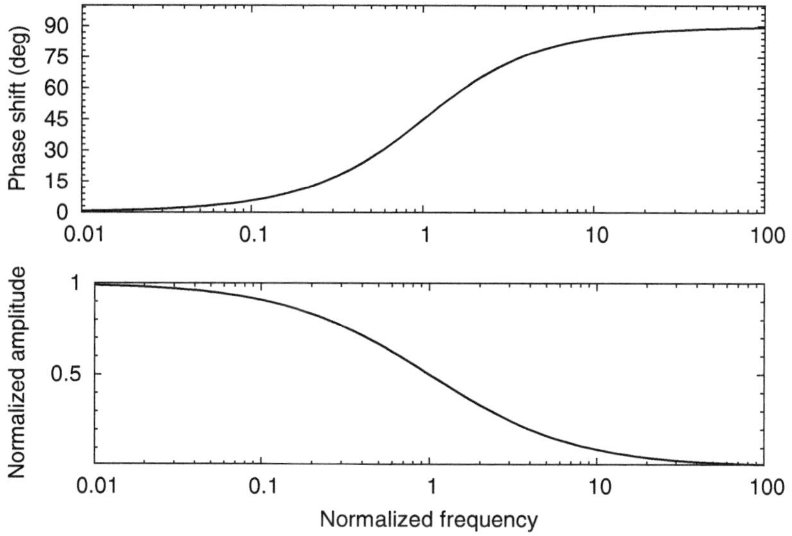

Fig. 6.1 Phase shift and modulation amplitude for single lifetime exponential decay undergoing sinusoidal excitation

This integral is readily solved by use of the Residue theorem [13], yielding

$$n(t) = n_0 e^{-\Gamma t} \tag{6.9}$$

This is a roundabout way to solve a simple first order differential equation, but the Laplace transform technique scales readily to solve some very difficult problems. Multi-exponential decay is handled as a straight-forward sum of the various single state terms. Bandwidth limitations in the detection system, which require a convolution in the time domain, become simple products of terms in the frequency domain. Equation 6.7 is the transfer function for simple exponential decay of an excited atomic state. It is also the transfer function of a first order low pass filter in electronics. The norm of the complex transfer function is the modulation depth, and the ratio of the imaginary part to the real part gives the tangent of the phase shift:

$$Mod = \sqrt{\frac{1}{\Gamma + i\omega}\frac{1}{\Gamma - i\omega}} = \sqrt{\frac{1}{\Gamma^2 + \omega^2}} = \frac{\tau}{\sqrt{1 + (\omega\tau)^2}} \tag{6.10}$$

$$tan\Phi = \frac{\omega}{\Gamma} = \omega\tau \tag{6.11}$$

where Φ is the phase shift, ω is the angular frequency, Γ is the transition rate, and the lifetime τ is the reciprocal or Γ. With some manipulation, the relationship between the *Mod* and Φ can be shown to be

$$Mod = \frac{cos\Phi}{\Gamma} = \frac{sin\Phi}{\omega} \tag{6.12}$$

6.2.2 Sinusoidal Driving

Another relevant example is the response of the system to a sine wave excitation function. By letting $g(t) = Asin(\omega t)$ the Laplace transform of Eq. 6.4 for a single decay rate is

$$N(s)(s + \Gamma) - n_0 = \mathscr{L}(Asin(\omega t)) \tag{6.13}$$

where $\mathscr{L}(sin(\omega t)$ represents the Laplace transform of the sine function. That transform is

$$\mathscr{L}(Asin(\omega t)) = A\frac{\omega}{\omega^2 + s^2} \tag{6.14}$$

and the expression to be transformed back into the time domain is

$$N(s) = \frac{n_0}{s + \Gamma} + \frac{A\omega}{(s + \Gamma)(s^2 + \omega^2)} \tag{6.15}$$

This integral has three simple poles (i.e., singularities where the function goes to infinity), so the Bromwich integral can be evaluated by use of the Residue theorem. The solution is

$$n(t) = n_0 e^{-\Gamma t} + A\left(\frac{\omega e^{-\Gamma t}}{\Gamma^2 + \omega^2} + \frac{e^{i\omega t}}{2i(\Gamma + i\omega)} - \frac{e^{-i\omega t}}{2i(\Gamma - i\omega)}\right). \tag{6.16}$$

The first term in Eq. 6.16, proportional to n_0, is a simple exponential decay of any population that happens to be present at time $t = 0$. Usually, the system starts in the ground state, such that $n_0 = 0$. The second term is a transient turn-on that damps away. It satisfies the $t = 0$ boundary condition, and it compensates for the steady state phase shift during the first few cycles of the response. The third and fourth terms in Eq. 6.16 are sines and cosines with phase shifts. To get a feel for them, let us look at two extremes of the driving frequency: low frequency where $\omega \ll \Gamma$, and high frequency where $\omega \gg \Gamma$. Concentrating on the steady state solution, when the transient terms have damped away, Eq. 6.16 becomes at low frequency ($\omega \ll \Gamma$) a sine function:

$$n(t) = \frac{A}{\Gamma}sin(\omega t) \tag{6.17}$$

and at high frequency ($\omega \gg \Gamma$) Eq. 6.16 becomes a cosine function:

$$n(t) = -\frac{A}{\omega}cos(\omega t) \tag{6.18}$$

so we see that by scanning the frequency of the drive the response transitions from a sine to a cosine, meaning that a phase shift from zero up to $\pi/2$ is accumulated in the response. The factor $1/\Gamma$ in front of the in-phase term means that the response is weakened by the fast decay rate, and that more drive power may be required to sample fast decay fluorophores. Also, for high driving frequencies the response diminishes by $1/\omega$, meaning that the modulation depth of the response dies away to a DC offset for higher and higher frequencies.

An important difference between the atomic system and the electronics filter analog is that there cannot be a negative population in an excited atomic state. As such, the response derived for a sinusoidal driving term either must be biased positively, or the drive term must be truncated at zero, and not allowed to go negative. This is not a real limitation; it simply means that some care must be taken to assure that any results appropriated from analog filter theory are physically realistic for fluorescence. In this case, a simple offset biasing of the sine wave positive suffices, and this is precisely what is done in the laboratory.

A positively biased sine wave can be written as $g(t) = A(1 - cos(\omega t))/2$. Its transform into s-space is

$$G(s) = \frac{A}{2}\left(\frac{1}{s} + \frac{s}{s^2 + \omega^2}\right) \tag{6.19}$$

and the s-space response function is

$$N(s) = \frac{A}{2}\left(\frac{1}{s} + \frac{s}{s^2 + \omega^2}\right)\left(\frac{1}{s + \Gamma}\right) \tag{6.20}$$

Solving the Bromwich integral via the Residue theorem to transform back to the time domain yields

$$n(t) = \frac{A}{2\Gamma}\left(1 - \frac{\left(\frac{\omega}{\Gamma}\right)^2 e^{-\Gamma t} - cos(\omega t) + \left(\frac{\omega}{\Gamma}\right) sin(\omega t)}{1 + \left(\frac{\omega}{\Gamma}\right)^2}\right). \tag{6.21}$$

Noting the appearance in Eq. 6.21 of both sine and cosine of the same frequency, this expression can be simplified further by use of the identity $cos(\alpha \pm \beta) = cos\alpha\, cos\beta \mp sin\alpha\, sin\beta$. Thus Eq. 6.21 can be written as

$$n(t) = \frac{A}{2\Gamma}\left(1 - \frac{\left(\frac{\omega}{\Gamma}\right)^2 e^{-\Gamma t}}{1 + \left(\frac{\omega}{\Gamma}\right)^2} - \frac{cos(\omega t + \Phi)}{\sqrt{1 + \left(\frac{\omega}{\Gamma}\right)^2}}\right) \tag{6.22}$$

where $tan\Phi = \omega/\Gamma$. Recognizing that Eq. 6.22 contains factors of $tan\Phi$ and $tan^2\Phi$ the result can be simplified even further.

$$n(t) = \frac{A}{2\Gamma}\left[1 - sin^2\Phi\, e^{-\Gamma t} - cos\Phi\, cos(\omega t + \Phi)\right] \tag{6.23}$$

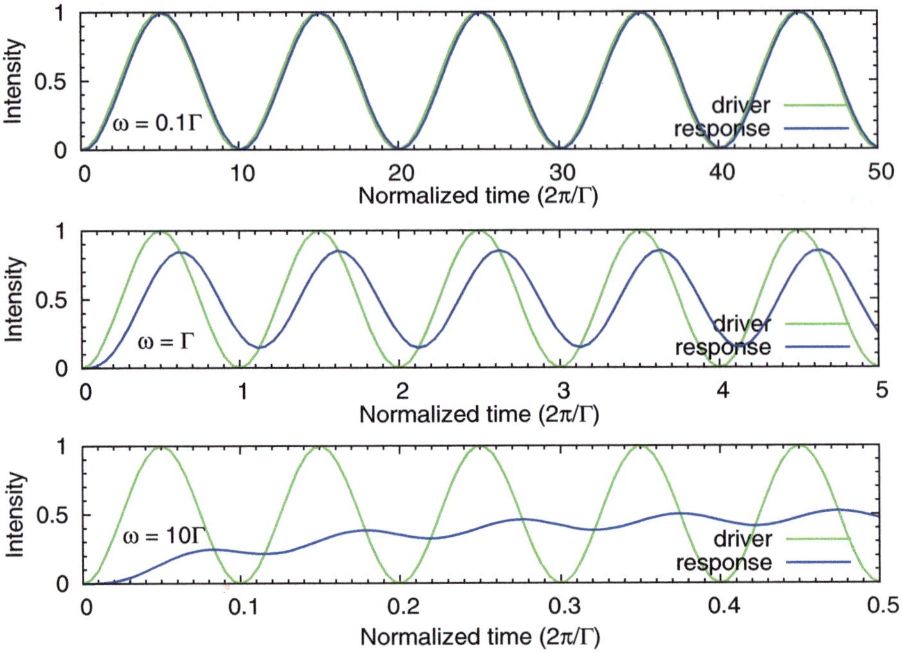

Fig. 6.2 These three plots of the modelled fluorescence signal driven by a positively biased sine wave show turn-on transients, DC offsets, and decreasing modulation as the drive frequency transitions through the fluorescence decay rate. The time axis for each plot is scaled by the fluorescence lifetime $1/\Gamma$. The plot at the *top* shows the signal for $\omega = 0.1\Gamma$, the *middle plot* is for $\omega = \Gamma$, and the *bottom plot* is for $\omega = 10\Gamma$

So now we see finally how the *Mod* and the phase shift are related to each other. The *Mod* behaves as the cosine of the phase shift times a scaling factor over the range of $\Phi = 0$ to $\pi/2$.

Modeled wave forms of the driver and response functions for several frequencies are shown in Fig. 6.2. In the figure the time axis for each plot is scaled by the fluorescence lifetime $1/\Gamma$. Multiplying the normalized frequency by $\Gamma/2\pi$ (or equivalently, by $1/(2\pi\tau)$) converts the frequency scale to Hz. One sees that at low frequency the response is in phase with the driver, and increasing frequency the response lags in phase and begins to demodulate to a DC level.

6.2.3 *Three Level System*

Fluorescence from a three level system can be represented by the diagram in Fig. 6.3. The driving term is Γ_{13}, and it excites population from level 1 to level 3. Level 3 quickly relaxes to level 2. Level 2 then relaxes more slowly back to level 1. Some

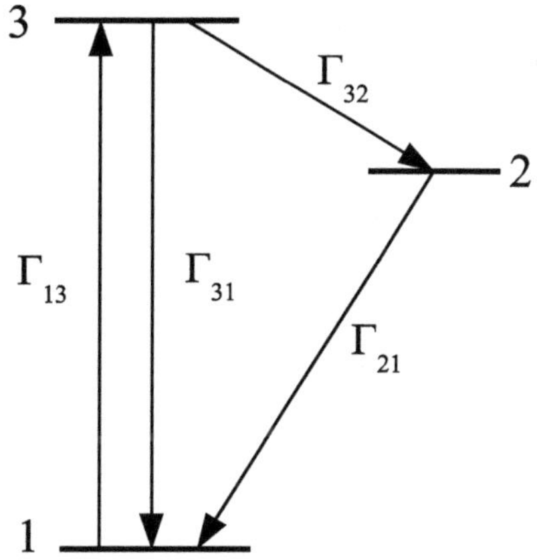

Fig. 6.3 Three level system diagram. The excitation Γ_{13} pumps population from level 1 into level 3, which then rapidly decays to level 2. The fluorescence emission is directly proportional to the population of level 2

portion of level 3 also relaxes directly back to level 1, but that transition is typically very weak compared to the relaxation from level 3 to level 2. The fluorescence signal is proportional to the population accumulated in level 2. We denote that time dependent population by $n_2(t)$, and correspondingly similar notation for the populations of the other levels.

The rate equations for the three levels are:

$$\frac{dn_1}{dt} = \Gamma_{31}n_3 + \Gamma_{21}n_2 - \Gamma_{13}n_1 \tag{6.24}$$

$$\frac{dn_2}{dt} = \Gamma_{32}n_3 - \Gamma_{21}n_2 \tag{6.25}$$

$$\frac{dn_3}{dt} = -\Gamma_{31}n_3 - \Gamma_{32}n_3 + \Gamma_{13}n_1 \tag{6.26}$$

The driving term Γ_{13} can be either a delta function for a pulsed time domain measurement, or a continuous wave form for a frequency domain measurement. If we assume that the pumping does not significantly deplete level 1 then we can readily solve for the transfer function for N_2 in terms of N_1. The Laplace transform solution for $N_2(s)$ is

$$N_2(s) = \frac{n_{2_0}}{s + \Gamma_{21}} + \frac{\Gamma_{32}\left(n_{3_0} + \Gamma_{13}N_1\right)}{\left(s + \Gamma_{21}\right)\left(s + \Gamma_{31} + \Gamma_{32}\right)} \tag{6.27}$$

The terms n_{2_0} and n_{3_0} are initial state populations for the levels 2 and 3. They are transients which die off exponentially, and those terms are usually equal to zero. If, however, the frequency of the driver is stepped discontinuously between data samplings then whatever steady state population is left from the previous frequency step evolves according to the transient terms (which are first order lowpass filters) to the next steady state solution.

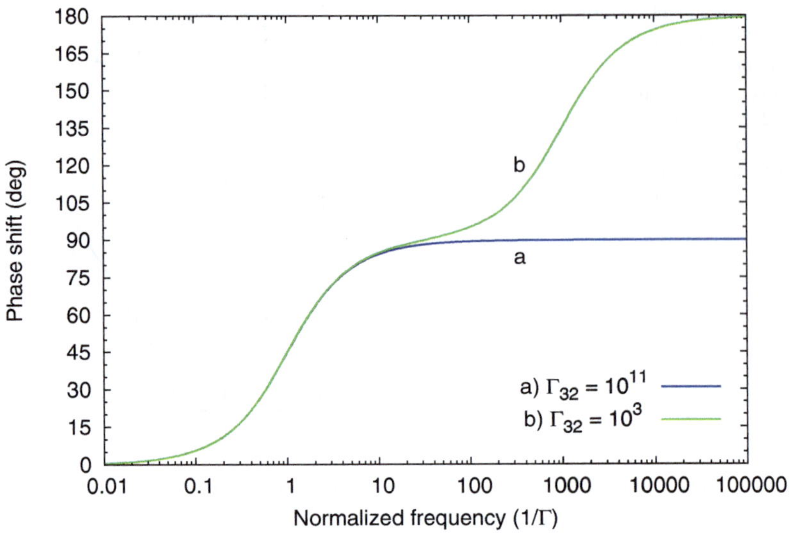

Fig. 6.4 Phase shift for the 3 level system with two sets of relaxation constants

Since Γ_{13} is the driver term we factor it out to arrive at a transfer function for the three level system. The transient terms are not part of the transfer function, since they are independent of the driver function. Figure 6.4 shows the modelled frequency responses for two sets of relaxation parameters. Curve *a* is the response for a system with a very fast relaxation from level 3 to 2 ($\Gamma_{32} \gg \Gamma_{21}$), and curve *b* is for comparable relaxation rates from 3 to 2 and 2 to 1 ($\Gamma_{32} \approx \Gamma_{21}$). The second pole at -($\Gamma_{32} + \Gamma_{31}$) in Eq. 6.27 lies beyond the range of relevant frequency scanning for a measurement, thus the wing of the pole approximately cancels the Γ_{32} in the numerator (assuming that $\Gamma_{32} \gg \Gamma_{31}$). This is the same response as the simple lowpass filter seen previously.

Curve *b* in Fig. 6.4 is for the situation where the relaxation from level 3 to 2 is comparable to that from 2 to 1. The transfer function is second order, and can be represented by two cascaded lowpass filters. The rolloff in frequency of the *Mod* is steeper, and the phase shift can extend up to 180°, since the phase shifts of the two stages are cumulative.

6.2.4 Multiple Decay Lifetimes

Multi-exponential decay can be modelled as two or more excited state decays, where the photons from each state decay are undiscriminated by the detection system. The transfer function for such a process is simply a weighted sum of the individual transfer functions accounting for each excited state. Thus Eq. 6.6 becomes

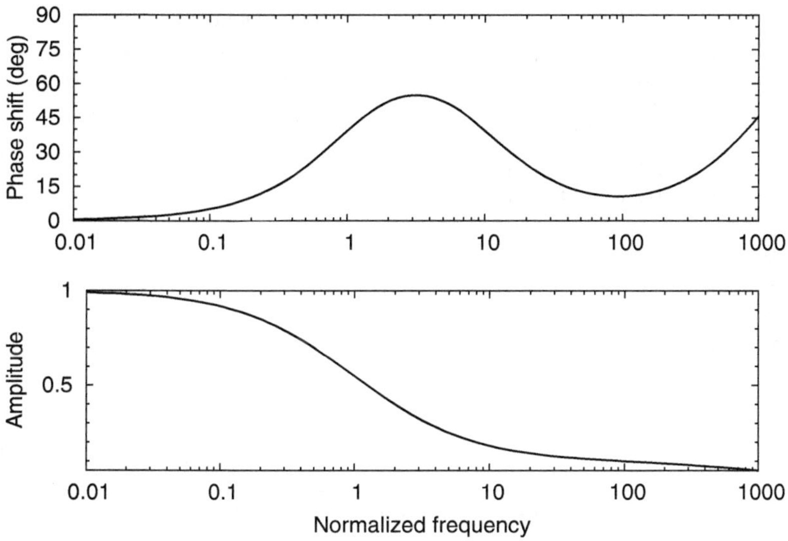

Fig. 6.5 Phase shift and modulation amplitude for double exponential decay undergoing sinusoidal excitation. The two decay rates used here are $\Gamma_1 = 1$ for 90 % of the signal amplitude, and $\Gamma_2 = 1000$ for 10 % of the signal amplitude

$$F(s) = \sum_k \frac{f_k}{s + \Gamma_k}. \tag{6.28}$$

where f_k is the weight of each component of the excited states.

Setting s to $i\omega$, inverting Γ_k to τ_k, and absorbing the extra τ_k of the numerator into the f_k we arrive at:

$$F(i\omega) = \sum_k \frac{f_k}{(1 + i\omega\tau_k)} \tag{6.29}$$

Figure 6.5 shows the phase shift and modulation amplitude of the response for a double exponential decay with sinusoidal driving. The two decay rates are $\Gamma_1 = 1$ ($\tau_1 = 1$), and $\Gamma_2 = 1000$ ($\tau_2 = 0.001$). The frequency scale is normalized to units of the transition rate. The amplitudes have been set so that at very low frequency the amplitude of the slow decay is 0.9, and the amplitude of the fast decay is 0.1. One sees that the phase shift peaks at a value less than 90° and then turns over. This occurs because the slow component dominates the response at low frequency, but for higher frequencies the magnitude of the modulation of the slow component dies away, leaving only the fast component to determine the response, and the phase shift. The *Mod* differs subtly from that of the single exponential case, and it could be difficult to distinguish experimentally.

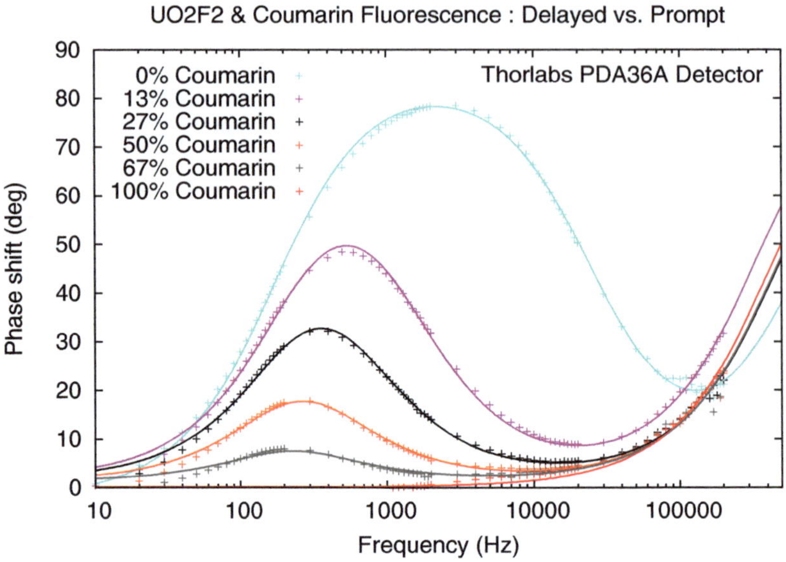

Fig. 6.6 This family of curves for mixtures of fast and slow fluorescence lifetimes was generated from a uranyl fluoride sample and coumarin dye excited by a 365 nm laser diode array, and the fluorescence light collected by a photodiode and measured with a lock-in detector

The behavior of the response phase shift with drive frequency is the most dramatic effect of multiple decay lifetimes. Figure 6.6 shows a family of experimental curves for known mixtures of fast and slow fluorophores. The two samples used were coumarin dye, with a picosecond lifetime, and uranyl fluoride, with a half millisecond lifetime. Varying the mixture of the two samples via f_k reveals that the fast lifetime fluorescence partially masks the full phase shift of the slow component. This is because the amplitude of the slow component diminishes as it is driven faster than its natural decay rate. Its fluorescence becomes a DC light level. The fast component, however, still oscillates in phase with the driver at full amplitude, so its signal tends to dominate the response as the drive frequency is increased. These experimental results were collected using a Nichia NCSU033AE LED array to provide excitation light at 365 nm. The signal waveform to drive the LED array was generated by an Agilent 33250A function generator and boosted by a Tegam 2348 power amplifier. The fluorescence light was filtered by an Andover 550FS40–50 bandpass filter to remove the driver light, and the photodiode detector was a Thorlab PDA36A set at 0 dB gain. The signal was measured by an EG & G Signal Recovery 7260 lock-in detector.

Time domain measurements of multi-component decay lifetimes are fraught with difficulty. In the simplest case of two unknown lifetimes, with unknown relative amplitudes, the fitting function is a sum of decaying exponentials with at least four unknown fitting parameters. Because of correlations between the fitting parameters, finding an optimum fit with lowest χ^2 is extremely difficult, and the error bars

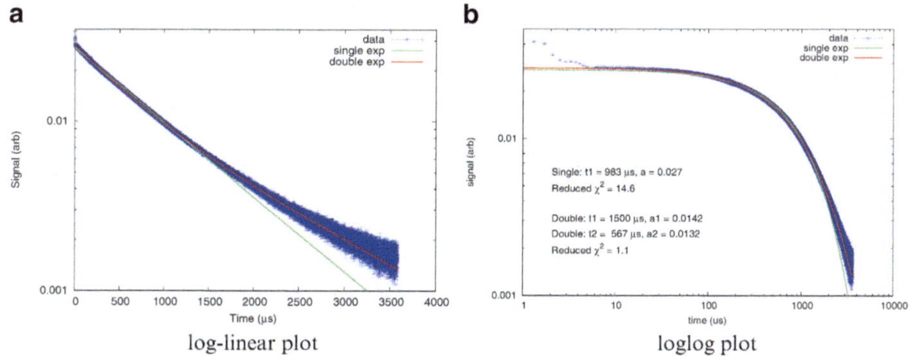

Fig. 6.7 Double exponential decay is a delicate measurement in the time domain. Plot (**a**) shows the standard log-linear plot of the data with fits to single and double decay lifetimes, and plot (**b**) shows the same data and fits on a loglog scale. The log-log plot in plot (**b**) illustrates two typical peculiar and pathological features of the time domain measurement. The features below 10 μs are artifacts from the detector electronics, and although the two fits only start to deviate from one another above 1000 μs, their estimates of the primary lifetime differ by 50 %

derived for the fit parameters tend to be bogus because those correlations are usually ignored. For a humorous admonition against doing this sort of thing please see the discussion on p. 253 in book by Acton [14]. In Fig. 6.7 we show data from a time domain measurement of a UO_2F_2 sample. Fits to single and double lifetimes are shown in the plots, and we note that adding a second lifetime drastically alters the estimate of the primary lifetime. The error bars for the parameters are in the last digit of the estimates, but the two estimates for τ_1 are different by about 50 %. Panel (b) of the figure shows the data and the fits on a loglog scale, and we note two peculiarities. The first is that the two fits only really deviate from one another in the times greater than 1000 μs. The other is that the data look pathological at times below 10 μs. The early time features are due to electronic artifacts from the receiver. The 10 ns laser pulse itself appears in the region from 1 to 2 μs feature, and the shoulder in the data up to 10 μs is probably electronic ringing of the detector that is being driven too fast at too high a gain.

6.2.5 Multiple Frequency Driving Functions

6.2.5.1 Square Wave

Driving fluorescence with a square wave excitation has several advantages over sine wave driving. The circuitry driving a laser diode with square wave driving is a simple switching circuit, and the diode's non-linear response for rapid turn-on and turn-off is not an issue. For sine wave driving, however, maintaining waveform fidelity close to laser threshold is quite tricky, and this requires more elaborate circuitry with feedback. Another advantage of the square wave is that the fundamental term in the harmonic series of the square wave is 27 % more intense than that of a single frequency sine wave operating from the same power supply. Finally, and most importantly, square wave driving allows multiple frequencies to be sampled simultaneously, and this feature can be exploited to perform faster data collections.

The Laplace transform for a positively biased square wave multiplied by the fluorescence low-pass filter transfer function is

$$R(s) = F(s)G(s) = \frac{1}{s+\Gamma}\left(\frac{1}{s} + \frac{1-e^{-s\pi/\omega}}{s\left(1+e^{-s\pi/\omega}\right)}\right) \tag{6.30}$$

where ω is the square wave's angular frequency. The first term in the parentheses, the $1/s$, is the positive bias. It produces an asymptotic exponential drift to the DC offset. The other term in the parentheses is the time dependent response, and when multiplied by the filter the transfer function produces both oscillatory and transient terms. It has three terms where poles occur. One is at the origin, but it is cancelled by a zero, also located at the origin. Another pole is at $-\Gamma$, and this produces a transient turn-on. Finally, an infinite set of poles occur at locations along the imaginary axis wherever the exponential term in the denominator is equal to -1. Those poles occur where

$$e^{s\pi/\omega} = e^{ik\pi} \tag{6.31}$$

for k odd. Since the complex variable s is located in an exponential, the poles of the expression should be evaluated with help from L'Hospital's rule. For the infinite set of poles along the imaginary axis the exponential term introduces a scaling factor of $k\omega/\pi$ in the denominator at each pole. Since k is an odd integer the oscillatory terms consist of odd harmonics weighted by the order of the harmonic. Transforming back to the time domain produces the following response function,

$$r(t) = \frac{1}{\Gamma}\left(1 - \frac{e^{-\Gamma t}}{1+e^{\pi\Gamma/\omega}}\right) + \frac{4}{\pi}\sum_{k=odd}^{\infty} \frac{\frac{\Gamma}{k}sin(k\omega t) - \omega cos(k\omega t)}{(k^2\omega^2 + \Gamma^2)} \tag{6.32}$$

which can be further simplified to

$$r(t) = \frac{1}{\Gamma}\left(1 - \frac{e^{-\Gamma t}}{1+e^{\pi\Gamma/\omega}}\right) + \frac{4}{\pi}\sum_{k=odd}^{\infty} \frac{cos\Phi_k}{\Gamma}\,\frac{sin(k\omega t - \Phi_k)}{k} \tag{6.33}$$

where $tan\Phi_k = k\omega/\Gamma$. The harmonic terms of the original square wave are recognizable with the expected $1/k$ decrease in amplitude with harmonic order. The filtering effect of the fluorescence introduces a phase lag, and further reduces the amplitude of each harmonic order by the cosine of the phase shift for that harmonic order, which is $tan\Phi_k = k\omega/\Gamma$, where k is the harmonic order, and k is odd and positive.

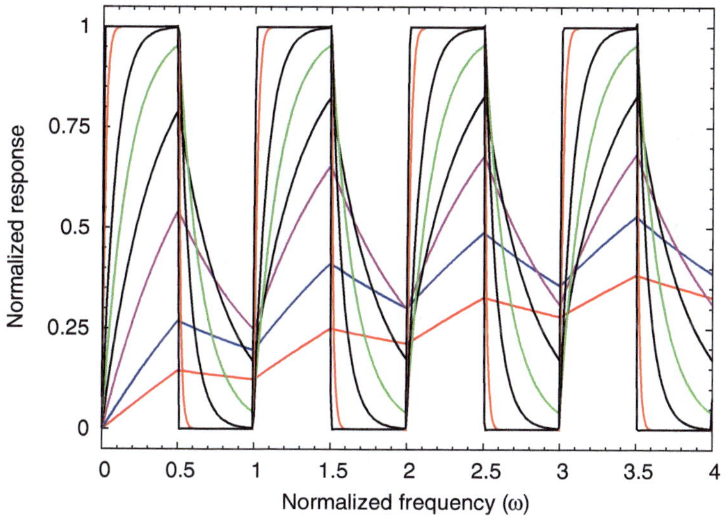

Fig. 6.8 The fluorescence response to a square wave drive signal has a shark's fin appearance at frequencies comparable to the decay rate. The family of curves shown here is for response wave forms ranging in frequency from $\omega = 0.01\Gamma$ to $\omega = 10\Gamma$

When the square wave drive frequency ω is slow compared to the fluorescence decay rate Γ the fluorescence response is a rounded square wave. When the frequency is comparable to the decay rate then the response has a shark's fin appearance. At higher frequencies the response becomes a triangle wave with smaller and smaller amplitude. Also at higher drive frequencies the turn-on transient persists over multiple cycles before settling down and oscillating about a constant DC offset. Examples of the response wave forms for drive frequencies ranging from $\omega = 0.01\Gamma$ to $\omega = 10\Gamma$ are shown in Fig. 6.8.

The harmonic content of a square wave can be exploited to sample the fluorescence response over an extended frequency range using only a few frequency steps of the driving term. In Fig. 6.9 the modelled phase response has been reproduced with only three frequency settings for the driver: $\omega = 0.2$, $\omega = 1$, and $\omega = 9$, where ω is in units of the fluorescence decay rate Γ.

Figure 6.10 shows experimental results from data collected with only three drive frequencies of the square wave, and exploitation of the phase shifts of harmonics up to harmonic 50. By comparison, the data in Fig. 6.6 were collected using a pure sine wave and 80 frequency samples for each curve. The experimental setup for Fig. 6.10 was identical to the setup used to collect the data shown in Fig. 6.6, except that the lock-in detector was replaced by a National Instruments USB-6251 fast waveform digitizer. The time domain signals from the driver and response waveforms were each averaged over a fixed number of cycles, typically 100, and then transformed into the frequency domain by Fast Fourier Transform (FFT). The FFT arrays were then averaged for mean and standard deviations of all complex frequency values

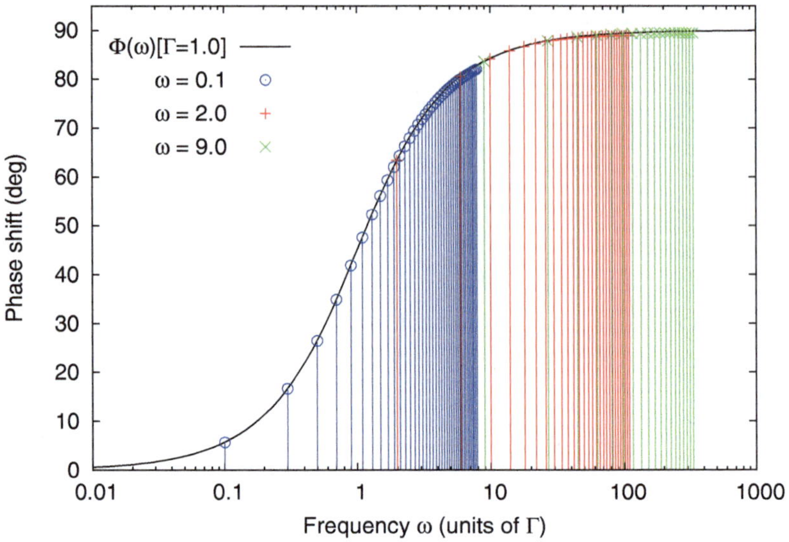

Fig. 6.9 The modelled phase of the fluorescence response is reproduced using only three frequency settings for the driver excitation: $\omega = 0.2$, $\omega = 1$, and $\omega = 9$, where ω is in units of the fluorescence decay rate Γ

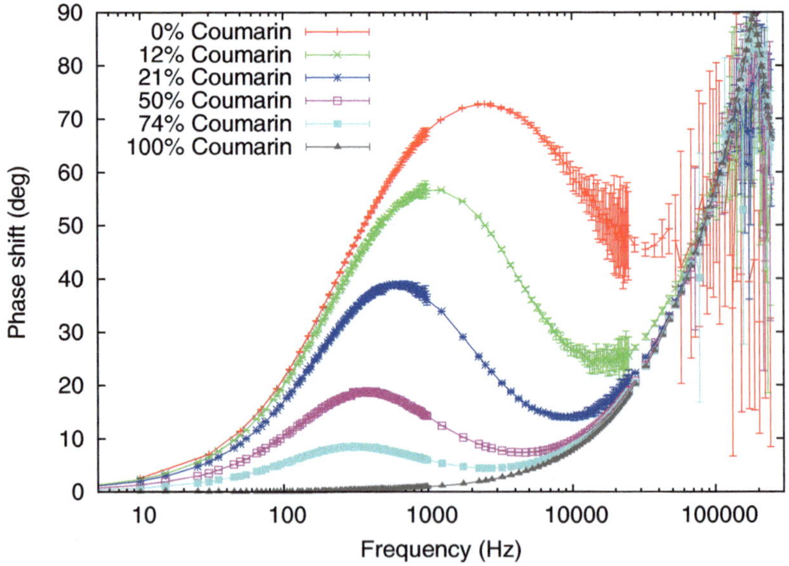

Fig. 6.10 This family of curves was generated using the same uranyl fluoride and coumarin fluorophores, the same laser, and the same photodiode as were used for Fig. 6.6, These curves were constructed using only three drive frequencies: 10 Hz, 250 Hz, and 2.5 kHz. The driver signal in this measurement was a square wave, and the data were collected by a fast digitizer. The data were processed for phase responses of the harmonic components up to 50 harmonics

of the harmonics. The phase delays between the response and driver signals of harmonics up to 50 were computed, and these are shown in Fig. 6.10. The error bars in the plot are the standard deviations of the computed phase shifts. The three frequencies used to reconstruct the family of phase shift curves were 10 Hz, 250 Hz, and 2.5 kHz. The various curves in the figure are for different mixtures of fast and slow lifetime fluorophores. The data collection time for each curve in Fig. 6.10, using three drive frequencies, was only a few seconds. By contrast, the data collection time in Fig. 6.6 required 6 min for each curve.

6.2.5.2 Sawtooth Wave

A sawtooth wave generator is another relatively easy driving circuit to implement in hardware, and it has the advantage that all even and odd harmonics are present in the wave form. Thus the fluorescence signal from a sawtooth excitation term contains twice the number of useful frequency samplings than that of a square wave. However, the harmonic terms alternate in sign, and thus the phases of the harmonics alternate from 0° to 180° from one harmonic to the next. Those phase shifts would be superimposed on the response phase shifts of the fluorescence, and the phase data would appear to be scattered or noisy. This is not a desirable situation for measurements in low signal-to-noise situations. Another draw back of the sawtooth wave is that the driving laser must respond linearly from lasing threshold up to full power, which requires complicated feedback circuitry of the laser driver.

6.2.5.3 Frequency Comb

The square wave has harmonic intensities proportional to the inverse of the harmonic order. Consequently, the power of the harmonics weakens rapidly as the harmonics probe higher frequencies. This situation is not desirable, since the *Mod* of the fluorescence response also falls off at higher frequency. Ideally, the driver's harmonic terms would have uniform intensity, or perhaps increasing intensities with harmonic order. An arbitrary waveform generator can synthesize any desired harmonic content up to the instrument's bandwidth limit, but the problem of linearity of the excitation laser's response would compromise performance. Also, incorporating an arbitrary waveform generator into an apparatus would add to the cost and complexity of the instrument.

A better alternative to an arbitrary waveform excitation driver is actually a rather simple solution. It is the Shah function, or the Dirac Comb. These wave forms are simple pulse trains. The Dirac comb is infinitely thin and intense, but it integrates to a finite value, and the Shah function is a discrete domain realization of the Dirac comb. The harmonic content of the Dirac comb has all harmonic orders of equal intensity, with all orders in phase with each other, thus the Dirac comb can also be called a frequency comb, or a phase comb. The experimental realization of the frequency comb is simply a short pulse laser with an adjustable repetition rate. The repetition rate is the fundamental frequency of the driver, and since the pulses are sharp, many harmonic orders of that repetition rate are present in the signal. In the limit that the pulse train is infinitely long and the individual pulses infinitely sharp the frequency comb contains an infinite number of harmonic orders.

A realization of the frequency comb can be approached by considering a periodic rectangle wave. If we let the duration of an individual pulse be T_1, and the pulse train repetition period be T, then the Laplace transform of a single rectangular pulse is

$$G_p(s) = \int_0^{T_1} e^{-st}\,\mathrm{d}t = \frac{1 - e^{-sT_1}}{s} \tag{6.34}$$

and to turn this into a pulse train we successively use the time delay feature of the Laplace transform, which is successive multiplication by e^{-sT}. The transform of the pulse train is therefore

$$G_T(s) = \sum_{k=0}^{\infty} \frac{1 - e^{-sT_1}}{s} e^{-sTk} = \frac{1 - e^{-sT_1}}{s(1 - e^{-sT})} \tag{6.35}$$

If we next define the duty cycle of the pulse train to be $D = T_1/T$ then the transformed function becomes

$$G_T(s) = \frac{1 - e^{-sTD}}{s(1 - e^{-sT})}. \tag{6.36}$$

This function has poles along the imaginary axis at $s = i2\pi k/T$, where k is an integer representing the harmonic order. The Residue theorem and the Bromwich integral have factors of $i2\pi$ that cancel, so the Residue at each pole is

$$Res = \frac{1 - e^{i2\pi kD}}{i2\pi k}. \tag{6.37}$$

The *Mod* of the function at each pole is therefore

$$Mod = D\sqrt{\left(\frac{sin(k\pi D)}{k\pi D}\right)^2} = D\left|\frac{sin\varphi_k}{\varphi_k}\right| \tag{6.38}$$

and the phase shift for each harmonic order is $\varphi_k = k\pi D$. The *Mod* of the transform of the pulse train is a frequency comb with an envelope function of $sin(x)/x$ (also called $sinc(x)$), where $x = k\pi D$, and wherever kD is integer valued the harmonic peak has zero amplitude. The phase shift for each harmonic peak is linear in the harmonic order with a slope of πD. In the special case of a square wave, where the duty cycle is $D = 1/2$, all even harmonics are null, and all of the phase shifts are identical, modulo $\pi/2$. For other duty cycle values, a harmonic peak is nulled whenever kD becomes an integer.

In the time domain the harmonic representation of the pulse train driver is

$$g(t) = D + 2D\sum_{k=1}^{\infty} \frac{sin\varphi_k}{\varphi_k} cos(k\omega t - \varphi_k) \tag{6.39}$$

where $\varphi_k = \pi kD$. Looking at the special case of $D = 1/2$, we see that the $\varphi_k = k\pi/2$ phases in both the sine and cosines of Eq. 6.39 yield results of ± 1 that cancel each other, so Eq. 6.39 reduces to the familiar harmonic expression of a positively biased square wave.

Looking at other duty cycles we find an interesting result. For example, with a duty cycle of $D = 1/10$ the harmonic peaks lie under the $sin(x)/x$ envelope where $x = \pi k/10$, and every tenth harmonic peak is null. This is illustrated in Fig. 6.11. The amplitudes of the harmonic peaks remain significant out to high order, thus the frequency comb allows many sample points of the fluorescence response function to be measured simultaneously with a single drive frequency. The phase of each harmonic is linearly delayed with the harmonic order. Figure 6.12 shows the phase delays versus harmonic order for a pulse train with a duty cycle of 1/10. One sees that every 10'th harmonic is missing, and the phase shifts of the harmonic peaks are $\Phi_k = k\pi/10$. The sawtooth appearance of the phase shifts in Fig. 6.12 is due to the modulo $\pm\pi/2$ convention of representing phase shifts of periodic signals.

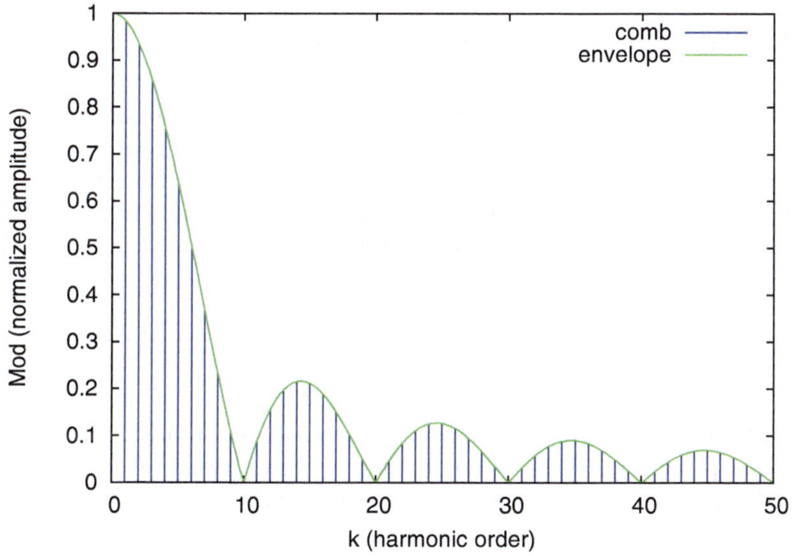

Fig. 6.11 The *Mod*, or the normalized power spectrum, for a 10 % duty cycle pulse train has nulls at every tenth harmonic. The amplitudes of the harmonic peaks follow an envelope function of $abs(sin(x)/x)$, with $x = \pi kD$, where k is the harmonic order and $D = 1/10$ is the duty cycle of the pulse train

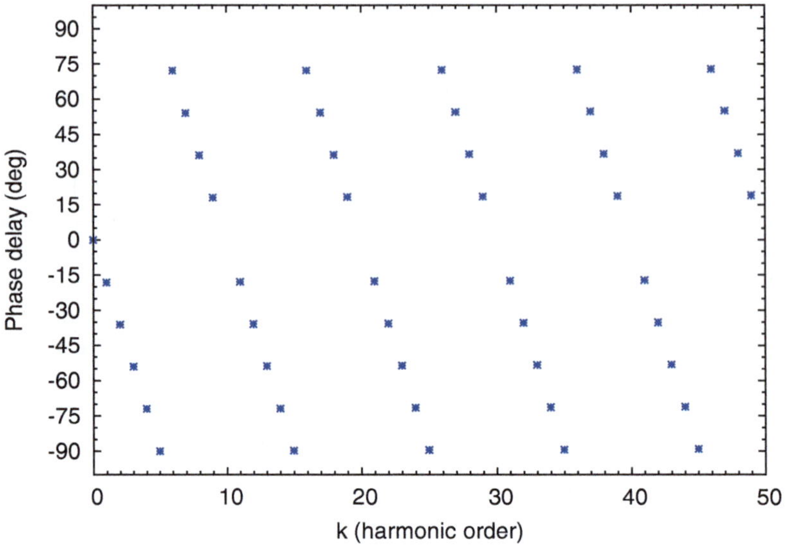

Fig. 6.12 The phase delays for the harmonic peaks of a 10 % duty cycle pulse train decrement linearly with a slope of 18° ($\pi/10$), and a null occurs at every tenth harmonic. The sawtooth appearance of the graph is due to the convention of mapping phase shifts into the range $[-\pi/2, \pi/2]$ to be consistent with the arc tangent function

6.2.6 Fluorescence From a Pulse Train

Having worked out the amplitude and phase offsets of the harmonic content of a pulse train driver, we can now proceed to analyze the low pass filtering effects from the fluorescence response. Multiplying the transfer function from Eq. 6.6 for the single fluorescence lifetime by the transform of the pulse train driver from Eq. 6.35 gives the frequency domain response function

$$R(s) = F(s)G(s) = \frac{1}{s+\Gamma}\sum_{k=0}^{\infty}\frac{1-e^{-sT_1}}{s}e^{-sTk} = \frac{1}{s+\Gamma}\frac{1-e^{-sT_1}}{s(1-e^{-sT})} \qquad (6.40)$$

To transform this into the time domain we evaluate the Bromwich integral as we have done previously. Using D for the duty cycle to replace T_1, as in Eqs. 6.36 and 6.39, the time domain function for the driver function is

$$r(t) = D + D\sum_{k=1}^{\infty}\left(\frac{sin\varphi_k}{\varphi_k}\right) cos(k\omega t - \varphi_k) \qquad (6.41)$$

and the time domain function for the fluorescence response is

$$r(t) = \frac{D}{\Gamma}\left(1 - \frac{1 - e^{(2\pi\Gamma D/\omega)}}{D\left(1 - e^{(2\pi\Gamma/\omega)}\right)} e^{-\Gamma t}\right)$$
$$+\frac{2D}{\Gamma}\sum_{k=1}^{\infty}\left(\frac{sin\varphi_k}{\varphi_k}\right) cos\Phi_{k\Gamma}\, cos(k\omega t - \varphi_k - \Phi_{k\Gamma}) \qquad (6.42)$$

where $\varphi_k = k\pi D$ is the initial phase offset of the $k'th$ harmonic of the driving pulse train, and $sin\varphi_k/\varphi_k$ is its amplitude (recall that $D = T_1/T$ is the duty cycle of the pulse train). The term $\Phi_{k\Gamma} = tan^{-1}(k\omega/\Gamma)$ is the phase shift introduced by the fluorescence lifetime, and $cos\Phi_{k\Gamma}/\Gamma$ is the amplitude, or *Mod*, of the fluorescence term. In the special case of the square wave, $D = 1/2$, the term $sin\varphi_k/\varphi_k$ is simply $1/\pi k$ for odd k, and zero for even k (the alternating signs of the sinc function are cancelled by the φ_k term in the cosine), thus all the even terms in k drop out of the sum.

Using a pulse train with a duty cycle different from 1/2 is somewhat more complicated than the special case of the square wave, but the amplitudes and phase shifts of the various harmonic terms of the response are easily determined, as long as one can keep track of the initial phases and amplitudes of the driver function's harmonic terms.

6.2.7 Nonlinear Detection Techniques

Nonlinear techniques of detection commonly used in spectroscopy, such as homodyne, heterodyne, and beat frequency detection require care when attempting an analysis with the analog filter approach. Homodyne and heterodyne techniques utilize either gain modulation, or signal mixing with a local oscillator, to remap phase and amplitude information into a lower frequency interval [15]. The beat frequency technique simply switches the detector gain on and off in order to perform an equivalent mix-down remapping of phase and amplitude. These techniques can be employed whenever an element in the detector chain does not have sufficient bandwidth to measure the response signal at the driver frequencies. Consequently, the response signal is mixed down into the appropriate frequency interval for measurement.

Nonlinear detection multiplies signals in the time domain, and unfortunately, the multiplication cannot be carried naively into the frequency domain. Recall that a multiplication in the frequency domain is equivalent to a convolution in the time domain, and vice versa. One approach to handling the mixing problem is simply to perform the convolution in the frequency domain, but this approach is awkward, since replacing the convolution integral with a multiplication is the primary reason for working in a frequency domain in the first place. The other approach is to work in the frequency domain up to the point of the mixing operation, then transform back into the time domain for the multiplication, and then transform back again into the frequency domain, if necessary, to finish the analysis. This sounds clunky, inefficient, and error prone, but often it can be done intuitively without difficulty. The trick is to remember a trigonometric identity showing that multiplication of a pair of sine or cosine functions is equivalent to the sum of sines and cosines whose frequencies are the sums and differences of the original frequencies. The trick works regardless of which transform one wishes to utilize. In other words, it works for Laplace transforms the same as it does for Fourier transforms.

Using the trigonometric identity

$$cos\alpha \, cos\beta = \frac{cos(\alpha - \beta) + cos(\alpha + \beta)}{2}, \tag{6.43}$$

we can investigate what happens when a periodic signal is modulated. Setting different frequencies and initial phase offsets in the above identity gives the following:

$$A\,cos(\omega_a t + \varphi_a)\,B\,cos(\omega_b + \varphi_b) = \frac{AB}{2}\left[cos(\omega_{\_}t + \Delta\varphi_{\_}) + cos(\omega_{+}t + \Delta\varphi_{+})\right] \tag{6.44}$$

where $\omega_{\pm} = \omega_a \pm \omega_b$, and $\Delta\varphi_{\pm} = \varphi_a \pm \varphi_b$. Now, if we call $A\,cos(\omega_a t + \varphi_a)$ the signal, and $B\,cos(\omega_b + \varphi_b)$ the modulation, then we see that a periodic gain modulation is equivalent to the summing of two other periodic functions of different frequencies, which are the sum and difference of the original frequencies. The original signal has thus been split into upper and lower side bands, with the energy being equally divided between the bands. The difference frequency is usually called the beat frequency. The modulating term (often called the local oscillator) multiplies the signal amplitude, therefore the strength of the side bands can be increased or decreased by adjusting the modulation amplitude. In this particular example, the input frequencies are suppressed in the output, and only the sum and difference frequencies appear. In electronics this type of mixing is called double balanced mixing. By contrast, if one the original frequencies is present in the output then the mixer is called a single balanced mixer. If both input frequencies are present in the output then the mixer is called an unbalanced mixer.

Unbalanced mixing sums input signals with their product. Often the terms "mixing" and "beating" are used to distinguish between multiplying and adding, and a Fourier transform of the signal can make this clear. If the original frequencies of the input are present in the output then the operation is a beating, whereas if components appear at the upper and lower side band frequencies then the operation is a mixing. Most realistic electronic mixing components do both operations.

Figure 6.13 shows an example of beating. For this figure two cosine waves of close, but slightly different, frequencies have been summed. The time domain plot

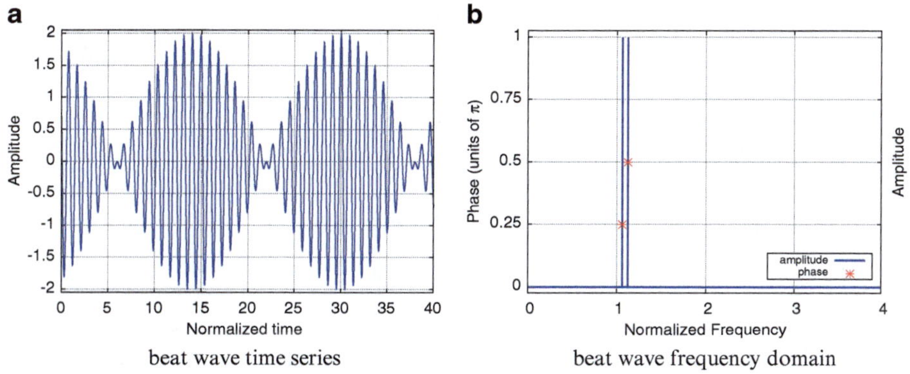

Fig. 6.13 Beating two waves together is the summing of the two waves. Although the beat frequency can clearly be seen in the envelope of the waveform shown in panel (**a**) there is no energy at the beat frequency, as is evident by the Fourier transform of this waveform shown in panel (**b**)

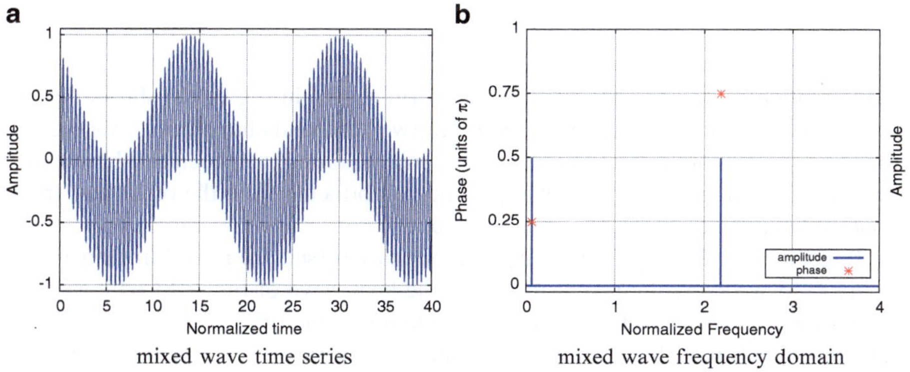

Fig. 6.14 Mixing two waves together is the multiplication of the two waves. The beat frequency, which is actually the difference frequency, is visible in panel (**a**), but its character is not the same as the as when beating the two waves together as in Fig. 6.13. The Fourier transform of this mixed waveform shown in panel (**b**) has energy at both the sum and difference frequency sidebands, and the energy has been split between the sidebands. Phase offsets are added in the upper sideband, and subtracted in the lower sideband

of the signal, shown in Fig. 6.13a, clearly shows a beat frequency in the amplitude envelope, but the Fourier transform, shown in Fig. 6.13b, has only components at the original two frequencies. There is no energy at the beat frequency, meaning that if this signal is passed through a low-pass filter to remove the original frequencies, then nothing survives at the beat frequency. The two cosine waves were given initial phase offsets of $\pi/2$ and $\pi/4$, and those phases were preserved by the addition, or beating, of the two waves.

Figure 6.14 shows an example of mixing, where the same cosine waves of Fig. 6.13 have been multiplied, rather than added. The beat frequency is visible with

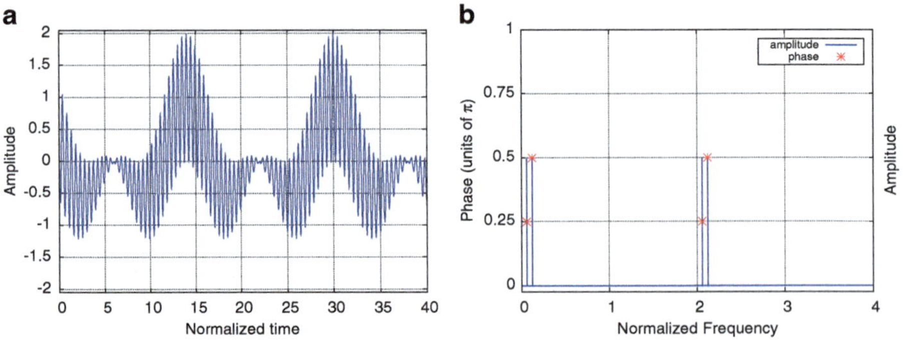

mixed beat wave time series

mixed beat wave frequency domain

Fig. 6.15 In panel (**a**) is the time domain beat wave of Fig. 6.13 mixed with a cosine wave with a phase offset of zero. In the Fourier transform shown in panel (**b**) the phase offsets of two components of the original beat wave are preserved when the mixing maps the information into the upper and lower sidebands. If the mixing cosine wave had an phase offset then it the phase have also been mapped into the sidebands (added for the upper sideband, and subtracted for the lower sideband)

a different character, and the Fourier transform shown in Fig. 6.14b shows that the original two frequencies are gone, and that two sidebands have been created. The lower sideband is at the beat frequency. The initial phases have been combined, such that in the lower sideband one phase has been subtracted from the other, and in the upper sideband the phases have been added.

To get a better feeling for the mapping of phase offsets from a mixing operation, in Fig. 6.15 we show the result of mixing the beat wave of Fig. 6.13 with a cosine wave of zero phase offset. The individual phases of the original two cosine waves (i.e., the signal) have been preserved by the mixing. However, if the mix wave had a non-zero phase offset then the phase of the mix wave would have been added to the other phases in the upper sideband, and subtracted from the others in the lower sideband. As long as a frequency component of the result of the mixing is properly identified as either an upper or lower sideband, and the initial phase of the mix wave is known, then the phases of the original signal components can be inferred. This becomes important when chopping a high frequency signal to beat it down for a low frequency measurement.

6.2.8 Chopping a Cosine Wave

Chopping a signal, or switching a detector on and off, does not, at first, appear to be a non-linear operation, but it really is. Mixing occurs, and signals of interest can be mapped into other frequency ranges where measurements are more easily made. Chopping a waveform is equivalent to multiplying the waveform by a pulse train whose amplitude alternates between 0 and 1. We can use Eq. 6.39 for the chopper and multiply it by the wave we wish to chop. Using a simple cosine for an example of the drive signal we have the following time domain representation of the chopped response signal,

$$g(t) = Dcos(\omega_d t + \varphi_d) + 2D \sum_{m=1}^{\infty} \frac{sin\varphi_m}{\varphi_m} cos(m\omega_c t - \varphi_m) cos(\omega_d t + \varphi_d). \quad (6.45)$$

where the subscript d represents the driver function, the subscript c represents the chopper function. The variable $\varphi_m = m\pi D$, where m is an integer for the harmonic order of the chopper function. The first term on the right of Eq. 6.45 is the original cosine signal mixed with the constant (DC) term of the chopper, and it remains at the original signal frequency. The second term is a product of cosines between the original signal term and the harmonics of the chopper function. As previously, we can replace the product of cosines with upper and lower sideband cosines. Focussing on the sideband terms (temporarily ignoring the mixing with the DC term of the chopper) we have

$$g(t) = D \sum_{m=1}^{\infty} \frac{sin\varphi_m}{\varphi_m} (cos((\omega_d - m\omega_c)t + \varphi_m + \varphi_d) + cos((\omega_d + m\omega_c)t - \varphi_m + \varphi_d)) \quad (6.46)$$

The result is that each harmonic of the chopper wave now has upper lower sidebands. The phase of the original cosine wave can be recovered by removing the phase of the chopper's harmonic, which has been added for the upper sideband, and subtracted for the lower sideband. In the special case of a square wave the phase of the chopper harmonic φ_m is always $\pi/2$. Doing the same for chopping a sum of cosines, we have in Fig. 6.16 the beat wave from Fig. 6.13 after it has been chopped at a frequency slightly different from the frequencies of the original signal components, and we see that the phases of the original two cosine waves behave as expected for the sidebands of the harmonic series of the chopper. The most useful peaks in the mixed signals of Fig. 6.16 are at the lowest order sideband, close to zero frequency. Since the signal wave and the chopper are approximately of the same normalized frequency of 1 all of the other harmonic sidebands fall at approximately even frequencies.

The pair of peaks at frequency 1 is the sideband derived from the DC offset of the chopper wave, which we ignored above. If the chopper had been bipolar (i.e., if instead of zeroing the signal during the "off" phase, it had inverted the signal, then the pair of peaks at frequency 1 would have been absent).

6.2.9 Chopping a Pulse Train

To represent a chopped pulse train we use Eq. 6.42 for the driver pulse train, and Eq. 6.39, at a slightly different frequency, to represent the chop wave. Multiplying the two together is a harmonic representation of the chopping. The term-by-term

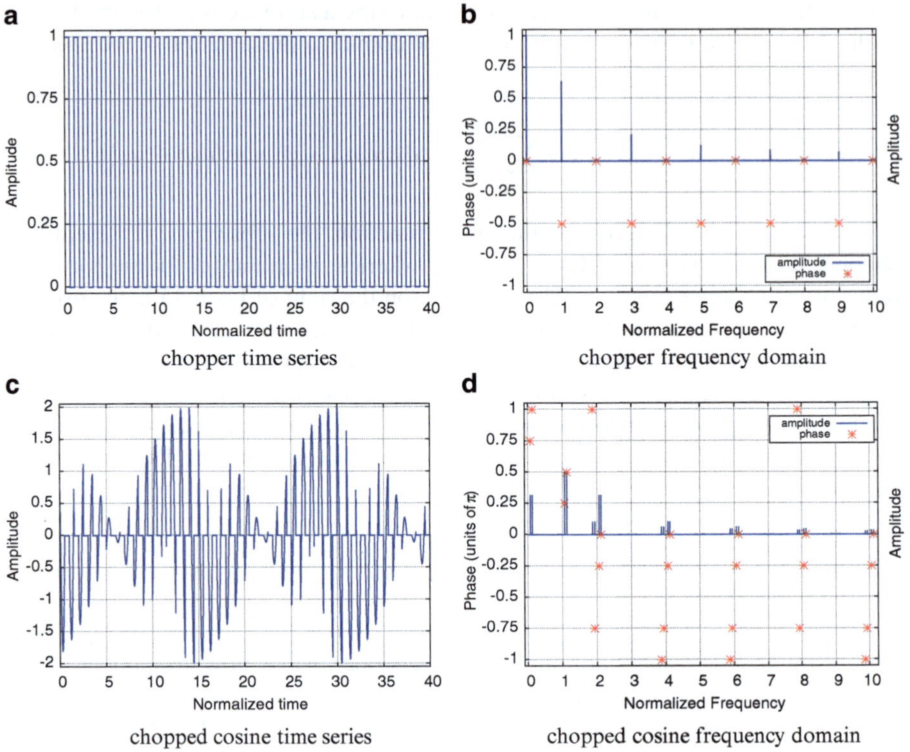

Fig. 6.16 Chopping a signal is a mixing operation, and here we see the result of chopping the beat wave of Fig. 6.13. The chopper signal is shown in panels (**a**) and (**b**) for the time domain signal and its Fourier transform. Panels (**c**) and (**d**) show the time domain waveform and the Fourier transform of the chopped beat drive signal. Upper and lower sidebands appear for each of the chopper's harmonic components. The phases (*red dots* in the figures) of the individual sideband peaks can be mapped back to the phases of the original signals by accounting for the phase offsets of the chopper's harmonics

products of multiplying the two infinite sums yields a dizzying array of sidebands, but if we look at an individual product term we can begin to sort out the identities, amplitudes, and phases of the terms. Ignoring the turn-on transients, the steady-state chopped response wave is

$$r_c(t) = \frac{4D_cD_d}{\Gamma} \sum_{k,m=1}^{\infty} \left(\frac{sin\varphi_{c_m}}{\varphi_{c_m}}\right)\left(\frac{sin\varphi_{d_k}}{\varphi_{d_k}}\right) cos\Phi_{d_k\Gamma}$$
$$\times\, cos(m\omega_c t - \varphi_{c_m})\, cos(k\omega_d t - \varphi_{d_k} - \Phi_{d_k\Gamma}) \qquad (6.47)$$

where we have added the subscripts *c* and *d* refer to the chopper and driver wave forms, respectively. The sum is understood to be a double sum with *k* and *m* summed independently. The variable *k* is the harmonic order of the driver function, and *m* is the harmonic order of the chopper function. Each product of cosines above can be replaced by a pair of sidebands for each harmonic pairing (m, k).

$$r_c(t) = \frac{2D_cD_d}{\Gamma} \sum_{k,m=1}^{\infty} \left(\frac{sin\varphi_{c_m}}{\varphi_{c_m}}\right)\left(\frac{sin\varphi_{d_k}}{\varphi_{d_k}}\right) cos\Phi_{d_k\Gamma}$$
$$\times \left(cos(\omega_{-}t - \varphi_{-} - \Phi_{d_k\Gamma}) + cos(\omega_{+}t - \varphi_{+} - \Phi_{d_k\Gamma})\right) \tag{6.48}$$

where $\omega_{\pm} = k\omega_d \pm m\omega_c$, and $\varphi_{\pm} = \varphi_{d_k} \pm \varphi_{c_m}$. If the duty cycles are different then the bookkeeping can be a bit tedious, but not intractable. Usually, however, the chopper and driver waves have the same duty cycle of $D = 1/2$ (i.e., both are square waves), and the frequencies are very close to one another, so that the lower sideband is mapped into a rather low frequency range for convenient measurement.

If we look at the frequency difference terms where $m = k$ (i.e., the lowest frequency harmonic comb), then that subseries is

$$r_{c_{bb}}(t) = 2D_cD_d \sum_{k=1}^{\infty} \left(\frac{sin\varphi_{c_k}}{\varphi_{c_k}}\right)\left(\frac{sin\varphi_{d_k}}{\varphi_{d_k}}\right)\left(\frac{cos\Phi_{d_k\Gamma}}{\Gamma}\right)$$
$$\times\, cos(k(\omega_d - \omega_c)t - \varphi_{d_k} + \varphi_{c_k} - \Phi_{d_k\Gamma}) \tag{6.49}$$

where we have added the subscript *bb* to indicate the response wave's lowest subseries, or its baseband. If we next make the duty cycles of the driver and chopper equal, then further simplification results,

$$r_{c_{bb}}(t) = \frac{2}{\pi^2} \sum_{k=odd}^{\infty} \left(\frac{cos\Phi_{d_k\Gamma}}{\Gamma}\right) \frac{cos(k(\omega_d - \omega_c)t - \Phi_{d_k\Gamma})}{k^2} \tag{6.50}$$

The baseband signal preserves the phase of the fluorescence response, but the amplitude terms have a $1/k^2$ dependence instead of $1/k$. If the signs of the sum alternated then the baseband waveform would appear to be a triangle wave, but those signs do not alternate. The response has a more complicated appearance, as we will see below. The treatment of the other sidebands is similar, and all of the sideband structures can be identified and analyzed if desired. The other sidebands have more complicated phase and amplitude dependences, but sorting them out is simply a matter of careful bookkeeping.

Time and frequency domain signals of a chopped square wave are shown in Fig. 6.17. Notice that the time domain signal has an accordion appearance. The beat frequency is not apparent in the amplitude, but it shows up as a modulation of the duty cycle of the wave form. The Fourier transform of that accordion wave, in Fig. 6.17b, shows multiple sideband structures. This frequency sideband starting at the beat frequency is the base band that we have been discussing, and the expected $1/k^2$ slope of the amplitude envelope for the harmonic peaks is plotted in red. The corresponding sideband structure for $m = k \pm 1$ is also shown in the plot with the magenta curve tracing its amplitude envelope. The upper sidebands have a frequency spacing of approximately 2 in normalized frequency units, and the fine structure of the frequency plot has separations in units of the beat frequency. We can gain more insight into the sideband structure at normalized frequency 2, and other frequencies, by letting $\omega_d = \omega_C + \delta_\omega$, where δ_ω is the beat frequency. Thus, from Eq. 6.48 for the lower sideband we have

$$cos(k\omega_d t - m\omega_c t - \varphi_{d_k} + \varphi_{c_m} - \Phi_{d_k\Gamma}) = cos((k-m)\omega_c t + k\delta_\omega t - \varphi_{d_k} + \varphi_{c_m} - \Phi_{d_k\Gamma}) \tag{6.51}$$

Recall that if the duty cycles of the driver and chopper are the same then $\varphi_{c_m} - \varphi_{d_k} = (m-k)\pi D$. Also, if $D = 1/2$ then the phase shifts of the driver and chopper simply convert cosines into sines for odd $(m-k)$ differences and are irrelevant for even differences. Consequently, the rightward extending harmonic band starting at normalized frequency 2 is the series for $k = m + 1$.

Other interesting features of Fig. 6.17b are the peaks at the odd frequencies 1,3,5, etc. Since the fundamental frequencies of both the driver and the chopper are approximately 1, the mixing of their harmonics should produce even frequencies at all orders. So what are these features at odd frequencies? They are not beating features, but they come from mixing of the harmonic terms of one wave with the DC offset term of the other wave. If the two square waves were without DC offsets then the features at the odd numbered harmonics would be absent from the spectrum.

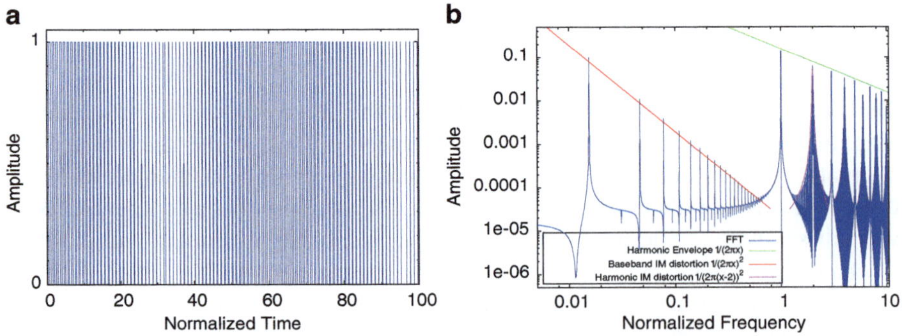

chopped pulse train time domain

chopped pulse train frequency domain

Fig. 6.17 Chopping a square wave does not show a beat wave in the amplitude envelope, as can be seen in plot (**a**), but it does show a modulation at the beat frequency of the apparent duty cycle. The Fourier transform in plot (**b**) confirms that a base band signal has been created at the beat frequency

6.2.10 Integrating Chopped Pulse Trains

Once a waveform has been chopped to produce the subharmonic baseband it makes sense to remove the higher order mixing terms to avoid aliasing effects in the measurements. This is often done by low pass filtering, but integrating the signal during the chopper's "on" phase can accomplish the same thing. In the measurement described by DiBenedetto [16] a camera was used to integrate photons from a microchannel plate intensifier switched on and off at a frequency slightly different from that of the laser driver. The frame rate of the camera was not fast enough to sample data at the frequencies necessary for measuring the fluorescence lifetime of their sample, so toggling the intensifier mixed the signal down to a frequency low enough for the camera to handle. The trigger to the camera was derived from the chopper signal, so integration period was synchronized with the chopper. The chosen beat frequency was 0.5 Hz, which allowed the camera to sample at least 40 points of the mixed down base band signal.

Not surprisingly, integrating the mixed signal introduces not only another amplitude contribution, but also an extra phase correction. Looking at a single harmonic term from Eq. 6.48 we perform an integration over half a cycle of the chopper wave.

$$\begin{aligned}
&\int_{nT_c}^{nT_c+T_c/2} Acos\left(\omega_{\pm}t+\varphi_{\pm}\right)\,\mathrm{d}t \\
&= \frac{A}{\omega_{\pm}}\left(sin((n+1/2)\omega_{\pm}T_c-\varphi_{\pm})-sin(n\omega_{\pm}T_c-\varphi_{\pm})\right) \\
&= \frac{2A}{\omega_{\pm}}cos\left(n\omega_{\pm}T_c+\omega_{\pm}T_c/4-\varphi_{\pm}\right)sin\left(\omega_{\pm}T_c/4\right) \\
&= \frac{AT_c}{2}\left(\frac{sin(\Omega_{\pm})}{\Omega_{\pm}}\right)cos\left(n\omega_{\pm}T_c+\Omega_{\pm}-\varphi_{\pm}\right)
\end{aligned} \tag{6.52}$$

where $T_c = 2\pi/\omega_c$, and

$$\Omega_{\pm} = \frac{\omega_{\pm}T_c}{4} = \frac{(k\omega_d \pm m\omega_c)}{\omega_c}\frac{\pi}{2} \tag{6.53}$$

Now if we let $\omega_d = \omega_c + \delta$, where δ is the beat frequency, then

$$\Omega_{\pm} = \frac{\pi}{2}\left((k \pm m) + k\frac{\delta}{\omega_c}\right) \tag{6.54}$$

Recall that k is the harmonic order of the driver, and m is the harmonic order of the chopper. To consider the base band we let $m = k$ and use the minus sign in the definition of $\Omega_{\pm}$ from Eq. 6.54. Consequently, for $m = k$, $\Omega_{-} = (k\pi/2)(\delta/\omega_c)$. This phase shift correction should normally be small, but at lower driving frequencies it may become significant.

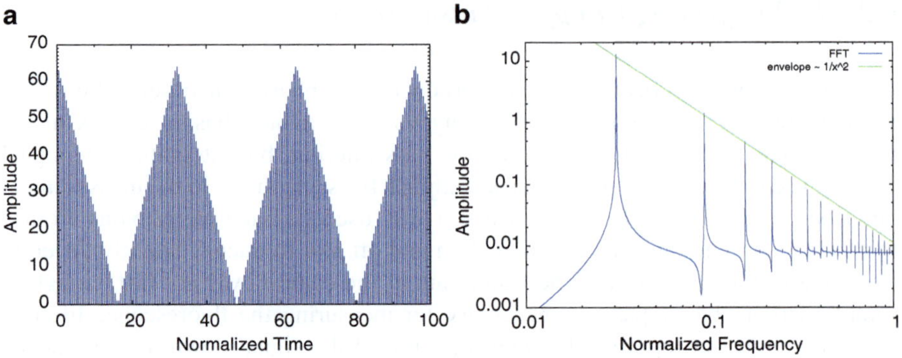

Fig. 6.18 Integrating a chopped square wave using a chopper at a slightly different frequency produces the triangle wave envelope as the signal and chopper waveforms come in and out of phase with each other. The frequency domain spectrum shows that the signal has been mixed down to base band, and that the upper bands have been eliminated. (**a**) Integrated chopped pulse train: time domain. (**b**) Integrated chopped pulse train: frequency domain

For the harmonic terms beyond the base band, $4n\Omega_{\pm}$ has multiple factors of 2π that modulo back to the base band. The extra factors of $(k \pm m)\pi/2$ in the $sin(\Omega_{\pm})/\Omega_{\pm}$ are correction terms that alternate in sign and damp out rapidly. The integration is thus a combination of aliasing and damping. In practice, it acts as a filter and produces in an additional phase shift related to the chopping frequency.

To visualize these results, Fig. 6.18 shows the time domain signal of a chopped and integrated square wave and its Fourier transform. The simulation data for this figure was prepared by simply chopping and summing a square wave and running it through an FFT, and the harmonic analysis that we have just discussed was not used for the figure. Thus this figure serves as a sanity check for the analysis. We note the triangle wave appearance of the integrated data, the falling off at second order of the harmonic peaks, and that the peaks occur only at odd harmonics, as expected.

The article by DiBenedetto [16] also discusses a quadrature phase detection (I/Q) method for measuring the beat wave. The I/Q method is simply an example of homodyne detection, in which the phase shifted fluorescence signal is mixed with a local oscillator signal for I, and with the quadrature of the local oscillator signal (which $\pi/2$ phase shifted) for Q. In this case the local oscillator was the waveform driving the laser. Looking at a single frequency cosine wave, the signal is $Acos(\omega t - \Phi)$, and the local oscillator is $Bcos(\omega t)$ for the in-phase component and $Bsin(\omega t)$ for the quadrature component. The I and Q components from the mixing are

$$I = Bcos(\omega t)Acos(\omega t - \Phi) = \frac{AB}{2}\left(cos(\Phi) + cos(2\omega t - \Phi)\right) \tag{6.55}$$

$$Q = Bsin(\omega t)Acos(\omega t + \Phi) = \frac{AB}{2}\left(sin(\Phi) + cos(2\omega t - \Phi)\right) \tag{6.56}$$

Passing the I and Q signals through a lowpass filter removes the double frequency components, so we can recover the original signal, such that the phase shift is $\Phi = tan^{-1}(Q/I)$, and the amplitude is $A = 2\sqrt{I^2 + Q^2}/B$.

6.2.11 Standoff Detection Phase Shifts

An easily overlooked contribution to a phase lag is the transit time of the light. In a laboratory setting the amount of time for light to travel from the excitation source to the fluorescent target and back to the receiver is usually negligible compared to the cycle time of the waveform, however, in a remote sensing application this is not the case. For example, if a is target located at 1 km then the round trip transit time for the light is 6.67 μs. This corresponds to a frequency of 150 MHz, meaning that if the driver is operating at 150 MHz then the phase lag from the transit time alone is a full 2π. Since the time lag is constant the phase lag is variable as the drive frequency changes. At 150 MHz the phase lag is 360°, and at about 4 MHz the phase lag is about 10°. So at each frequency the transit time phase lag due to the transit time needs to be properly accounted for in the data analysis.

6.2.12 Numerical Methods for Simulating the Frequency Comb

The numerical simulations for most of the figures were performed with Gnuplot's [17] "set table" function and a buffer size of 2^{22} points. Conversion into the frequency domain was computed by the complex Fast Fourier Transform (FFT) function `gsl_fft_complex_radix2_forward` from the Gnu Scientific Library [18]. Real and imaginary parts of the converted data were used to calculate the *Mod*s and the phase shifts of the harmonic peaks. The sampling density and the number of cycles of the numerical treatment were sufficient to show excellent agreement with the analytical results. Since harmonic orders up to 50 are shown in Figs. 6.11 and 6.12 the minimum number of samples per cycle for 1° phase accuracy at orders up to 50 would be 18,000 points per cycle. The waveforms used to generate those particular figures actually had 41,943 samples per cycle. When collecting fluorescence data similar considerations of adequate sampling and aliasing of harmonics are necessary in order to avoid distortion and errors of the measured phases and *Mod*s. Examples of the gnuplot scripts and C programs used for the numerical simulations are listed in the appendices.

6.2.13 Circuit Representations

Perhaps the most interesting heuristic to be found in the similarity of the fluorescence problem to analog filter theory is the ability to use the tools of electronics to solve the fluorescence equations. After moderate experience in circuit analysis one can draw circuits of most arbitrary transfer functions simply by inspection. The key concepts are that poles are represented by integrators, and zeros by differentiators. The transfer function can be factored and rearranged algebraically in several ways, and sometimes it is convenient to convert a zero acting on an input into a pole acting on the output and feed it back into the circuit. Usually, several different circuit realizations can be made to represent a single transfer function.

Once a transfer function has been modelled as a circuit then SPICE [11] can generate phase plots and Bode plots (which are plots of the *Mod*). Transient and steady state responses to arbitrary driving functions can be solved easily, and the analysis of the circuit can verify a fluorescence solution obtained analytically. Insights into issues of adequate sampling and of responses to exotic and complicated driving functions can be obtained from a SPICE analysis.

The transfer function of Eq. 6.27 can be represented by the circuit shown in Fig. 6.19 with: the initial values of n_{2_0} and n_{3_0} have been set to zero, but they could easily have non-zero starting voltages. The driving term Γ_{13} is the voltage source, and the factored terms in the denominator of Eq. 6.27 are cascaded low pass filters. Since each low-pass can introduce a phase shift up to 90° the cascade of two of low-pass filters can phase shift the input by up to 180°.

Figure 6.20 shows the phase response calculated by Tina-TI SPICE [19] of the circuit in Fig. 6.19 for two sets of *RC* time constants of the cascaded low pass filters. The curve labeled *a* is for a very fast relaxation from level 2 to 3 (refer to Fig. 6.3), and this response is for the usual situation found in fluorescence. Curve *b* in Fig. 6.20 is for a slower relaxation from level 2 to 3, and one sees that the phase shift of the response can extend beyond 90° when the driving frequency reaches the vicinity of the level 2-to-3 relaxation transition rate. For both curves the $2\pi RC$ lifetime of

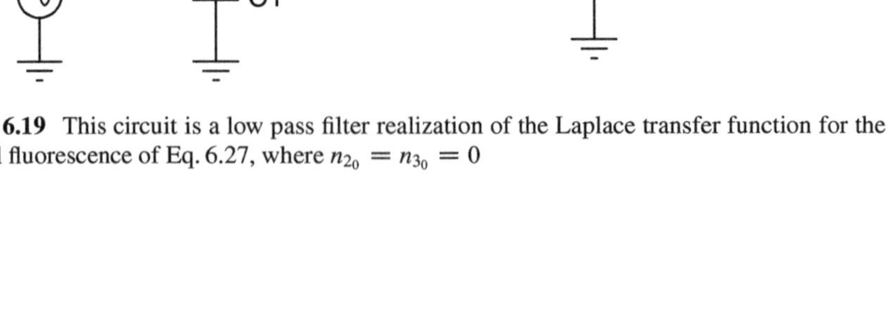

Fig. 6.19 This circuit is a low pass filter realization of the Laplace transfer function for the three level fluorescence of Eq. 6.27, where $n_{2_0} = n_{3_0} = 0$

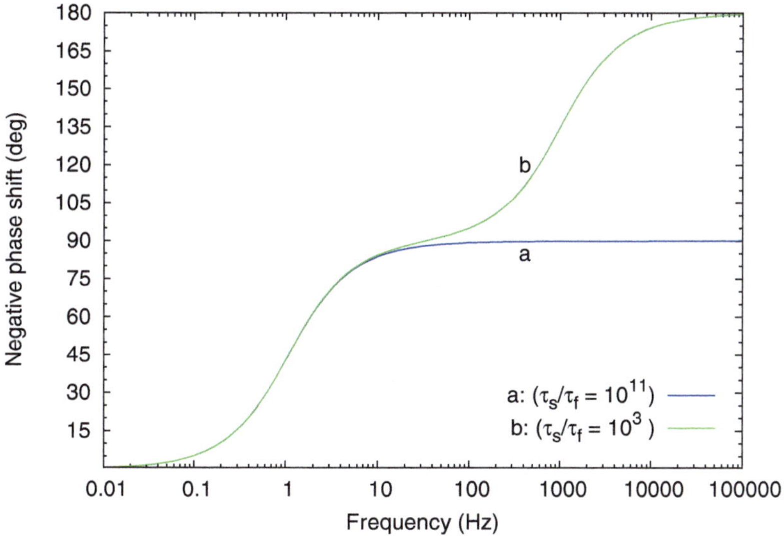

Fig. 6.20 The AC transfer characteristic of the Tina-TI SPICE implementation was used to generate the phase shift plotted here. These phase shifts correspond to two sets of fast and slow *RC* time constants for the double low pass circuit shown in Fig. 6.19. These results are identical to that of the calculated Laplace solutions shown in Fig. 6.4. The $2\pi RC$ lifetime of slow component (stage 1 of the circuit) for the both SPICE calculations is $\tau_S = 1$. Curve (*a*) above shows the phase shift for $\gamma_F = 1/\tau_F = 10^{11}$, and curve (*b*) for $\gamma_F = 1/\tau_F = 10^3$

the first stage (slow component) was set at 1 s. The electronic component values were $R_1 = 159k\Omega$ and $C_1 = 1\,\mu F$. For curve (a) the second stage electronic components values were $R_2 = 1.6k\Omega$ and $C_2 = 100\,nF$. For curve (b) the second stage component values were $R_2 = 0.159\Omega$ and $C_2 = 10pF$.

The circuit in Fig. 6.19 was drawn by inspection of the transfer function of Eq. 6.27, but it could been drawn directly from the three level diagram in Fig. 6.3 without having to solve the algebraic equations in Laplace space. In the electronic analogue each level population of Fig. 6.3 represents a voltage, and the transition rates represent either drive terms Γ_{13}, or filter terms, Γ_{31}, Γ_{23}, Γ_{21}. Each transition term of Fig. 6.3 is a circuit element joined to the other circuit elements in the figure. The correspondence is quite useful and powerful. Even if a hypothetical fluorescence system appears to be too onerous to solve in Laplace space then it should still be possible to draw the electronic equivalent and solve the problem with SPICE. Certain simplifying assumptions, such as negligible initial state depletion, are not necessary. SPICE can calculate the response behavior for arbitrary initial conditions, even when the system is driven hard.

The equivalent circuit for the transfer function with two lifetime components of Eq. 6.29 is shown in Fig. 6.21. This circuit was loaded into Tina-TI SPICE using two fixed *RC* time constants for slow and fast components. In SPICE different summing resistor values were substituted for R3 and R4 in order to generate the family of mixture curves shown in Fig. 6.22. The actual frequency values used for the slow

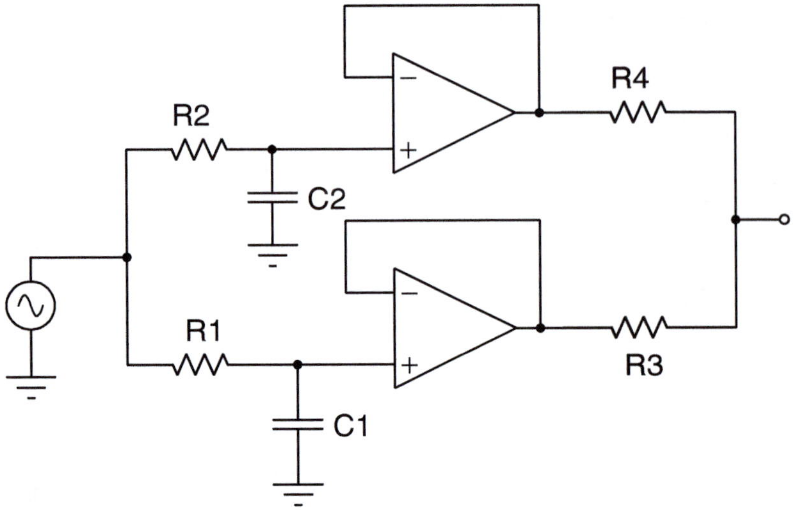

Fig. 6.21 This circuit is a realization of the transfer function for the two lifetime fluorescence of Eq. 6.29

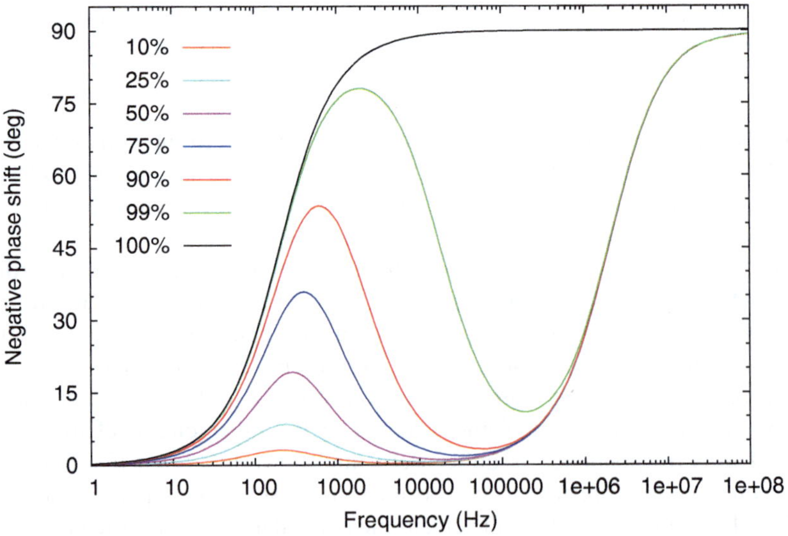

Fig. 6.22 This family of curves of phase shifts for different mixtures of two lifetimes was calculated by the Tina-TI SPICE. The slow and fast decay rates are separated by 4 decades, and the actual frequencies used in the simulation were 100 Hz and 1 MHz. However, the frequencies can be scaled appropriately. The ratios of intensity of the slow component to the fast component ranges from 10 % to 100 %. This figure should be compared to Figs. 6.6 and 6.10

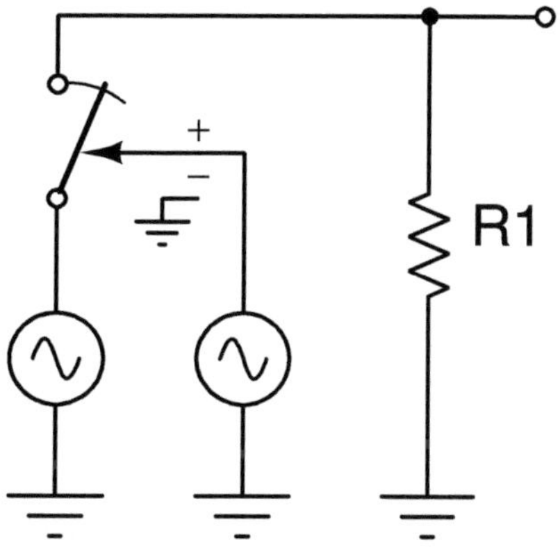

Fig. 6.23 This circuit represents the frequency detection technique for a chopped square wave

and fast transition rates in the circuit were 100 Hz and 1 MHz, but those frequencies can be rescaled arbitrarily to represent other decay rate regimes. The four decades difference between the two decay rates scales appropriately, so the simulation could equally apply to a problem with GHz decay rates.

For both the low-pass circuit and the fluorescence response the phase shift is, in fact, a phase lag. In electronics it is customary for a phase lag to be negative. The plots here show the phase lag as a positive phase shift, which is the convention in the literature on frequency domain fluorescence.

The circuit in Fig. 6.23 is a representation of the beat frequency technique without integration. The SPICE circuit consists of only a square wave signal source, a temporal switch driven by another square wave source at a slightly different frequency, and a pull-down resistor at the point of measurement. The frequencies chosen for Fig. 6.23 are 10 kHz for the chopper and 10.01 kHz for the signal. Figure 6.24 shows the pulse train in the time domain as calculated by SPICE of the circuit in Fig. 6.23, and the Fourier transform of that pulse train. Figure 6.24 should be compared to Fig. 6.17, which shows the analytic solution. The result of the SPICE calculation is similar, but there is additional hashing seen above first few harmonics of the base band. This is due to aliasing of frequencies above the Nyquist limit. The Tina-TI version of SPICE used here does not allow user specified time increments, so the Nyquist frequency is not readily selectable. Another difference between the figures is the shorter and broader lowest order baseband peak in the SPICE solution. This is due to a smaller number of cycles of the beat frequency in the SPICE solution. Not only is the Nyquist frequency lower in the SPICE solution, the frequency increment is larger, so the SPICE solution is at a lower resolution compared to the calculation with the analytical model used for Fig. 6.17.

One caveat of using a circuit to model a fluorescence problem is that SPICE may calculate more detail than would appear in the circuit taken at face value. The op-amp models in SPICE have varying degrees of realistic behavior, and at certain frequencies the solution may be dominated by the filter characteristics of the op-amp model itself, rather than by the other circuit components. For example op-amp models in Tina-TI SPICE include a selectable slew rate, realistic impedances,

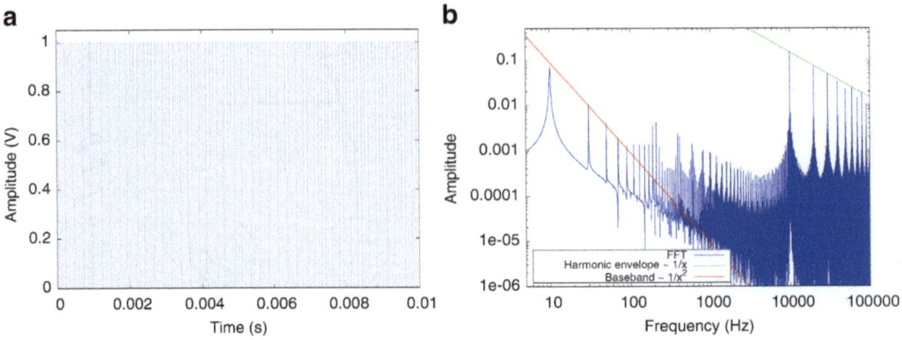

Fig. 6.24 This SPICE solution of the chopped square wave circuit of Fig. 6.23 shows essentially the same harmonic behavior of the analytical solution from Fig. 6.17. The differences in the figures are due to different temporal sampling and aliasing effects of the SPICE calculation, since the version Tina-TI version of SPICE used here automatically selects the time increment and does not allow the user to select it. (**a**) Time domain. (**b**) Frequency domain

and two poles for phase responses. One must be mindful of these features when converting a transfer function into a circuit. It is prudent to choose an appropriate scaled frequency range and to move the poles of the op-amp into regions where they do not affect the expected response.

6.3 Conclusion

Frequency domain fluorescence measurements in atomic and molecular physics can be modelled as electronic analog low-pass filters and processed with the tools of electrical engineering. The mathematical equivalence of fluorescence to analog filters permits a unified treatment of the entire fluorescence chain by cascading transfer functions. Driver terms, fluorescence response, and detection electronics can all be modelled on equal footing. Phase shifts and amplitude distortion from high gain and bandwidth limited detection electronics can be readily accounted for, and complicated fluorescence rate equations in both time and frequency domains can be handled elegantly. Equivalent circuits can be constructed to model fluorescence equations, and tools such as SPICE can be used to generate Bode plots, phase plots, and transient response signals. Frequency combs permit the sampling of many response points with just a few driver frequencies, so data collection can be accelerated considerably by use of driver functions rich in harmonic structure. This is advantageous in situations where measurement dwell times are limited. Nonlinear mixing techniques can be used to map fluorescence phase shifts into frequency ranges where measurements are most convenient, and can be analyzed with the linear techniques discussed in this chapter. The beat frequency technique, where a detector signal is chopped, is simply an example of mixing. Integrating the signal during the chop acts as another filter, introducing an additional phase shift. The techniques of modelling and analysis described in this chapter should make it possible to design more compact instrumentation for remote sensing applications.

Acknowledgements We wish to thank Howard Bender, John DiBenedetto, and Ian McKenna for their insights and support. This work was performed under the auspices of the U.S. Department of Energy, National Nuclear Security Administration by National Security Technologies, LLC, under Contract No. DE-AC52-06NA25946. Financial support was provided by the *Site Directed Research and Development Program* (SDRD) of National Security Technologies, LLC.

Appendices

We list here two C programs, several input scripts to Gnuplot, and an example of a shell script used to generate many of the figures in this chapter.

Program Listing of gsl_mod_phase.c

```
/****************************************************************/
/*  gsl_mod_phase.c : Calculate complex FFT using Gnu
    Scientific Library                                         */
/*  To compile:                                                */
/*  gcc  -O2 -o gsl_mod_phase.out gsl_mod_phase.c -lgsl
    -lgslcblas -lm                                             */
/****************************************************************/
#include <stdio.h>
#include <stdlib.h>
#include <math.h>
#include <gsl/gsl_errno.h>
#include <gsl/gsl_fft_complex.h>

#define Pi M_PI
#define Square(X)  ((X)*(X))
#define REAL(z,i)  ((z)[2*(i)])
#define IMAG(z,i)  ((z)[2*(i)+1])

#define NPTS 1048576

int main(int argc, char *argv[])
{
  static double data[2*NPTS];
  double Xmax=0, Xmin=0, scale=NPTS, step, x, y;
```

```
  char *ptr, buffer[256];
  FILE *Ch1;
  int i, Npts=0;

  Ch1 = fopen(argv[1], "r");
  while (fgets(buffer, 255, Ch1) != NULL) {
    if ((buffer[0] == '#') || (buffer[0] == '\n')) continue;
    ptr = buffer;
    x = strtod(ptr, &ptr);
    REAL(data, Npts) = strtod(ptr, &ptr);
    IMAG(data, Npts) = 0;
    if (x > Xmax) Xmax = x;
    if (x < Xmin) Xmin = x;
    if (++Npts >= NPTS) break;
  }
  fclose(Ch1);

  step = (Xmax - Xmin)/(NPTS-1);
  printf("Npts=%d, Xmin=%f, Xmax=%f, step=%f\n", Npts,
          Xmin, Xmax, step);

  Ch1=fopen("wave_input.txt", "w");
  for (i=0; i<NPTS; i++) fprintf(Ch1, "%e %e %e\n", i*step,
      REAL(data,i), IMAG(data,i));
  fclose(Ch1);

  gsl_fft_complex_radix2_forward(data, 1, NPTS);
  step = 1.0/step/(NPTS);

  Ch1=fopen("fft_output.txt", "w");
  for (i=0; i<NPTS; i++) {
     x = REAL(data,i)/scale;
     y = IMAG(data,i)/scale;
     fprintf(Ch1, "%e %e %e\n", i*step, x, y);
  }
  fclose(Ch1);

  return 0;
}
```

Program Listing of Make_integrated_accordion.c

```
/******************************************************************/
/*  make_integrated_accordion.c                                   */
/*  To compile:                                                   */
/*  gcc -O2 -o make_integrated_accordion.out
    make_integrated_accordion.c                                   */
/******************************************************************/
#include <stdio.h>
#include <stdlib.h>

#define CHOP 128
#define FREQ 130
#define SAMP 1048576

inline int toggle(int x, int f)
{
   if ((x % f) < f/2) return 0;
   else return 1;
}

int main(int argc, char *argv[])
{
   int Sum=0, tmp, i;

   for (i=0; i<SAMP; i++) {
      if ((i % CHOP) == 0) {
          printf("%d %d\n", i, Sum);
          Sum = 0;
      }
      if ((tmp = toggle(i, FREQ) > 0) && (toggle(i, CHOP) > 0)) {
         Sum += tmp;
      }
   }

   return 0;
}
```

Gnuplot Script to Generate a Chopped Square Wave

```
#!/usr/bin/gnuplot
# chopper_table-ON_OFF_Square_Wave_Switching.gnu

if (!exists("datafile")) datafile="chopper_table.dat"
if (!exists("samp")) samp = 1048576
if (!exists("chop")) chop = 127
if (!exists("freq1")) freq1 = 129

print "Datafile = ".datafile
rate = samp/32
chop = floor(rate/chop)
omega = floor(rate/freq1)
Toggle(t,f) = ((floor(t+0.5) % f) < f/2) ? 0 : 1
Chopper(t) = ((floor(t) % chop) < chop/2) ? 0 : Toggle(t,omega)
set samples samp
set format x "%.0f"
set table datafile
plot [0:samp-1] Chopper(x)
unset table
```

Gnuplot Script to Plot the Time Domain Chopped Square Wave

```
#!/usr/bin/gnuplot
set terminal postscript eps enhanced defaultplex \
   leveldefault color colortext \
   solid dashlength 1.0 linewidth 1.0 butt noclip \
   nobackground \
   palfuncparam 2000,0.003 \
   "Helvetica" 18  fontscale 1.0
set output 'ON_OFF_Square_Wave_Time_Series.eps'
set key title ""
set key inside right top vertical Right noreverse enhanced
    autotitles nobox
set key noinvert samplen 4 spacing 1 width 0 height 0
set key maxcolumns 0 maxrows 0
set key noopaque
set mxtics 2
set mytics 10
set ytics 0, 1
```

```
# set title "Detector ON/OFF Switching of a Square Wave
     (Accordian detection)"
set xlabel "Normalized Time"
set ylabel "Amplitude"
plot [0:100][0:1.05]  "wave_input.txt" using ($1/256):2 w l lt
     3 not
```

Gnuplot Script to Plot the Frequency Domain Chopped Square Wave

```
#!/usr/bin/gnuplot
set terminal postscript eps enhanced defaultplex \
   leveldefault color colortext \
   solid dashlength 1.0 linewidth 1.0 butt noclip \
   nobackground \
   palfuncparam 2000,0.003 \
   "Helvetica" 16  fontscale 1.0
set output 'ON_OFF_Square_Wave_Switching-log_log.eps'
set key title ""
set key inside left bottom vertical Right noreverse enhanced
    autotitles box opaque
set key noinvert samplen 4 spacing 1 width 0 height 0
set key maxcolumns 0 maxrows 0
set key noopaque
set logscale y 10
set logscale x 10
set mxtics default
set mytics 10
set xlabel "Normalized Frequency"
set ylabel "Amplitude"
plot [0.005:10.5][5E-7:0.5] \
     "fft_output.txt" using (32786*($1*1.00/128)):
         (sqrt($2*$2 + $3*$3)) w l lc 3 t "FFT", \
     (1/(2*pi*x)) w l lc 2 t "Harmonic Envelope
         ~ 1/(2{/Symbol p}x)", \
     (x > 0.8 ? 1/0 : (0.00078125/(4*pi*pi*x*x))) lc 1 t
         "Baseband IM distortion ~ 1/(2{/Symbol p}x)^2", \
     (((x < 1.2) || (x > 2.8)) ? 1/0 : (0.00078/
         (4*pi*pi*(x-2)*(x-2)))) lc 4 t "Harmonic IM distortion \
     ~ 1/(2{/Symbol p}(x-2))^2"
```

Gnuplot Script to Generate Beat and Mix Wave Cosine Data

```
#!/usr/bin/gnuplot
if (!exists("samp")) samp = 1048576
if (!exists("chop")) chop = 256
if (!exists("freq1")) freq1 = 272
if (!exists("freq2")) freq2 = 288

inv_duty = 2
rate = (samp/32)
omega1 = 2.0*pi*freq1/rate
omega2 = 2.0*pi*freq2/rate
omega_m = 2.0*pi*chop/rate
ccycle = floor(rate/chop)
phi1 = 0.25*pi
phi2 = 0.5*pi

S1(t,omega,phi) = cos(omega*t + phi)
Mixer(t,omega) = cos(omega*t)
Unmodulated(t) = S1(t,omega1,phi1)+S1(t,omega2,phi2)
Modulated(t)   = S1(t,omega1,phi1)*S1(t,omega2,phi2)
Mixdown(t) = Unmodulated(t)*cos(omega_m*t)
Chop_func(t) = ((floor(t) % ccycle) > ccycle/inv_duty) ? 0 : 1
Chopper(t)   = ((floor(t) % ccycle) > ccycle/inv_duty) ? 0 :
    Unmodulated(t)

set samples samp
set format x "%.0f"
set table "unmodulated_cosine_table.dat"
plot [0:samp-1] Unmodulated(x)
unset table
set table "modulated_cosine_table.dat"
plot [0:samp-1] Modulated(x)
unset table
set table "chopper_cosine_table.dat"
plot [0:samp-1] Chopper(x)
unset table
set table "chopper_alone_table.dat"
plot [0:samp-1] Chop_func(x)
unset table
set table "mixdown_cosine_table.dat"
plot [0:samp-1] Mixdown(x)
unset table
```

Linux Shell Script to Generate Fig. 6.17

```
#!/bin/bash
# create numerical data and plot figures for chopped square wave
gnuplot chopper_table-ON_OFF_Square_Wave_Switching.gnu
./gsl_fft_mod_phase.out chopper_table.dat
gnuplot ON_OFF_Square_Wave_Time_Series.gnu
gnuplot ON_OFF_Square_Wave_Switching-log_log.gnu
```

References

1. Trainham R, O'Neill M, McKenna IJ (2105) An analog filter approach to frequency domain fluorescence spectroscopy. J Fluoresc 25:1801–1812. doi: 10.1007/s10895-015-1669-z
2. McGLynn SP, Smith JK (1961) The electronic structure, spectra, and magnetic properties of actinyl ions. Part 1. The uranyl ion. J Mol Spectrosc 6:164–187
3. Glinka YD, Krak TB (1995) Luminescence spectra of uranyl ions adsorbed on disperse SiO_2 surfaces. Phys Rev B 52:14985–14995
4. Lakowicz JR (2006) Principles of fluorescence spectroscopy. Springer, Berlin
5. Eaton DF (1990) Recommended methods for fluorescence decay analysis. Pure Appl Chem 62:1631–1648
6. Lakowicz JR, Laczko G, Cherek H, Gratton E, Limkeman M (1984) Analysis of fluorescence decay kinetics from variable-frequency phase shift and modulation data. Biophys J 46:463–477
7. Gafni A, Modlin RL, Brand L (1975) Analysis of fluorescence decay curves by means of the Laplace transformation. Biophys J 15:263–280
8. Cooper GR, McGillem CD (1967) Methods of signal and system analysis. Holt, Renehart and Winston, New York
9. Lacanette K (1991) A basic introduction to filters-active, passive, and switched-capacitor, application note 779. http://www.national.com/an/AN/AN-779.pdf
10. Rader CM, Gold B (1967) Digital filter design techniques in the frequency domain. Proc IEEE 55:149–171
11. Nagel LW (1975) Spice2: a computer program to simulate semiconductor circuits. Ph.D. thesis, EECS Department, University of California, Berkeley. http://www.eecs.berkeley.edu/Pubs/TechRpts/1975/9602.html
12. Arfken G (1970) Mathematical methods for physicists, 2nd edn. Academic Press, New York
13. Greenberg MD (1978) Foundations of applied mathematics. Prentice-Hall, Englewood
14. Acton FS (1970) Numerical methods that work. Harper and Row, New York, p 253
15. Horowitz P, Winfield H (1989) The art of electronics, chapter 13, 2nd edn. Cambridge University Press, Cambridge
16. John DiBenedetto J, Capelle GA, O'Neill M (2012) Time-resolved hyperspectral fluorescence spectroscopy using frequency-modulated excitation. J Appl Phys 112:013109
17. Gnuplot is a freely available computer program for data fitting and plotting (Jun 2016). http://www.gnuplot.info/
18. The Gnu Scientific Library software is freely available for download (Jun 2016). http://www.gnu.org/software/gsl
19. Tina-TI is the Spice implementation from Texas Instruments (2015). http://www.ti.com/tool/tina-ti

Chapter 7
Synthesis and Characterization of Europium Doped Zirconium Based Phosphor for Display Applications

Neha Dubey (Tiwari) and Vikas Dubey

Keywords Zirconium based phosphors • Europium doping ion concentration • Solid state lightening application • Display devices

7.1 Introduction

Zirconium is an important ceramic with excellent mechanical, electrical, thermal and optical properties. These properties have made the n-type semiconductor attractive for several applications including electrolytes for oxide fuel cells, catalysts, sensors, thermal barrier coatings and photonics [1–3]. Zirconium oxide (ZrO_2) has three polymorphs such as monoclinic (m-ZrO_2), tetragonal (t-ZrO_2) and cubic (c-ZrO_2) phases. The m-ZrO_2 has lattice symmetry of C_s and is thermodynamically stable at temperatures below 1170 °C while the t-ZrO_2 structure with symmetry of D_{2d} stabilizes at temperatures in the range 1170 to 2370 °C. The c-ZrO_2 is stable at temperatures beyond 2370 °C [4]. Among these polymorphs, the cubic and tetragonal phases are the most stable and symmetric, and are considered crucial for technological applications.

In general stable and high symmetric ZrO_2 phase can be achieved by heat treatment at high temperatures, however, low temperature phase stabilization is possible when the oxide is doped with trivalent rare earths such as Pr^{3+}, Y^{3+}, Ce^{3+} and Eu^{3+} [5, 6]. Incorporating these rare earth ions into the ZrO_2 lattice introduces oxygen vacancies which favours eight-fold coordination of oxygen ions around the Zr^{4+} cation [7] that stabilizes the tetragonal and cubic phases at low temperatures. Eu^{3+} ion has the $^5D_0$ – $^7F_2$ transition which is hypersensitive to the crystal field symmetry and has been utilized to probe the morphological changes in a host crystal.

N. Dubey (Tiwari) (✉)
Department of Physics, Govt. V.Y.T.PG. Auto. College, Durg, Chhattisgarh 491001, India
e-mail: tiwarineha1441@gmail.com

V. Dubey
Department of Physics, Bhilai Institute of Technology, Raipur, Chhattisgarh, India

C.D. Geddes (ed.), *Reviews in Fluorescence 2016*, Reviews in Fluorescence,
DOI 10.1007/978-3-319-48260-6_7

Report has shown that the crystal lattice of zirconia tend to transform into a more symmetrical structure with the addition of rare earths into zirconia lattice [6].

7.1.1 Rare Earths (RE) and Luminescence in Rare Earths

Rare earths are group of 17 metallic elements that includes the 15 lanthanides, yttrium and scandium.The rare–earth elements (REE) have a wide variety of lighting applications. The principal sources of the REE are the four minerals, bastnasite, loparite, monazite and xenotime. Production of REE has been primarily occurred in China Australia, Malaysia, India, Thailand and Russia with China being the leading REI producer in the world [7–13].

Rare earths are family of 17 elements with atomic number 57–71 and 21, 39. Elements with atomic number 57, lanthanum has no free electron in the 4f shell, while with atomic number 71; lutetium has a completely filled ,but is completely monitored by the outer 5s and 5p sub shell, which are completely filled. The optical and electromagnetic properties of 13 rare earth elements are essentially due to the screening of this incompletely filled 4f shell, which is generally forbidden by quantum mechanical spin and parity prohibition rules. Efficient luminescence can still occur in these ions under conditions where such ions do not occupy a position having a centre of symmetry in a crystalline lattice. Some of the RE^{3+} ions such as europium, terbium and dysprosium are good luminescent emitters- a factor depending on the number of electrons in its 4f shell. The ions, which are inert to luminescent emission , are those of yttrium,lanthanum gadolinium and lutetium.Some common luminescent ions are shown in Fig. 7.1.

Common Luminescent Ions

Ion	Excited State	Ground State	λ_{max} Emission
Mn^{2+} $(3d^5)$	$t_2^4e^1$ $(^4T_1)$	$t_2^3e^2$ $(^6A_1)$	Green-Orange-Red*
Sb^{3+} $(5s^2)$	$5s^15p^1$	$5s^2$	Blue*
Ce^{3+} $(4f^1)$	$4f^05d^1$	$4f^15d^0$	Near UV to Red*
Eu^{2+} $(4f^7)$	$4f^65d^1$	$4f^75d^0$	Near UV to Red*
Tm^{3+} $(4f^{12})$	$^1G_4$	$^3H_6$	450 nm (Blue)
Er^{3+} $(4f^{11})$	$^4S_{3/2}$	$^4I_{15/2}$	545 nm (Green)
Tb^{3+} $(4f^8)$	$^5D_4$	$^7F_5$	545 nm (Green)
Pr^{3+} $(4f^2)$	$^3P_0$	$^3H_5$ $(^3F_2)$	605 (635) nm (Red)
Eu^{3+} $(4f^6)$	$^5D_0$	$^7F_2$	611 nm (Red)

*The energy of transitions involving d, p or s orbitals is very sensitive to the crystal field splitting induced by the lattice.

Fig. 7.1 Rare earth ions

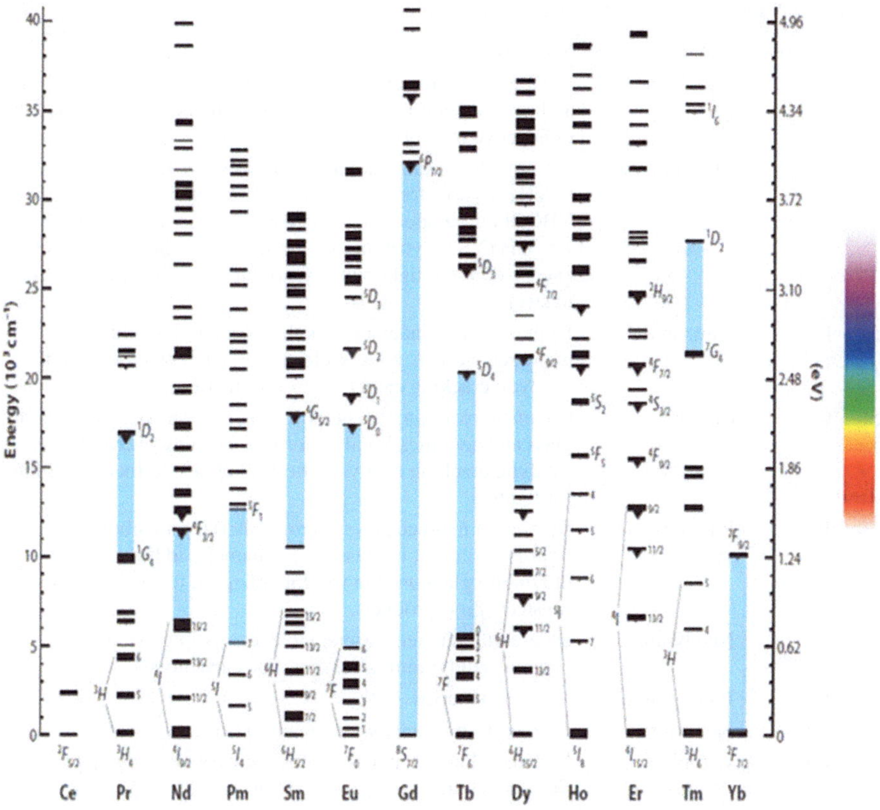

Fig. 7.2 Rare earth spectroscopy

The rare earth spectroscopic diagram for the trivalent rare earth from Ce^{3+} to Yb^{3+} is shown in Fig. 7.2.

Significant improvement has been made by introduction of rare earth ions as luminescent species in the field of luminescent materials. The use of Eu^{3+} activated materials as the red component for colour television screen was an important step. After few years later the use of rare earth activated phosphors gave up considerable developments in solid state lighting (SSL), luminescent lamps and X-ray intensifying screen. Trivalent rare earth complexes ,which featured line-like high luminescent efficiency, emissions, high quantum efficiency and easy synthesis, have been introduced as the emission materials into the field of organic and inorganic light-emitting diodes [14–17].

A Table 7.1 listing the seventeen rare earth elements, their atomic number, symbol and their main usages are listed here.

Table 7.1 Various rare earth elements and its applications

Z	Symbol	Name	Selected applications
21	Sc	Scandium	Light aluminium-scandium alloys for aerospace components, additive in metal-halide lamps and mercury-vapor lamps, radioactive tracing agent in oil refineries
39	Y	Yttrium	Yttrium aluminium garnet (YAG) laser, yttrium vanadate (YVO_4) as host for europium in TV red phosphor, YBCOhigh-temperature superconductors, yttria-stabilized zirconia (YSZ), yttrium iron garnet (YIG) microwave filters, energy-efficient light bulbs, spark plugs, gas mantles, additive to steel
57	La	Lanthanum	High refractive index and alkali-resistant glass, flint, hydrogen storage, battery-electrodes, camera lenses, fluid catalytic cracking catalyst for oil refineries
58	Ce	Cerium	Chemical oxidizing agent, polishing powder, yellow colors in glass and ceramics, catalyst for self-cleaning ovens, fluid catalytic cracking catalyst for oil refineries, ferrocerium flints for lighters
59	Pr	Praseodymium	Rare-earth magnets, lasers, core material for carbon arc lighting, colorant in glasses and enamels, additive in didymium glass used in welding goggles, ferrocerium firesteel (flint) products.
60	Nd	Neodymium	Rare-earth magnets, lasers, violet colors in glass and ceramics, didymium glass, ceramic capacitors
61	Pm	Promethium	Nuclear batteries
62	Sm	Samarium	Rare-earth magnets, lasers, neutron capture, masers
63	Eu	Europium	Red and blue phosphors, lasers, mercury-vapor lamps, fluorescent lamps, NMR relaxation agent
64	Gd	Gadolinium	Rare-earth magnets, high refractive index glass or garnets, lasers, X-ray tubes, computer memories, neutron capture, MRI contrast agent, NMR relaxation agent, magnetostrictive alloys such as Galfenol, steel additive
65	Tb	Terbium	Green phosphors, lasers, fluorescent lamps, magnetostrictive alloys such as Terfenol-D
66	Dy	Dysprosium	Rare-earth magnets, lasers, magnetostrictive alloys such as Terfenol-D
67	Ho	Holmium	Lasers, wavelength calibration standards for optical spectrophotometers, magnets
68	Er	Erbium	Infrared lasers, vanadium steel, fiber-optic technology
69	Tm	Thulium	Portable X-ray machines, metal-halide lamps, lasers
70	Yb	Ytterbium	Infrared lasers, chemical reducing agent, decoy flares, stainless steel, stress gauges, nuclear medicine
71	Lu	Lutetium	Positron emission tomography , lutetium tantalate hosts

7.1.2 Phosphors

Phosphors are solid luminescent materials that emit photons when excited by an external energy source, such as Ultra-violet light (Photoluminescence) or an electron beam (Cathodoluminescence). Phosphors are composed of an inert host lattice and an activator. Host lattice is transparent to the excitation radiation and an activator, typically a 3d or 4f electron metal, which is excited under energy bombardment. The process of luminescence occurs by adsorption of energy at the activator site, relaxation and subsequent emission of photon, then return to a ground state. The efficiency of a phosphor depends on the amount of relaxation that occurs during the activation and emission. Relaxation is the process in which energy is lost to the lattice as heat; it needs to be minimized in order to extract the highest luminous efficiency. The ratio of the energy emitted to the energy absorbed is known as luminous efficiency. In recent years, fields of application of phosphors have expanded greatly to include lighting, display and medicine. For these purpose, thousands of different phosphors have been synthesized. However, approximately 50 of them have properties (luminescence efficiency, emitted light colour, chemical and physical stability, etc.) satisfactory for practical use [18–24].

7.1.3 Significance of Phosphors

Most phosphors consist of a host composition plus activator. Activator should be added in carefully controlled quantities. The activator itself is a substitution defect and is subject to lattice phonon perturbations. Therefore, it is essential that the charge on the substitution cation is equal to that of the host lattice cation, otherwise, an inefficient phosphor does not result. We denote a phosphor as: M_aYO_b:N_x, where M_a is the cation , YO_b is the anion, and N_x is the activator.

7.1.4 Properties of Zirconium Based Phosphors

Zirconium oxide (ZrO_2) is one of the widely studied oxide materials over the last two decades because of its excellent electrical and optical properties, such as high dielectric constant (about 23–29), good thermal stability, high melting point, and wide band gap (5–7 eV) [25, 26]. It is widely used as an essential material in the optical fields including broadband interference filters and active electro- optical devices, due to its excellent optical properties such as high refractive index, large optical band gap, low optical loss and high transparency in the visible and near infrared region [27, 28]. It has wide band gap and short wavelength Photoluminescence (PL) emission property [29]. Oxide phosphors were found to be optimal candidates in field emission display (FED) and plasma display panel (PDP) devices as they were sufficiently conductive to release electric charges, stored on the phosphor particle surfaces [30]. Because, these potential applications are still very much in the design phase, further fundamental research in the field remains a challenge. Among these oxide phosphors, zirconium oxide/zirconia is a material that deserves attention in the field of photonics. Zirconia was a technologically important material due to its high melting point, high thermal and mechanical resistance, high thermo chemical resistance, high thermal expansion coefficient, low thermal conductivity, high corrosion resistance, high dielectric constant, photo thermal stability etc. [31–34]. Most of these applications make use of size-dependent properties of zirconia nanocrystals. It was an important task to synthesize the crystallite size as small as possible. When zirconia particles are reduced to nano scale size, novel characteristics may be found due to quantum size effects, which occur when cluster size was smaller than the Bohr radius [35]. Therefore, the precise control of the crystallite size, crystalline phase, morphology, textural properties and lattice defects was absolutely required [36]. Zirconia exists in three crystalline phases: monoclinic (below 1170 °C), tetragonal (1170–2370 °C) and cubic (above 2370 °C) [32, 33]. Among these, tetragonal/cubic phase was most desirable for technical applications [31, 37, 38]. In general, the increasing temperature stabilizes more symmetric phases [39–43]. The assignment of tetragonal and cubic phase solely from the XRD can be misleading, since these two phases have similar XRD patterns [31]. The problem was more pronounced when the peaks are broadening due to small sizes of the crystals [42]. The incorporation of trivalent rare earth (RE) ions such as Y^{3+}, Pr^{3+} and Eu^{3+} promotes the formation of oxygen vacancies for charge compensation and favours sevenfold coordinated oxygen ions around the 'Zr′ cations and stabilize tetragonal or cubic phases at low temperature [44–49]. The stabilization of the tetragonal or cubic phase from monoclinic phase is very crucial as these two phases are considered to be more important in fundamental and technological applications than the low temperature phase. Norris and co-workers described the important role of the crystal phase in the efficiency of transition metal doped semiconducting nanoparticles [50].

7.1.5 $SrZrO_3$ Phosphors

Compounds with perovskite structure have been the subject of many researches for their important physical characteristics, chemical stability, and nicer mechanical properties. The alkaline earth zirconate $SrZrO_3$, as one of the perovskite type compounds with a high melting point (over 2600 °C), can be used as high-temperature materials [51]. Its proton conductivity at elevated temperatures makes it a candidate material for electrochemical devices [52]. To the best of our knowledge, however, few researchers have been done concerning the photoluminescence of

$SrZrO_3$. The number of luminescent zirconates is small, especially of those with visible emission. The bulk material properties and structures of $SrZrO_3$ are well known [52–55]. However, it is now widely accepted that materials exhibit novel properties when their sizes decreased to nanometer scales, and the performance of materials depends largely on the synthesis procedure. Here, we aim to prepare nano-scaled $SrZrO_3$ crystallites using a simple and mild chemical method and to reveal its luminescence property at room temperature. Several methods have been reported to prepare $SrZrO_3$ powders: conventional solid state reaction of $SrCO_3$ and ZrO_2, co-precipitation, hydrothermal [56], sol–gel [57]. In order to get homogeneous $SrZrO_3$ nanocrystals, Kashinath and co-workers adopted a sol–gel combustion method, and introduced citric acid in the synthesis process and as fuel during the combustion process [58].

To date, many methods have been developed to prepare ZrO_2: Eu^{3+} luminescent phosphors in the form of powders, thin films or colloidal solutions by sol–gel method [59–66], co-precipitation method [67] etc. Traditionally solid-state method was used for the preparation of phosphor materials for Solid State Lightening (SSL) applications. The conventional SSL method is less considered due to (i) inhomogeneity of the final product, (ii) low-surface area, formation of large agglomerated particles and (iii) presence of various defects (surface, impurities) which are harmful to luminescence [68, 69]. These problems were overcome by the use of solution combustion method [70]. This technique produces uniform product, high surface area, fine particles, lesser processing time and cost effective method.

7.1.6 $CaZrO_3$ Phosphors

Calcium zirconate (CaZrO3) due to its valuable properties such as high melting point (2340 °C), high dielectric permittivity and low dissipation factor, is a ceramic material that is currently being used in a wide range of applications: multilayer ceramic capacitors, solid electrolyte, crystalline host for phosphor materials, moderate temperature thermal barrier catalyst, etc. [71–73]. Also, as a new application, Robert Iano and co-workers has used this ceramic material in production of a potential red shade pigment, chromium-doped calcium zirconate [74]. There are several methods for the synthesis of this material. $CaZrO_3$ powders is conventionally synthesized via a high temperature (1500 °C) solid state reaction of powdered CaO (or $CaCO_3$) and zirconia (ZrO_2) (conventional mixed oxide synthesis (CMOS)). As the reactions are generally controlled by slow diffusion mechanisms, highly reactive raw materials, high temperatures and long times have to be used for the reactions to achieve completion.

In recent year search for the new materials with desirable mechanical and optical properties has become important. In particular, there is a need of the compound with the improved luminescence properties. Considerable improvement in the field of luminescent materials has been made by the introduction of rare earth ion doped in the material, the optical properties of material is improved. So rare earth activated phosphors played the way of development of optical amplifiers. There has been constant interest in trying to develop luminescent materials with higher radiative efficiency. This has led to increased research activities on the study of the synthesis and characterization of rare-earth doped materials where the host materials used have a high energy band gap, low phonon energy and chemical stability over photon and electron radiations [73–77].

7.2 Synthesis and Characterization

Phosphor of $SrZrO_3$, $CaZrO_3$, $Y_2Zr_2O_7$ and $La_2Zr_2O_7$ doped with Eu^{3+} ions with variable molar concentration of Eu^{3+} (0.1–2.5 %) was prepared by solid state reaction method. The composition of each chemical weighed in proper stoichiometric ratio then mixed thoroughly for 45 minutes using mortar and pestle. The grinded sample was placed in an alumina crucible and subsequently fired at 1000 °C for 1 h for calcinations and then at 1250 °C for 3 h for sintering in a muffle furnace. Every heating was followed by intermediate grinding. Finally the samples were cooled slowly to room temperature in the furnace and ground again into powder for subsequent characterization.

The XRD measurements were carried out using Bruker D8 Advance X-ray diffractometer. The X-rays were produced using a sealed tube and the wavelength of X-ray was 0.154 nm (Cu K-alpha). The X-rays were detected using a fast counting detector based on Silicon strip technology. The FTIR spectrum of a sample was recorded at room temperature in the wave number range of 4000-400 cm^{-1} on a Bruker spectrophotometer. The photoluminescence (PL) emission and excitation spectra were recorded at room temperature by use of a Shimadzu RF-5301 PC spectrofluorophotometer. The excitation source was a Xenon lamp [78–83].

7.3 Results and Discussion

7.3.1 Luminescence and Structural Analysis of $CaZrO_3$ Phosphor

7.3.1.1 XRD Analysis

For obtaining the lattice parameters of the sample the powder diffraction pattern was analyzed by Rietveld fitting method [84]. Lattice parameters were refined until the observed and calculated XRD spectra were fairly in good agreement. The pattern is characterized by few prominent peaks found at different glancing angles. The crystallite size was computed from the full width half maxima (FWHM) of maximum intense peak using the Scherer's formula [85, 86].

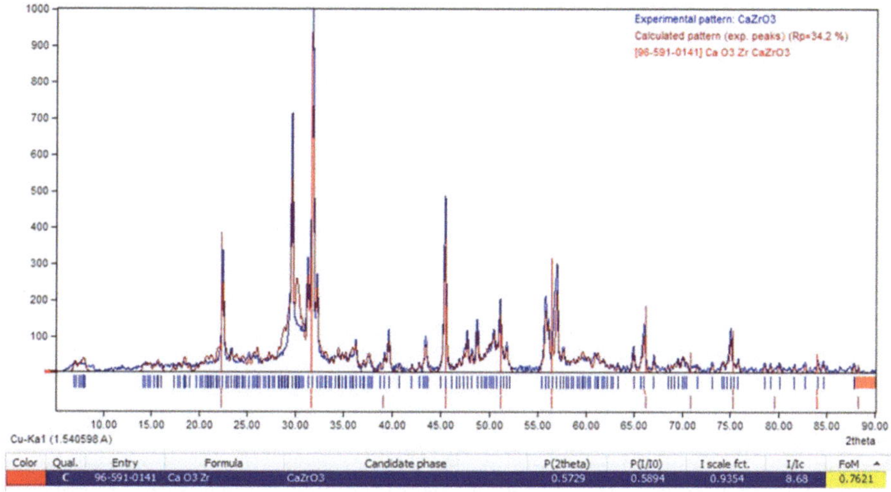

Fig. 7.3 XRD pattern of $CaZrO_3{:}Eu^{3+}$ doped phosphor matched with COD 96-591-0141 having FOM 76 %

The Scherer's formula is given by:

$$D = 0.9\lambda/\beta\cos\theta$$

Where,

D = crystallite size;

λ = wavelength of X-rays

β = FWHM

θ = diffraction angle

Calculated crystallite size found 76 nm. Figure 7.3 represents the XRD pattern of of $CaZrO_3{:}Eu^{3+}$ doped phosphor matched with COD 96-591-0141 having FOM 76 %. Celref v. 3 software is used for indexing and refinement parameters.

7.3.1.2 FTIR Results for $CaZrO_3$ Phosphor Doped with Europium

FTIR Spectrum of $CaZrO_3{:}Eu^{3+}$ (0.5 %) phosphor is shown in the Fig. 7.4. This spectrum expresses strong broad peaks at 473 cm^{-1} to 721 cm^{-1} which are the characteristics of Zr-O vibrations. Presence of Ca-O gives rise to an IR peak within a range of 1437–1824 cm^{-1}. All these observed peaks found together confirm the formation of $CaZrO_3$ phosphor. From the FTIR spectrum it is observed that the peak at ~2163–3014 cm^{-1} are assigned to intermolecular C–C stretching. The specimen might burn some contents due to high temperature synthesis method.

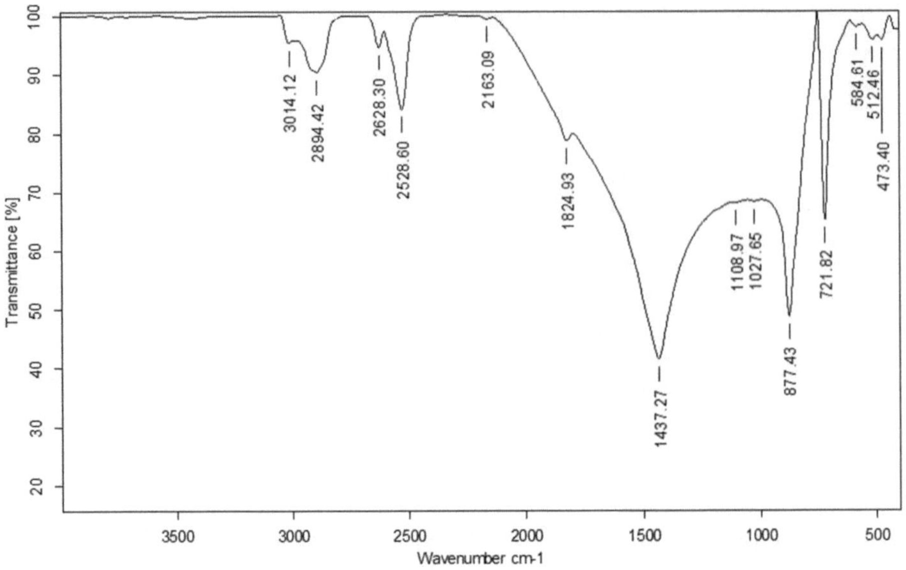

Fig. 7.4 FTIR analysis of $CaZrO_3$: Eu^{3+} doped phosphor

7.3.1.3 HRTEM Study

The images revealed that the Eu^{3+} doped $CaZrO_3$ phosphor was a collection of nanorods 55–58 nm long and 2–4 nm thick. The synthesised sample showed the nanorod shapes. The nano-rod formation is responsible for the very good PL spectra from prepared phosphor. It is due to because the procedure of sample preparation liquid grinding occurs using acetone in intermediate grinding as well as after sample preparation the specimen was thoroughly grounded (Fig. 7.5).

7.3.1.4 PL Spectra of $CaZrO_3$ Phosphor

Charge-Transfer (CT) Transition

The accurate position of the energy σ of Eu-O CT band in different host matrix can be calculated via the Jorgensen formula [87]:

$$\sigma = \left[\chi\left(O^{2-}\right) - \chi\left(Eu^{3+}ion\right)\right] \times 30,000\ \text{cm}^{-1} \tag{7.1}$$

where, χ (O^{2-}) and χ (Eu^{3+} ion) are the electronegativity of the ligand anion O^{2-} and central metal ion Eu^{3+} respectively. In general the Eu^{3+}- O^{2-} band located at ~250 nm [91]. However, a red shift in the band position is observed in the present case which can be explained on the basis of two different factors as mentioned by X. Liu et al. [88]. These two factors comprise the electronegetivity and bond characteristics of Eu-O bond.

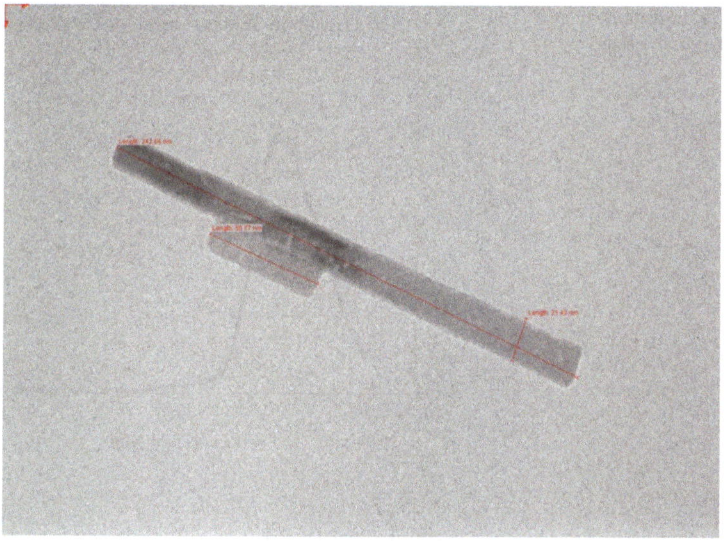

Fig. 7.5 TEM micrograph of $CaZrO_3$:Eu^{3+} doped phosphor

Previously, Parchur et al. [89] explained the large shift of the Eu–O CT band on the basis of the difference in electronegetivity by using the Jørgensen equation. According to their proposed model, the change in the crystal field environments can modify the difference in optical electronegativity between Eu^{3+} and O^{2-}. The position of energy is dependent on the difference in optical electronegativity between Eu^{3+} and O^{2-}. For the present host matrix, the electronegativity difference χ (O^{2-}) – χ (Eu^{3+} ion), is observed to be 1.25. The present value of electronegativity difference is much lower than the reported value [40]. Such lower electronegativity difference indicates the lesser value of CT energy σ. Therefore, the red shift of CT band can be attributed to the lowering of CT energy.

The type of Eu–O bond in a host matrix is an indicative of the cause of shifting in the position of CT band. The higher value of ionic bond strength of the Eu-O bond indicates the presence of greater CT energy owing to which the CT band shift towards lower wavelength. However, in the present case, the position of CT band shifts towards higher wavelength indicating the Eu-O ionic bond energy weakened. For this purpose the nephelauxetic ratio (β^-), bonding parameter ($b^{1/2}$) and covalence (δ) parameters in the present host matrix were calculated as [90]

$$\beta^- = \frac{1}{n}\sum \frac{\nu_{complex}}{\nu_{freeion}} \tag{7.2}$$

$$\delta = \frac{(1-\beta^-)}{\beta^-} \tag{7.3}$$

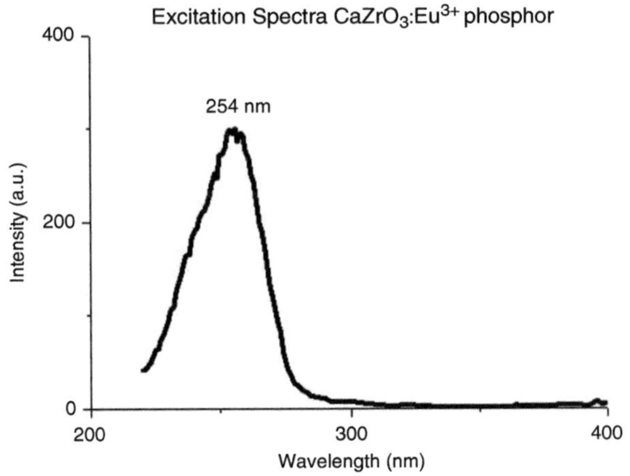

Fig. 7.6 PL emission spectra of $CaZrO_3:Eu^{3+}$ doped phosphor

$$b^{1/2} = \left[\frac{(1-\beta^{-})}{2}\right]^{1/2} \tag{7.4}$$

where 'ν' is the wave number of an absorption transition of Eu^{3+} and 'n' is the number of observed absorption transitions. The values of β^{-}, δ and $b^{1/2}$ in the present host matrix were calculated as 0.996, 0.004 and 0.045. According to the detail literature survey, the current values fulfil the possibility of covalent bonding between the Eu^{3+} ion and oxygen atom. Furthermore, the currently obtained values are lesser than the same reported earlier indicating the presence of more covalencey in the Eu-O bond (Fig. 7.6). Such outcome indicates the lesser ionic strength which is responsible for the shift in the CT band position towards higher wavelength.

Figure 7.6 shows the emission spectra of $CaZrO_3:Eu^{3+}$ phosphor with different concentration of Eu^{3+} ions. The emission spectra are characteristic $^4f_6$ energy level transition emission. They are mainly due to two dipole transitions. One is $^5D_0$ to $^7F_1$ magnetic dipole transition, and another is $^5D_0$ to $^7F_2$ forced electric dipole transition. The intensity ratio of $^5D_0$ to $^7F_2$ to $^5D_0$ to $^7F_1$ can be viewed as a clue concerning the nature of the chemical surroundings of the luminescent center and its symmetry [25].

The strong emission peak of $CaZrO_3:Eu^{3+}$ phosphor is due to forced electric dipole transition of $^5D_0$ to $^7F_2$ centered at 613 nm. It is characteristic red emission. The $CaZrO_3$ phase crystallizes in monoclinic system with space group P21/c. Then Eu^{3+} ions with C2h point symmetry are in the strict inversion center. Therefore, the phosphor should mainly exhibit the orange $^5D_0$–$^7F_1$ emission (for 594 nm). However, the strong emission peak of $CaZrO_3:Eu^{3+}$ crystalline phosphor is $^5D_0$–$^7F_1$ transition (centered at 594 nm) and $^5D_0$–$^7F_2$ transition (centered at 613 nm). It is possible that some Eu^{3+} ions deviate from the inversion center of the crystal lattice (Fig. 7.7).

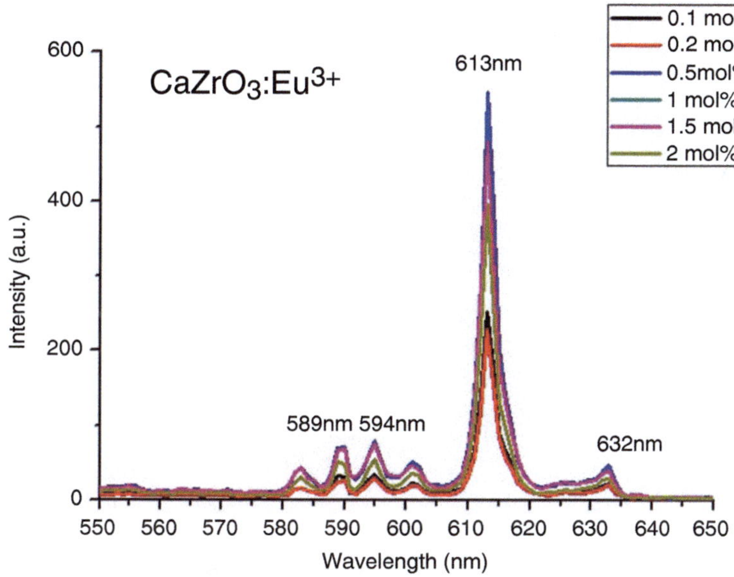

Fig. 7.7 PL emission spectra of $CaZrO_3:Eu^{3+}$ doped phosphor

The CIE coordinates were calculated by Spectrophotometric method using the spectral energy distribution of the $CaZrO_3:Eu^{3+}$ sample (Fig. 7.8). The color co-ordinates for the Eu^{3+} doped sample are x = 0.638 and y = 0.312 (these coordinates are very near to the orange-red light emission). Hence this phosphor having excellent color tenability from orange-red light emission.

7.3.2 Luminescence and Structural Analysis of $Y_2Zr_2O_7$ Phosphor Doped with Europium

7.3.2.1 FTIR Spectra of $Y_2Zr_2O_7$ Phosphor Doped with Europium

Figure 7.9 shows the FTIR spectra of $Y_2Zr_2O_7:Eu^{3+}$ phosphor prepared by solid state reaction method. A strong absorption peak at 443.63 & 649.1 cm^{-1} corresponds to Zr-O vibrational modes of ZrO_2 phase in $Y_2Zr_2O_7$ host. The band at 784.58 attributed to the Eu-O stretching and band at 848.25, 891.07 cm^{-1} corresponds to Y-O stretching. Moreover, the absorption band centered at 3787 cm^{-1} corresponds to OH stretching vibrations and peak centered at 1461.54 cm^{-1} corresponds to bending vibration of O-H in H_2O. Other peaks at 2924, 2866 cm^{-1} which is due to C-O bending representing very weak peak may be some phosphor material burn during the sample preparation.

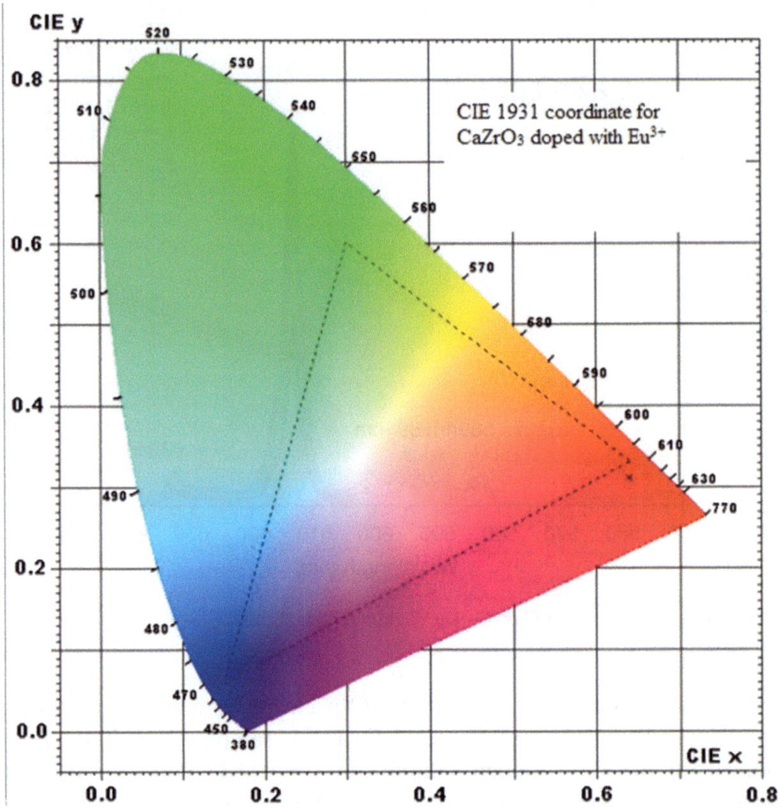

Fig. 7.8 CIE 1931 coordinate for $CaZrO_3$ doped with europium ion

7.3.2.2 TEM Analysis

Figure 7.10 shows TEM image of prepared phosphors for fixed concentration (2 mol%) of doping ions (Eu^{3+}). As this figures shows, the particle size of $Y_2Zr_2O_7$: Eu^{3+} Seems in the range of nano-meter. The surface morphology is uniform, some spherical shape was found in surface morphology of prepared phosphors which shows the cube like structure and good agreement for luminescence intensity enhancement.

7.3.2.3 Photoluminescence Study of Europium Doped $Y_2Zr_2O_7$ Phosphor

As seen, the excitation spectrum of Eu^{3+} in $Y_2Zr_2O_7$ host consists with a broad band at around 252 nm attributed to the charge-transfer (CT) transition from the oxygen 2p orbital to the empty 4f orbital of Eu^{3+} (Fig. 7.11). Whereas in the corresponding

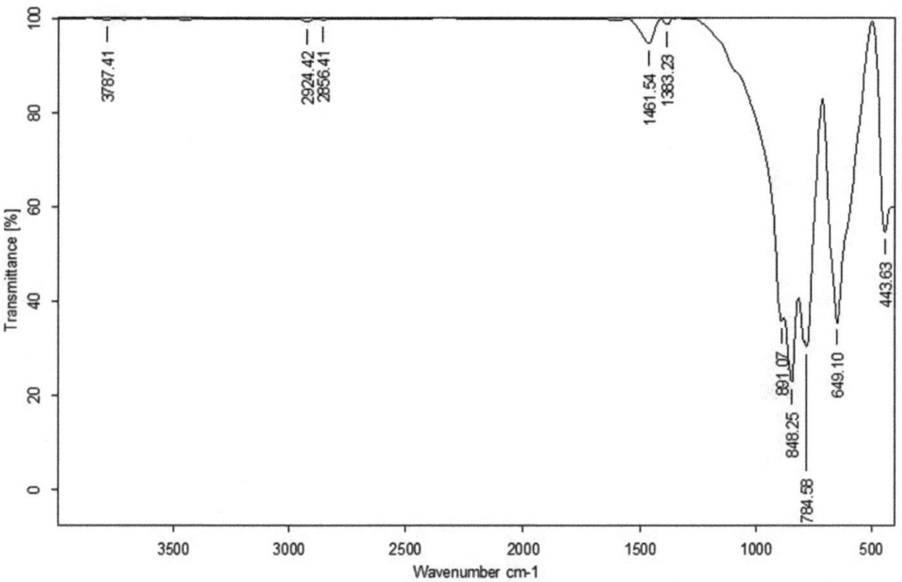

Fig. 7.9 FTIR spectra of europium doped $Y_2Zr_2O_7$ phosphor

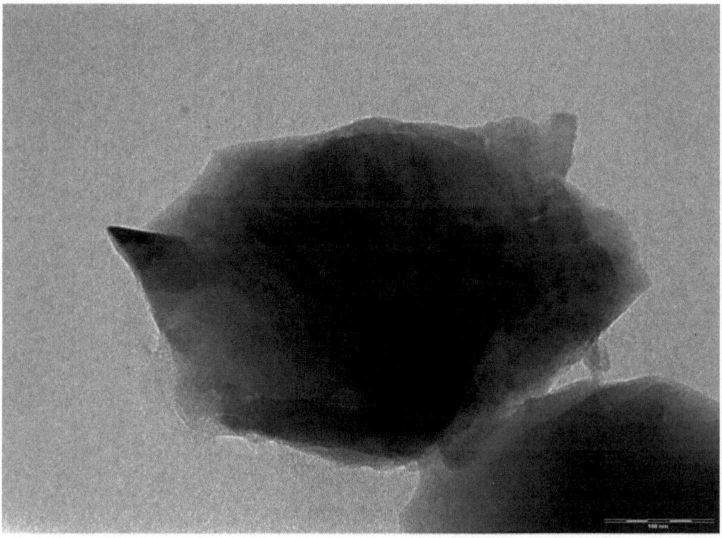

Fig. 7.10 TEM image of europium doped $Y_2Zr_2O_7$ phosphor

emission spectra the observed sharp peaks around ~ 379, 396, 416, 483 and 544, 616 and 629 nm are attributed to the $^5F_J \rightarrow {}^7F_0$, $^5L_6 \rightarrow {}^7F_0$, $^5D_0 \rightarrow {}^7F_0$ and $^5D_0 \rightarrow {}^7F_1$, $^5D_0 \rightarrow {}^7F_2$ electron transitions of Eu^{3+} ions, respectively (Fig. 7.12).

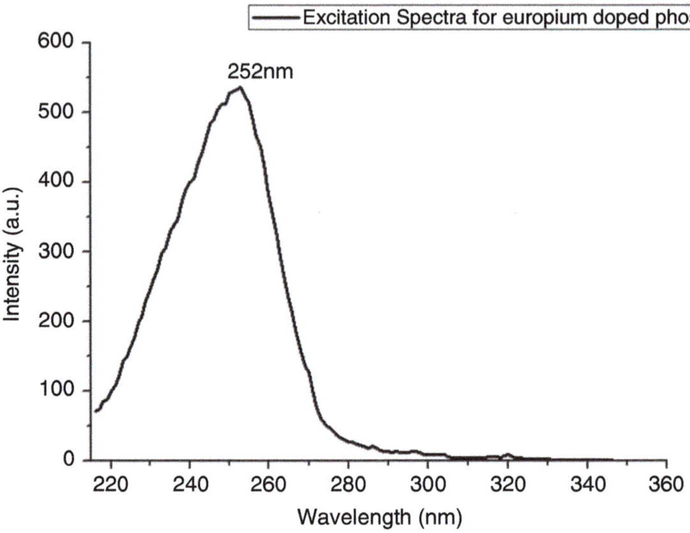

Fig. 7.11 Excitation spectra of Eu^{3+} in $Y_2Zr_2O_7$ host

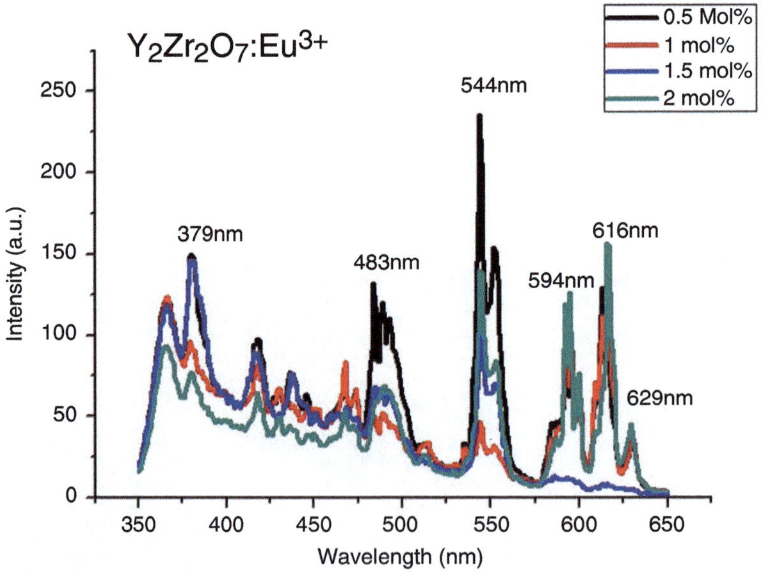

Fig. 7.12 PL emission spectra of $Y_2Zr_2O_7$:Eu^{3+} doped phosphor

From PL studies these prepared phosphors are suitable for compact florescent lamp (Ex. – 252) with excitation at 252 nm, several emission peaks corresponding to $^5D_0$-^7F_j (j = 0, 1, 2) transitions of Eu^{3+} ions are observed. The PL spectra of the nano spherical particles showed intense blue (469 and 483 nm), green (having highest intensity 544 nm), orange (594 nm) and red (616, 629 nm) luminescence with main peaks between 350-650 nm corresponding to magnetic dipole transition, $^5D_0$-$^7F_1$, of Eu^{3+} and the peak at 616 nm is attributed to structurally sensitive electric dipole transition, $^5D_0$-$^7F_2$, of Eu^{3+}. The photoluminescence study shows that the emission intensity of electric dipole transition ($^5D_0 \rightarrow {}^7F_2$) [616 nm] ($^5D_0 \rightarrow {}^7F_2$) dominates over that of magneic dipole transition ($^5D_0 \rightarrow {}^7F_1$) [594 nm]. The optimum concentration of Eu^{3+} (2 %) high for red emission but 0.5 mol% concentration of europium shows intense green emission (Fig. 7.12).The results indicated that present phosphor could find application for white light emitting diode (WLED) application because in this host matrix $Y_2Zr_2O_7$ europium doping ions gives their all possible transition in visible region.

7.3.2.4 CIE Coordinate for $Y_2Zr_2O_7$ Phosphor Doped with Europium Ion

Commission Internationale de l'e´ clairage (CIE) co-ordinate of the nano-phosphor is calculated. The CIE co-ordinate is found to be (0.32, 0.33), which is closed to that of the ideal white light (0.33, 0.33). The emission of white light from this nano-spherical shaped phosphor can be attributed to comparable intensities of the vacancy and Eu^{3+} emissions in the visible range of electromagnetic spectrum. To identify technical applicability of this white light, correlated color temperature (CCT) is determined from CIE coordinate [43]. Figure 7.13 shows Planckian locus in the CIE chromaticity diagram for calculation of CCT. The lines perpendicular to the Planckian locus is called isothermal lines and they nearly converge towards a point on the chromaticity diagram. A slope line can be drawn by joining the point and the coordinate (x, y) in quest. With this slope, CCT at the coordinate (x, y) can be calculated by using the approximation:

$$\mathrm{CCT}\,(x;y) = -449n^3 + 3525n^2 - 6823.3n + 5520.33 \tag{7.5}$$

Where $n = (x - x_e)/(y - y_e)$ is inverse of the slope line and ($x_e = 0.33$, $y_e = 0.19$) is the convergent point [44, 45]. The calculated CCT value of this nanocrystal is 5964 K which corresponds to vertical daylight (5500–6000 K) [44]. Therefore, the white light emitted from this nano-phosphor can be used for artificial production of vertical daylight of the Sun.

7.3.3 *Luminescence and Morphological Analysis of $La_2Zr_2O_7$ Phosphor Doped with Europium*

7.3.3.1 FTIR Analysis of $La_2Zr_2O_7$ Phosphor Doped with Europium

Figure 7.14 shows FTIR spectra of lanthanum zirconate phosphor doped with europium ion. In the spectrum the absorption band of zirconium groups are clearly evident. The intense broad band centered at 457.73, 591.11, 723 and 765 cm^{-1} is

Fig. 7.13 CIE coordinate for $Y_2Zr_2O_7{:}Eu^{3+}$ doped phosphor

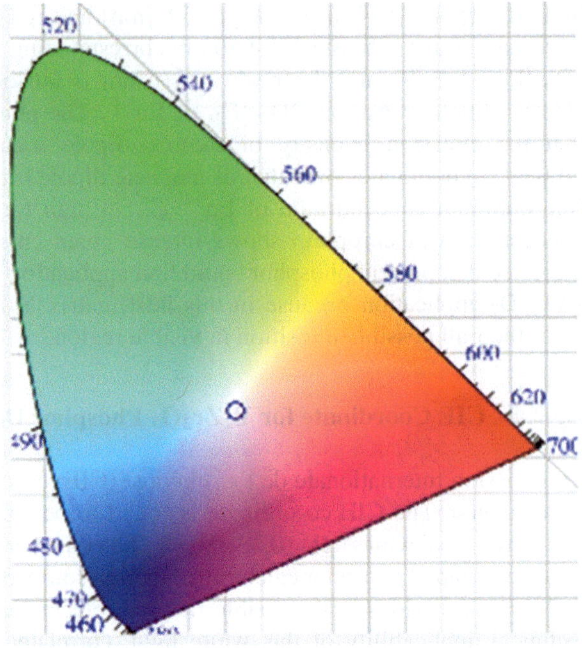

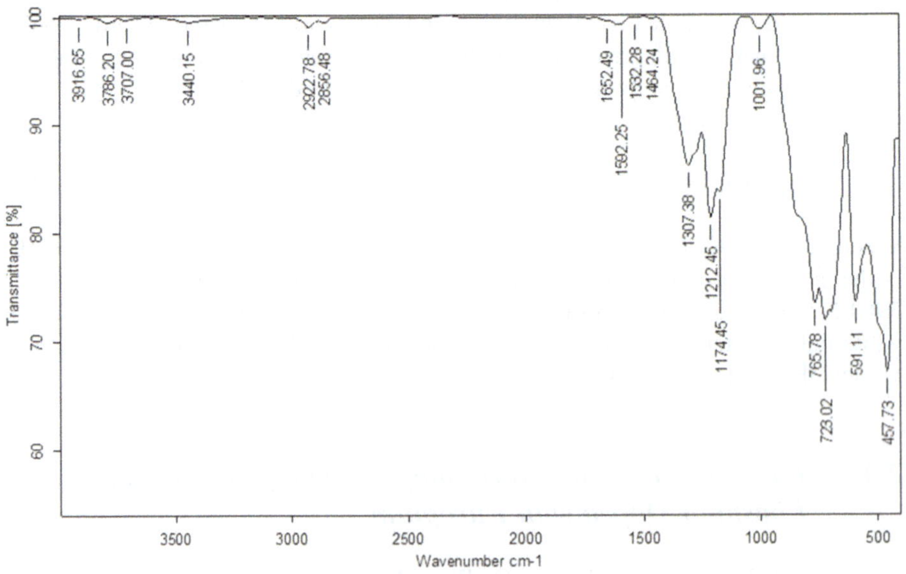

Fig. 7.14 FTIR spectra of La2Zr2O7 phosphor doped with europium for optimized concentration of europium ion (1.5 mol%)

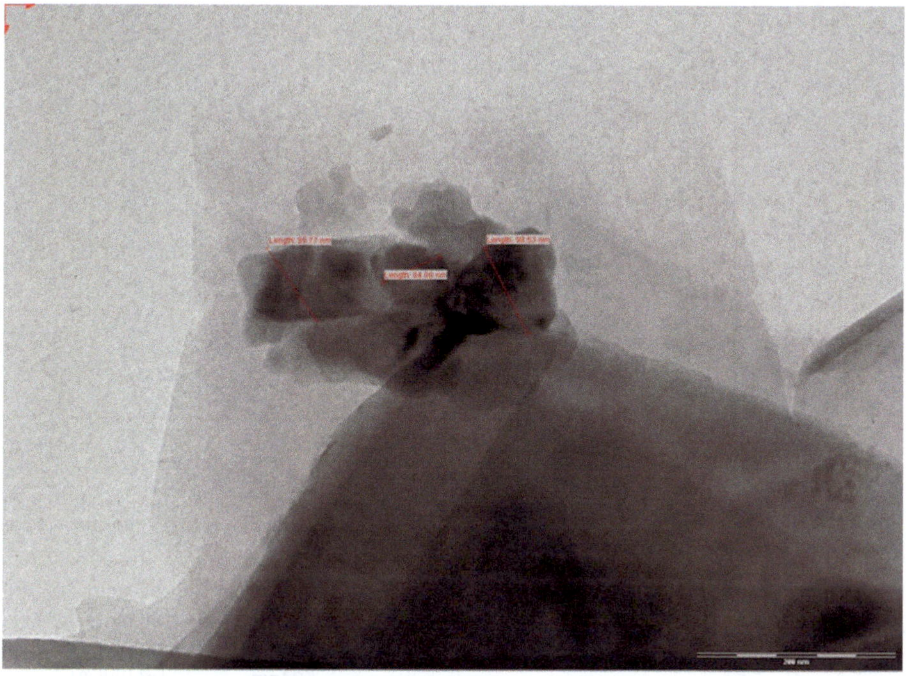

Fig. 7.15 TEM image of $La_2Zr_2O_7{:}Eu^{3+}$phosphor

assigned to the presence of Zr-O vibration. The band at 1174, 1307, 1001.98 cm^{-1} are assigned to La-O vibration mode. The band at 1592 cm^{-1} is due to Eu-O vibration. The band at 1652 cm^{-1} is assigned to the Eu-O vibration. It also confirms the formation of $La_2Zr_2O_7$: Eu^{3+} phosphor. Some peaks in 3440-3916 is may be H-O stretching due to some atmospheric moisture absorbed my materials.

7.3.3.2 TEM Image of $La_2Zr_2O_7$ Phosphor Doped with Europium

To investigate exact particle size and surface morphology of phosphor transmission electron microscopy images was obtained (Fig. 7.15). It is found that from TEM analysis of prepared LZO:Eu^{3+} sample the particles are uniformly distributed and some plate like structure was found. Surface morphology was clustered and particle size was found nearly 67–97 nm range.

7.3.3.3 PL Excitation and Emission Spectra

Figure 7.16 shows the excitation and emission spectra of $La_2Zr_2O_7{:}Eu^{3+}$phosphor sintered at 1300 °C by the solid state reaction method. The excitation spectra (Fig. 7.16) consist of a broad band extending from 215 to 300 nm. The former is

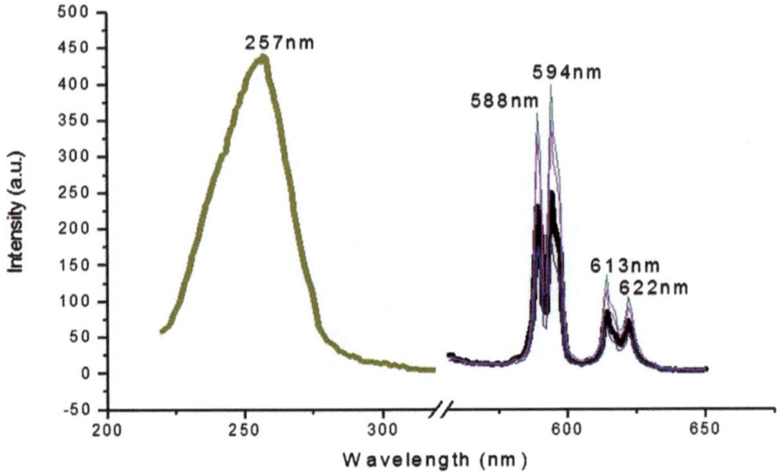

Fig. 7.16 PL excitation and corresponding emission spectra of $La_2Zr_2O_7$:Eu^{3+}phosphor (550–650 nm range emission)

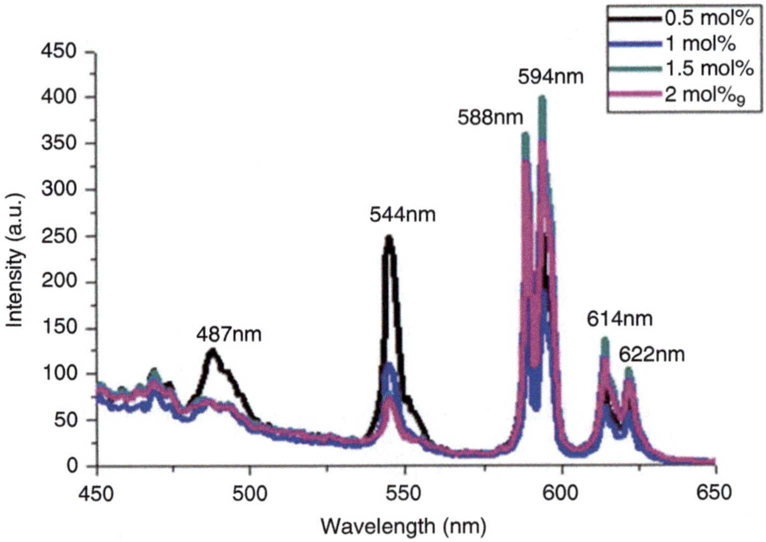

Fig. 7.17 PL emission spectra of $La_2Zr_2O_7$:Eu^{3+} doped phosphor

due to part of the charge transfer band (CTB) of Eu^{3+}–O^{2-} bond, and the latter is from the f–f transitions with in $4f^6$ electron of Eu^{3+}. Upon excitation in to the CTB at 257 nm, we obtained the emission spectrum (Fig. 7.17).

The emission spectrum of phosphors was recorded by excitation with 257 nm. The emission spectrum are shown in Fig. 7.17 which is composed of $^5D_0$–^7F_J (J = 1, 2, and 4, as labeled in the figure) emission lines of Eu^{3+}. In general, when the Eu^{3+} ion is located at crystallographic site without inversion symmetry, its hypersensitive forced electric-dipole transition $^5D_0$–$^7F_2$ red emission dominates in the emission spectrum. If the Eu^{3+} site possesses an inversion center, $^5D_0$–$^7F_1$ orange emission is dominant [10]. The distinct emission lines lying between 450–650 nm are observed due to transitions from excited $^5D_0$ to the ^7F_J (J = 0–3) levels of Eu^{3+} ions. The origin of these transitions (electric dipole or magnetic dipole) from emitting levels to terminating levels depend upon the location of Eu^{3+} ion in $La_2Zr_2O_7$ lattice and the type of transition is determined by selection rule [11]. The most intense peak in the vicinity of 594 nm is ascribed to the magnetic dipole transition of $^5D_0$ and $^7F_1$ levels. The weak emission at 614 and 622 nm corresponds to the hypersensitive transition between the $^5D_0$ and $^7F_2$ levels due to forced electric dipole transition mechanism.

Emission spectrum consists of emission peaks in the range of 450–650 nm, which result from $^5D_0 \rightarrow ^7F_J$ (J = 1, 2) transitions of Eu^{3+} ion, respectively. For an excitation wavelength of 395 nm, the emission spectrum of $La_2Zr_2O_7$:Eu^{3+} consists of sharp emission lines at 594 nm, 614 nm and 622 nm. The emission at 594 nm originates from the allowed magnetic dipole (MD) transition $^5D_0 \rightarrow ^7F_1$ which is having highest intensity. The peaks observed at 614 and 622 nm are due to the electric dipole $^5D_0 \rightarrow ^7F_2$ transition has the half the intensity when compared to highest emission intensity peak which is 594 nm. The ratio of 594 nm peak intensity to 612 and 628 nm peaks are one third. This result is interesting when compared the earlier work where the 594:612:628 nm peak intensities are nearly same [12]. From the emission spectrum it is clearly observed that the emission intensity of magnetic dipole was lower than that of electric dipole transition, due to this Eu^{3+} ions occupy a low symmetry site in $La_2Zr_2O_7$ host [14, 15]. Both magnetic dipole transition and electric dipole transition transitions are shown in the emission spectra. If the magnetic dipole transition $^5D_0 \rightarrow ^7F_1$ having the highest intensity then Eu^{3+} ions in host lattice occupies an inversion centre. If the emission intensity of magnetic dipole transition was lower than that of electric dipole transition, which indicates that Eu^{3+} ions occupied without an inversion symmetric centers in the host. The increasing concentrations of Eu^{3+} ions have no change on the peak position of emission spectra. The concentration quenching occurs at 1.5 mole% of Eu^{3+} ions concentration. This concentration quenching was associated with the energy transfer occurs from one Eu^{3+} ions to another ions and finally quenches the emission intensity [16, 17]. The results indicate that $La_2Zr_2O_7$:Eu^{3+} (1.5 %)phosphors can be selected as a potential candidate for LED (Light Emitting Diode) application (Ex.395) as well as for FL (Fluorescent Lamp) and Compact Fluorescent Lamp (CFL) (Ex.254).

Generally, the emission spectrum of $La_2Zr_2O_7$:Eu^{3+} consists of three sharp lines at 594, 614, and 621 nm, which are associated with the transitions from the excited $^5D_0$ level to ^7F_J ($J^{1/4}$ 1, 2) levels of Eu^{3+} ions [35, 36]. It is reported that the red emission (R) at 614 nm from the $^5D_0$–$^7F_2$ transition is a typical electric-dipole transition, whereas the orange emission (O) at 594 nm from the $^5D_0$–$^7F_1$ transition is a typical magnetic-dipole transition. In addition, it can clearly be seen that the emission intensity of the samples rapidly increase with the K/C concentration [34]. Our sample shows similar response with reported studies [79–82].

7.3.3.4 CIE Coordinate for $La_2Zr_2O_7$:Eu^{3+} Doped Phosphor (0.1–2 mol%)

The sharp peak of the vacancy state and Eu^{3+} peaks within the visible range (400–650 nm) disclose that this phosphor will emit white light (figure). Therefore, Commission Internationale de l'é clairage (CIE) co-ordinate of the nanophosphor is calculated. The CIE co-ordinate is found to be (0.30, 0.23), which is closed to that of the ideal white light (0.22, 0.21). The emission of white light from this phosphor can be attributed to comparable intensities of the vacancy and Eu^{3+} emissions in the visible range of electromagnetic spectrum. To identify technical applicability of this white light, correlated color temperature (CCT) is determined from CIE coordinate [43]. Figure 7.18 shows Planckian locus in the CIE chromaticity diagram for calculation of CCT. The lines perpendicular to the Planckian locus is called isothermal lines and they nearly converge towards a point on the chromaticity diagram. A slope line can be drawn by joining the point and the coordinate (x, y) in quest. With this slope, CCT at the coordinate (x, y) can be calculated by using the approximation:

$$\mathrm{CCT}\,(x;y) = -449n^3 + 3525n^2 - 6823.3n + 5520.33 \tag{7.6}$$

Where $n = (x-x_e)/(y-y_e)$ is inverse of the slope line and ($x_e = 0.33$, $y_e = 0.19$) is the convergent point [44, 45]. The calculated CCT value of this nanocrystal is 5894 K which corresponds to vertical daylight (5500–6000 K) [44]. Therefore, the white light emitted from this nanophosphor can be used for artificial production of vertical daylight of the Sun.

7.3.4 Luminescence and Morphological Analysis of $SrZrO_3$ Phosphor Doped with Europium

7.3.4.1 FTIR Analysis of $SrZrO_3$ Phosphor Doped with Eu^{3+}

The phase formation and purity of the products are further confirmed by FTIR spectroscopy, and results are shown in Fig. 7.19. The strong absorption peak 452.24, 645.28 cm^{-1} is ascribed to the stretching vibration of the Zr–O bond. The absorption peak at 3787 cm^{-1} is due to O–H stretching vibration of H_2O. This is typical of most of nano-materials, since their high specific areas results in significant adsorption of atmospheric water vapour. Some peaks 778.80, 847.43 is due to Sr-O vibration found in $SrZrO_3$ host. There are some impurity peaks of europium ion 1384.77, 1458.95 cm^{-1} ascribed Eu-O vibration.

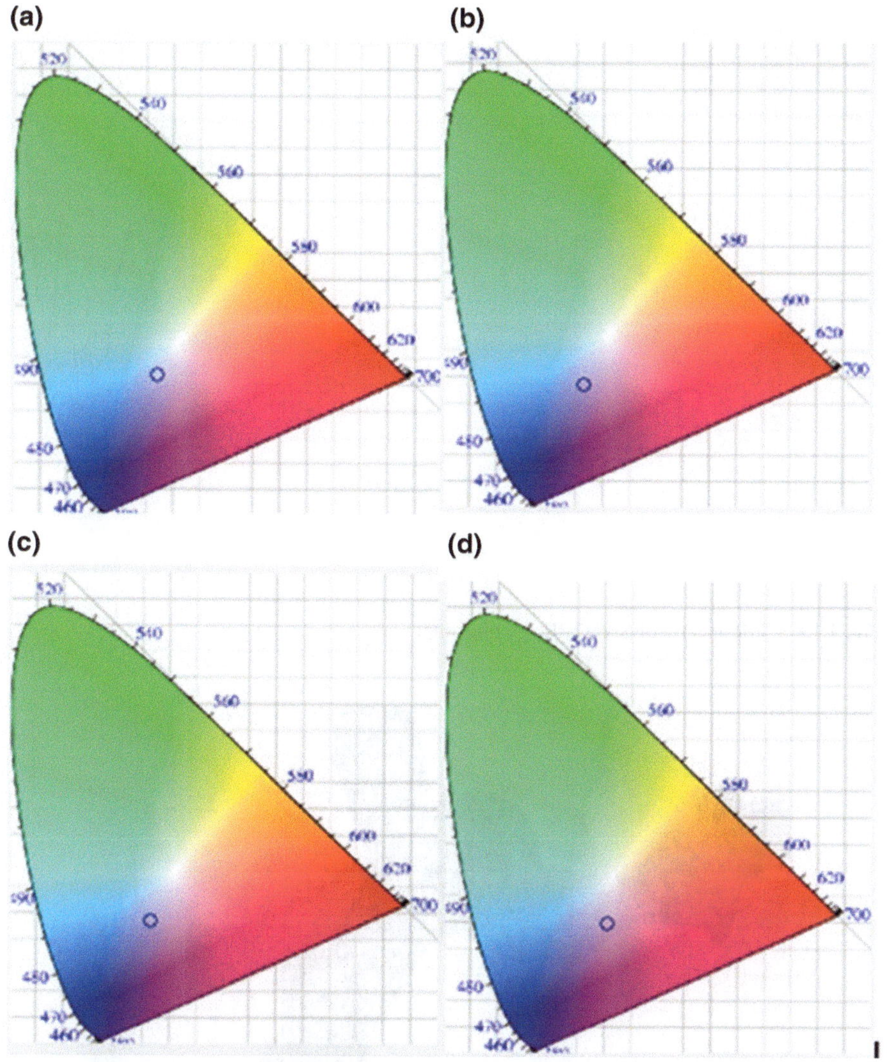

Fig. 7.18 **(a–d)**. CIE diagram showing the coordinate (0.30, 0.23), Planckian locus and point of convergence (0.22, 0.21)

7.3.4.2 TEM Image of $SrZrO_3$ Phosphor Doped with Europium Ion

Figure 7.20 shows representative TEM images of $SrZrO_3$ phosphor doped with 1.5 % of Eu^{3+}, ions prepared using solid state reaction method at 1300 °C for 4 h. Particles are evidently clustered/aggregated with sizes in the range of 70–90 nm for the 1300 °C heated sample. The sample is having nearly uniform particle size distributions.

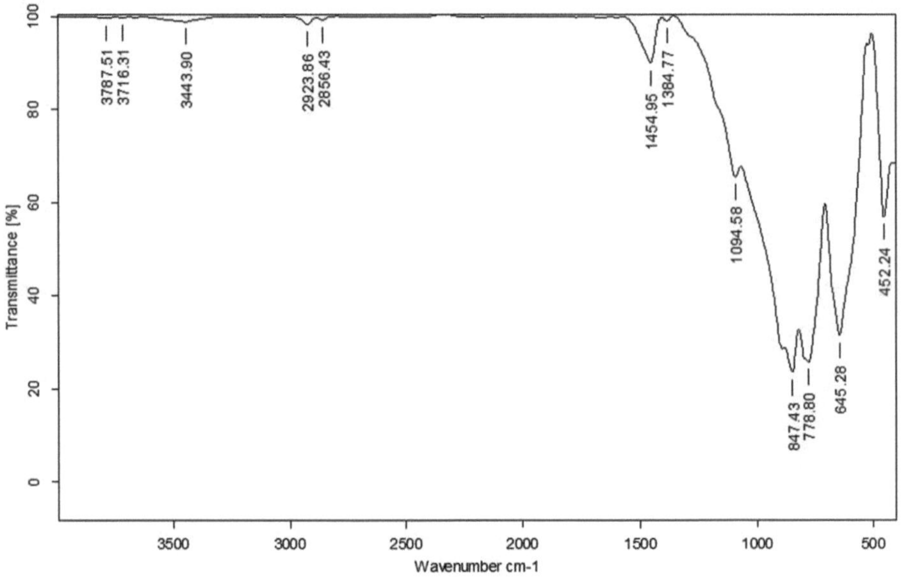

Fig. 7.19 FTIR spectra of $SrZrO_3$:Eu^{3+} phosphor

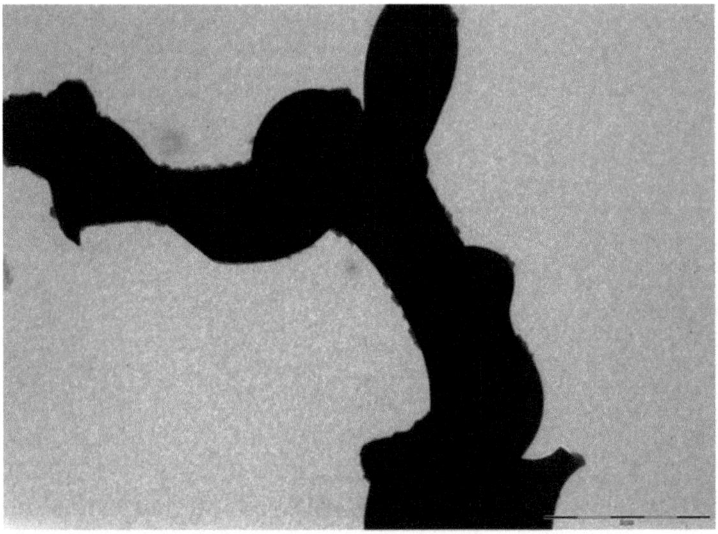

Fig. 7.20 TEM image of $SrZrO_3$ phosphor doped with Eu^{3+} (1.5 mol%)

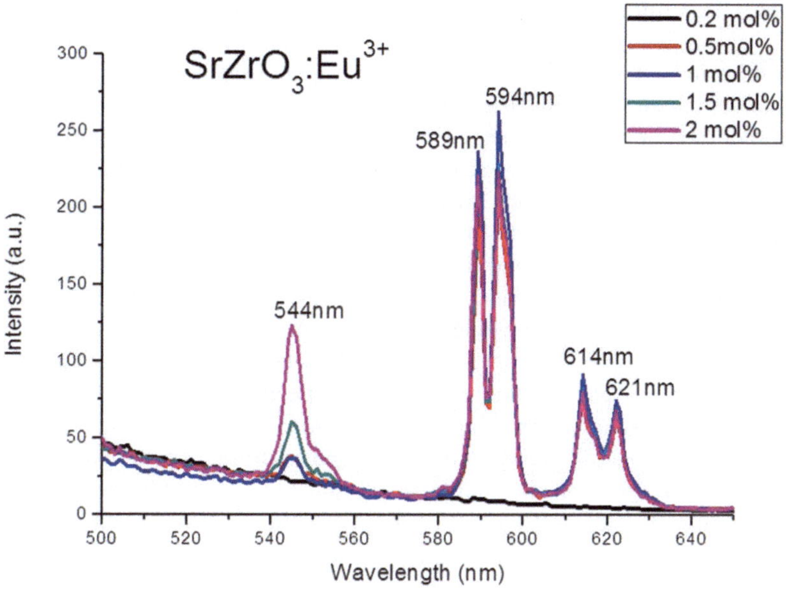

Fig. 7.21 PL emission spectra of $SrZrO_3:Eu^{3+}$ doped phosphor

7.3.4.3 PL analysis of $SrZrO_3$ Phosphor Doped with Europium Ion

The emission spectra of $SrZrO_3:Eu^{3+}$ phosphor contain a series of sharp lines in the regions of 400–650 nm, as shown in (Fig. 7.21). The lines belong to transitions between the energy levels of the $4f^6$ configuration of Eu^{3+} ion [92]. In addition, we can see that there are main emission lines at 544, 589, 594, 614 and 621 nm due to the $^5D_0$–$^7F_1$, $^5D_0$–$^7F_2$ and $^5D_0$–$^7F_3$ transitions, respectively. The peaks from $^5D_0$–$^7F_2$ (electric–dipole transition) are weaker than those from $^5D_0$–$^7F_1$ (magnetic–dipole transition). It is known that the $^5D_0$–$^7F_2$ transition of Eu^{3+} belongs to hypersensitive transitions, which is strongly influenced by crystal field outside surroundings. When Eu^{3+} is located at a low-symmetry local site (without an inversion center), this emission transition is often dominated in their emission spectra. Therefore, the Eu^{3+} ions occupy then on-inversion symmetric Zr^{4+} sites [93]. It is well known that $SrZrO_3$ lattice. There are one Sr^{3+} sites in $SrZrO_3$ also having Zr^{4+} sites, 75 % of these sites are non-centro symmetric with C_2 symmetry and the remaining 25 % are centro symmetric having S_6 symmetry [38]. When the Eu^{3+} ion is located at a low- symmetry local site without inversion center, this forced-electric dipole transition is often dominant in the emission spectrum. So the strongest $^5D_0$–$^7F_2$ transition (614 nm, 622 nm) and nearly all of other features in the spectrum are due to the Eu^{3+} on C_2 site. Here the forced magnetic dipole transition is dominant (594 nm) having excellent orange emission and 544 nm also dominant peaks which covers overall emission of europium ions.

7.3.4.4 CIE Coordinates for $SrZrO_3:Eu^{3+}$ Phosphor

The results indicate that $SrZrO_3:Eu^{3+}$ (1.5 %) phosphors can be selected as a potential candidate for LED (Light Emitting Diode) application as well as for FL (Fluorescent Lamp) and Compact Fluorescent Lamp (CFL) (Ex.254). However, the relative intensity of the emission bands which provide the fundamental colors balance for white-light emission was achieved with the 0.1 mol% sample with the spectrum (Fig. 7.22a–e) providing the CIE 1931 chromaticity coordinates much closer to the equal-energy white-light. If one increases the activator concentration even further, the emission intensity commences to decrease owing to concentration quenching. This concentration quenching is due to the increase in the ion–ion interaction provoked by the shorter distance between interacting activators as the concentration increases. The fluorescence light spectral profile as a function of the activator concentration was examined and the results indicated that the chromaticity coordinates of the overall emission light changed resulting in different colors of the overall emission light, for different concentrations as can be observed. It is found that the optimized concentration is 1.5 mol% for white light emission in the prepared phosphor which touches nearer white line boundary having excellent color tenability for white light emission.

7.4 Conclusion

In this work, $CaZrO_3$, $Y_2Zr_2O_7$ (YZO), $LaZr_2O_7$ (LZO) and $SrZrO_3$ doped with various amount of Eu^{3+} ions was synthesized by the solid state reaction method. The photoluminescence (PL) emission measurements upon excitation in ultraviolet range showed the characteristics emission lines of Eu^{3+} ions within 400–650 nm due to the $^5D_0 \rightarrow {^7F_J}$ (J = 0, 1, 2, 3, 4) transitions of Eu^{3+} ions. The occurrence of charge transfer band in the excitation spectra has been explained in details. The PL intensity was increased with the increase of Eu^{3+} concentrations up to a certain extent above which the PL intensity decreased owing to the concentration quenching. This has been explained on the basis of electron phonon coupling and multipolar interaction. Important spectral parameters were calculated via the Judd Ofelt approach. This observation concludes that the prepared phosphor is good emitting phosphor for the visible light excitations. Here for LZO doped with europium phosphor shows white light emission and having excellent color tenability for WLED applications. Similar observation was also found for the $Y_2Zr_2O_7$ (YZO) and $SrZrO_3$ phosphor but for $CaZrO_3$ phosphor doped with europium having red light emission so the application point of view the all prepared host matrix doped with a single doping ion i.e. europium give the all possible transition in visible region and found suitable for display devices application. The purity of the red emission has been checked and the obtained results indicate the suitability of Eu^{3+} doped zirconium based phosphor as a promising candidate for the lighting and display applications.

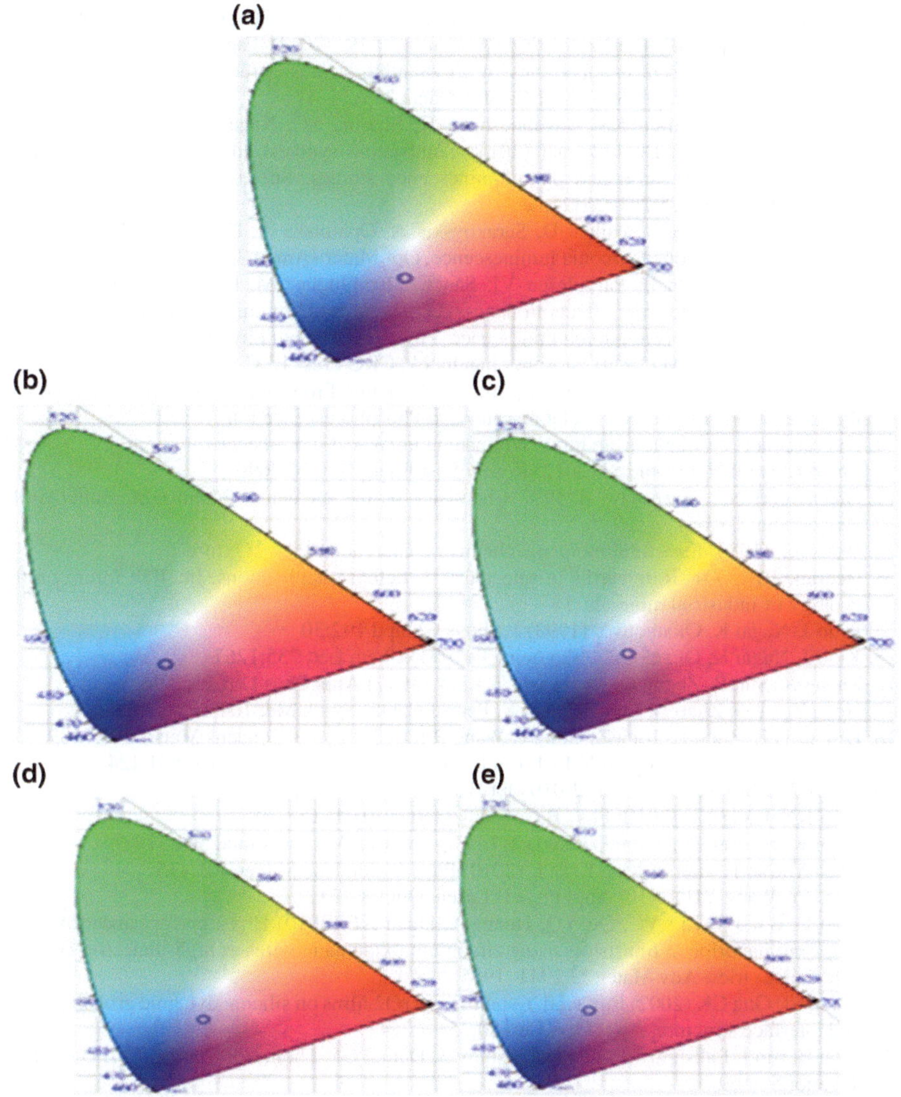

Fig. 7.22 (**a–e**) CIE coordinates for $SrZrO_3:Eu^{3+}$ (0.1, 0.2, 0.5, 1.5 and 2 mol%)

References

1. Tiwari N, Kuraria RK, Tamrakar RK (2014) Thermoluminescence glow curve for UV induced ZrO_2:Ti phosphor with variable concentration of dopant and various heating rate. J Radiat Res Appl Sci 7(4):542–549
2. Meetei SD, Singh SD, Sudarsan V (2012) Polyol synthesis and characterizations of cubic ZrO2:Eu3+ nanocrystals. J Alloys Compd 514:174–178

3. Gurushantha K, Anantharaju KS, Nagabhushana H, Sharma SC, Vidya YS, Shivakumara C, Nagaswarupa HP, Prashantha SC, Anilkumar MR (2014) Facile green fabrication of iron-doped cubic ZrO2 nanoparticles by Phyllanthus acidus: structural, photocatalytic and photoluminescent properties. J Mol Catal A Chem 397:36–47
4. Vidya YS, Anantharaju KS, Nagabhushana H, Sharma SC, Nagaswarupa HP, Prashantha SC, Shivakumara C, Danithkumar (2015) Combustion synthesized tetragonal ZrO2: Eu3+ nanophosphors: structural and photoluminescence studies. Spectrochim Acta Part A Mol Biomol Spectrosc 135:241–251
5. Smits K, Grigorjeva L, Millers D, Sarakovskis A, Opalinska A, Fidelus JD, Lojkowski W (2010) Europium doped zirconia luminescence. Opt Mater (Amst) 32(8):827–831
6. Tiseanu C, Cojocaru B, Parvulescu VI, Sanchez-Dominguez M, Primus PA, Boutonnet M (2012) Order and disorder effects in nano-ZrO2 investigated by micro-Raman and spectrally and temporarily resolved photoluminescence. Phys Chem Chem Phys 14(37):12970
7. Shionoya S, Yen WM (1999) Phosphor handbook. CRC Press LLC, Boca Raton
8. Fluorescence and Phosphorescence, [Online]. Available From http://www.Physik.Unibas.Ch/Praktikum/VPII/Fluorescence/Fluorescenceand Phosphorescence.Pdf
9. Fluorescence. Available from http://Wikipedia.Org/Wiki/Fluorescence
10. Thejo Kalyani N, Dhoble SJ (2012) Renew Sust Energ Rev 16:2696
11. Greenwood NN, Earnshaw A 1997 Chemistry of the elements, 2ndnd edn. Butterworth-Heinemann, Oxford
12. Wells AF (1984) Structural inorganic chemistry. Clarendon Press, Oxford
13. Hedrick JB (2004) Rare earths in selected U.S. defense applications. In: 40th forum on the geology of industrial minerals
14. Kido J, Nagai K, Okamoto Y (1993) J Alloys Compd 192:30
15. Kido J, Nagai K, Okamoto Y, Skotheim T (1991) Chem Lett 235:1267
16. Zhang X, Sun R, Zheng Q, Kobayashi T, Li W (1997) Appl Phys Lett 71:2596
17. Hong Z, Li W, Zhao D, Liang C, Liu X, Peng J (1995) Synth Met 104:165
18. Lan Y, Yi L, Zhou L, Tong Z, Gong F, Wang R (2010) Phys B Condens Matter 405:3489
19. Xie A, Yuan X, Wang F, Shi Y, Li J, Liu L, Mu Z (2010) J Alloys Compd 501:124
20. Kuo TW, Liu WR, Chen TM (2010) Opt Express 18:1888
21. Jung KY, Kang YC (2010) Phys B Condens Matter 405:1615
22. Shioi K, Hirosaki N, Xie RJ, Takeda T, Li YQ (2010) J Alloys Compd 504:579
23. Chen Y, Cao F, Tian Y, Xiao L, Li L (2010) Phys B Condens Matter 405:435
24. Jiao HY, Wang YH (2010) Appl Phys B Lasers Opt 98:423
25. Liao L, Bai JW, Lin YC, Qu YQ, Huang Y, Duan XF (2010) High- performance top-gated graphene-nanoribbon transistors using zirconium oxide nano wires as high-dielectric-constant gate dielec- trics. Adv Mater 22:1941–1945
26. Wang SJ, Ong CK (2002) Epitaxial Y-stabilized ZrO2 films on silicon: dynamic growth process and interface structure. Appl Phys Lett 80:2541–2543
27. John Berlin I, Lakshmi JS, SujathaLekshmy S, Daniel GP, Thomas PV, Joy K (2011) J gel Sci Technol 58:669
28. John Berlin I, Maneeshya LV, Jijimon K, Thomas PV, Joy K (2012) J Lumin 132:3077
29. John Berlin I, Anitha VS, Thomas PV, Joy K (2012) J Sol-Gel Sci Technol 64:289
30. Zhang H, Fu X, Niu S, Xin Q (2008) J Non Cryst Solids 354:1559–1563
31. Meetei SD, Singh SD, Sudarshan V (2012) J Alloys Compd 514:174–178
32. Gillani R, Ercan B, Qiao A, Webster TJ (2010) Int J Nanomedicine 5:1–11
33. Khatatbeh YA, Lee KKM, Kiefer B (2010) Phys Rev B 81:214102
34. Harrison HDE, Mclamed NT, Subbarao EC (1963) J Electrochem Soc 110:8–23
35. Khosravi AA, Kundu M, Kuruvilla BA (1995) Appl Phys Lett 67:2506–2508
36. Gu F, Lu MK, Wang SF, Qi YX, Song CF, Zhou GJ, Xu D, Yuan DR (2004) Appl Phys A Mater Sci Process 78:1059–1061
37. Lopez BJ, Luz VDL, Gonell F, Cordonicillo E, Haro ML, Calvino JJ, Escribano P (2012) J Alloy Compd 519:29–36
38. Chen L, Liu Y, Li Y, Alloys J (2004) J Alloy Compd 381:266–271

39. Kuratani K, Mizuhata M, Kajinami A, Deki S (2006) J Alloys Compd 408(412):711–716
40. Moon BK, Kwon IM, Jeong JH, Kim CK, Yi SS, Kim PS, Choi H, Kim JH (2007) J Lumin 122(123):855–857
41. Benyagoub A (2005) Phys Rev B 72:094114
42. Joo J, Yu T, Kim YW, Park HM, Wu F, Zhang JZ, Hyeon T (2003) J Am Chem Soc 125:6553–6557
43. Ayyub P, Palkar VR, Chattopadhyay S, Multani M (1995) Phys Rev 851:6135–6138
44. Soares MRN, Soares MJ, Fernandes AJS, Rino L, Costa FM, Monteiro TJ (2011) J Mater Chem 21:15262–15265
45. Brito FR, Armenta CA, Hipolito MG, Camarillo E, Hernandez AJ, Murrieta SH, Falcony C (2008) Opt Mater 30:1840–1847
46. Carcoache MC, Rivas PC, Pasquevich AP, Garcia ARI (1993) J Mater Res 8:605–607
47. Kim DJ, Jung HJ (1960) J Am Ceram Soc 43:234–239
48. Morell G, Katiyar RS, Torres D, Paje SE, Llopis J (1997) J Appl Phys 81:2830–2834
49. Mumpton FA, Roy R (1993) J Am Ceram Soc 76:2157–2160
50. Erwin SC, Zu L, Haftel MI, Efros AL, Kennedy TA, Norris D (2005) J Nat 43:91–94
51. Slonimskaya EA, Belyakov AV (2001) Glas Ceram 58(1–2):54
52. Davies RA, Islam MS, Gale JD (1999) Solid State Ionics 126:323
53. Osaka T, Numako C, Koto K (1999) Mater Res Bull 34(1):11
54. Matsuda T, Yamanaka S, Kurosaki K, Kobayashi S (2003) J Alloys Compd 351:43
55. Huntelaar ME, Cordfunke EHP, van der Laan RR (1996) Thermochim Acta 274:101
56. Zheng W, Pang W, Meng G (1998) Solid State Ionics 108:37
57. Pullar RC, Taylor MD, Bhattacharya AK (1998) J Mater Sci 33(13):3229
58. Kashinath C, Patil ST, Aruna T (2002) Mimani. Curr Opin Solid State Mater 6:507
59. Shen Y, Clarke DR (2009) Effects of reducing atmosphere on the luminescence of Eu 3+ doped yttria-stabilized zirconia sensor layers in thermal barrier coatings. J Am Ceram Soc 92:125–129
60. Ninjbadgar T, Garnweitner G, Borger A, Goldenberg LM, Sakhno OV, Stumpe J (2009) Synthesis of luminescent ZrO2: Eu3+ nano- particles and their holo graphic sub-micrometer patterning in polymer composites. Adv Funct Mater 19:1819–1825
61. Hobbs H, Briddon S, Lester E (2009) The synthesis and fluorescent properties of nanoparticulate ZrO2 doped with Eu using continuous hydrothermal synthesis. Green Chem 11:484–491
62. Gu F, Wang SF, Lu MK, Zhou GJ, Liu SW, Xu D, Yuan DR (2003) Chem Phys Lett 380:185
63. Torres LAD, la Rosa ED, Salas P, Romero VH, Chavez CA (2008) J Solid State Chem 181:75
64. Akiyama M, Xu CN, Nonaka K (2002) Appl Phys Lett 81:457–459
65. Chandra BP (2010) J Lumin 130:2218–2222
66. Torres LAD, Meza O, Solis D, Salas P, De la Rosa E (2011) Opt Lasers Eng 49:703–708
67. Giu F, Wang SF, Lu MK, Zhou GJ, Liu SW, Xu D, Yuan DR (2003) Chem Phys Lett 380:185–189
68. Kang YC, Park SB (2000) Mater Res Bull 35:1143–1151
69. Shilpa CJ, Dhananjaya N, Nagabhushana H, Sharma SC, Shivakumara C, Sudheerkumar KH, Nagabhushana BM, Chakradhar RPS (2014) Spectrochem Acta A 128:730–739
70. Patil KC, Hegde MS, Rattan T, Aruna ST (2008) Chemistry of nanocrystalline oxide materials, combustion synthesis, properties and applications. World Scientific Publishing Co. Pvt. Ltd., Singapore
71. Ro'g G, Dudek M, Kozlowska-Ro'g A, Buc'ko M (2002) Calcium zirconate: preparation, properties and application to the solid oxide galvanic cells. J Electrochim Acta 47:4523–4529
72. Yu T, Chen CH, Chen XF, Zhu W, Krishnan RG (2004) Fabrication and characterization of perovskite CaZrO3 oxide thin films. Ceram Int 30:1279–1282
73. Lee WJ, Wakahara A, Kim BH (2005) Decreasing of CaZrO3 sintering temperature with glass frit addition. Ceram Int 31:521–524
74. Iano R, Lazău R (2014) Chromium-doped calcium zirconate – a potential red shade pigment: preparation, characterization and testing. Dyes Pigments http://dx.doi.org/10.1016/j.dyepig.2014.02.007

75. Moon BK, Kwon IM, Jeong JH, Kim CK, Yi SS, Kim PS, Choi H, Kim JH (2007) J Lumin 122(123):855–857
76. Benyagoub A (2005) Phys Rev B 72:094114
77. Gillani R, Ercan B, Qiao A, Webster TJ (2010) Int J Nanomedicine 5
78. Dubey V, Kaur J, Agrawal S, Suryanarayana NS, Murthy KVR (2013) Optik Int J Light Electron Opt. doi:10.1016/j.ijleo.2013.03.153
79. Dubey V, Suryanarayana NS, Kaur J (2010) Kinetics of TL glow peak of limestone from Patharia of CG basin (India). J Miner Mater Charact Engin 9(12):1101–1111
80. Dubey V, Kaur J, Suryanarayana NS, Murthy KVR (2012) Res Chem Intermed. doi:10.1007/s11164-012-0872-7
81. Dubey V, Kaur J, Agrawal S (2013) Res Chem Int. doi:10.1007/s11164-013 1201-5
82. Dubey V, Kaur J, Agrawal S, Suryanarayana NS, Murthy KVR (2014) Superlattice Microst 67:156–171
83. Kaur J, Parganiha Y, Dubey V (2013) Luminescence studies of Eu^{3+} doped calcium bromofluoride phosphor. Phys Res Int 2013 . doi:10.1155/2013/4948075 pages
84. Rietveld HM (1969) J Appl Crystallogr 2:65–71
85. Qi Z, Shi C, Zhang G, Han Z, Hung HH (2013) Phys Stat Sol (a) 195(2):311–316
86. Lee WG, Lee DH, Kim YK, Kim JK, Park JW (2008) J Nucl Sci Tech Suppl 4:572–574
87. Huang K-W, Chen W-T, Chu C-I, Hu S-F, Sheu H-S, Cheng B-M, Chen J-M, Liu R-S (2012) Chem Mater 24:2220–2227
88. Liu X, Li L, Noh HM, Moon BK, Choi BC, Jeong JH (2014) Dalton Trans 43:8814–8825
89. Parchur AK, Ningthoujam RS (2012) RSC Adv 2:10859–10868
90. Condon SU, Shortley GH (1963) The theory of atomic spectra. Cambridge University Press, Cambridge
91. Som S, Kunti AK, Kumar V, Kumar V, Dutta S, Chowdhury M, Sharma SK, Terblans JJ, Swart HC (2014) J Appl Phys 115:193101
92. Yang J, Quan ZW, Kong DY, Liu XM, Lin J (2007) Cryst Growth Des 7:730
93. Ravichandran D, Roy R, Chakhovskoi AG, Hunt CE, White WB, Erdei S (1997) J Lumin 71:291

Chapter 8
Noninvasive Cancer Diagnostics Using Native Fluorescence Analysis of Biological Fluids

Zuzana Šteffeková, Terézia Kisková, Anna Birková, Mária Mareková, Zuzana Kostecká, Zdenka Bedlovičová, and Miroslava Smrčová

Keywords Cancer • Diagnostics • Metabolomics • Native fluorescence • Biological fluids

8.1 Cancer

Worldwide, cancer is today a major public health problem. It is currently the second leading cause of death in the United States and is expected to overtake heart diseases in mortality in the next few years [1]. Accordingly, up to 1.4 million Europeans are projected to die of cancer in 2015: 766,200 men and 592,900 women [2].

Cancer is basically defined as an uncontrolled dividing of transformed or abnormal cells that form a tissue called a tumor. From this point of view it is logical that molecular changes have first occurred in the evaluated cells. The uncontrolled dividing of cells takes place due to the altered cell cycle. In normal cells, this cycle cannot proceed without a series of successive events occurring [3]. The cycle is controlled at defined checkpoints, and if an event fails, then the cell arrests. However, transformed cells exhibit unregulated growth and genetic changes in cyclins and cyclin-dependent kinases as the checkpoints of the cell cycle. Because cancer cells have shown an altered p53 and RB protein (apoptosis triggers and therefore homeostasis controllers), cells can proliferate [3].

Z. Šteffeková (✉) • Z. Kostecká • Z. Bedlovičová • M. Smrčová
Department of Chemistry, Biochemistry and Biophysics, University of Veterinary Medicine and Pharmacy, Košice, Slovakia
e-mail: steffekova.zuzana@gmail.com

T. Kisková
Faculty of Sciences, Department of Animal Physiology, Institute of Biology and Ecology, Pavol Jozef Šafárik University, Košice, Slovakia

A. Birková • M. Mareková
Faculty of Medicine, Department of Medical and Clinical Biochemistry, Pavol Jozef Šafárik University, Košice, Slovakia

C.D. Geddes (ed.), *Reviews in Fluorescence 2016*, Reviews in Fluorescence,
DOI 10.1007/978-3-319-48260-6_8

The somatic mutation (molecular) theory deals with changes in DNA that can alter the characteristics of the cell so that it undergoes clonal expansion [4]. Changes in DNA may lead to altered genes, namely protooncogenes and tumor-suppressor genes. Protooncogenes encode many proteins that help to control cell growth and proliferation. However, mutations in these genes can contribute to the development of cancer. Conversion or activation of a protooncogene into an oncogene generally involves a mutation (point mutation, gene amplification, chromosomal translocation and/or epigenetic changes). On the other hand, tumor-suppressor genes encode proteins that are involved in the control of cell proliferation. The mutation of these genes contributes to the development of many cancers [5].

Tumor cells differ from normal cells in many mechanisms: gene expression, protein secretion, morphology, intracellular interactions, membrane properties, cytoskeletal structure, as well as growth control [5]. Growth control is regulated by growth factors and growth factor receptors providing communication between cells in a tissue. Under physiological conditions, cells receive a variety of signals from their tissue surroundings, primarily in the form of polypeptide growth factors. Integration of these extracellular signals underlies tissue homeostasis. Growth factors are the main regulators of all steps of tumor progression as well as metastasis [6].

Tracking the changes in tissue level creates a space to discuss the role of the tumor microenvironment, because it plays an important role in cancer development, epithelial-mesenchymal transition of cancer cells, angiogenesis, metastasis and the development of drug resistance [7]. An abnormal extracellular matrix affects cancer progression by directly promoting cellular transformation and metastasis [8]. Changes in the tumor microenvironment also include metabolic changes, e.g. lactate accumulation or a decrease in pH.

Monitoring such a wide range of molecules and metabolites draws the attention of researchers and clinicians to identify the proper cancer markers and work further with the most relevant of them. This is why biomarkers have many potential applications in oncology. They may be used in risk assessment, screening, determination of prognosis, prediction of response to treatment and monitoring disease progression [9]. Currently, metabolomics research is being used to discover relevant cancer biomarkers in the clinic, to determine cancer pathways that could be used for new targets and to monitor metabolic biomarkers during therapeutic intervention [10].

8.2 Cancer Metabolomics

All biological *in vivo* systems include a genome, a transcriptome, a proteome and a metabolome. However, only the latter represents the organism's phenotype [11]. It has been known for several decades that cancer metabolism differs from that of normal tissue [12]. Nevertheless, cancer metabolic reprogramming was only first recognized by Hanahan et Weinberg a few years ago [13]. Currently, metabolomics research is being used to discover and to determine relevant cancer biomarkers in clinical practice [10]. Moreover, the profiling of metabolic biomarkers is much cheaper and sample preparation simpler when compared with transcriptomics and proteomics. Data can be compared over several conditions and used in analysis between organisms [14].

The metabolome is defined as the quantitative complement of low-molecular-weight metabolites present in an organism under physiological or pathophysiological conditions [15]. Because thousands of metabolites are altered during cancer, there is a need to find them and define them. However, the questions still remain: which ones are relevant to be used as the essential markers and if and how do they differ in regard to cancer type?

There are five main metabolic pathways involved in the formation of the unique metabolomics of cancer: upregulation of glycolysis, glutaminolysis, pentose phosphate pathway, mitochondrial biogenesis and lipid metabolism [16].

The key finding in cancer metabolomics was the fact that cancer cells preferentially convert pyruvate to lactate rather than undergo the Krebs cycle under both aerobic and anaerobic conditions [17]. In normal cells, glucose is predominantly catabolized to pyruvate for proper functioning of the mitochondrial respiratory chain, followed by the Krebs cycle generating NADH and $FADH_2$ for energy production [18]. However, cancer cells compensate the lower efficacy of energy generation by intensive glucose uptake via glucose transporters upregulation [19]. Moreover, AKT1 (a serine/threonine kinase), hypoxia inducing factor and tumor suppressor protein p53 together cause an increased flux of glucose through glycolysis and downregulation of flux through the Krebs cycle [20]. The reactive by-products of sugar metabolism – advanced glycation end products (AGEs) – support alteration of cell signaling events and gene expression profiles. The nonenzymatic, spontaneous glycosylation of proteins, lipids and DNA causes many molecular consequences that can directly affect tumor biology. Glycative stress thus contributes to the development and progression of cancer [21, 22]. In addition to glycolysis, many cancer cells rely on glutaminolysis for a constant supply of nutrients and to fuel metabolism [23]. Glutaminolysis is supported by c-Myc, a factor indirectly regulating glutaminase. Glutaminase is a mitochondrial enzyme that converts glutamine to glutamate [24]. Thus, glutaminolysis builds a key source of energy and precursors for the Krebs cycle [25]. The intermediates of Krebs cycle can further be used for the synthesis of lipids, cholesterol, amino acids and other essential metabolites [16].

Another common metabolic pathway is the pentose phosphate pathway (PPP), which generates NADPH used for glutathione production in the oxidative arm of the PPP. On the other hand, the metabolites of the non-oxidative arm of the PPP (fructose-6-phosphate and glyceraldehyde-3-phosphate) could participate in glycolysis and downstream metabolic pathways. Importantly, the PPP builds a "defense" counteracting oxidative stress and supporting nucleotide synthesis [16]. Because tumor cells proliferate rapidly, they demand nucleotides and other material (predominantly lipids) for biosynthesis by requiring NADPH and pentose phosphate. Importantly, the PPP could promote tumor resistance to chemotherapy and radiation by providing NADPH, a potent antioxidative agent. NADPH protects cancer cells from the reactive oxygen species (ROS) induced cell death caused by chemotherapy and radiation [26].

Another change in cancer metabolism is the enhanced mitochondrial biosynthesis. Mitochondria are the main source of energy in most eukaryotic cells. They play a vital role in ATP production via oxidative phosphorylation and the Krebs cycle. Fuel catabolism produces NADH and $FADH_2$ via mitochondrial dehydrogenases by transferring electrons to oxygen via the electron transport chain [27]. While energy production is a major process performed by the mitochondrion, many other important metabolic pathways (e.g. fatty acid and amino acid oxidation, biosynthesis of pyrimidines, steroids, etc.) are also present [28]. Moreover, mitochondria control the redox balance and cellular homeostasis via Ca^{2+} concentration [29].

Lipidomics is a specialized subset of metabolomics that evaluates lipid profiles [30, 31]. Lipids play many important roles in cancer processes, including invasion, migration and proliferation [10]. Fatty acid synthesis strongly depends on glucose through the generation of acetyl-Co-A, a central metabolic precursor [32]. In tumor cells, most fatty acids are synthesized *de novo* by fatty acid synthase to arrange the intensive bioenergetics and structural changes. Indeed, increased levels of fatty acid synthase in breast cancer tissue indicate a poor prognosis [33]. Fatty acid oxidation is a repeated process of the production of acetyl-Co-A, NADH and $FADH_2$. Acetyl-Co-A is important for the Krebs cycle to further generate more NADH and $FADH_2$. NADPH derived from fatty acid oxidation is important for cancer cells because of its antioxidative effects [16].

The intense catabolism of sugar, lipid and protein stores as well as body weight loss and weakness lead to cancer cachexia syndrome [34]. Considering the intense lipid catabolism, high fasting triglycerides and low serum HDL-cholesterol were significantly associated with an increased breast cancer risk in one or more models [35]. On the other hand, an increase of total cholesterol in plasma could accelerate the development of tumors and is associated with the aggressiveness of the disease [36], though low total cholesterol levels have been considered as a risk marker for future cancer [37].

Hundreds of metabolites are known to get altered during cancer, and as seen below, many of them are native fluorophores. Thus, metabolic reprogramming is a key cancer hallmark. Cancer metabolites provide cancer cells with energy and necessary materials used in biosynthesis, proliferation, invasion and metastasis, and could cause a resistance to therapies [16]. Therefore, tracking alterations at different levels could yield promising results.

8.3 Techniques in Metabolomics

Altered metabolism plays a crucial role in carcinogenesis. The best way to explore the cancer metabolome is to reveal the phenotypic changes relative to biological function [11]. Therefore, employing the features of cancer metabolism for cancer detection and monitoring is a promising trend in cancer detection, treatment or even prevention [16].

The most effective and noninvasive way is to monitor metabolic changes in biological fluids, such as urine, blood, saliva, etc. However, when talking about metabolome analysis, there has to be a sufficient amount of liquid volume to evaluate the spectrum of biomarkers. Thus, most clinicians and researchers are used to working with blood and urine as the most affordable biological fluids.

Measurement of a complete set of metabolites requires rapid analysis due to the quenching of metabolism. Moreover, the method is time-consuming, the extraction and separation procedure is inadequate, and the metabolites could be present at widely different concentrations [15]. Therefore, development of suitable methods for metabolomics evaluation is currently a major challenge.

Nuclear magnetic resonance spectroscopy (NMR spectroscopy) is a technique that exploits the magnetic properties of atomic nuclei. NMR can provide detailed information about the structure, dynamics, reaction state and chemical environment of molecules. NMR is highly reproducible and can quantify the compounds in complex mixtures precisely due to the direct relationship between the peak area and the concentration. The main limitation of NMR technique is its sensitivity [38].

Mass spectrometry (MS) is probably the most commonly used platform in cancer metabolomics. It is an analytical technique that ionizes chemical species and sorts the ions based on their mass-to-charge ratio. Gas chromatography – mass spectrometry (GC–MS) combines GC with MS to identify different substrates within a sample. GC–MS is applied to the analysis of low molecular weight metabolites [39]. Hydroxyl, carboxyl and amino functional groups readily react with the chemical derivatisation agents that are added to convert these polar groups into derivatives with increased volatility. In the process of GC–MS, chemical samples are fractioned in the interaction with the stationary phase (analytical column) and mobile phase (carrier gas) [11]. Fragments are then detected, producing a spectrum for every eluent peak of the chromatogram. GC–MS is highly sensitive, quantitative and reproducible, and the latter has made more advanced compound identification possible compared with other techniques in the field [40]. On the other hand, liquid chromatography – MS (LC–MS) is a highly sensitive technique oriented towards the separation and potential identification of chemicals or particular masses in the presence of other chemicals or natural products. LC–MS may be prefixed with ultra or ultra-high performance and is particularly useful for non-volatile chemicals; therefore, it is complementary to GC–MS [39].

Positron emission tomography (PET) is a nuclear medicine technique used to observe metabolic processes in the body. PET is a widely used method that measures the signals of radio-labeled tracers (e.g. ^{18}F-fluoro-2-deoxyglucose, 3′-deoxy-^{18}F-fluorothymidine or 16-α-^{18}F-fluoro-17β-estradiol) taken up by cancer cells. For this reason, PET is commonly used for determining cancer stages as well as assessing the response of tumor to anti-cancer therapy [16].

Magnetic resonance imaging (MRI/MRSI) is an imaging method that provides spectroscopic information in addition to the image that is generated by MRI alone. MRI can provide morphological data, such as size, shape, number of lesions, presence of edema and necrosis, relationship with adjacent structures and characteristics related to intravenous contrast enhancement [41]. MRSI can be used to evaluate the activity of cells as well as metabolic activity occurring in the tumor. Therefore, MRI and MRSI are often combined in clinical oncology diagnostics [42]. In recent years MRSI has become a reliable imaging method [16]. Commonly found altered metabolites are lactate, citrate, choline, creatine, N-acetyl aspartate and ATP [43].

Autofluorescence analysis using fluorescence spectroscopy has been considered as one of the potential diagnostic tools for the detection of various cancers and even precancerous tissues. This technique is fast, highly sensitive and cheap compared to other techniques suitable for metabolomics approaches [44]. Nowadays, there are two main directions of endoscopic fluorescence diagnoses: autofluorescence and exogenous (fluorophore-enhanced) fluorescence [45]. While autofluorescence is generated by endogenous molecules, exogenous fluorescence diagnosis is based on the use of exogenous fluorophores revealing certain tumor selectivity (chlorins, phthalocyanines, porphyrins, or 5-aminolevulinic acid) [45].

Different analytical platforms are suited to different types of samples as well as experiments with respect to the metabolic information required from the sample. In addition to choosing the right analytical platform for the application, choosing the correct way to prepare a sample based on its type and the instrument to be employed for the experiment can enhance the results [16].

8.4 Native Fluorophores and Cancer

The pathological process changes the concentrations of various metabolites, and thus the light scattering and absorption properties of the tissues and body fluids. Atypical nuclei in abnormal cells lead to more expressive scattering. Optical methods provide new options for characterizing the physical and chemical changes occurring in tissues and cells, which results in interesting possibilities for new diagnostic and therapeutic procedures. There are a number of biologically important molecules with a system of conjugated bonds which are native or potential fluorophores. Native fluorescence is especially a property of conjugated polycyclic and aromatic molecules, like some vitamins (folic acid), coenzymes (NAD(P)H, FAD), structural proteins (collagen, elastin), lipids, porphyrins or aromatic amino acids and their derivates. Autofluorescence of collagen and elastin is related to structural arrangements of cells and tissues. Other fluorophores, such as tryptophan, tyrosine, phenylalanine, pyridoxine derivatives, NAD(P)H, FAD and porphyrins, are related to cellular metabolic and functional processes [46]. Among native fluorophores which are closely related to neoplastic transformation are the amino acids tryptophan and tyrosine [47], the structural proteins collagen and elastin [48], the coenzymes NAD(P)H and FAD [49] and advanced glycation end products [21] and porphyrins [46]. In addition to these fluorophores, a number of other fluorescent compounds occur in blood serum or urine. The characteristics of the fluorescence of tryptophan metabolites, catecholamines, bilirubin, phospholipids, lipoproteins and vitamins are well-known. Many other fluorophores have not yet been identified. The table below (Table 8.1) provides an overview of some fluorophores, present in blood and urine, which are considered as potential tumor markers.

Table 8.1 Representative fluorophores found in blood and urine of cancer patients. AGEs - advanced glycation end products

Fluorophore	Type of cancer	Bio-fluid	References
AGEs	Prostate cancer	Blood	Yang et al. [50]
Coproporphyrine	Ovarian cancer	Urine	Lu et al. [51]
	Cervical cancer	Urine	Lu et al. [51]
Elastin	Lung cancer	Plasma	Al Salhi et al. [52]
FAD, flavins	Colorectal cancer	Blood	Lawaetz et al. [53]
	Ovarian cancer	Urine	Lu et al. [51]
	Cervical cancer	Urine	Lu et al. [51]
		Urine	Masilamani et al. [54], [55]
	Different etiology	Urine	Masilamani et al. [49]
Isoxanthopterine	Stomach cancer	Urine	Wan et al. [56]
	Bladder cancer in rat	Urine	Huang et al. [57]
NAD(P)H	Colorectal cancer	Blood	Lawaetz et al. [53]
	Cervical cancer	Urine	Masilamani et al. [54], [55]
	Different etiology	Urine	Masilamani et al. [49]
			Rajasekaran et al. [44]
Porphyrine	Breast cancer	Blood	Kalaivani et al. [46]
	Colorectal cancer	Blood	Lawaetz et al. [53]
	Gastric cancer	Blood components	Masilamani et al. [58]
	Hodgkin's lymphoma	Blood components	Masilamani et al. [58]
	Different etiology	Blood	Xu et al. [59]
		Serum	Wenchong [60]
		Urine	Masilamani et al. [61]
Pteridine	Ovarian cancer	Urine	Zvarik et al. [62]
Tryptophan	Breast cancer	Blood	Bell et al. [63]
	Colorectal cancer	Blood	Lawaetz et al. [53]
	Lung cancer	Plasma	Al Salhi et al. [52]
	Bladder cancer in rat	Urine	Huang et al. [57]
	Colorectal cancer in rat	Urine	Steffekova et al. [64]
Tyrosine	Colorectal cancer	Blood	Lawaetz et al. [53]
Xanthopterin	Stomach cancer	Urine	Wan et al. [56]
	Bladder cancer in rat	Urine	Huang et al. [57]

8.5 Fluorescence of Biological Fluids

The composition of biological fluids results from metabolic processes. Significant changes in the concentration of fluorophores, in their ratios or the presence of atypical fluorophores, may be related to various pathological processes of an organism. Biological fluids have an interpretation disadvantage (compared to tissues) due to their higher variability. There are many various interactions between fluorophores and the environment, and many of them are still unknown. The most common changes in fluorescence spectral matrices may include: a reduction in fluorescence intensity – energy transfer to another fluorophore; an increase of fluorescence intensity – energy transfer from another fluorophore; shifts of excitation – emission maxima; the formation of new excitation – emission maximums; or other effects (e.g. non-linear summation spectra). Despite the complications in interpretation, the autofluorescence of biological fluids seems to be very beneficial.

8.6 Blood

One of the most common biological materials used for clinical examinations in human medicine is blood. However, it requires minimal invasive sampling. It contains many substances, such as proteins, whose composition is changed in various diseases. Pilot studies focused on blood fluorescence date from the end of the last century. Bell et al. [63] studied alterations in the concentration of degradation products of tryptophan in blood and their relationship to steroid hormone in regard to breast cancer. Xu et al. [59] studied porphyrins fluorescence at 635 nm in the serum of cancer patients. Wencheng [60] published a study on the use of porphyrins as tumor markers in blood and tissues. Leiner et al. [65] and Hubmann et al. [66] studied total serum fluorescence properties in detail. Masilamani et al. [61] first demonstrated the relationship between porphyrins and cell proliferation in an animal model, and a year later, Masilamani et al. [58] extended the same study to human blood. The work focused on the possibility of diagnostics of different cancer types by measuring porphyrins autofluorescence in serum, and it confirmed alterations in cancer patients, but different malignancies could not be distinguished from each other. They detected a dependence between a high concentration of porphyrins in blood and the grade of cancer [58]. In following years, many studies established the importance of the fluorescence technique used for blood analysis either in healthy [67] or cancer patients. The technique has become more popular, and the fluorophores present in many types of cancer have been established. Blood components of patients with gastric cancer, breast cancer and Hodgkin's lymphoma showed enhanced fluorescence around 630 nm due to the porphyrin fluorophore [58]. In breast cancer, the most reliable fluorophores that could be used as a tumor

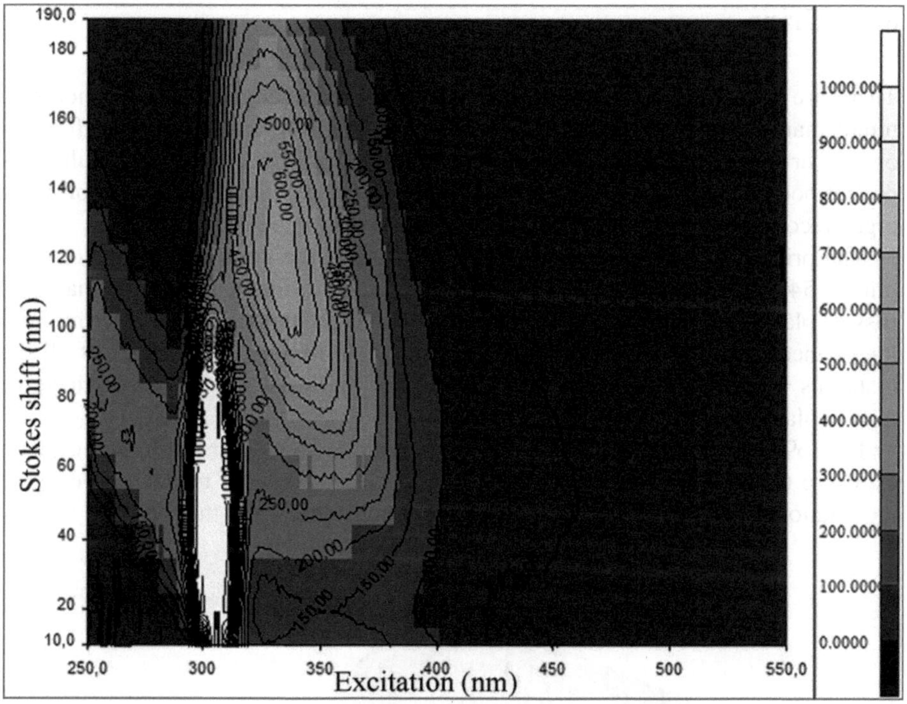

Fig. 8.1 Synchronous matrix (Δλ 30) nm of four-times diluted blood plasma of a healthy human

marker are the porphyrins seen at 639 nm as a neutral porphyrin and at 597 nm as a basic form of porphyrin [46]. Another group of researchers has been evaluating the fluorophores in prostate cancer using animal model – nude mice [68]. The blood porphyrin fluorophore showed enhanced fluorescence at 632 nm following tumor growth. Lawaetz et al. [53] revealed that fluorescence peaks differ when comparing the spectra from undiluted and diluted (100 times) samples. While in both samples the major peak is in the region with fluorescence from the aromatic amino acids tryptophan and tyrosine, in the undiluted sample a distinct peak is seen with the emission maximum at a higher wavelength. The peak structure could indicate the mixture of molecules, such as NAD(P)H, FAD and riboflavin. Blood obtained from cervical cancer patients again indicated the presence of porphyrin. However, the porphyrin of cervical cancer samples has approximately 1.6-times longer fluorescence time than that from the normal control [69]. In lung cancer, fluorescence peaks from tryptophan and elastin were found in the blood plasma of cancer patients [52]. Plasma fluorescent AGEs were associated with an increased risk of prostate cancer [50]. Figure 8.1 shows a synchronous matrix (Δλ 30 nm) of four-times diluted blood plasma of a healthy human.

8.7 Urine

There are more studies that describe in detail blood endogenous fluorophores as tumor markers using fluorescence spectroscopy. Much fewer studies have focused on cancer diagnosis by using the autofluorescence of urine. Urine is a multicomponent body system –a solution of many inorganic salts with the presence of many organic compounds including a lot of fluorescence compounds (Fig. 8.2.)

The principal advantage of urinalysis is basically the non-invasiveness of sampling. [54, 55] compared the fluorescence of four different biological materials (tissue, plasma, urine and pleural effusion) obtained from one patient suffering from lung cancer. Despite the diversity of materials, the fluorescent features of all the materials were specific and typical. In urine, this finding was even more detectable than in plasma and pleural effusion.

In 1869 Jaffe [70] noted that normal urine exhibited a blue-green fluorescence on exposure to ultraviolet radiation. Subsequently attempts were made to investigate the relation between the output of fluorescent substances and that of other urinary

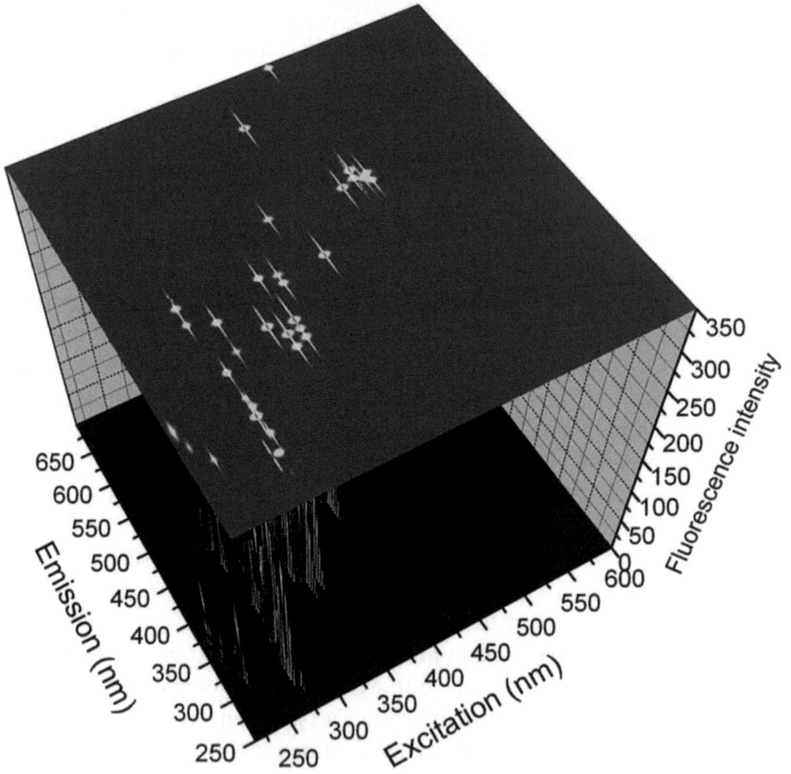

Fig. 8.2 Fluorescence metabolome of urine

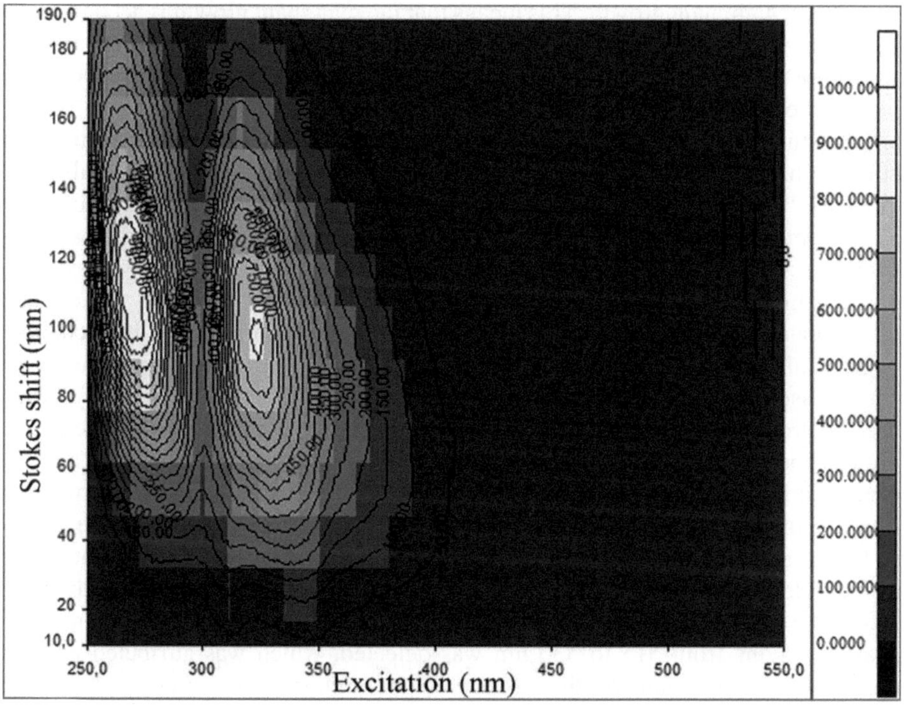

Fig. 8.3 Synchronous matrix (Δλ 30) nm of five-times diluted urine of a healthy human

constituents in human urine. In this context Squires and Jeffree [71] investigated the fluorescence of urine by means of a simple form of ultraviolet fluorophotometer. Within certain limits of dilution, the intensity of fluorescence is proportional to the concentration of the fluorescent substance. Based upon this principle, a series of extended observations on eight individuals was completed. The results indicated that the daily output of fluorescent material was independent of the volume of urine excreted and bore no close relation to that of the familiar urinary constituents. Figure 8.3 shows the synchronous matrix (Δλ 30 nm) of five-times diluted urine of a healthy human.

The first oncological application includes the research work of Rabinowitz [72], who reports on studies of fluorescent intensity of blue fluorescent compounds (B values) and red fluorescent compounds (R values), respectively, present in human urine, their relation to each other and their significance in normal, benign and malignant growth. The B and R values and the B/R ratios on samples of 24-h urines were determined for three groups of cases: (a) controls comprising cases free of benign or malignant growth in good health and those free of benign or malignant growth with associated ailments; (b) proven cases of benign growth, including pregnancy; (c) proven cases of malignant growth. Compared to the control group, the B/R ratio was increased in the group with benign growth and decreased in the group with malignant growth. This means that the malignant growth is accompanied by an increase in the red fluorescent compounds and a decrease of compounds with blue fluorescence. The result opened up a new field of interest in the study of fluorescent porphyrins as related to the genesis of cancer.

The study of native fluorescence of urine in a context with cancers of different etiology has been mentioned in several papers. Masilamani et al. [73] published the work "Fluorescence and synchronous spectral characteristics of urine for early cancer detection", which reports the result of investigation on optical diagnosis of cancer from urine analysis. They compared the fluorescence emission spectra and synchronous fluorescence spectra of 125 urine samples of healthy donors and of 185 cancer-diseased patients of different etiology. By employing the technique on specially prepared urine, they were able to show the distinctive features of cancer-diseased urine. In 2010 Masilamani et al. published similar study. They used fluorescence emission spectra and Stokes' shift spectra to analyze the native fluorescence of the first voided urine samples of 100 healthy controls and 50 cancer patients of different etiology. They showed that flavoproteins and porphyrins released into the urine can act as generic biomarkers of cancer with a specificity of 92 %, a sensitivity of 76 % and an overall accuracy of 86.7 %. Patients were also divided into those who had early stage ($n = 16$) and those who had advanced stage ($n = 34$) cancer. In patients with advanced cancer a shift of the emission peak of flavins from 515 to 530 nm was detected, which was attributed to their excessive concentration. At the same time the authors stated that the advanced stages of various cancers have similar spectra. Porphyrins and flavins (perhaps bilirubin) are increased, and NADH is decreased. Rajasekaran et al. [44] compared the autofluorescence features of human urine samples using excitation–emission matrices over a range of excitation and emission wavelengths, and emission spectra at 405 nm excitation, to discriminate cancerous patients of different etiologies and stages from a healthy control group. The fluorescence spectra of urine samples of cancer patients exhibited considerable spectral differences in both excitation–emission matrices and emission spectra with respect to the control group. The authors calculated different ratios using the fluorescence intensity values of the emission spectra. The discriminant analysis classified 94.7 % of the original grouped cases and 94.1 % of the cross-validated grouped cases correctly. Based on the fluorescence emission characteristics of urine and statistical analysis, the authors consider NADH and flavins as cancer markers.

Most works about cancer diagnosis by analyzing urine native fluorescence are dedicated to gynecological malignancies. Among other studies, Lu et al. [51] studied the intrinsic fluorescence spectra of urine in relation to ovarian cancer. The three groups of analyzed urine samples (ovarian cancer, cervical carcinoma groups and healthy donors) showed different emission spectra peak and fluorescence intensity. The ovarian cancer group had the largest displacement of maximum spectra peak at 380–400 nm excitation wavelength. The fluorescence intensity from the ovarian cancer group was significantly higher with the healthy donors and the cervical carcinoma group in comparison to the short wave excitation region of 320–360 nm, while the cervical carcinoma group had higher fluorescence intensity than the ovarian cancer group at 380–500 nm excitation. Fluorescence spectra at 440 nm excitation were characterized by three peaks, which originated from coproporphyrin, riboflavin and p-hydroxyphenol derivatives existing in the urine. In 2011 Masilamani et al. presented the results of a study to distinguish cervical cancer patients ($n = 50$) from healthy subjects ($n = 50$). They analyzed fluorescence emission spectra of urine obtained by excitation at 400 nm and the Stokes' shift spectra ($\Delta\lambda$ 70 nm). In the Stokes' shift spectra, the ratio of intensity of the two bands due to flavin and NADH was 0.2 for normal controls and 0.8 for cancer patients. Combining fluorescence analysis of blood and urine showed a sensitivity of 80 % and a specificity of 78 %.

Zvarík et al. [62] dealt with the fluorescence characteristics of urine of ovarian cancer patients using fluorescence excitation-emission matrices. They detected the most significant difference at emission wavelengths from 400 to 460 nm, when excited at 310–390 nm. Statistical analyses of their data showed a fivefold reduction in the intensity of the peak at 330/420 nm (excitation/emission wavelength) for urine samples of cancer patients as compared with the control group. Moreover, the ratio of intensities of the peaks at 370/440 nm and at 330/420 nm was 18-times elevated in the urine of ovarian cancer patients. They considered the detected changes as a reduction of the presence of pyridoxic acid, whereas blue-fluorescing pteridines become dominant in the excitation-emission matrices of cancer urine samples. They suggested pteridines, which are related to cellular metabolism, as suitable candidates for neoplasia-associated fluorescent markers in human urine. Birková et al. [74] compared the synchronous spectrum ($\Delta\lambda$ 30 nm) of undiluted urine of three groups of subjects: healthy women without cancer anamnesis, patients with a benign ovarian tumor and patients with malignant ovarian cancer. They detected significant differences in several areas of the spectrum between patients with malignant ovarian cancer and healthy women (at 315, 425 and 540 nm) or those with benign histology (350 and 540 nm). The spectra from women suffering from ovarian cancer differed, with the highest significance in the part of spectra from 505 to 545 nm. They did not detect significant differences between the healthy group and women with benign ovarian tumors.

In 2010 Wan et al. studied the xanthopterin and isoxanthopterin content in human urine using synchronous fluorescence spectroscopy in the context with stomach cancer. The results showed that the average level of isoxanthopterin was significantly elevated in urine excreted by stomach cancer patients ($P < 0.01$), while no significant change of xanthopterin level was found between stomach cancer patients and healthy individuals.

Some studies have also examined the native fluorescence of urine of animals, such as rats and dogs, in connection with a cancer diagnosis. Šteffeková and Birková [75] compared the fluorescence fingerprints of diluted urines of healthy dogs and dogs with various diseases, including cancer. The urine of dogs with cancer differs from healthy dogs and those with other disorders.

Huang et al. [57] compared three potential cancer biomarkers – tryptophan, isoxanthopterin and xanthopterin – in the urine of rats with bladder cancer and a control group using synchronous fluorescence spectra ($\Delta\lambda$ 70 nm). Results showed that the average level of tryptophan, isoxanthopterin and xanthopterin shows different changing trends with the growth of the tumor. Šteffekova et al. [64] studied the fluorescence fingerprints of the urine of rats with induced colorectal cancer and control rats. The study was focused on the possibility of detection of early lesions of colorectal cancer. They detected significant differences ($P < 0.05$) between the two groups at the 300 nm wavelength of the fluorescence fingerprints of rat urine.

References

1. Llaverias G, Danilo C et al (2011) Role of cholesterol in the development and progression of breast cancer. Am J Pathol 178:402–412
2. Richardson AD, Yang C et al (2008) Central carbon metabolism in the progression of mammary carcinoma. Breast Cancer Res Treat 110:297–307
3. Henry NL, Hayes DF (2012) Cancer biomarkers. Mol Oncol 6(2):140–146
4. Griffiths EJ, Rutter GA (2009) Mitochondrial calcium as a key regulator of mitochondrial ATP production in mammalian cells. Biochim Biophys Acta 1787:1324–1333
5. De Berardinis RJ, Sayed N et al (2008) Brick by brick: metabolism and tumor cell growth. Curr Opin Genet Dev 18:54–61
6. De Berardinis RJ, Mancuso A et al (2007) Beyond aerobic glycolysis: transformed cells can engage in glutamine metabolism that exceeds the requirement for protein and nucleotide synthesis. Proc Natl Acad Sci USA 104:19345–19350
7. Van QN, Veenstra TD (2009) How close is the bench to the bedside? Metabolic profiling in cancer research. Genome Medicine 1:5
8. Yang S, Pinney SM et al (2015) Impact of oxidative stress biomarkers and carboxymethyllysine (an advanced glycation end product) on prostate cancer: a prospective study. Clin Genitourin Cancer 13:347–351
9. Leiner MJ, Schaur RJ et al (1986) Fluorescence topography in biology. III. Characteristic deviations of tryptophan fluorescence in sera of patients with gynecological tumors. Clin Chem 32:1974–1978
10. Hubman MR, Leiner MJ et al (1990) Ultraviolet fluorescence of human sera: I. Source of characteristic differences in the ultraviolet fluorescence spectra of sera from normal and cancer-bearing humans. Clin Chem 36:1880–1883
11. Guimaraes MD, Schuch A et al (2014) Functional magnetic resonance imaging in oncology: state of the art. Radiol Bras 47(2):101–111
12. Lin JA, Wu CH et al (2016) Glycative stress from advanced glycation end products (AGEs) and dicarbonyls: an emerging biological factor in cancer onset and progression. Mol Nutr Food Res doi. doi:10.1002/mnfr.201500759
13. Wang HW, Willis J et al (1999) Quantitative laser scanning confocal autofluorescence microscopy of normal, premalignant and malignant colonic tissues. IEEE Trans Biomed Eng 46:1246–1252
14. Nagrath D, Caneba C et al (2011) Metabolomics for mitochondrial and cancer studies. Biochim Biophys Acta 1807:650–663
15. Huang L, Guo L et al (2014) Simultaneous determination of three potential cancer biomarkers in rat urine by synchronous fluorescence spectroscopy. Spectrochim Acta A Mol Biomol Spectrosc 120:595–601
16. Masilamani V, Al-Zhrani K et al (2004) Cancer diagnosis by autofluorescence of blood components. J Lumin 109:43–154
17. Masilamani V, Kalaivani R et al (2009) Fluorescence and synchronous spectral characteristics of urine for early cancer detection. J King Saud Univ 21:75–83

18. Kalaivani R, Masilamani V et al (2014) Cervical cancer detection by time-resolved spectra of blood components. J Biomed Opt 19:057011
19. Masilamani V, Vijmasi T et al (2010) Cancer detection by native fluorescence of urine. J Biomed Opt 15:057003-1-057003-9
20. Malvezzi M, Bertuccio P et al (2015) European cancer mortality predictions for the year 2015: does lung cancer have the highest death rate in EU women? Ann Oncol 26:779–786
21. Kim JW, Dang CV et al (2006) Cancer's molecular sweet tooth and the Warburg effect. Cancer Res 66:8927–8930
22. De Oliveira Silva FR, Bellini MH et al (2010) Intrinsic fluorescence of protoporphyrin IX from blood samples can yield information on the growth of prostate tumours. J Fluoresc 20:1159–1165
23. Boros LG, Torday JS et al (2000) Transforming growth factor beta2 promotes glucose carbon incorporation into nucleic acid ribose through the nonoxidative pentose cycle in lung epithelial carcinoma cells. Cancer Res 60:1183–1185
24. Squires BT, Jeffree JH (1928) A simple ultra-violet fluorophotometer. J Sci Instr 5:273–277
25. Wallace D (2012) Mitochondria and cancer. Nat Rev Cancer 12:685–698
26. Wenchong L (1989) Some fluorescence observation on the canceration tissue and the blood of cancer patients. Proc SPIE 1054:196–199
27. Warburg O, Wind F, Negelein E (1927) The metabolism of tumours in the body. J General Physiol 8:519–530
28. Mountford C, Ramadan S et al (2009) Proton MRS of breast in the clinical setting. NMR Biomed 22:54–64
29. Gao P, Tchernyshyov I et al (2009) cMyc suppression of miR23a/b enhances mitochondrial glutaminase expression and glutamine metabolism. Nature 458:762–765
30. Gonzalez HC, Esteban L et al (2010) The role of magnetic resonance imaging in oncology. Clin Transl Oncol 12:606–613
31. Riganti C, Gazzano E et al (2012) The pentose phosphate pathway: an antioxidant defense and a crossroad in tumor cell fate. Free Radic Biol Med 53:421–436
32. Vermeersch KA, Styczynski MP (2013) Applications of metabolomics in cancer research. J Carcinog 12:9
33. Wan YQ, Tang LJ et al (2010) Simultaneous determination of xanthopterin and isoxanthopterin in human urine by synchronous fluorescence spectroscopy. J Fluoresc 20:1191–1198
34. Han X, Gross RW (2003) Global analyses of cellular lipidomes directly from crude extracts of biological samples by ESI mass spectrometry a bridge to lipidomics. J Lipid Res 44:1071–1079
35. Lodish H, Berk A et al (2000) Proto-oncogenes and tumor-suppressor genes. Molecular cell biology, 4th edn. W. H. Freeman, New York; Section 24.2
36. Armitage EG, Barbas C (2014) Metabolomics in cancer biomarker discovery: current trends and future perspectives. J Pharm Biomed Anal 87:1–11
37. Ravnskov U, McCully KS, Rosch PJ (2011) The statin-low cholesterol-cancer conundrum. QJM 105:383–388
38. Birková A, Grešová A et al (2014) Changes in urine autofluorescence in ovarian cancer patients. Neoplasma 61:724–731
39. Al-Salhi M, Masilamani V et al (2011) Lung cancer detection by native fluorescence spectra of body fluids-a preliminary study. J Fluoresc 21:637–645
40. Šteffeková Z, Birková A (2015) Pilot study of canine urine analysis using fluorescent fingerprint. Spectrosc Lett 48:447–453
41. Xu X, Meng J et al (1988) The characteristic fluorescence of the serum of cancer patients. J Lumin 40–41:219–220
42. Tsai MJ, Chang WA et al (2014) Tumor microenvironment: a new treatment target for cancer. ISRN Biochemistry 2014:351959
43. Rabinowitz HM (1949) A correlation of fluorescence of human urine with benign and malignant growth. Cancer Res 9:672–676

44. Roede JR, Park Y et al (2012) Detailed mitochondrial phenotyping by high resolution metabolomics. PLoS One 7:e33020
45. Siegel RL, Miller KD, Jemal A (2015) Cancer statistics, 2015. A Cancer J Clinicians 65:5–29
46. Masilamani V, Sivakumar N et al (2003) Diagnosis of cancer from blood by native fluorescence. Asian J Phys 12:125–132
47. Ganesan S, Sackc PG et al (1998) Native fluorescence spectroscopy of normal and malignant epithelial cells. Cancer Biochem Biophys 16:365–373
48. Garcia A, Barbas C (2011) Gas chromatography–mass spectrometry (GC–MS)- based metabolomics. Methods Mol Biol 708:191–204
49. Witsch E, Sela M, Yarden Y (2010) Roles for growth factors in cancer progression. Physiology (Bethesda) 25:85–101
50. Masilamani V, Vijmasi T et al (2011a) A new lung cancer biomarker – a preliminary report. Photomed Laser Surg 29:161–170
51. Lu J, Gao S et al (2008) Study on characteristic intrinsic fluorescence spectra of urine from ovarian cancer patients. Proc. SPIE 6826:682620
52. Li BH, Zhang ZX et al (2006) Fluorescence spectral characteristics of human blood and its endogenous fluorophores. Guang Pu Xue Yu Guang Pu Fen Xi 26:1310–1313
53. Jaffe M (1869) Zur Lehre von den Eigenschaften und der Abstammung der Harnpigmente. Virchows Arch f path Anat 47:405–427
54. Kalaivani R, Masilamani V et al (2008) Fluorescence spectra of blood components for breast cancer diagnosis. Photomed Laser Surg 26:251–256
55. Agnoli C, Berrino F et al (2010) Metabolic syndrome and postmenopausal breast cancer in the ORDET cohort: a nested case-control study. Nutr Metab Cardiovasc Dis 20:41–48
56. Masilamani V, Vijmasi T et al (2011b) Detection of cervical cancer by fluorescence emission and stokes' shift spectra of blood and urine. Proc SPIE 7895:78950A
57. Wenk MR (2010) Lipidomics: new tools and applications. Cell 143:888–895
58. Zvarik M., Martinicky D et al (2013) Fluorescence characteristics of human urine from normal individuals and ovarian cancer patients. Neoplasma 60:533–537
59. Vaux D (2011) In defense of the somatic mutation theory of cancer. Bioessays 33:341–343
60. Amoedo ND, Valencia JP et al (2013) How does the metabolism of tumour cells differ from that of normal cells. Biosci Rep 33. doi:10.1042/BSR20130066
61. Bell EM, Mainwaring WI et al (1971) Relationships between excretion of steroid hormones and tryptophan metabolites in patients with breast cancer. Am J Clin Nutr 24:694–698
62. Hanahan D, Weinberg RA (2011) Hallmarks of cancer: the next generation. Cell 144(5):646–674
63. Lu P, Weaver VM, Werb Z (2012) The extracellular matrix: a dynamic niche in cancer progression. J Cell Biol 196:395–406
64. Anastasiou D, Cantley LC (2012) Breathless cancer cells get fat on glutamine. Cell Res 22:443–446
65. Halket JM, Waterman D et al (2005) Chemical derivatization and mass spectral libraries in metabolic profiling by GC/MS and LC/MS/MS. J Exp Bot 56:219–243
66. Kell DB, Brown M et al (2005) Metabolic footprinting and systems biology: the medium is the message. Nat Rev Microbiol 3:57–565
67. Sieron A, Sieron-Stołtny K et al (2013) The role of fluorescence diagnosis in clinical practice. Onco Targets Ther 6:977–982
68. Phan LM, Yeung SCJ, Lee MH (2014) Cancer metabolic reprogramming: importance, main features, and potentials for precise targeted anti-cancer therapies. Cancer Biol Med 11:1–19
69. De Lima C, Alves LE et al (2008) Anaerobic exercise reduces tumor growth, cancer cachexia and increases macrophage and lymphocyte response in Walker 256 tumor-bearing rats. Eur J Appl Physiol 104:957–964
70. Stephens F, Aigner K (2009) Basics of oncology. Springer, Berlin
71. Lawaetz AJ, Bro R et al (2012) Fluorescence spectroscopy as a potential metabolomic tool for early detection of colorectal cancer. Metabolomics 8:111–121
72. Beger RD (2013) A review of application of metabolomics in cancer. Metabolites 3:552–574

73. Šteffeková Z, Birková A et al (2014) Early diagnosis of colorectal cancer in rats with dmh induced carcinogenesis by means of urine autofluorescence analysis. Photochem Photobiol 90:682–685
74. Rajasekaran R, Aruna PR et al (2013) Characterization and diagnosis of cancer by native fluorescence spectroscopy of human urine. Photochem Photobiol 89:483–491
75. Turner DP (2015) Advanced glycation end-products: a biological consequence of lifestyle contributing to cancer disparity. Cancer Res 75:1925–1929

Chapter 9
Cellular Applications of Semiconductor Quantum Dots at the U.S. Naval Research Laboratory: 2006–2016

Clare E. Rowland, Kimihiro Susumu, Michael H. Stewart, Lauren D. Field, Ajmeeta Sangtani, Igor L. Medintz, and James B. Delehanty

Keywords Quantum dot • Nanocrystal • Fluorescence • Bioconjugate • Peptide • Cell • Organelle • Sensor

9.1 Introduction

9.1.1 The U.S. Naval Research Laboratory, Nanotechnology, and Quantum Dots and Cells

The U.S. Naval Research Laboratory (NRL), which began operation in 1923, is the corporate research laboratory of the Navy and Marine Corps, and its primary

C.E. Rowland
Center for Bio/Molecular Science and Engineering, Code 6900, U.S. Naval Research Laboratory, Washington, DC 20375, USA

National Research Council, Washington, DC 20036, USA

K. Susumu
Optical Sciences Division, Code 5600, U.S. Naval Research Laboratory, Washington, DC 20375, USA

Sotera Defense Solutions, Inc., Columbia, MD 21046, USA

M.H. Stewart
Optical Sciences Division, Code 5600, U.S. Naval Research Laboratory, Washington, DC 20375, USA

L.D. Field • A. Sangtani
Center for Bio/Molecular Science and Engineering, Code 6900, U.S. Naval Research Laboratory, Washington, DC 20375, USA

Fischell Department of Bioengineering, University of Maryland, College Park, MD 20742, USA

I.L. Medintz • J.B. Delehanty (✉)
Center for Bio/Molecular Science and Engineering, Code 6900, U.S. Naval Research Laboratory, Washington, DC 20375, USA
e-mail: james.delehanty@nrl.navy.mil

C.D. Geddes (ed.), *Reviews in Fluorescence 2016*, Reviews in Fluorescence,
DOI 10.1007/978-3-319-48260-6_9

role is to fulfill the Department of Defense's (DoD) mission of developing new materials and technologies to benefit the warfighter and, more generally, society as a whole. In meeting this mission, the NRL performs both basic and applied research and development across a wide range of disciplines encompassing sonar, radar, satellite technology, electronics, and advanced functional materials. Since the inception of the National Nanotechnology Initiative (NNI) in 2001, DoD has been directly tasked to pursue the vision of "*a future in which the ability to understand and control matter at the nanoscale leads to a revolution in technology and industry that benefits society*" [1]. One research frontier that emerged from this pursuit is the interfacing of novel nanomaterials with biological systems (biological molecules, cells and tissues) and the development of a fundamental understanding of the potential benefits offered by novel nanomaterials for new diagnostic and therapeutic reagents.

In keeping with this theme, researchers in the Center for Bio/Molecular Science and Engineering at NRL became interested in novel photoluminescent semiconductor nanocrystals or quantum dots (QDs) as the basis for various biological applications. Given their small size, extreme brightness, and photostability, the working hypothesis was that QDs and QD-bioconjugates held enormous promise as the basis for robust sensing platforms and could potentially be used for cellular labeling, imaging and sensing. Initial studies were focused on methodologies for the attachment of protein recognition elements (*e.g.*, antibodies) to QDs [2–4] which facilitated a number of QD-bioconjugate-based sensors for various analytes [5–7]. This, coupled with some of the first pioneering demonstrations of the potential power of QDs for cellular labeling and tracking [8–9] led directly to the more advanced and intricate cellular studies that are detailed chronologically in this chapter.

We begin with a brief discussion of the use of fluorescence in cellular biology followed by a survey of the pertinent photophysical attributes of QDs that make them attractive alternatives to traditional fluorescent materials for cellular applications. In Sects. 9.2, 9.3, and 9.4, we detail three key enabling technologies that have facilitated the transition of QDs into the cellular realm: 1-surface-capping ligands for rendering QDs stable in biological/cellular media; 2-bioconjugation chemistries for the attachment of peptides, proteins and other biologicals to the QD surface; and 3-techniques for the delivery of QDs to cellular and subcellular locations with fine control. In the remaining Sects. 9.5, 9.6, and 9.7 we detail, in chronological fashion, the progression of the use of QDs for a variety of cellular applications in our and collaborators' laboratories over the past 10 years, including a consideration of the potential toxicity of QD materials. Finally, we offer a brief forward-looking perspective of what can be expected to develop going forward into the future.

9.1.2 Fluorescence in Cellular Biology

Fluorescence has found application in recent years as a literal and proverbial means of illuminating the cell. From labelling and imaging cellular components to more

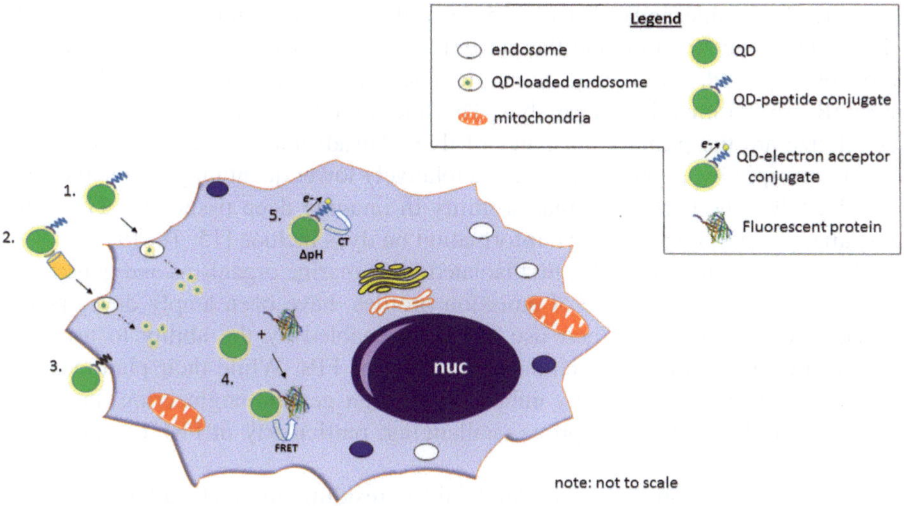

Fig. 9.1 Representative cellular applications of QD bioconjugates realized at NRL (2006–2016). Shown are a variety of cellular structures and processes for which we have implemented QDs for labeling, imaging, sensing, or cargo delivery. The various applications are: endocytosis and endosomal escape of (*1*) QD-peptide and (*2*) QD-peptide-cargo conjugates; (*3*) peptide-mediated QD membrane tethering; (*4*) Förster resonance energy transfer (FRET)-based monitoring of QD-protein assembly; and (*5*) charge transfer (CT)-based pH sensing. Other biological processes (not shown) that we have successfully monitored with QDs are: multicolor labeling of disparate cellular structures, tracking of embryonic brain development, and deep-tissue electrophysiology/imaging of live mouse brain

complex paradigms for sensing processes and interactions and even delivering cargoes into cells, advances in fluorophores and in instrumentation for fluorescence microscopy have contributed to a plethora of reports. The use of traditional fluorophores, which include small molecule organic dyes and fluorescent proteins, has, for example, recently been augmented by QDs, robust quantum-confined semiconductors that have found extensive use in cellular applications [10–12]. By conjugating them to peptides, for example, they have been used for endosomal tracking and in delivering cargoes into cells [13–14]. Within cells, they have been used not only in direct imaging but also in electron transfer and energy transfer-based applications [15–17]. Here we discuss the advances in cellular applications of QDs that have come out of our group, from the optimization of these materials for use in biological environments through the wide range of cellular labeling and sensing applications that they have enabled (see Fig. 9.1).

9.1.3 Traditional Fluorophores

Traditional fluorophores used in cellular biology include organic dyes, fluorescent proteins (FPs), and metal chelates. *Organic dyes* constitute a broad range of organic small molecules built on one of a number of basic scaffolds including fluorescein, cyanine, rhodamine, and coumarin [15, 18–20]. Addition of substituents changes the dye's electronic structure and therefore its spectral properties. These substitutions are well-explored, and high quantum yield derivatives emitting from the UV to the near-IR are commercially available. This fact coupled with their brightness and small size are the primary attributes of dyes. Disadvantages include toxicity, low molar absorptivity, photobleaching, and relatively lower quantum yields in the near-IR (2–25 %), the region of greatest utility in imaging deep tissue [18, 21]. Other resources with more substantive information on dyes include [15, 18–19, 22].

Fluorescent proteins (FPs), first isolated from marine organisms and genetically engineered for a wide range of emission profiles, have been amply developed in recent decades [23–28]. Their use has been enabled by the ability to use gene transfection to prompt intracellular generation of FPs. While their photophysical properties have been markedly enhanced through genetic engineering, the broad spectral profiles of FPs can prove challenging, particularly in FRET applications [29].

Metal chelates are typically lanthanides residing in a chelating pocket of an organic complex that sensitizes the metal's emission [30]. Unlike the other fluorophores discussed here, the emission of a metal chelate relies on the electronic properties of a single atom, resulting in very well-defined transitions. The need for a sensitizer arises from the low molar extinction coefficients of the lanthanides, and the chelating molecule doubles as a means of excluding water, which quenches lanthanide emission, from the lanthanide's inner coordination sphere; it must also serve as a point of bioconjugation in systems where that is desired [15]. These dyes are long-lived, with millisecond lifetimes [30].

9.1.4 Properties of Quantum Dots

More recently, quantum dots have gained traction as fluorophores for cellular imaging. While the term 'nanomaterials' covers a broad range of sizes, compositions, and morphologies, 'quantum dot' refers specifically to an optically nanoscale or quantum confined spherical semiconductor [31]. This definition encompasses materials that range in size from several nm to over a hundred (*e.g.*, ZnS at 4.4 nm [32], InSb at 108 nm [33]). The electronic structure of QDs falls between that of a bulk semiconductor and a large molecule; that is, they possess the discrete energy levels associated with molecular orbitals, but the spacing of these energy levels is sufficiently small that band structure is a suitable approximation; and indeed this is the terminology used with QDs (see Fig. 9.2). Typical of a semiconductor, the Fermi level of a QD lies between the valence and conduction bands, and photoexcitation by a photon of sufficient energy will result in the promotion of an electron from the valence to the conduction band, leaving behind a ground state hole. The energy of the requisite photon for photoexcitation depends on the size of the bandgap, which is greater than that of the bulk semiconductor as a result of the quantum confinement, a phenomenon originating from the QD diameter's being smaller than the material's

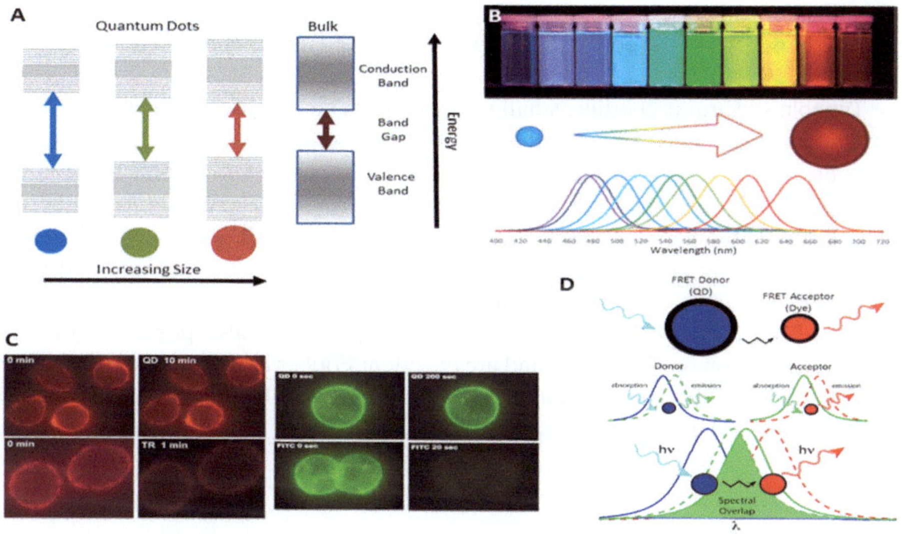

Fig. 9.2 Photophysical properties of QDs. (**a**) The band structure of QDs gives rise to the size-dependent absorption and emission profiles. Increasing the size of the particle brings the valence and conduction band closer in energy to the bulk semiconductor. (**b**) The size-dependent band gap of QDs results in size-dependent spectral properties, with a decrease in band gap (and therefore red-shift in emission) occurring with increasing particle size (Reproduced with permission from Ref. [47]). (**c**) QDs exhibit much higher photostability than other fluorophores such as organic dyes. Here QDs are compared to Texas Red (TR) and fluorescein isothiocyanate (FITC). While similar in brightness at the zero time-point, both dyes show substantial degradation after illumination of a minute or less, while the QDs show no evidence of photobleaching after illumination ten times as long (Reproduced from Ref. [42]). (**d**) QDs function as excellent donors in FRET systems, in which a photo-excited donor non-radiatively transfers energy to a ground state acceptor species, resulting in emission from the acceptor (Reproduced with permission from Ref. [48])

Bohr exciton diameter and the energetic cost of creating an electron/hole pair (overcoming the electron/hole attraction) within that space [34]. This electronic configuration – a feature that defines QDs - gives rise to their most celebrated property: their absorption and emission properties are size-dependent. The addition of a shell to the QD and the influence of the bandgap of the shell on the confinement of the electron or hole gives rise to an added dimension of manipulation in these materials. CdSe/ZnS core/shell materials have arguably become the preferred QD platform of choice for biological applications as their luminescence is very stable, spanning a wide range of the visible spectrum (~480–650 nm), with core sizes that range from 1 to 10 nm in diameter. At present, CdSe/ZnS remain the most widely studied QD material for interfacing with biological molecules and cells [35]. Here, the ZnS shell plays a critical role as it passivates the CdSe core and minimizes the leaching of Cd^{2+} ions into the surrounding media as well as mediating the bioconjugation/self-assembly of histidine-containing biologicals to the QD (*vide infra*). For more detail about core/shell QDs and the factors that contribute to the classification of these materials as Type I or Type II – assignments based on the relative bandgaps of the core and shell materials – the reader is referred to reference [36].

Tunable spectral properties, small size, and high molar absorption cross sections have made QDs interesting to the bio-imaging community, while resistance to photobleaching, low toxicity, and general physical robustness overcome some of the principal problems associated with organic dyes [37–42]. The ability to synthesize QDs with high quantum yields (20–70%) in the near-IR marks an additional improvement over many traditional fluorophores [18, 43]. Further, the ability to bioconjugate QDs has facilitated their integration into a variety of biological systems [44]. Suited to more specialized techniques, QDs also possess high two-photon absorption cross sections and are excellent Förster resonance energy transfer (FRET) and electron transfer donors [17, 45–46]. In combination, these properties have resulted in the application of QDs to cellular imaging with remarkable success.

9.2 Ligand Technology

Surface ligands are indispensable for the synthesis of colloidal QDs. They lend QDs their colloidal stability and protect the inorganic core from the ambient environment. Moreover, surface functionalization of QDs is critical for biomedical applications, as most high quality (*i.e.,* bright) QDs are synthesized using hydrophobic surface ligands at high temperature and thus require additional surface modification for use in aqueous media.

Water-solubilizing ligands possess two primary units: an anchoring or surface-interacting group that secures the ligand to the QD and a hydrophilic group that endures the solubility of the QD/ligand complex in aqueous media. From these basic requirements, further design criteria of the surface ligands are tailored to the conditions of the aqueous media and the desired application of the QD. For biological applications, desired properties for surface ligands and the QD/ligand complexes they form include (i) strong affinity to the QD surface, (ii) colloidal stability in a wide range of pH conditions and in high salt concentration, (iii) small hydrodynamic size, (iv) ability to conjugate with biological molecules, (v) minimal nonspecific adsorption in biological environments, (vi) minimal toxicity, and (vii) low-cost commercial availability or straightforward large-scale synthesis. The first two criteria are essential, as the ligand's continued presence on the QD surface is critical to the QD's continued colloidal stability, and the ligand's ability to withstand changes in pH and high salt environments is what preserves that colloidal stability under diverse cellular pH conditions and in biological fluids rich in electrolytes, peptides and proteins [49–51]. For example, endocytosis is one of the common pathways for cellular uptake [52], and thus pH stability under the weakly acidic conditions typical of an endosome is crucial for reliable cellular tracking and imaging. Moreover, many aqueous coupling reactions frequently occur under slightly acidic or basic conditions [53].

Synthesizing a QD/ligand complex with a small hydrodynamic radius is important in certain size-critical functions such as ensuring efficient renal clearance [54–56] and sensing based FRET [57–59]. FRET efficiency is correlated to the distance between the donor and acceptor species, and minimizing the length of the ligand and therefore the distance between a QD FRET donor and its energy transfer acceptor aids in maximizing FRET efficiency. Both in FRET systems and elsewhere, surface ligands that possess functional groups accessible for *in situ* chemistry can allow the QD to be coupled to fluorescent dyes or biological molecules. Conjugation with certain biological molecules can be achieved simply by modifying the QD surface with target entities. Such targeting can help reduce nonspecific interactions, which often lead to unnecessary background signals in cellular labeling and imaging studies. Also important to any *in vitro* or *in vivo* studies, the surface-functionalized QDs should be minimally toxic and should not disturb any essential biological functions of the target environments.

The final concern – the availability or ease of synthesis of the ligand – is a practical matter. It is most convenient if the ligand is commercially available, and indeed most instances of QDs used in biological applications have employed commercially available materials as the surface ligands. However, many of the aforementioned criteria are not satisfied by commercially available ligands, a fact that has required the design and synthesis of new ligands. Since the surface coating procedures of QDs are normally mass-driven reactions and thus require large excess of the ligand, straightforward synthetic schemes are desirable. While each of the requirements mentioned here is important to the synthesis of an excellent QD/ligand complex, no single ligand exists that fulfills each one. Selecting a ligand therefore requires balancing the needs of a project with the practical limitations of the available ligands.

Approaches to the surface modification of hydrophobic QDs include ligand exchange, encapsulation, and silica coating. Ligand exchange is the method to replace the native hydrophobic surface ligands on the QD surface with hydrophilic ligands. Encapsulation involves overcoating the QDs and their hydrophobic ligands with extraneous amphiphilic ligands. The hydrophobic chains of the amphiphilic ligands are interdigitated into the QD's native ligands, and the hydrophilic segments form the new surface of the complex and ensure its aqueous solubility. Silica coating refers to the formation of a siloxane network on the QD surface through hydrolysis of silane molecules and the subsequent crosslinking process. Initial ligand exchange with (3-mercaptopropyl)trimethoxysilane forms nucleation sites on QD surface, and hydrolysis and condensation of silane molecules such as tetraethoxysilane result in shell growth. [60–62] Alternatively, a water-in-oil reverse microemulsion method has been used [63–65]. Encapsulation and silica coating both generate QD/ligand complexes with large hydrodynamic diameters, and the importance of retaining a small size for many biological sensing applications limit the utility of these techniques. Large nanoparticles can be more difficult to deliver to cellular environments and show poor biodistribution. In the case of single particle tracking studies, larger materials are less mobile and may hamper the cellular processes to be monitored. FRET-based sensing works only when short distances between the donor and acceptor can be achieved. In clinical studies, a small hydrodynamic size is critical for efficient renal clearance [55–56]. Since our major research goals have been the development of FRET based QD biosensing systems and targeted cellular delivery of QD bioconjugates, ligand exchange has been the natural choice for our studies.

In ligand exchange, thiol groups are frequently utilized as anchors on QD surfaces due to their strong affinity to Cd and Zn, metals commonly available on QD surfaces. While other anchoring groups such as amines and carboxylic acids have also been explored in the preparation of hydrophilic QDs, their use has been limited due to their relatively lower affinity with QD surfaces and their sensitivity to pH, a potential problem in biological aqueous media. Mercaptoacetic acid (MAA) [66–69], mercaptopropionic acid (MPA) [70–74], and mercaptoundecanoic acid (MUA) [75–77] are among the frequently used, commercially available single thiol ligands. The aqueous solubility of QDs coated with these ligands relies on deprotonation of the carboxyl groups, meaning that the QDs are only soluble under basic conditions. These single thiol ligands modified with poly(ethylene glycol) (PEG) have been explored to expand the stable pH windows for hydrophilic QDs [78–80]. Nevertheless, the long-term colloidal stability of QDs coated with the single thiol ligands is questionable.

To improve upon the poor colloidal stability of monodentate thiol ligands, a variety of multidentate thiol ligands have been developed, including the frequently studied bidentate ligand dihydrolipoic acid (DHLA) [2]. DHLA is prepared from the commercial precursor thioctic acid (TA) in one step *via* a ring opening reduction of the disulfide unit using $NaBH_4$. Its dithiol anchoring group significantly enhances the colloidal stability of QDs and has expanded their utility in a variety of biological applications including *in vitro* assays and cellular imaging. The ligand's compact size facilitates use in FRET-based sensing. However, the aqueous solubility of DHLA-coated QDs still relies on deprotonation of the carboxyl group; thus colloidal stability remains limited to basic media. To improve pH stability, our group has developed a series of PEG-appended DHLA derivatives, in which a terminal hydroxy group of a PEG with different chain lengths is connected with DHLA *via* an ester bond [81]. PEG groups have been widely explored as aqueous solubilizing segments for QDs and other nanoparticles [82]. Because they are composed of flexible and neutral ethylene oxide repeating units, PEGylated QDs experience minimal nonspecific adsorption and aggregation in biological environments [83], and they have excellent colloidal stability in a wide range of pHs and under high salt concentration. As a result, PEG is one of the most popular hydrophilic segments for use in biological applications. To synthesize the DHLA-PEG ligands, the commercial PEGs are directly attached to the thioctic acid *via* a carbodiimide coupling using *N,N'*-dicyclohexylcarbodiimide (DCC). The subsequent ring opening reduction of the dithiolane ring of TA unit with $NaBH_4$ yields the DHLA-PEG-OH ligands. Synthesis of DHLA-PEG-OH requires only two steps from the commercially available precursors, which helps mitigate the laborious purification procedures and makes large scale synthesis feasible. The ligand exchange of as-prepared hydrophobic QDs with DHLA-PEG-OH results in colloidally stable QDs between pH ∼5 and 12. These ligands do suffer from limitations, however: the ester bond which connects the PEG and DHLA units tends to be hydrolyzed and cleaved in acidic and basic media, and the conjugation chemistry of the terminal hydroxy group is limited. The simple DHLA-PEG-OH ligands prepared in this study preclude easy implementation of common QD bioconjugation techniques such as avidin-biotin binding and covalent conjugation chemistry (*via* 1-ethyl-3-(3-dimethylaminopropyl)carbodiimide (EDC) coupling) due to the absence of accessible functionalities at the ligand terminus.

In order to further improve the ligand stability and access common bioconjugation methods, a series of modular PEG-appended DHLA derivatives was designed and synthesized [84]. In this configuration, each modular ligand has a central PEG segment with tunable chain length to ensure water solubility, DHLA as a bidentate terminal group for strong anchoring to the QD surface, and a terminal functional group that promotes conjugation with target biomolecules. The PEG and DHLA units are connected *via* an amide bond, which is more robust than an ester bond under physiological and common bioconjugation conditions. The terminal functional groups that have thus far been introduced in this ligand configuration include methoxy, amino, carboxyl, biotin, azide, nitrilotriacetic acid (NTA) and a few others [84–89]. The methoxy group is chemically inert, which helps simplify the ligand synthesis and purification steps. Amino and carboxyl groups can be used to form covalent bonds with a variety of organic dyes and biological receptors using common bioconjugation chemistry, including EDC coupling. Biotin groups permit the use of ubiquitous and robust avidin-biotin binding chemistry. Azide groups can be useful for bioorthogonal coupling with alkyne groups *via* click chemistry. NTA groups have been commonly used for binding with oligohistidine-tagged peptides and proteins *via* metal affinity interactions. These modular PEG ligands have demonstrated enhanced colloidal stability of QDs in weakly acidic to strongly basic aqueous media while possessing functional groups for a wide variety of bioconjugation chemistries [81, 84–85, 90–91].

With the demonstrated success of bidentate DHLA in improving colloidal stability of QDs in aqueous media, subsequent research has explored the use of tridentate and tetradentate thiol ligands [92–96]. For example, our group has incorporated two bidentate DHLA groups to a molecular backbone *via* modification of the terminal amine of the hydrophilic unit to construct a tetradentate ligand [93]. QDs coated with the tetradentate PEG ligands exhibited remarkable long-term colloidal stability over a substantial pH range (1.1 ∼ 13.9) and in the high salt concentration. These results clearly indicate the utility of multidentate structures in the design of surface coating ligands to enhance the QD colloidal stability for their long-term use in harsh biological environments.

As with all ligands, PEG does, however, have some drawbacks. In particular, aqueous solubility of PEGylated QDs requires relatively long PEG chain lengths, resulting in large hydrodynamic volumes. Zwitterionic ligands have recently been explored as smaller alternative surface coating materials for QDs [97]. A zwitterionic ligand possesses both a positively and negatively charged functional group in the single molecule. Amines, including quaternary amines, are common as cationic functionalities, and carboxylic, phosphoric, and sulfonic acids are the most common anionic groups. Net neutrality of the ligand helps to minimize nonspecific adsorption in biological environments. While sulfobetaine is one of the most common zwitterions and its utility as a hydrophilic surface has been well demonstrated, its structure does not allow further bioconjugation. Our group therefore focused on developing a series of DHLA-based ligands incorporating a zwitterion containing tertiary amines and hydroxyl and/or carboxyl substituents; in the latter case, the carboxyl groups of the zwitterionic ligands were successfully used to conjugate acceptor dyes *via* EDC coupling [98]. The benefits that these compact zwitterionic ligands provide for biological applications are substantial and include (i) enhanced colloidal stability under wide pH ranges, high salt concentrations, and in cellular environments; (ii) the ability to directly conjugate His-tagged proteins and peptides onto QD surfaces, which is often hampered in case of long PEG ligands; [99] (iii) the availability of carboxyl groups for chemical modification of QD surfaces; (iv) a small hydrodynamic size for the QD/ligand complex, resulting in efficient FRET; and (v) low toxicity.

The use of multidentate capping ligands has also been extended beyond DHLA and its derivatives to include a variety of anchoring groups incorporated onto polymer chains as pendant groups. Examples include phosphine [100], phosphine oxide [101], amine [102–104], carboxylic acid [105–106], imidazole [107–108], pyridine [109] and thiol [110–115] used in conjunction with polymer backbones such as poly(acrylic acid) [110, 113–114, 116], poly(methacrylic acid) [111, 115], poly(maleic anhydride) [117], poly(glutamic acid) [118] and poly(aspartic acid) [112]. Among the anchoring groups, thiols have been the most extensively explored, a fact that is not surprising given their stable coordination to Cd and Zn surfaces. One of their limitations, however, is the propensity for thiols to photooxidize, and imidazole-appended polymeric ligands have been developed as a photooxidation-resistant alternative. Bawendi's group demonstrated radical addition-fragmentation chain transfer (RAFT) polymerization to incorporate a series of functional groups, including imidazole as anchor group, PEG as a water solubilization segment, and either amine or biotin groups as functionalization units on the polymer backbone [107]. The QDs coated with the imidazole-appended polymeric ligands exhibited small hydrodynamic volumes, high fluorescence quantum yields, reasonable colloidal stability under physiological conditions, and low nonspecific adsorption.

Our group has also developed short poly(acrylic acid) (PAA, MW $\sim$1,800: $\sim$25 average repeating units) ligands appended with pyridines and PEGs as anchoring and hydrophilic groups, respectively, in order to prepare compact and biocompatible QDs with enhanced photostability [109]. Pearson's Hard and Soft Acid and Base (HSAB) Principle categorizes Zn^{2+} and pyridine as borderline acid and base, respectively, suggesting pyridine's potential utility as an anchoring group on the Zn surface of QDs [119]. PAA has repeating carboxylic acid groups within its linear chain, which were simply modified with carbodiimide coupling chemistry to attach amine-terminated pyridines and PEGs as the polymer pendant groups with desired grafting ratios. Multidentate pyridine coordination onto QD surfaces was robust under air and light, which imparted extended photostability to QDs and benefited a variety of biological applications. The QDs coated with these polymeric ligands still allowed us to demonstrate two different types of conjugation to the QD surface, both covalent EDC conjugation to carboxyl groups displayed on the polymers with dyes and peptides as well as metal-affinity coordination of proteins to the QD surface. Our cellular labeling studies have shown the QDs to be well-dispersed in cellular environments without aggregation following microinjection, and the cells remained unperturbed over the 6 h experimental time frames utilized. Alternative delivery by cell penetrating peptide (CPP)-mediated uptake followed by cytoviability assays confirmed minimal QD toxicity under our experimental conditions. Most importantly, the photostability and small hydrodynamic sizes along with both colloidal stability and a lack of aggregation in the cellular environments observed for the pyridine polymer coated QDs enabled us to perform intracellular single QD tracking studies.

Clearly, as the QD materials themselves have progressed, so too have the capabilities of the solubilizing capping ligands. Still, it is important to appreciate that each ligand comes with its own inherent benefits and liabilities. For example, the earliest DHLA ligands, while short and compact, mediated only modest colloidal stability in aqueous media (limited to pH $>\sim 8.5$). Later, ligand species such as the compact ligand series (based on the same DHLA backbone) opened up a wider pH range. It is also worth noting that these ligands possess dual applicability as they have been used to effectively cap other nanoparticles (NPs), such as gold NPs [87, 120]. Collectively, this is an area of continuing research and surely new ligand designs will evolve as new applications/needs present themselves (Table 9.1).

9.3 Bioconjugation

QDs can be coupled to biomolecules using bioconjugation strategies to impart additional functionality to the QD constructs. Common biomolecules for this purpose include proteins, peptides, and nucleic acids, with the selection dependent on the intended application. An ideal bioconjugation strategy would result in reproducible control over the number and orientation of biomolecules on the nanocrystal, yield stable conjugates that could be rapidly and easily assembled without labor intensive chemistries or purifications, and preserve the intrinsic properties of both the nanocrystal and the biomolecule [121].

Bioconjugation can rely on either covalent bonding or non-covalent interactions such as coordination (or dative) bonding. The most general covalent coupling strategy is through amide bond formation between amino and carboxyl groups *via* carbodiimide reagents, such as EDC. However, amino and carboxyl groups are ubiquitous in biomolecules, leading to complications such as cross linking and random conjugation and orientation. Furthermore, excess EDC is required because the reaction intermediates are easily hydrolyzed, resulting in low coupling efficiency. This leads to poor control over the number of biomolecules per QD and necessitates additional purification steps that can result in precipitation of the QDs.

Table 9.1 Capping ligands developed in our group and their properties

Structure and features	M.W.	Approx. length (nm)	Charge	References
DHLA : compact, stable at basic pHs 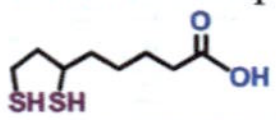	208	1	Negative	[2]
DHLA-PEG-OH : ester bond, stable between weakly acidic and basic pHs 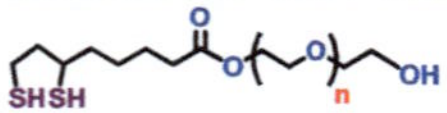	n = 3:385	2.4	Neutral	[81]
	n = 5:473	3.2		
	n = 8:605	4.2		
	n = 12:781	5.7		
	n = 21:1178	8.9		
DHLA-PEG-OCH$_3$: amide bond, stable between weakly acidic and basic pHs 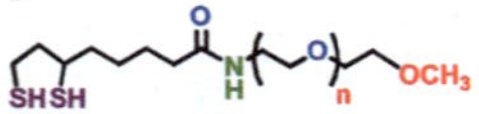	n = 11:750	5.4	Neutral	[85, 90]
	n = 15:926	6.9		
DHLA-PEG-NH$_2$: conjugation ready with carboxyl counterpart 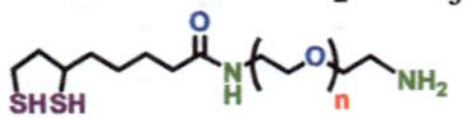	n = 8:603	4.2	Positive	[84, 91]
	n = 12:779	5.7		
DHLA-PEG-COOH: conjugation ready with amine counterpart 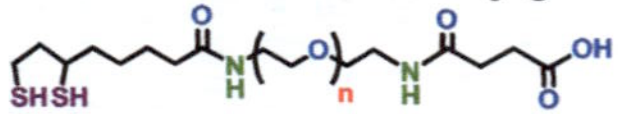	n = 8:703	4.8	Negative	[84, 91]
	n = 12:879	6.3		
DHLA-PEG-biotin: avidin-biotin coupling 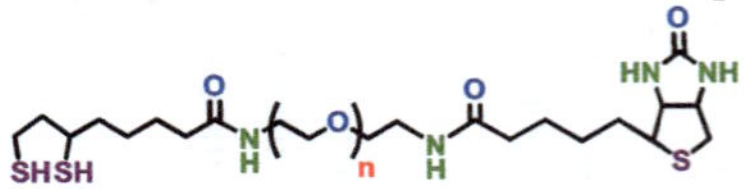	n = 8:829	5.2	Neutral	[84, 86, 91]

DHLA-PEG-NTA: metal affinity interaction with oligohistidine-tagged peptides and proteins 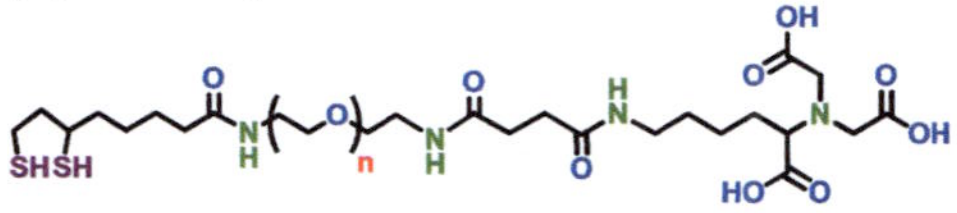	n = 12:1123	7.4	Negative	[88]
(DHLA)$_2$-PEG-OCH$_3$: stable at pH between 1.1 and 13.9 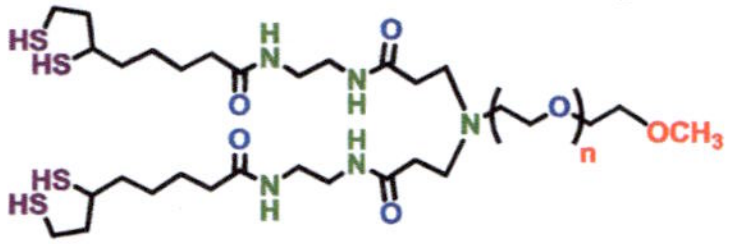	n = 15:1345	7.8	Neutral	[93]
CL1: compact zwitterionic ligand 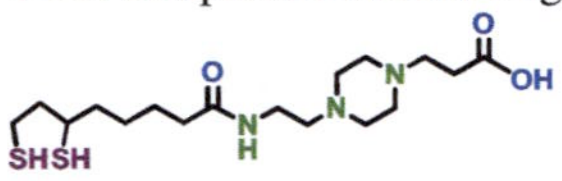	392	2.0	Negative (pH 8.3)	[98]
			Positive (pH 4)	
CL2 : compact zwitterionic ligand 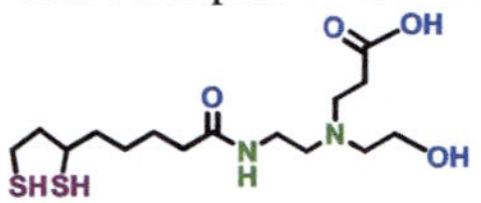	367	1.8	Negative (pH 8.3)	[98]
			Positive (pH 4)	
CL3 : compact ligand 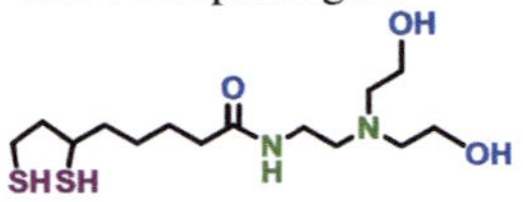	339	1.6	Negative (pH 8.3)	[98]
			Positive (pH 4)	

(continued)

Table 9.1 (continued)

Structure and features	M.W.	Approx. length (nm)	Charge	References
CL4 : compact zwitterionic ligand 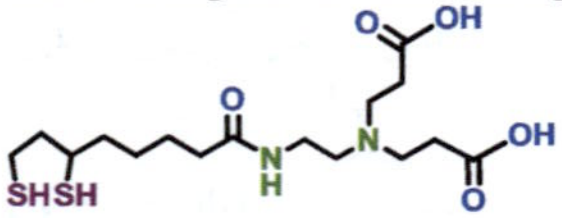	395	1.8	Negative (pH 8.3)	[98]
			Negative (pH 4)	
Pyridine-appended polymer ligands 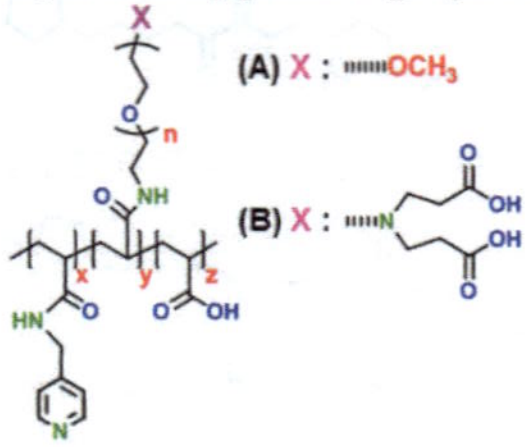	(A) (x:y:z = 45:45:10, n ∼ 15) ∼ 11,000	6 (PAA)	Neutral	[109]
	(B) (x:y:z = 45:50:5, n ∼ 8) ∼ 10,000		Negative	

While covalent conjugation with EDC is not ideal, a number of refined click and bioorthogonal covalent coupling chemistries have been developed that overcome the shortcomings of carbodiimide chemistry [44, 122–123]. These generally require the incorporation onto the biomolecule and nanocrystal of exotic functional groups that form exclusive covalent bonds, a synthetically intensive task.

Another approach to creating QD-biomolecule conjugates is through the exploitation of non-covalent bonds such as electrostatic interactions, avidin-biotin chemistry, and dative thiol bonding. Additionally, our group has developed the use of polyhistidine tags (His_n) for the self-assembly of biomolecules to QDs, which has proven to be a versatile and powerful method for bioorthogonal conjugation. This strategy was inspired by protein purification *via* immobilized metal affinity chromatography, which relies on the ability of multiple histidine residues (His_n) on expressed proteins to chelate divalent metal cations (*e.g.*, Ni^{2+}, Cu^{2+},Co^{2+}, Zn^{2+}) attached to a solid support. Importantly, zinc ions are located in the inorganic surface of CdSe/ZnS core/shell QDs, and we have found that polyhistidine groups (typically His_6) on proteins or peptides will coordinate to those ions *via* self-assembly at room temperature. Benefits to this bioconjugation process include strong affinity (dissociation constants (K_d) on the order of $\sim 10^{-9}$ M) and rapid, near-stoichiometric assembly, which eliminates the need for conjugate purification to remove excess reagents [124]. Moreover, more than one type of His-tagged biomolecule can be self-assembled to the same QD surface while each retains its functionality. Examples include two different peptides or a mixture of peptides and proteins [14, 125–126]. Furthermore, the His_6 interacts directly with the inorganic QD surface, while the QD ligands can still provide colloidal stability and be used for additional conjugation [125]. The His_6 tag can also be assembled to carboxyl groups in the ligand coating through a divalent metal cation, such as Ni^{2+} [127]. The types of biomolecules used for cellular applications by the implementation of this bioconjugation strategy in tandem with others will be reviewed in the following sections.

9.4 Quantum Dot Delivery Techniques

One of the continuing challenges of interfacing QD materials with cells for labeling, imaging, and sensing applications is the development of robust and noninvasive methods for their controlled cellular delivery. In a 2010 investigation, we undertook an exhaustive, comparative analysis of a range of methods for the delivery of QDs to mammalian cells [128]. The chief means of the passive cellular internalization of QDs (and NPs in general) is *via* the vesicle-based endolysosomal system which results in the sequestration of the materials with endocytic vesicles [129]. Thus, our primary goal was to ascertain the ability of various active and facilitated methods to deliver QDs to the cellular cytosol where they would not be entrapped in endosomes. In this study, we compared a variety of active and facilitated means of QD delivery (summarized in Table 9.2). Active delivery techniques

Table 9.2 Survey of cellular delivery strategies for quantum dots

Delivery method	Peptide sequence	Mechanism of uptake	Uptake efficiency/ localization	QD fate	Toxicity
Active delivery					
Electroporation		Membrane pore formation	Poor uptake	QDs aggregated	Toxic
Nucleofection		Membrane pore formation	Poor uptake	QDs aggregated	Toxic
Facilitated delivery					
Polymer-mediated					
Lipofectamine 2000		Endocytosis	Endosomal	QDs punctate	Toxic
Polyethyleneimine		Endocytosis	Endosomal	QDs punctate	Toxic
PULSin™		Endocytosis	Cytosolic	QDs disperse	Toxic
Peptide-mediated					
JB434	R_9GGLAA(Aib)SGWKH_6	Endocytosis	Endosomal	QDs punctate	Minimally toxic
JB577	WG(Dap)PalVKKIKKP_9GGH_6	Endocytosis	Cytosolic	QDs disperse	Minimally toxic
JB858	WG(Dab)PalVKKIKKP_9GGH_6	Membrane binding; minimal uptake	Cellular Membrane	QDs membrane-located	Minimally toxic
Nuclear localization sequence peptide	H_6WGLA(Aib)SGPKKKRKV	Endocytosis	Endosomal	QDs punctate	Minimally toxic

Note: Table is adapted with modification from Ref. [125]

Abbreviations: *Aib* alpha-amino isobutyric acid, *Dap* diaminopropionic acid, *Dab* diaminobutyric acid, *Pal* palmitoyl. All peptides are N-terminal acetylated/C-terminal amidated (*e.g.*, electroporation, nucleofection) delivered only modest amounts of QDs and the materials were often present largely as aggregates within the cytosol. Facilitated delivery modalities (*e.g.*, cationic amphiphiles (PULSin^{TM}, Lipofectamine), polymers or cell-penetrating peptides (CPPs)) yielded mixed results ranging from complete endosomal entrapment of QDs to the time-dependent release of modest amounts of QD materials to the cytosol. One common drawback shared by the polymer- and lipid-based facilitated delivery modalities is the concomitant cytotoxicity which must always be taken into consideration and balanced against efficient QD uptake and delivery. Two salient discoveries regarding peptide-mediated QD delivery were made in this study. First, the His-based assembly of the peptide to the QD surface is remarkably stable intracellularly for several days, even within the slightly acidic environment of endocytic compartments. Second, we identified the ability of a rationally-designed, multidomain, multifunctional peptide to mediate QD endocytosis followed by a slower sustained release of the QDs to the cytosol over a ~48 h time course. Importantly, we observed minimal cytotoxicity using this peptidyl platform. Originally referred to as Palm-1 [128] and later renamed peptide JB577 [130], this peptide became the subject of a comprehensive structure/function analysis to understand the contribution of each domain to the peptide's full function (*vide infra*). As a result of this study, we came to appreciate the robust and flexible nature of the peptidyl platform as a controllable and tractable QD delivery modality and it became one of our preferred QD delivery methods and the basis of multiple follow-on studies.

9.5 Specific Cellular Applications of Quantum Dot Bioconjugates

9.5.1 Cellular Delivery of QDs and QD-Appended Cargos Using Cell-Penetrating-Peptides

In our first detailed study of the cellular delivery of QDs, we had several goals. We sought to design a modular cell-penetrating peptide (CPP) that displayed separate domains for facile assembly to the QD surface and cellular uptake, mediated efficient internalization of the QD-peptide conjugates *via* the endocytosis pathway, exhibited specificity in terms of QD uptake, and elicited minimal cytotoxicity under conditions necessary for efficient cellular labeling [13]. At that time, the number of constructs that met these criteria listed were very limited. Thus, we designed a modular peptide containing two functional domains: a polyarginine (Arg_8 or R_8) motif derived from the HIV-1 Tat peptide and a His_8 tract that self-assembles the peptide onto DHLA-capped CdSe/ZnS QDs *via* metal-affinity coordination (Fig. 9.3a). A short spacer separates the two functional domains.

Once assembled to the QD surface, the resulting bioconjugates were efficiently internalized by HEK 293T/17 and COS-1 cells *via* endocytosis where most of the

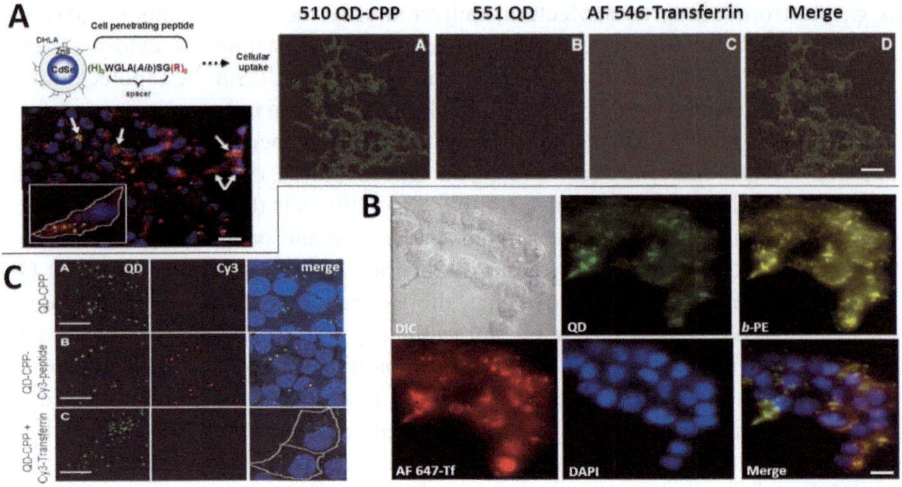

Fig. 9.3 Cellular labeling with QD-peptide conjugates. (**a**) CPP-mediated QD uptake. DHLA-capped QDs appended with polyarginine (R_8) displaying peptides self-assembled onto the QD surface *via* metal affinity coordination. Lower micrograph shows the QDs (*green*) in endosomes co-localized with AlexaFluor-conjugated transferrin (Tf) marker (*red*). Nuclei are stained with DAPI (*blue*). Scale bar, 10 μm. Multipanel micrograph (*right*) shows cells labeled with 510 nm-emitting QD-CPP conjugates while unlabeled (no CPP) 551 nm-emitting QDs show no cellular labeling. Transferrin panel shows endosomes tracked with AlexaFluor-Tf. Scale bar, 10 μm (Modified from Ref. [13] © ACS). (**b**) CPP-mediated QD-cargo delivery. 540 nm-emitting QDs appended with 40 CPP were decorated with ∼2 *b*-phycoerythrin (*b*-PE) protein complexes. AlexaFluor-transferrin (AF 647-Tf) conjugate was used to label endosomes. Respective channels are as indicated in each panel. Scale bar, 10 μm (Adapted from Ref. [14] © ACS). (**c**) Two photon FRET imaging of QD-CPP assemblies. 510 nm QDs conjugated to CPP (A, C) and CPP plus 2 Cy3-labeled peptides (**b**) were delivered to HEK 293T/17 cells and imaged by two-photon fluorescence microscopy. In (**c**) QD-CPP conjugates were also incubated with Cy3-labeled transferrin (not bound to QDs). Row B (*middle panel*) shows efficient FRET from QDs to Cy3-peptides while no FRET is seen for QDs delivered to cells with Cy3-transferrin. In one-photon mode, significant direct excitation of the Cy3 acceptor was observed (not shown) (Image from Ref. [131] © Wiley-VCH)

QD materials remained sequestered within endocytic vesicles as determined by colocalization with dye-labeled transferrin, a marker for the endocytic pathway. In addition, when the experiment was conducted at 4 °C, where endocytosis is inhibited, there was no internalization of the complexes, further implicating the endocytic pathway as the route of QD-CPP entry. Examination of internalization of the QD-CPP assembly across varying CPP:QD ratios showed a direct correlation between the amount of peptide conjugated to the QD surface and the efficiency of uptake. Further, QD uptake was shown to be dependent on the presence of the CPP, demonstrating the low degree of nonspecific binding/uptake in the absence of the CPP (Fig. 9.3a). We further showed that the efficiency of QD-peptide uptake was dependent not only on the number of peptides assembled onto the QD surface but also on the overall concentration of the resulting conjugates incubated on the cells. This was one of the first QD delivery papers to characterize this concept of "peptide valence" wherein the number of peptides assembled onto the QD surface (conjugate ratio) ultimately determines efficiency of uptake. Finally, cytotoxicity analysis of the QD-CPP complexes showed distinct differences that were dependent not only on time of incubation but also on the cell line tested. For short-term, acute incubation times (1 h), the cell viability of HEK 293T/17 and COS-1 cells survival was greater than 85 %, even at the highest QD concentrations tested (~250 nM). For longer, chronic incubations (24 h), the QD alone and QD-CPP complex cytotoxicity, approached 50% viability, a result largely attributable to the DHLA ligand (*vide infra*).

Cumulatively, this initial study demonstrated the successful implementation of CPPs for the intracellular uptake of hydrophilic QDs while highlighting the facile formation of QD-CPP assemblies by His-mediated metal affinity interactions. It also provided a firm foundation for future studies aimed at the generation of mixed surface QDs; QDs decorated with multiple biologicals (peptides, proteins) for value-added cellular delivery, labeling and tracking.

In a follow-on study to this work, we next used the QD-CPP platform not just for cellular labeling but additionally for the delivery of structurally diverse fluorescent protein cargos [14]. Here, monomeric yellow fluorescent protein (YFP, 27 kDa) or the multichromophore *b*-phycoerythrin light-harvesting complex (*b*-PE, 240 kDa), were attached to QDs using either His-driven self-assembly or biotin-streptavidin binding, respectively. QDs with mixed surfaces comprised of ~40 CPPs and ~1–2 (*b*-PE) or ~10 (YFP) fluorescent proteins were incubated with COS-1 or HEK 293T/17 cells. The cellular uptake of assembled complexes was dependent on the presence of the CPP as QD-fluorescent protein conjugates lacking the CPP showed minimal uptake (Fig. 9.3b). Consistent with our initial findings for QD-CPP internalization, the QD-peptide-protein assemblies were localized mostly within endolysosomal compartments, indicating that intracellular delivery of both QD assemblies was primarily driven by endocytotic uptake. Two color tracking confirmed the structural integrity of the bioconjugate assembly. This study clearly demonstrated the utility and ease afforded by the His-driven assembly of multiple, disparate biologicals to the QD surface while retaining the innate biological function of the various species; in this case, cellular uptake of QDs driven by the CPP.

Finally, we utilized the QD-CPP/protein platform to demonstrate the effectiveness of QDs as the basis for two-photon excitation FRET-based fluorescence imaging [131]. Relative to organic fluorophores and fluorescent proteins, QDs have significantly larger (10^3–10^4 times larger) two-photon action cross sections [132]. In a QD-fluorophore donor-acceptor FRET system, the negligible two-photon absorbance of the acceptor (dye or fluorescent protein) effectively eliminates the direct excitation of the dye or protein acceptor. This concept was demonstrated in COS-1 and HEK 293T/17 cells where 510 nm-emitting QDs appended with ~60 CPPs and ~2 Cy3-peptides showed significant QD donor-sensitized emission of the Cy3 acceptor under two-photon excitation. In this example, the two-photon action cross sections of the QD and Cy3 were 8500 and <1 Goeppert-Mayer units, respectively. QD-CPP conjugates co-delivered with Cy3-labeled transferrin (Cy3 not attached to the QD), on the other hand, showed no sensitization or direct Cy3 excitation (Fig. 9.3c). This example clearly demonstrates the value-added benefit of using the QD as the donor in a FRET configuration in two-photon excitation mode where the lack of direct excitation of the Cy3 acceptor significantly improves and simplifies imaging and data interpretation.

9.5.2 *Identification of Diverse Peptidyl Motifs for Specific Subcellular QD Delivery*

As our initial studies on cellular labeling and cargo delivery, we came to appreciate the utility of the peptidyl platform as a robust means for the cellular delivery of QDs (*vide supra*). At the same time, we also realized the significant technical hurdle posed by the endocytic pathway as a route of QD uptake; particularly when one employs CPPs based on tracts of polyarginine or polylysine residues which result in sequestration of the QD cargos within endosomes. Thus, we sought to overcome this technical roadblock by rationally designing a multidomain, multifunctional peptide capable of mediating *both* QD endocytosis coupled with the endosomal escape of the QDs to the cytosol [130]. As shown in Fig. 9.4a, the peptide (known as peptide JB577) comprises four functional modules. The first domain is a C-terminal His_6 tract to provide self-assembly of the peptide to the QD surface. The second domain is a polyproline (Pro_9) tract that is thought to adopt a rigid, type II helical structure that displays away from the QD surface to facilitate cell uptake/endosomal escape motifs. The first of these motifs is a lysine-rich domain to drive initial binding of the QD-peptide complex to the cell surface. The second motif is an aliphatic palmitoyl moiety that is attached to the peptide backbone by a non-hydrolyzable amide linkage. Initial characterization of cellular QD delivery using JB577 revealed both time- and peptide valence-dependent nature of the endocytosis and subsequent endosomal escape of QDs to the cytosol (Fig. 9.4b). QDs were rapidly endocytosed (within 1 h) and this was followed by a slow, sustained release of the QDs to the cytosol over a 48 h period where they adopted a diffuse staining pattern throughout the entire cell volume. It was also clear that a minimum number peptides on the QD were necessary to mediate this effect. In A549 cells, $\sim$40 peptides were required to achieve the robust cytosolic delivery shown in Fig. 9.4b. Equally as important as the efficient endosomal escape mediated by JB577, we also showed that the His_n-based assembly of the peptide to the QD surface was stable upon release of the QDs to the cytosol for up to 72 h (Fig. 9.4c). A chemoselectively synthesized variant of JB577 (peptide JB829-26) along with an AlexaFluor594 dye-labeled reporter peptide were assembled onto a 550 nm-emitting QD donor. FRET between the QD donor and Alexa dye acceptor confirmed the stable, intact nature of the QD-peptide complex. Along with confirmation of the cytosolic delivery of QDs, we also demonstrated the robust delivery of a range of disparate nanomaterials including the proteins maltose-binding protein ($\sim$40 kDa) and the multimeric phycobilliprotein beta-phycoerythrin ($\sim$250 kDa), 8.2 nm hydrophilic AuNPs, and a generation 5 dye-labeled dendrimer.

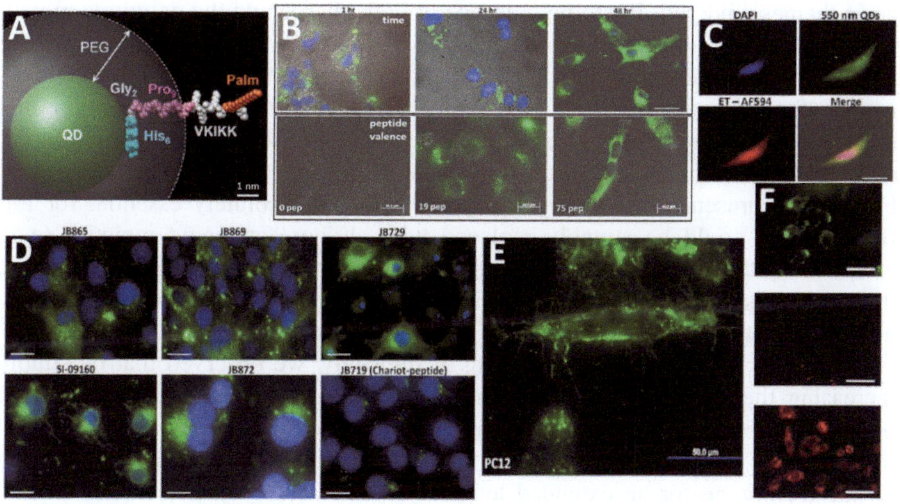

Fig. 9.4 Diverse peptidyl motifs for specific subcellular QD delivery. (**a**) Model of peptide JB577 assembled onto the surface of 550 nm-emitting QD shows assembly of the peptide to the QD surface *via* a polyhistidine tract (*blue*). A polyproline helix (*magenta*) forms a rigid spacer that extends the rest of the peptide beyond the PEG layer (*dashed line*). This results in the presentation of the lysine-rich sequence (VKIKK, *white*) and an aliphatic palmitoyl moiety (*orange*) accessible to the solvent and the cell surface. (**b**) JB577 mediates time- and peptide valence-dependent endocytosis and subsequent endosomal release to the cytosol in A549 cells. QDs were rapidly endocytosed within 1 h and were released to the cytosol over a 48 h period. (*top panel*). The cellular uptake and endosomal escape required a minimum of ~40 JB577 peptides on the QD surface. Scale bars: 50 μm. (**c**) Peptide assembly to the QD surface is stable in A549 cells for up to 72 h after delivery in A549 cells. Cells were labeled with 550 nm-emitting QDs assembled with peptides JB829-26 (75/QD) and AlexaFluor (AF) 594-labeled JB780 (16/QD) as a FRET acceptor. QD-peptide complexes were incubated on cells for 3 h followed by 72 h culture period prior to fixation. The ET-A594 image shows the QD-dye FRET signal denoting the intact QD-peptide complex. (**d**) Multiple structural variants of peptide JB577 showed a comparable ability to deliver QDs to the cellular cytosol. JB865 (longer proline tract), JB869 (amphipathic [VRLP$_3$]$_3$ substituted for VKIKK), JB729 (C_8F_{17} substituted for palmitoyl), SI-09160 (cholesterol replacing palmitoyl) and JB872 (farnesyl replacing palmitoyl) all displayed a greater degree of QD endosomal escape compared to the commercial Chariot peptide. (**e**) JB858-mediated plasma membrane delivery and tethering of QDs. PC12-Adh cells labeled with 550 nm-emitting QDs show distinct membranous labeling 48 h after initial QD delivery. Scale bar 50 μm. (**f**) Enhanced photostability of the rhodamine-labeled phosphoethanolamine afforded by FRET excitation from QDs appended to the plasma membrane by peptide JB858. PC12-Adh cells co-labeled with 550 nm-emitting QDs (*green*) and Rhod-PE (*red*). *Middle panel* shows dye photobleaching when excited directly while *bottom panel* shows photostable signal when illuminated for 40 sec in a FRET configuration. Scale bar: 50 μm (All images from Ref. [130] © ACS)

These results clearly demonstrate the flexibility and utility of the JB577 platform for the cellular delivery of a range of nanomaterials.

A detailed structure/function analysis was performed on the parent JB577 peptide wherein each domain was iteratively altered to determine the functional contribution of each domain in the ensemble peptide (Fig. 9.4d). From this examination, several basic themes emerged. First, the presence of both charge and the palmitoyl moiety were critical for fully efficient endocytosis and endosomal escape (*e.g.*, peptide JB869 displaying repeats of [VLRP$_3$] along with the palmitoyl showed native JB577 peptide function). Further, the Pro$_9$ domain was determined to be the minimal helical extension needed for full peptide function; shortening the proline linker abrogated function while lengthening the domain (as in JB865) maintained it. Finally, the presence of the aliphatic palmitoyl was absolutely essential for full function and could be suitably replaced with various fatty acid moieties (e.g., multifluorinated C_8F_{17} (JB729), cholesterol (SI-09160), or farnesyl (JB869). Also, it was a requisite that the fatty acid moiety be attached to the peptide backbone by a non-hydrolyzable linkage.

In perhaps the most striking finding of the study, we observed that by simply increasing the length of the linker that covalently attaches the palmitoyl to the peptide backbone by one carbon from diaminopropionic to a diaminobutyric acid linkage, the QD cargo was no longer endocytosed but rather remained localized on the membrane for an extended length of time (Fig. 9.4e). In some instances, the attached QD payload remained membrane-localized for up to 48 h with only minimal internalization. This remarkable feature allowed us to perform FRET sensitization of membrane-localized dye-labeled phospholipids wherein we showed the ability to impart substantial photostability to the dye acceptor when excited using the QD as energy donor (Fig. 9.4f). Importantly, all of these peptide-QD conjugates displayed remarkable biocompatibility where cellular viabilities of >80 % were observed across six different cultured cell lines and primary human dermal fibroblasts. Cumulatively, this study showed the wide utility of the JB577 peptidyl platform and the tractability of its multidomain design in overcoming a substantial roadblock in the delivery of QDs to the cellular cytosol by escaping the endolysosomal system.

9.5.3 Chemoselectively-Ligated Peptides for Site-Specific Cellular Labeling

The specific labeling of discrete, targeted regions of the cell using QDs has been a continuing goal of nanotechnology as the synthesis of increasingly more biocompatible materials has progressed. In an extension of the polyhistidine-based assembly of peptides and other biologicals to the Zn^{2+}-rich shell of CdSe-ZnS core-shell QDs, we demonstrated the ability to make functional chimeric peptides consisting of a "starter" His$_n$ peptide that can be ligated to other peptidyl or DNA sequences that can be used in cellular delivery, light harvesting and nucleic acid hybridization [133]. This chemistry relied on the chemoselective ligation (catalyzed by aniline) of hydrazines to aldehydes (Fig. 9.5a). To extend this work, we designed a series of chemoselectively ligated peptides for targeting various subcellular locales [134]. A common His$_6$-bearing starter peptide was covalently

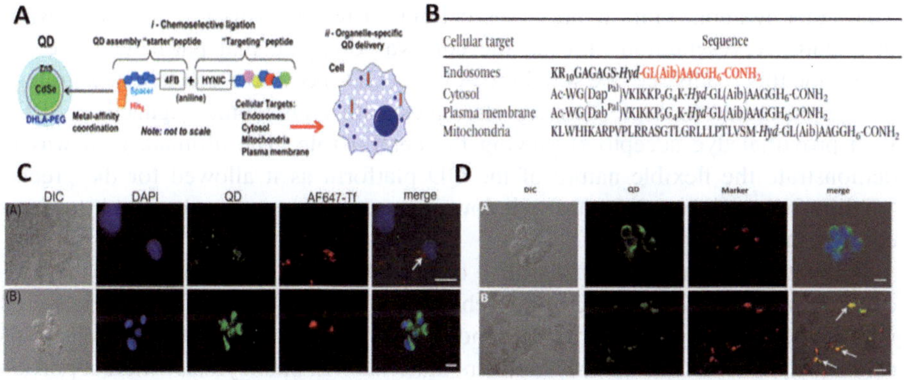

Fig. 9.5 Subcellular targeting peptides generated by chemoselective ligation. (**a**) Novel peptides generated by aniline-catalyzed coupling of 4-formylbenzyl (4FB)-bearing starter peptides to 2-hydrazinonicotinyl (HYNIC)-terminated targeting peptides for self-assembly to and specific delivery of QDs to subcellular locations. (**b**) Peptidyl motifs for targeting QD to endosomes, the cytosol, plasma membrane and mitochondria. The base starter peptide comprising the sequence 4-formylbenzyl-GL(Aib)AAGGH6-CONH$_2$ is shown in *red*. Ac is an acetyl group blocking the C-terminal; CONH$_2$ is an amide blocking the N-terminal. *Hyd* is hydrazone bond. *Dap* diaminopropionic acid, *Dab* diaminobutyric acid, *Pal* palmitoyl group. (**c**) QD targeting to endosomes (*top*) and to the cellular cytosol (*bottom*). QD (*green*), nuclei (DAPI, *blue*) and AlexaFluor647-transferrin (Tf) (endosomes, *red*). Scale bar is 50 μm. (**d**) QD delivery to the plasma membrane and mitochondria using chemoselective ligated peptides. Micrographs show the specific labeling of QDs (*green*) at the plasma membrane (*top*, costained with AlexaFluor647-Tf) and in mitochondria (*bottom*, costained with MitoTracker Red). Scale bar is 50 μm. In (**c**) and (**d**), *arrows* indicate localization of QDs with respective cellular markers (All images from Ref. [134] © Royal Society of Chemistry)

coupled to various targeting peptides for specific delivery to the plasma membrane, endosomes, cytosol, and mitochondria (Fig. 9.5b). Peptides generated in this manner successfully delivered QDs to the various targeted locations, as confirmed using specific cellular markers (Fig. 9.5c, d). Notably, the mitochondria-directing peptide used in this study was 43 amino acids in length, which is considerably longer than peptides typically generated by standard peptide chemistry. This approach provides a flexible platform for the synthesis of peptidyl structures and motifs that might otherwise be challenging to synthesize using traditional approaches.

9.5.4 *QD-Based Multicolor Labeling of Cells*

The simultaneous labeling of multiple, disparate intra- and extracellular structures is a powerful tool for understanding the complex biochemical and spatiotemporal processes that regulate cellular homeostasis. Typically, this requires the combinatorial use of numerous targeting probes (*e.g.*, antibodies), different delivery regimes, and often necessitates cellular fixation and membrane permeabilization.

This latter feature, while it preserves cellular structure, kills the cell and results in a static representation of what the cell was doing at that particular time. To overcome this roadblock, we sought to devise a QD-based live-cell labeling strategy that incorporated multiple delivery strategies with QDs (some engaged in FRET with proximal dye acceptors) playing the central role. The ultimate goal was to demonstrate the flexible nature of the QD platform as it allowed for the precise "dialing in" of discrete spectral windows that could be assigned to each targeted cellular structure.

The rationale for the multi-day live cell labeling strategy is depicted schematically in Fig. 9.6a. On day one of the scheme, 520 nm-emitting QDs were delivered with PULSin^{TM} reagent to target late endosomes and the cytosol of A549 cells. This cationic lipid-based formulation delivers materials *via* endocytosis where a portion of the materials escape the endosome and reach the cytosol. In this instance, the QDs were capped with a mixture of DHLA and DHLA-PEG-OCH_3 ligands to realize simultaneous electrostatic assembly of the PULSin^{TM} reagent while maintaining

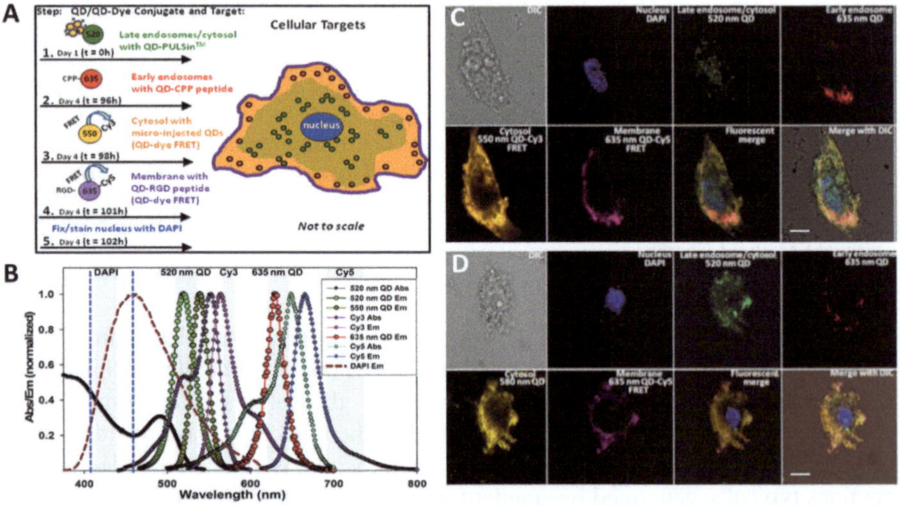

Fig. 9.6 QD-enabled multicolor cellular labeling. (**a**) Multi-day live cell labeling strategy. On day 1, 520 nm-emitting QDs were delivered with PULSin^{TM} reagent to target late endosomes and the cytosol of A549 cells. Four days later, early endosomes were labeled with 635 nm-emitting QDs appended with CPPs. 550 nm-emitting QDs appended with a peptide-Cy3 acceptor (FRET configuration for a *yellow/orange* emission window) were then microinjected. Finally 635 nm-emitting QDs decorated with RGD peptides (to target membrane integrins) and Cy5-peptides as a FRET acceptor labeled the plasma membrane. After delivery, cells were fixed and DAPI-stained to yield a fifth color. (**b**) Spectra and spectral windows of the QDs and dyes used in the delivery scheme depicted in *A*. The excitation lines used for DAPI (402 nm) and QD-dye excitation (457 nm) are depicted by the blue dotted lines. Images of A549 cells after the sequential delivery of the indicated color QDs to specific cellular locations (as indicated). (**c**) Cells were microinjected with QD-C3 conjugates engaged in FRET. (**d**) 580 nm-emitting QDs were injected into the cytosol (non FRET configuration). Scale bar in (**c**) and (**d**) is 5 μm (All images from Ref. [135] © ACS)

colloidal stability of the QD-lipid assembly. Four days later, early endosomes were labeled with 635 nm-emitting QDs appended with CPPs. The cytosol was subsequently microinjected with 550 nm-emitting QDs that were decorated with a peptide-Cy3 acceptor. This QD donor-dye acceptor FRET configuration produced a shifted yellow/orange emission window from the QD donor's native green. Finally, 635 nm-emitting QDs decorated with RGD peptides (to target membrane integrins) and Cy5-peptides (as a FRET acceptor) were used to label the plasma membrane. After this sequential delivery was done on live cells, the cells were then fixed with paraformaldehyde and DAPI-stained to yield a fifth color window. Laser scanning confocal microscopy was used where the DAPI and QD-dye systems were excited with 402 and 457 nm laser lines, respectively, while the emission of the various channels was acquired per the spectral output shown in Fig. 9.6b. Fluorescence imaging revealed the fidelity of both the delivery and imaging strategies as the QDs and QD-peptide-dye conjugates targeted their respective structures/cellular locations (Fig. 9.6c, d).

This report detailed a facile method for the delivery to and labeling of various cellular structures while incorporating multiple delivery modalities (lipid-based delivery, receptor-ligand interactions, direct microinjection). Further, it demonstrated the ability to create discrete spectral windows to which targets could be assigned by simply employing QD donor-dye acceptor constructs where the peptide-dye acceptors functioned as relays to create discrete shifted emission windows that could be easily resolved without the need for complicated spectral deconvolution. Additionally, it highlighted the compatibility of the QD materials with a range of delivery techniques while bringing to the forefront the inherent limitations of some delivery modalities. Microinjection, for example, while affording direct access to the cytosol, is rather limited in its throughput. Cationic polymer-mediated QD delivery, meanwhile, induced considerable toxicity that was not observed when delivering QDs using CPPs or RGD peptides. Clearly, a balance must be struck between efficacy of delivery and cellular viability when one employs these mixed delivery approaches.

9.5.5 Monitoring Realtime Intracellular QD-Protein Assembly Using QD-Based Energy Transfer

Understanding the spatiotemporal nature of nanomaterial-protein interactions *in vivo* is critical to the development of NP-based medicines and diagnostics and fluorescence-based techniques continue to play a critical role in discerning the details of these interactions. Specifically, the realtime fluorescence tagging of targeted proteins with bright, highly photostable materials is one such approach to realize this goal. In one example, the FlAsH/ReAsH system employs biarsenical fluorophores which react with vicinal tetracysteine motifs on targeted proteins [136]. To demonstrate the feasibility and utility of integrating QD materials into a similar approach, we targeted commercially prepared QDs to self-assemble with His_n-tagged mCherry fluorescent proteins in the cytosol of mammalian cells [137]. First, we demonstrated that 565 nm-emitting QDs coated with ITK-carboxyl groups (Invitrogen), when loaded with exogenous Ni^{2+}, mediated the coordination of His_6-mCherry proteins to the QD surface. We then used FRET to evaluate the ability of microinjected QDs to assemble to His_6-mCherry proteins expressed in the cytosol of COS-1 cells. As shown in Fig. 9.7a, only cells expressing His-tagged mCherry mediated energy transfer from the donor QD to the acceptor protein; no FRET was observed in QD-injected cells expressing non His-tagged mCherry. Notably, these experiments demonstrated the rapid (within minutes) assembly of the QD-protein complexes and the existence of a ~6 h experimental window over which the assemblies were stable as evidenced by no loss in the FRET signal over this time course. Further, the results showed the enhanced photostability of the QD-sensitized mCherry acceptor compared to when the protein was directly excited with the UV lamp.

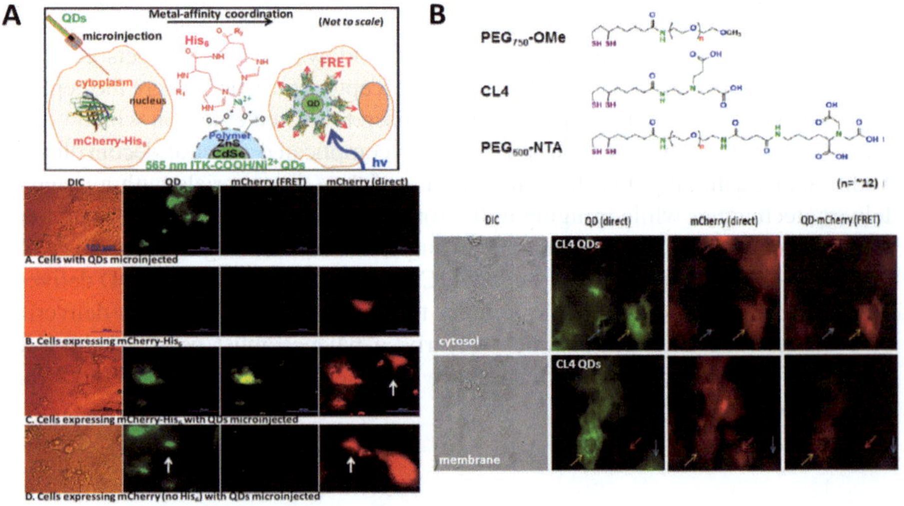

Fig. 9.7 Monitoring realtime intracellular QD-protein assembly using QD-based energy transfer. (**a**) Commercial 565 nm-emitting QDs coated with Ni^{2+}-loaded, carboxyl-terminated ITK ligands were microinjected in to COS-1 cells transfected to express His_6-tagged mCherry proteins in the cytosol (*top*). Successful QD-mCherry assembly was confirmed by FRET where excitation of the QD donor resulted in FRET-sensitized emission of the mCherry acceptor only in cells expressing His tagged mCherry (compare *panel C* and *D*) (Image from Ref. [137] © ACS). (**b**) In-house QDs capped with various solubilizing ligands (*top*) were microinjected into COS-1 cells expressing His6-mCherry proteins in the cytosol or appended to the inner leaflet of the plasma membrane (as indicated). Analysis showed the most efficient FRET was mediated by the combination of CL4-capped QDs assembled with mCherry within the cytosol. Assemblies of QDs with mCherry proteins at the inner leaflet of the plasma membrane showed less efficient FRET (Images adapted from Ref. [138] © MDPI Publishers)

In a recent follow-on study, this QD-fluorescent protein FRET system was extended to determine the role played by the nature of the coating ligand on the QD surface and the intracellular location of assembly on FRET efficiency [138]. In house-synthesized QDs capped with either short, zwitterionic compact ligands (CL4) or longer PEGylated ligands (DHLA-PEG-OCH_3 or DHLA-PEG-NTA) were compared for their ability to mediate efficient FRET to His_6-mCherry acceptor proteins assembled to the QD surface. Several salient features of these systems were demonstrated. While DHLA-PEG-NTA ligands mediated the most efficient FRET in test tube/plate-based assays, QDs capped with CL4 ligands displayed the most efficient FRET when injected into the cytosol of COS-1 cells where they assembled to mCherry proteins expressed in the cytosol. In contrast, assembly of the QDs to His_6-mCherry expressed at the inner leaflet of the plasma membrane as a fusion to a transmembrane-spanning receptor showed the least efficient FRET, likely due to steric hindrance of the assembly process within the close proximity to the membrane. Going forward, such QD-fluorescent protein assemblies could serve as the basis for realtime sensors and this study points to the careful consideration that must be given to the design of these intracellular FRET-based systems, both with respect to QD architecture and location of the assembled complexes.

9.5.6 Intracellular Realtime pH Sensing with QDs

The use of QDs for "active" cellular sensing has continued to grow at an almost exponential pace (reviewed in [21]). This growth has been accompanied by an ever-increasing number of new configurations in which the QD plays a pivotal role in the sensing platform. In addition to their utility in FRET-based sensors, increasing numbers of QD-based sensors take advantage of the QD's ability to engage in charge transfer with proximal electron acceptors. Our laboratory developed an intracellular pH sensing platform based on the charge transfer inherent in QD-peptide-dopamine bioconjugates [139]. The scheme takes advantage of the redox properties of the peptide-appended dopamine ligand (Fig. 9.8a). Dopamine is characterized by two inherent redox properties: a Nernstian dependence of formal potential on pH and oxidation of the hydroquinone to quinone at basic pH. Under acidic conditions, the dopamine appended to the QD *via* the His-peptide bridge is reduced and is a poor electron acceptor and the proximal photoexcited QD luminescence is bright. In basic media, however, dopamine is oxidized to the quinone form in a pH-dependent manner, making it an excellent electron acceptor. Here, the photoexcited QD donates electrons to the dopamine acceptor, resulting in a pH-dependent quenching of QD PL. When microinjected into the cytosol of COS-1 cells induced to undergo alkalosis using the drug nystatin, the QD PL tracked with increasing pH in a linear manner (from pH 6.5 to 12.8) over a 1 h observation window (Fig. 9.8b). While such high ranges of intracellular pH values are unlikely to be encountered in nature, the experimental format showed that the QD-peptide-dopamine conjugates could be

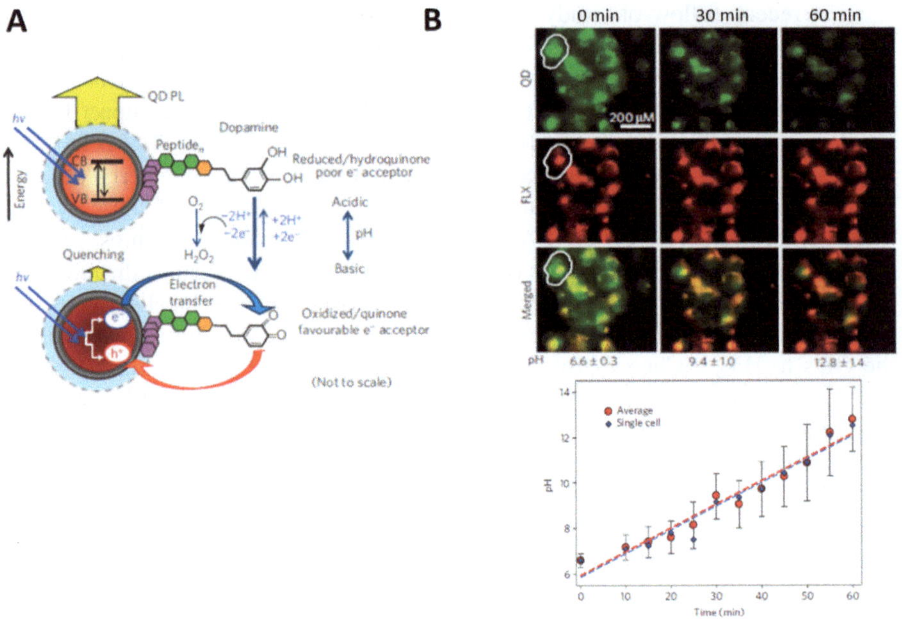

Fig. 9.8 QD-dopamine charge transfer for realtime pH sensing in live cells. (**a**) QD-peptide-dopamine bioconjugate for intracellular pH sensing. At acidic pH, the reduced/hydroquinone form of the dopamine ligand (appended to the QD surface *via* a His peptide) is a poor electron acceptor and the photoexcited QD emits luminescence (*top*). At basic pH, the oxidized/quinone form of dopamine is an efficient electron acceptor. Electron from the excited state QD are attracted to the quinone acceptor, resulting in pH-dependent quenching of QD photoluminescence (PL) (*bottom*). (**b**) 550 nm-emitting QDs assembled with peptide-dopamine conjugates along with 20 nm 680 nm red fluorescent Fluorophorex (FLX) nanospheres (used as internal control) were microinjected into the cytosol of COS-1 cells. pH of the injection mixture was 6.5. The extracellular medium was switched to PBS (pH 11.5) supplemented with nystatin to form membrane pores and promote cytosolic alkalosis. Fluorescence images were collected at the indicated times from both channels (*top*). Scale bar 200 μm. Average pH values were extracted from the PL data using a calibration curve (not shown) and plotted for total image or a single cell (*bottom*) (Images adapted from Ref. [139] © MacMillan Publishers Ltd)

stably introduced into cells and sense broad pH ranges. Further, this was one of the first experimental uses of charge transfer for realtime, live cell sensing.

9.5.7 Use of QDs in Neuroscience Applications

As our QD materials, surface/ligand coatings and bioconjugation chemistries have evolved and matured, so too have the myriad uses and applications of the QD-bioconjugates. Perhaps no better example bears this out than in the multiple recent demonstrations of our QDs used in neuroscience research. One of the research

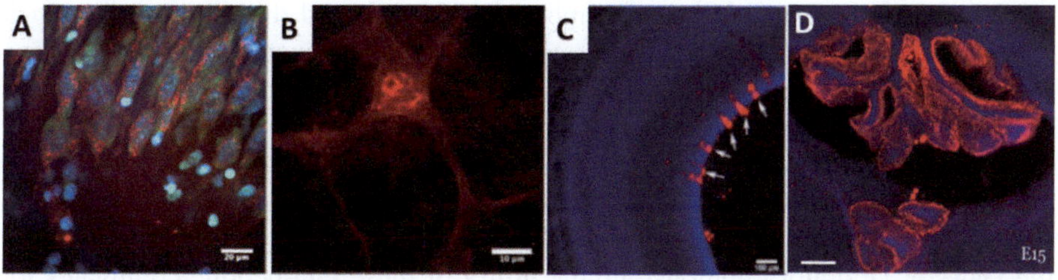

Fig. 9.9 Neuroscience applications of QDs and QD-peptide conjugates (1). (**a**) QD-peptide conjugates for tracking peptide localization in neuronal tissue slice. 625 nm-emitting QDs capped with CL4 ligands and JB577 peptides label the cytosol of CA3 layer cells in rat hippocampal slice. QDs (*red*), Nissl (neuronal marker, *green*) and DAPI staining showing a punctate, perinuclear staining pattern. Scale bar 20 μm. (**b**) Pyramidal neuron of the CA3 layer showing diffuse distribution of QD-JB577 peptide conjugate 24 h after delivery. Scale bar 10 μm. *Panels A* and *B* adapted from Ref. [141] © the authors). (**c**) Section of midbrain showing QD-CL4-JB577 conjugates at E8 (4 days after embryo injection). Image shows uptake of QDs from ventricles and long distance migration along glial tracks (*arrows*). Scale bar is 100 μm in each panel. (**d**) QD-CL4-JB577 labeling of the choroid plexus at E15. The choroid plexus has not developed at E4 when the QD-peptides are injected. Scale bar is 200 μm. *Panels C* and *D* adapted from Ref. [143] © ACS)

foci of the Dawson Laboratory at the University of Chicago is the development of peptidyl-based therapies for the treatment of a variety of diseases in which protein misfolding plays a key role (*e.g.*, Batten disease, a lysosomal storage disease [140]). One of the challenges, particularly in experimental tissue models and *in vivo*, is the robust visualization and tracking of the delivered therapeutic peptide. To address this, these investigators employed commercial 625 nm-emitting QDs that were cap-exchanged and the native, commercial ligand coating replaced with our custom ligands (either DHLA-PEG_{750}-OCH_3 or the shorter, zwitterionic CL4 ligand (see Table 9.1)). The goal was to use the QD's brightness and photostability in tracking the tissue penetration and localization of peptide JB577 in a rat hippocampal slice tissue culture system [141]. Their initial findings showed that the CL4-capped QDs appended with JB577 peptides localized efficiently to the cells of the CA3 layer in the slice model (Fig. 9.9a). Importantly, these conjugates showed specificity for pyramidal neuronal cell uptake as they were largely excluded from astrocytes, microglia, and oligodendrocytes. They subsequently noted distinct differences in cellular uptake by CA3 layer cells afforded by the capping ligand in the absence of the JB577 peptide. Specifically, the net negatively-charged zwitterionic CL4 ligands promoted uptake of the QDs while the uncharged PEGylated QDs remained on the periphery of the cell bodies, largely excluded from the cell interior. In a follow-on study, they determined that the balance between the negatively charged CL4 ligands mediating neuronal cell uptake combined with the negatively-charged chondroitin sulfate of the extracellular matrix restricting uptake was critical for maintaining specific QD uptake [142].

In another experimental platform, the Dawson laboratory used the same 625 QDs capped with our DHLA-PEG_{750}-OCH_3 or CL4 ligands to track peptide delivery and migration in the developing chick embryo [143]. Injected directly into the chick spinal cord at embryonic day 4 (E4), the QD-JB577 conjugates allowed tracking of the spinal cord extension into the ventricles, migratory neuroblasts, maturing brain cells, and complex architectures such as the choroid plexus (Fig. 9.9c, d). Importantly, no abnormalities in embryonic patterning or embryo survival were observed and mRNA expression analysis confirmed the proper spatiotemporal expression of key developmental genes. Cumulatively, these studies conducted by the Dawson laboratory demonstrate the utility of these QD-peptide conjugates for long-term tracking of embryonic development and suggest the exciting possibility of their use in neural stem cell tracking and migration and drug delivery to the developing brain.

In another demonstration of the implementation of our QD materials in critical neuroscience experimental formats, researchers at the Janelia Farm Research Campus (Howard Hughes Medical Institute) employed our as-synthesized, hydrophobic QDs to coat patch clamp pipettes and took advantage of the enormous two-photon action cross section absorption of the nanocrystal to facilitate deep tissue imaging and electrophysiological manipulations in live mouse brain [144]. As an example, patch pipettes whose surface was coated with 625 nm-emitting QDs were loaded with DsRed-encoding plasmid. Two-photon excitation of the QD-coated pipet allowed the targeting of deep-layer EGFP-expressing neurons in brain of live mice (Fig. 9.10a). The QD-labeled patch pipet successfully electroporated the plasmid

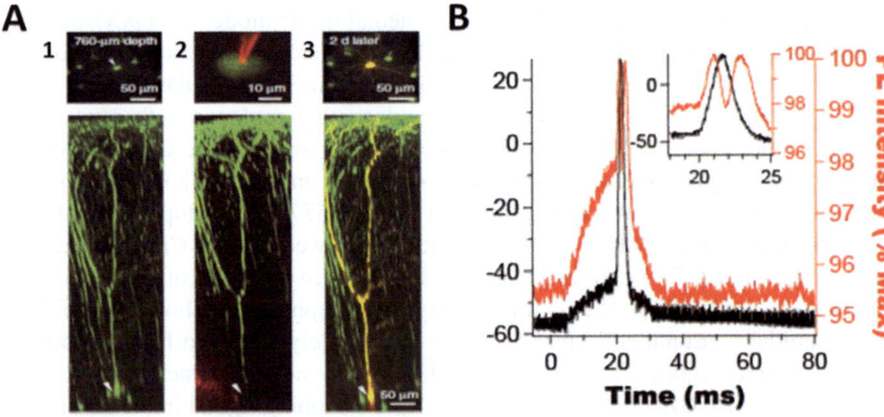

Fig. 9.10 Neuroscience applications of QDs and QD-peptide conjugates (2). (**a**) Deep layer targeting of electroporation in live mouse brain guided by 625 nm-emitting QD-coated patch pipettes. (*1*) *top*, z-projection showing EGFP-expressing targeted neuron soma; *bottom*, orthogonal view showing ~800 μm depth. (*2*) *top*, targeted neuron during electroporation (arrow denotes the site of electroporation); *bottom*, QD shown in *red*. (*3*) *top*, targeted neuron expressing DsRed and EGFP 2 d after electroporation (*yellow*) (Image adapted from Ref. [144] © Nature America, Inc). (**b**) Action potential tracing using the inherent opto-electrical properties of QDs. Quasi-Type II QDs were subjected to a time-resolved electric field intensity profile emulating that of firing neuron (*black trace*). The QD PL response (*red trace*) tracks with the field strength with msec time resolution. Note the repeated QD emission maxima at ~21 and 23 ms corresponding to the action potential passing through zero (Image from Ref. [145] © ACS)

repetitively at multiple target points along the length of the neuron covering a total tissue depth of ~760 μm while being excited with a 940 nm laser. This depth of imaging was completely facilitated by the QD's inherent advantages afforded by the unique photophysical properties of the QD materials to enable guided imaging at depths not possible using conventional organic dyes.

Finally, in our most recent work, we sought to ascertain the ability to function as robust imaging probes for visualizing changes in the membrane potential in live neurons (and potentially other electrically active cells) [145]. Here, the goal was to take advantage of QD photostability combined with their known PL modulation when subjected to an applied electric field. Previous reports demonstrated that QDs exhibit a quantum confined Stark effect in the presence of an electric field that manifests as a concomitant decrease in QD PL, a red shift in maximum emission wavelength, and broadening of the emission profile [146–147]. However, these previous studies were performed on single QD particles and did not address the QD response to an applied modulated electric field resembling that of a firing neuron. Thus, we performed steady-state and time-correlated measurements of QD PL responses to electric fields (in ensemble QDs cast in layer of dielectric polymer) in both Type I and Quasi-Type II QD materials and found distinct differences between the two materials. Namely, the Quasi-Type II QDs exhibited a significantly greater decrease in PL in response to an applied field, a response attributed to an increase in the number of charged QDs in the ensemble population. Most importantly for potential utility in neuronal cell imaging, we showed for the first time the ability of QDs to track in realtime the time-resolved electric field profile representative of that in a firing neuron (Fig. 9.10b). Here, the QD PL tracks, with millisecond fidelity, the tracing of the action potential while clearly reflecting the transition of the potential trace through zero field strength. Coupled with the inherent photostability and superior two-photon action cross sections of QDs compared to conventional fluorophores, these findings offer the exciting promise of the integration of QD materials into robust neuronal cell imaging platforms for *in vitro* and *in vivo* use.

9.6 Quantum Dot Toxicity

Clearly, the examples detailed in the preceding section demonstrate our Laboratory's continued effort to integrate QDs and QD-bioconjugates into living systems (cells, tissues, organisms) with ever-increasing control and sophistication for a variety of applications. Yet, no discussion of the use of QDs in biology would be complete without giving some consideration to the impact of the QD materials with biological systems. As the CdSe/ZnS core/shell platform has been a key player in much of our work (they represent one of the more robust semiconductor composites in terms of control over their photophysical properties) concerns about the long-term and toxicity of Cd^{2+}-containing materials persist. While these concerns are valid, it is worth pointing out that in all of our studies, we perform accompanying cytotoxicity assays to assess the impact on cellular viability for the application being pursued.

Indeed, in one of the few studies of its kind, we performed a head-to-head comparison of the cytotoxicity of our in-house synthesized QDs (capped with various ligands) delivered to cells *via* a number of delivery modalities (*e.g.*, peptides, cationic lipids) [41]. In this study we showed that CdSe/ZnS QDs delivered to a number of cell lines (epithelial, endothelial, monocyte) *via* peptides were remarkably well-tolerated compared to when delivered using traditional transfection reagents such as Lipofectamine^{TM}. Further, when toxicity was observed, it was often in conjunction with incubation conditions that were not particularly representative of those needed for robust labeling needed. For example, exposure of HEK 293T/17 cells to ~200 nM QD-CPP for 1 h, an incubation time that affords robust labeling of a large percentage of cells, corresponds to ~95 % cellular viability. Incubation of these same complexes on cells for 24 h (a more chronic scenario that is beyond what is necessary for efficient labeling) lowers cellular viability to ~70 %. More importantly, this study showed that many commercially available organic dyes, when used at the manufacturer's recommended concentrations, elicited toxicities that far surpassed those of the QD materials. Further evidence for the compatibility of our QDs is provided by the very elegant work of the Dawson and Janelia groups whose findings clearly show the ability to integrate these materials at the tissue and organismal level with minimal impact on the system being studied over the time course of the experiment. Finally, a recent meta-analysis performed on published toxicity data on cadmium-containing QDs showed that multiple factors correlate well with cytotoxicity, including the presence of a shell, the nature of the capping ligand, and the (bio)modifications made to the QD surface [148]. Clearly, this subject is more complex than initially anticipated.

In sum, the impact of the QD materials on cellular viability will always need to be assessed in the context of the experimental platform being interrogated. To date, amongst the various applications we have demonstrated, our observations point to minimal impact on cellular health and homeostasis during experiments and we have found the CdSe/ZnS QD platform to be a remarkably versatile system in which to learn basic fundamental principles (*e.g.*, bioconjugation, cellular delivery, intracellular FRET) that can ultimately be transferred and integrated into non-Cd^{2+}-containing QD platforms such as InP [149] and CuInS [150].

9.7 Conclusions

In this chapter, we have highlighted our efforts over the past decade to utilize QDs and QD bio-conjugates with living cells, tissues and organisms. QDs, with their remarkable size-dependent photophysical properties, have clearly moved beyond a mere scientific curiosity and have become a value-added material with labeling, imaging, long-term fate tracking and cargo delivery capabilities that are not afforded by many other nanomaterials. In our work with these materials, several common themes have emerged. First, the histidine-driven self-assembly is an invaluable tool for creating QD-biological conjugates that are stable in cells in tissues over extended periods of time (hours to days). Second, we have employed the QD as an efficient FRET donor to proximal fluorophore acceptors for a variety of applications including the confirmation of intracellular bioconjugate stability [118, 120] and the successful assembly of QDs to targeting proteins in realtime [127, 129]. Going forward, we expect that this type of realtime "readout" will serve as the basis for a multitude of active sensing, assembly and multistage drug/cargo release platforms for controlling and modulating the behavior of cells and tissues. Clearly, the basic research that will facilitate these efforts will continue and we are excited to see what new sensing and drug delivery capabilities evolve.

Acknowledgement The authors acknowledge the Defense Threat Reduction Agency, NRL Base Funding Program and NRL Institute for Nanoscience for financial support. C.E.R is supported by a National Research Council Postdoctoral Research Associateship. L.D.F. and A.S. are graduate students in the Fischell Department of Bioengineering, University of Maryland College Park, MD. USA.

References

1. National Nanotechnology Initiative Strategic Plan. Washington DC, 2014
2. Mattoussi H, Mauro JM, Goldman ER, Anderson GP, Sundar VC, Mikulec FV, Bawendi MG (2000) Self-assembly of CdSe-ZnS quantum dot bioconjugates using an engineered recombinant protein. J Am Chem Soc 122(49):12142–12150
3. Goldman ER, Anderson GP, Tran PT, Mattoussi H, Charles PT, Mauro JM (2002) Conjugation of luminescent quantum dots with antibodies using an engineered adaptor protein to provide new reagents for fluoroimmunoassays. Anal Chem 74(4):841–847
4. Goldman ER, Balighian ED, Mattoussi H, Kuno MK, Mauro JM, Tran PT, Anderson GP (2002) Avidin: a natural bridge for quantum dot-antibody conjugates. J Am Chem Soc 124(22):6378–6382
5. Goldman ER, Clapp AR, Anderson GP, Uyeda HT, Mauro JM, Medintz IL, Mattoussi H (2004) Multiplexed toxin analysis using four colors of quantum dot fluororeagents. Anal Chem 76(3):684–688
6. Medintz IL, Clapp AR, Mattoussi H, Goldman ER, Fisher B, Mauro JM (2003) Self-assembled nanoscale biosensors based on quantum dot FRET donors. Nat Mater 2(9):630–638
7. Medintz IL, Goldman ER, Lassman ME, Mauro JM (2003) A Fluorescence Resonance Energy Transfer Sensor Based on Maltose Binding Protein. Bioconjug Chem 14(5):909–918
8. Jaiswal JK, Mattoussi H, Mauro JM, Simon SM (2003) Long-term multiple color imaging of live cells using quantum dot bioconjugates. Nat Biotechnol 21(1):47–51
9. Voura EB, Jaiswal JK, Mattoussi H, Simon SM (2004) Tracking metastatic tumor cell extravasation with quantum dot nanocrystals and fluorescence emission-scanning microscopy. Nat Med 10(9):993–998
10. Arya H, Kaul Z, Wadhwa R, Taira K, Hirano T, Kaul SC (2005) Quantum dots in bio-imaging: revolution by the small. Biochem Biophys Res Co 329(4):1173–1177
11. Li J, Zhu J-J (2013) Quantum dots for fluorescent biosensing and bio-imaging applications. Analyst 138(9):2506–2515
12. Medintz IL, Uyeda HT, Goldman ER, Mattoussi H (2005) Quantum dot bioconjugates for imaging, labelling and sensing. Nat Mater 4(6):435–446
13. Delehanty JB, Medintz IL, Pons T, Brunel FM, Dawson PE, Mattoussi H (2006) Self-assembled quantum dot-peptide bioconjugates for selective intracellular delivery. Bioconjug Chem 17(4):920–927

14. Medintz IL, Pons T, Delehanty JB, Susumu K, Brunel FM, Dawson PE, Mattoussi H (2008) Intracellular delivery of quantum dot-protein Cargos mediated by cell penetrating peptides. Bioconjug Chem 19:1785–1795
15. Sapsford KE, Berti L, Medintz IL (2006) Materials for fluorescence resonance energy transfer analysis: beyond traditional donor–acceptor combinations. Angew Chem Int Ed 45(28):4562–4589
16. Medintz IL, Stewart MH, Trammell SA, Susumu K, Delehanty JB, Mei BC, Melinger JS, Blanco-Canosa JB, Dawson PE, Mattoussi H (2010) Quantum-dot/dopamine bioconjugates function as Redox coupled assemblies for in vitro and intracellular pH sensing. Nat Mater 9(8):676–684
17. Rowland CE, Brown CW, Medintz IL, Delehanty JB (2015) Intracellular FRET-based probes: a review. Methods Appl Fluoresc 3(4):042006
18. Resch-Genger U, Grabolle M, Cavaliere-Jaricot S, Nitschke R, Nann T (2008) Quantum dots versus organic dyes as fluorescent labels. Nat Methods 5(9):763–775
19. Yuan L, Lin W, Zheng K, Zhu S (2013) FRET-based small-molecule fluorescent probes: rational design and bioimaging applications. Acc Chem Res 46(7):1462–1473
20. Mishra A, Fischer MKR, Bäuerle P (2009) Metal-free organic dyes for dye-sensitized solar cells: from structure: property relationships to design ules. Angew Chem Int Ed 48(14):2474–2499
21. Delehanty JB, Susumu K, Manthe RL, Algar WR, Medintz IL (2012) Active cellular sensing with quantum dots: transitioning from research rool to reality; A review. Anal Chim Acta 750(0):63–81
22. Haugland RP (2002) Handbook of fluorescent probes and research products. Molecular Probes, Eugene
23. Day RN, Davidson MW (2009) The fluorescent protein palette: tools for cellular imaging. Chem Soc Rev 38(10):2887–2921
24. Lam AJ, St-Pierre F, Gong Y, Marshall JD, Cranfill PJ, Baird MA, McKeown MR, Wiedenmann J, Davidson MW, Schnitzer MJ, Tsien RY, Lin MZ (2012) Improving FRET dynamic range with bright green and red fluorescent proteins. Nat Methods 9(10):1005–1012
25. Zhang J, Campbell RE, Ting AY, Tsien RY (2002) Creating new fluorescent probes for cell biology. Nat Rev Mol Cell Biol 3(12):906–918
26. Frommer WB, Davidson MW, Campbell RE (2009) Genetically encoded biosensors based on engineered fluorescent proteins. Chem Soc Rev 38(10):2833–2841
27. Chudakov DM, Lukyanov S, Lukyanov KA (2005) Fluorescent proteins as a toolkit for in vivo imaging. Trends Biotechnol 23(12):605–613
28. Tsien RY (2009) Constructing and exploiting the fluorescent protein paintbox (nobel lecture). Angew Chem Int Ed 48(31):5612–5626
29. Piston DW, Kremers G-J (2007) Fluorescent protein FRET: the good, the bad and the ugly. Trends Biochem Sci 32(9):407–414
30. Selvin PR (2002) Principles and biophysical applications of lanthanide-based probes. Ann Rev Biophys Biomolecular Struct 31(1):275–302
31. Klimov VI (2007) Spectral and dynamical properties of multiexcitons in semiconductor nanocrystals. Annu Rev Phys Chem 58:635–673
32. Zhong X, Feng Y, Knoll W, Han M (2003) Alloyed $Zn_xCd_{1-x}S$ nanocrystals with highly narrow luminescence spectral width. J Am Chem Soc 125(44):13559–13563
33. Wise FW (2000) Lead salt quantum dots: the limit of strong quantum confinement. Acc Chem Res 33(11):773–780
34. Reimann SM, Manninen M (2002) Electronic structure of quantum dots. Rev Mod Phys 74(4):1283–1342
35. Barroso MM (2011) Quantum dots in cell biology. J Histochem Cytochem 59(3):237–251
36. Kim S, Fisher B, Eisler H-J, Bawendi M (2003) Type-II quantum dots: CdTe/CdSe(Core/Shell) and CdSe/ZnTe(Core/Shell) heterostructures. J Am Chem Soc 125(38):11466–11467

37. Clapp AR, Medintz IL, Mattoussi H (2006) Förster resonance energy transfer investigations using quantum-dot fluorophores. Chem Phys Chem 7(1):47–57
38. Rowland CE, Liu W, Hannah DC, Chan MKY, Talapin DV, Schaller RD (2013) Thermal stability of colloidal InP nanocrystals: small inorganic ligands boost high-temperature photoluminescence. ACS Nano 8(1):977–985
39. Rowland CE, Schaller RD (2013) Exciton fate in semiconductor nanocrystals at elevated temperatures: hole trapping outcompetes exciton deactivation. J Phys Chem C 117(33):17337–17343
40. Rowland CE, Hannah DC, Demortière A, Yang J, Cook RE, Prakapenka VB, Kortshagen U, Schaller RD (2014) Silicon nanocrystals at elevated temperatures: retention of photoluminescence and diamond silicon to β-silicon carbide phase transition. ACS Nano 8(9):9219–9223
41. Bradburne CE, Delehanty JB, Boeneman Gemmill K, Mei BC, Mattoussi H, Susumu K, Blanco-Canosa JB, Dawson PE, Medintz IL (2013) Cytotoxicity of quantum dots used for in vitro cellular labeling: role of QD surface ligand, delivery modality, cell type, and direct comparison to organic fluorophores. Bioconjug Chem 24(9):1570–1583
42. Smith AM, Nie S (2004) Chemical analysis and cellular imaging with quantum dots. Analyst 129(8):672–677
43. Guo Y, Rowland CE, Schaller RD, Vela J (2014) Near-infrared photoluminescence enhancement in Ge/CdS and Ge/ZnS core/shell nanocrystals: utilizing IV/II–VI semiconductor epitaxy. ACS Nano 8(8):8334–8343
44. Sapsford KE, Algar WR, Berti L, Gemmill KB, Casey BJ, Oh E, Stewart MH, Medintz IL (2013) Functionalizing nanoparticles with biological molecules: developing chemistries that facilitate nanotechnology. Chem Rev 113(3):1904–2074
45. Andrasfalvy BK, Galinanes GL, Huber D, Barbic M, Macklin JJ, Susumu K, Delehanty JB, Huston AL, Makara JK, Medintz IL (2014) Quantum dot-based multiphoton fluorescent pipettes for targeted neuronal electrophysiology. Nat Methods 11(12):1237–1241
46. Clapp AR, Pons T, Medintz IL, Delehanty JB, Melinger JS, Tiefenbrunn T, Dawson PE, Fisher BR, O'Rourke B, Mattoussi H (2007) Two-photon excitation of quantum-dot-based fluorescence resonance energy transfer and its applications. Adv Mater 19(15):1921–1926
47. Zrazhevskiy P, Sena M, Gao X (2010) Designing multifunctional quantum dots for bioimaging, detection, and drug delivery. Chem Soc Rev 39(11):4326–4354
48. Rowland CE, Fedin I, Zhang H, Gray SK, Govorov AO, Talapin DV, Schaller RD (2015) Picosecond energy transfer and multiexciton transfer outpaces Auger recombination in binary CdSe nanoplatelet solids. Nat Mater 14(5):484–489
49. Geisow MJ, Evans WH (1984) pH in the endosome: measurements during pinocytosis and receptor-mediated endocytosis. Exp Cell Res 150:36–46
50. Mellman I (1992) The importance of being acid: the role of acidification in intracellular membrane traffic. J Exp Biol 172:39–45
51. Varkouhi AK, Scholte M, Storm G, Haisma HJ (2011) Endosomal escape pathways for delivery of biologicals. J Control Release 151:220–228
52. Delehanty JB, Mattoussi H, Medintz IL (2009) Delivering quantum dots into cells: strategies, progress and emaining issues. Anal Bioanal Chem 393:1091–1105
53. Hermanson GT (2008) Bioconjugate techniques, 2 edn. Academic Press, Inc., San Diego, p. 1202
54. Liu W, Choi HS, Zimmer JP, Tanaka E, Frangioni JV, Bawendi M (2007) Compact cysteine-coated CdSe(ZnCdS) quantum dots for in vivo applications. J Am Chem Soc 129:14530–14531
55. Choi HS, Liu W, Misra P, Tanaka E, Zimmer JP, Ipe BI, Bawendi MG, Frangioni JV (2007) Renal clearance of quantum dots. Nat Biotechnol 25:1165–1170
56. Choi HS, Liu W, Liu F, Nasr K, Misra P, Bawendi MG, Frangioni JV (2010) Design considerations for tumor-targeted nanoparticles. Nat Nanotechnol 5:42–47
57. Clapp AR, Medintz IL, Mauro JM, Fisher BR, Bawendi MG, Mattoussi H (2004) Fluorescence resonance energy transfer between quantum dot donors and dye-labeled protein acceptors. J Am Chem Soc 126(1):301–310

58. Clapp AR, Medintz IL, Mattoussi H (2006) Förster resonance energy transfer investigations using quantum-dot fluorophores. Chem Phys Chem 7:47–57
59. Medintz IL, Mattoussi H (2009) Quantum dot-based resonance energy transfer and its growing application in biology. Phys Chem Chem Phys 11:17–45
60. Bruchez M Jr, Moronne M, Gin P, Weiss S, Alivisatos AP (1998) Semiconductor nanocrystals as fluorescent biological labels. Science 281(5385):2013–2016
61. Gerion D, Pinaud F, Williams SC, Parak WJ, Zanchet D, Weiss S, Alivisatos AP (2001) Synthesis and properties of biocompatible water-soluble silica-coated CdSe/ZnS semiconductor quantum dots. J Phys Chem B 105(37):8861–8871
62. Nann T, Mulvaney P (2004) Single quantum dots in spherical silica particles. Angew Chem Int Ed 43:5393–5396
63. Selvan ST, Tan TT, Ying JY (2005) Robust, non-cytotoxic, silica-coated CdSe quantum dots with efficient photoluminescence. Adv Mater 17:1620–1625
64. Darbandi M, Thomann R, Nann T (2005) Single quantum dots in silica spheres by microemulsion synthesis. Chem Mater 17:5720–5725
65. Selvan ST, Patra PK, Ang CY, Ying JY (2007) Synthesis of silica-coated semiconductor and magnetic quantum dots and their use in the imaging of live cells. Angew Chem Int Ed 46:2448–2452
66. Chan WCW, Nie SM (1998) Quantum dot bioconjugates for ultrasensitive nonisotopic detection. Science 281(5385):2016–2018
67. Kim JH, Morikis D, Ozkan M (2004) Adaptation of inorganic quantum dots for stable molecular Beacons. Sens Actuators B 102:315–319
68. Algar WR, Krull UJ (2006) Adsorption and hybridization of oligonucleotides on mercaptoacetic acid-capped CdSe/ZnS quantum dots and quantum dot−oligonucleotide conjugates. Langmuir 22:11346–11352
69. Wilson R, Spiller DG, Beckett A, Prior IA, Sée V (2010) Highly stable dextran-coated quantum dots for biomolecular detection and cellular imaging. Chem Mater 22:6361–6369
70. Mitchell GP, Mirkin CA, Letsinger RL (1999) Programmed assembly of DNA functionalized quantum dots. J Am Chem Soc 121:8122–8123
71. Freeman R, Gill R, Shweky I, Kotler M, Banin U, Willner I (2009) Biosensing and probing of intracellular metabolic pathways by NADH-sensitive quantum dots. Angew Chem Int Ed 48:309–313
72. Haldar KK, Sen T, Patra A (2010) Metal conjugated semiconductor hybrid nanoparticle-based fluorescence resonance energy transfer. J Phys Chem C 114:4869–4874
73. Ghadiali JE, Lowe SB, Stevens MM (2011) Quantum-dot-based FRET detection of histone acetyltransferase activity. Angew Chem Int Ed 50:3417–3420
74. Xia Y, Song L, Zhu C (2011) Turn-on and near-infrared fluorescent sensing for 2,4,6-trinitrotoluene based on hybrid (gold nanorod)-(quantum dots) assembly. Anal Chem 83:1401–1407
75. Hoshino A, Fujioka K, Oku T, Suga M, Sasaki YF, Ohta T, Yasuhara M, Suzuki K, Yamamoto K (2004) Physicochemical properties and cellular toxicity of nanocrystal quantum dots depend on their surface modification. Nano Lett 4:2163–2169
76. Jiang W, Mardyani S, Fischer H, Chan WCW (2006) Design and characterization of lysine cross-linked mercapto-acid biocompatible quantum dots. Chem Mater 18:872–878
77. Hering VR, Gibson G, Schumacher RI, Faljoni-Alario A, Politi MJ (2007) Energy transfer between CdSe/ZnS core/shell quantum dots and fluorescent proteins. Bioconjug Chem 18:1705–1708
78. Åkerman ME, Chan WCW, Laakkonen P, Bhatia SN, Ruoslahti E (2002) Nanocrystal targeting in vivo. Proc Natl Acad Sci U S A 99(20):12617–12621
79. Hong R, Fischer NO, Verma A, Goodman CM, Emrick T, Rotello VM (2004) Control of protein structure and function through surface recognition by tailored nanoparticle scaffolds. J Am Chem Soc 126:739–743
80. Kang E-C, Ogura A, Kataoka K, Nagasaki Y (2004) Preparation of Water-soluble PEGylated Semiconductor Nanocrystals. *Chem Lett 33*:840–841

81. Uyeda HT, Medintz IL, Jaiswal JK, Simon SM, Mattoussi H (2005) Synthesis of compact multidentate ligands to prepare stable hydrophilic quantum dot fluorophores. J Am Chem Soc 127(11):3870–3878
82. Karakoti AS, Das S, Thevuthasan S, Seal S (2011) PEGylated Inorganic Nanoparticles. *Angew Chem Int Ed 50*:1980–1994
83. Bentzen EL, Tomlinson ID, Mason J, Gresch P, Warnement MR, Wright D, Sanders-Bush E, Blakely R, Rosenthal SJ (2005) Surface modification to reduce nonspecific binding of quantum dots in live cell assays. Bioconjug Chem 16:1488–1494
84. Susumu K, Uyeda HT, Medintz IL, Pons T, Delehanty JB, Mattoussi H (2007) Enhancing the Stability and Biological Functionalities of Quantum Dots via Compact Multifunctional Ligands. *J Am Chem Soc 129*:13987–13996
85. Mei BC, Susumu K, Medintz IL, Delehanty JB, Mountziaris TJ, Mattoussi H (2008) Modular poly(ethylene glycol) ligands for biocompatible semiconductor and gold nanocrystals with extended pH and ionic stability. J Mater Chem 18:4949–4958
86. Susumu K, Uyeda HT, Medintz IL, Mattoussi H (2007) Design of biotin-functionalized luminescent quantum dots. J Biomed Biotechnol 2007:90651
87. Oh E, Fatemi FK, Currie M, Delehanty JB, Pons T, Fragola A, Lévêque-Fort S, Goswami R, Susumu K, Huston AL, Medintz IL (2013) PEGylated luminescent gold nanoclusters: synthesis, characterization, bioconjugation, and application to one- and two-photon cellular imaging. Part Part Syst Charact 30:453–466
88. Field LD, Walper SA, Susumu K, Oh E, Medintz IL, Delehanty JB (2015) Modulation of intracellular quantum dot to fluorescent protein Förster resonance energy transfer via customized ligands and spatial control of donor–acceptor assembly. Sensors 15:30457–30468
89. Brown CW III, Oh E, DA H Jr, Walper SA, Susumu K, Stewart MH, Deschamps JR, Medintz IL (2015) Kinetic enhancement of the diffusion-limited enzyme beta-galactosidase when displayed with quantum dots. RSC Adv 5:93089–93094
90. Mei BC, Susumu K, Medintz IL, Mattoussi H (2009) Polyethylene glycol-based bidentate ligands to enhance quantum dot and gold nanoparticle stability in biological media. Nat Protoc 4:412–423
91. Susumu K, Mei BC, Mattoussi H (2009) Multifunctional ligands based on dihydrolipoic acid and polyethylene glycol to promote biocompatibility of quantum dots. Nat Protoc 4:424–436
92. Dif A, Boulmedais F, Pinot M, Roullier V, Baudy-Floc'h M, Coquelle FM, Clarke S, Neveu P, Vignaux F, Le Borgne R, Dahan M, Gueroui Z, Marchi-Artzner V (2009) Small and stable peptidic PEGylated quantum dots to target polyhistidine-tagged proteins with controlled stoichiometry. J Am Chem Soc 131:14738–14746
93. Stewart MH, Susumu K, Mei BC, Medintz IL, Delehanty JB, Blanco-Canosa JB, Dawson PE, Mattoussi H (2010) Multidentate poly(ethylene glycol) ligands provide colloidal stability to semiconductor and metallic nanocrystals in extreme conditions. J Am Chem Soc 132:9804–9813
94. Thiry M, Boldt K, Nikolic MS, Schulz F, Ijeh M, Panicker A, Vossmeyer T, Weller H (2011) Fluorescence properties of hydrophilic semiconductor nanoparticles with tridentate polyethylene oxide ligands. ACS Nano 5:4965–4973
95. Gravel E, Tanguy C, Cassette E, Pons T, Knittel F, Bernards N, Garofalakis A, Ducongé F, Dubertret B, Doris E (2013) Compact tridentate ligands for enhanced aqueous stability of quantum dots and in vivo imaging. Chem Sci 4:411–417
96. Zhan N, Palui G, Safi M, Ji X, Mattoussi H (2013) Multidentate Zwitterionic ligands provide compact and highly biocompatible quantum dots. J Am Chem Soc 135:13786–13795
97. García KP, Zarschler K, Barbaro L, Barreto JA, O'Malley W, Spiccia L, Stephan H, Graham B (2014) Zwitterionic-coated "Stealth" nanoparticles for biomedical applications: recent advances in countering biomolecular corona formation and uptake by the mononuclear phagocyte system. Small 10:2516–2529
98. Susumu K, Oh E, Delehanty JB, Blanco-Canosa JB, Johnson BJ, Jain V, Hervey WJ, Algar WR, Boeneman K, Dawson PE, Medintz IL (2011) Multifunctional compact Zwitterionic ligands for preparing robust biocompatible semiconductor quantum dots and gold nanoparticles. J Am Chem Soc 133:9480–9496

99. Dennis AM, Sotto DC, Mei BC, Medintz IL, Mattoussi H, Bao G (2010) Surface ligand effects on metal-affinity coordination to quantum dots: implications for nanoprobe self-assembly. Bioconjug Chem 21:1160–1170
100. Kim S, Bawendi MG (2003) Oligomeric ligands for luminescent and stable nanocrystal quantum dots. J Am Chem Soc 125:14652–14653
101. Kim S-W, Kim S, Tracy JB, Jasanoff A, Bawendi MG (2005) Phosphine oxide polymer for water-soluble nanoparticles. J Am Chem Soc 127:4556–4557
102. Nann T (2005) Phase-transfer of CdSe@ZnS quantum dots using amphiphilic hyperbranched polyethylenimine. Chem Commun 2005:1735–1736
103. Duan H, Nie S (2007) Cell-penetrating quantum dots based on multivalent and endosome-disrupting surface coatings. J Am Chem Soc 129:3333–3338
104. Wang M, Felorzabihi N, Guerin G, Haley JC, Scholes GD, Winnik MA (2007) Water-soluble CdSe quantum dots passivated by a multidentate diblock copolymer. Macromolecules 40:6377–6384
105. Xie M, Liu H-H, Chen P, Zhang Z-L, Wang X-H, Xie Z-X, Du Y-M, Pan B-Q, Pang D-W (2005) CdSe/ZnS-labeled carboxymethyl chitosan as a bioprobe for live cell imaging. Chem Commun 2005:5518–5520
106. Duong JT, Bailey MJ, Pick TE, McBride PM, Rosen EL, Buonsanti R, Milliron DJ, Helms BA (2012) Efficient polymer passivation of ligand-stripped nanocrystal surfaces. J Polym Sci Part A Polym Chem 50:3719–3727
107. Liu W, Greytak AB, Lee J, Wong CR, Park J, Marshall LF, Jiang W, Curtin PN, Ting AY, Nocera DG, Fukumura D, Jain RK, Bawendi MG (2010) Compact biocompatible quantum dots via RAFT-mediated synthesis of imidazole-based random copolymer ligand. J Am Chem Soc 132:472–483
108. Zhang P, Liu S, Gao D, Hu D, Gong P, Sheng Z, Deng J, Ma Y, Cai L (2012) Click-functionalized compact quantum dots protected by multidentate-imidazole ligands: conjugation-ready nanotags for living-virus labeling and imaging. J Am Chem Soc 134:8388–8391
109. Susumu K, Oh E, Delehanty JB, Pinaud F, Gemmill KB, Walper S, Breger J, Schroeder MJ, Stewart MH, Jain V, Whitaker CM, Huston AL, Medintz IL (2014) A new family of pyridine-appended multidentate polymers as hydrophilic surface ligands for preparing stable biocompatible quantum dots. Chem Mater 26:5327–5344
110. Smith AM, Nie S (2008) Minimizing the hydrodynamic size of quantum dots with multifunctional multidentate polymer ligands. J Am Chem Soc 130:11278–11279
111. Yildiz I, Deniz E, McCaughan B, Cruickshank SF, Callan JF, Raymo FM (2010) Hydrophilic CdSe−ZnS core−shell quantum dots with reactive functional groups on their surface. Langmuir 26:11503–11511
112. Jana NR, Erathodiyil N, Jiang J, Ying JY (2010) Cysteine-functionalized polyaspartic acid: a polymer for coating and bioconjugation of nanoparticles and quantum dots. Langmuir 26:6503–6507
113. Liu L, Guo X, Li Y, Zhong X (2010) Bifunctional multidentate ligand modified highly stable water-soluble quantum dots. Inorg Chem 49:3768–3775
114. Palui G, Na HB, Mattoussi H (2012) Poly(ethylene glycol)-based multidentate oligomers for biocompatible semiconductor and gold nanocrystals. Langmuir 28:2761–2772
115. Giovanelli E, Muro E, Sitbon G, Hanafi M, Pons T, Dubertret B, Lequeux N (2012) Highly enhanced affinity of multidentate versus bidentate Zwitterionic ligands for long-term quantum dot bioimaging. Langmuir 28:15177–15184
116. Sun M, Yang L, Jose P, Wang L, Zweit J (2013) Functionalization of quantum dots with multidentate Zwitterionic ligands: impact on cellular interactions and cytotoxicity. J Mater Chem B 1:6137
117. Duan H, Kuang M, Wang YA (2010) Quantum dots with multivalent and compact polymer coatings for efficient fluorescence resonance energy transfer and self-assembled biotagging. Chem Mater 22:4372–4378

118. Cui Y, Gong X, Zhu S, Li Y, Su W, Yang Q, Chang J (2012) An effective modified method to prepare highly luminescent, highly stable water-soluble quantum dots and its preliminary application in immunoassay. J Mater Chem 22:462–469
119. Pearson RG (1968) Hard and soft acids and bases, HSAB, part 1. Fundamental principles. J Chem Educ 45:581–587
120. Oh E, Delehanty JB, Sapsford KE, Susumu K, Goswami R, Blanco-Canosa JB, Dawson PE, Granek J, Shoff M, Zhang Q, Goering PL, Huston A, Medintz IL (2011) Cellular uptake and fate of PEGylated gold nanoparticles is dependent on both cell-penetration peptides and particle size. ACS Nano 5(8):6434–6448
121. Medintz I (2006) Universal tools for biomolecular attachment to surfaces. Nat Mater 5(11):842–842
122. Sletten EM, Bertozzi CR (2009) Bioorthogonal chemistry: fishing for selectivity in a sea of functionality. Angew Chem Int Ed 48(38):6974–6998
123. Algar WR, Prasuhn DE, Stewart MH, Jennings TL, Blanco-Canosa JB, Dawson PE, Medintz IL (2011) The controlled display of biomolecules on nanoparticles: a challenge suited to bioorthogonal chemistry. Bioconjug Chem 22(5):825–858
124. Sapsford KE, Pons T, Medintz IL, Higashiya S, Brunel FM, Dawson PE, Mattoussi H (2007) Kinetics of metal-affinity driven self-assembly between proteins or peptides and CdSe-ZnS quantum dots. J Phys Chem C 111(31):11528–11538
125. Delehanty JB, Bradburne CE, Boeneman K, Susumu K, Farrell D, Mei BC, Blanco-Canosa JB, Dawson G, Dawson PE, Mattoussi H, Medintz IL (2010) Delivering quantum dot-peptide bioconjugates to the cellular cytosol: escaping from the endolysosomal system. Integr Biol-UK 2(5–6):265–277
126. Delehanty JB, Bradburne CE, Susumu K, Boeneman K, Mei BC, Farrell D, Blanco-Canosa JB, Dawson PE, Mattoussi H, Medintz IL (2011) Spatiotemporal multicolor labeling of individual cells using peptide-functionalized quantum dots and mixed delivery techniques. J Am Chem Soc 133(27):10482–10489
127. Boeneman K, Delehanty JB, Susumu K, Stewart MH, Medintz IL (2010) Intracellular bioconjugation of targeted proteins with semiconductor quantum dots. J Am Chem Soc 132(17):5975–5977
128. Delehanty JB, Bradburne CE, Boeneman K, Susumu K, Farrell D, Mei BC, Blanco-Canosa JB, Dawson G, Dawson PE, Mattoussi H, Medintz IL (2010) Delivering quantum dot-peptide bioconjugates to the cellular cytosol: escaping from the endolysosomal system. Integr Biol 2(5–6):265–277
129. Nazarenus M, Zhang Q, Soliman MG, Del Pino P, Pelaz B, Carregal-Romero S, Rejman J, Rothen-Rutishauser B, Clift MJ, Zellner R, Nienhaus GU, Delehanty JB, Medintz IL, Parak WJ (2014) In vitro interaction of colloidal nanoparticles with mammalian cells: what have we learned thus far? Beilstein J Nanotechnol 5:1477–1490
130. Boeneman K, Delehanty JB, Blanco-Canosa JB, Susumu K, Stewart MH, Oh E, Huston AL, Dawson G, Ingale S, Walters R, Domowicz M, Deschamps JR, Algar WR, DiMaggio S, Manono J, Spillmann CM, Thompson D, Jennings TL, Dawson PE, Medintz IL (2013) Selecting improved peptidyl motifs for cytosolic delivery of disparate protein and nanoparticle materials. ACS Nano 7(5):3778–3796
131. Clapp AR, Pons T, Medintz IL, Delehanty JB, Melinger JS, Tiefenbrunn T, Dawson PE, Fisher BR, O'Rourke B, Mattoussi H (2007) Two-photon excitation of quantum dot-based fluorescence resonance energy transfer and its applications. Adv Mater I19:1921–1926
132. Resch-Genger U, Grabolle M, Cavaliere-Jaricot S, Nitschke R, Nann T (2008) Quantum dots vVersus organic dyes as fluorescent labels. Nat Methods 5(9):763–775
133. Prasuhn DE, Blanco-Canosa JB, Vora GJ, Delehanty JB, Susumu K, Mei BC, Dawson PE, Medintz IL (2010) Combining chemoselective ligation with polyhistidine-driven self-assembly for the modular display of biomolecules on quantum dots. ACS Nano 4:267–278
134. Delehanty JB, Blanco-Canosa JB, Bradburne CE, Susumu K, Stewart MH, Prasuhn DE, Dawson PE, Medintz IL (2013) Site-specific cellular delivery of quantum dots with chemoselectively-assembled modular peptides. Chem Commun 49(72):7878–7880

135. Delehanty JB, Bradburne CE, Susumu K, Boeneman K, Mei BC, Farrell D, Blanco-Canosa JB, Dawson PE, Mattoussi H, Medintz IL (2011) Spatiotemporal multicolor labeling of individual cells using peptide-functionalized quantum dots and mixed delivery techniques. J Am Chem Soc 133(27):10482–10489
136. Adams SR, Tsien RY (2008) Preparation of the membrane-permeant biarsenicals FlAsH-EDT2 and ReAsH-EDT2 for fluorescent labeling of tetracysteine-tagged proteins. Nat Protoc 3(9):1527–1534
137. Boeneman K, Delehanty JB, Susumu K, Stewart MH, Medintz IL (2010) Intracellular bioconjugation of targeted proteins with semiconductor quantum dots. J Am Chem Soc 132(17):5975–5977
138. Field L, Walper S, Susumu K, Oh E, Medintz I, Delehanty J (2015) Modulation of intracellular quantum dot to fluorescent protein Förster resonance energy transfer via customized ligands and spatial control of donor–acceptor assembly. Sensors 15(12):29810
139. Medintz IL, Stewart MH, Trammell SA, Susumu K, Delehanty JB, Mei BC, Melinger JS, Blanco-Canosa JB, Dawson PE, Mattoussi H (2010) Quantum-dot/dopamine bioconjugates function as Redox coupled assemblies for in vitro and intracellular pH sensing. Nat Mater 9(8):676–684
140. Parenti G (2009) Treating lysosomal storage fiseases with pharmacological chaperones: from concept to clinics. EMBO Mol Med 1(5):268–279
141. Walters R, Kraig RP, Medintz I, Delehanty JB, Stewart MH, Susumu K, Huston AL, Dawson PE, Dawson G (2012) Nanoparticle targeting to neurons in a at hippocampal slice culture model. ASN Neuro 4(6):383–392
142. Walters R, Medintz IL, Delehanty JB, Stewart MH, Susumu K, Huston AL, Dawson PE, Dawson G (2015) The role of negative charge in the delivery of quantum dots to neurons. ASN Neuro 7(4)
143. Agarwal R, Domowicz MS, Schwartz NB, Henry J, Medintz I, Delehanty JB, Stewart MH, Susumu K, Huston AL, Deschamps JR, Dawson PE, Palomo V, Dawson G (2015) Delivery and tracking of quantum dot peptide bioconjugates in an intact developing Avian brain. ACS Chem Neurosci 6(3):494–504
144. Andrasfalvy BK, Galinanes GL, Huber D, Barbic M, Macklin JJ, Susumu K, Delehanty JB, Huston AL, Makara JK, Medintz IL (2014) Quantum dot-based multiphoton fluorescent pipettes for targeted neuronal electrophysiology. Nat Methods 11(12):1237–1241
145. Rowland CE, Susumu K, Stewart MH, Oh E, Mäkinen AJ, O'Shaughnessy TJ, Kushto G, Wolak MA, Erickson JS, AL E, Huston AL, Delehanty JB (2015) Electric field modulation of semiconductor quantum dot photoluminescence: insights into the design of robust voltage-sensitive cellular imaging probes. Nano Lett 15(10):6848–6854
146. Park K, Deutsch Z, Li JJ, Oron D, Weiss S (2012) Single molecule quantum-confined stark effect measurements of semiconductor nanoparticles at room temperature. ACS Nano 6(11):10013–10023
147. Empedocles SA, Bawendi MG (1997) Quantum-confined stark effect in single CdSe nanocrystallite quantum dots. Science 278(5346):2114–2117
148. Oh E, Liu R, Nel A, Gemill KB, Bilal M, Cohen Y, Medintz IL (2016) Meta-analysis of cellular toxicity for cadmium-containing quantum dots. Nat Nanotechnol Advance online publication
149. Thomas A, Nair PV, George Thomas K (2014) InP quantum sots: an environmentally friendly material with resonance energy transfer requisites. J Phys Chem C 118(7):3838–3845
150. Das A, Snee PT (2016) Synthetic developments of nontoxic quantum dots. Chem Phys Chem 17(5):598–617

Chapter 10
Fluorescence Applications for Structural and Thermodynamic Studies of Membrane Protein Insertion

Alexander Kyrychenko, Yevgen O. Posokhov, Mauricio Vargas-Uribe, Chiranjib Ghatak, Mykola V. Rodnin, and Alexey S. Ladokhin

Keywords Fluorescence spectroscopy • Fluorescence quenching • Fluorescence correlation spectroscopy • FRET • Distribution analysis of depth-dependent quenching • Membrane protein • Thermodynamics • Annexin • Diphtheria toxin • Bcl-xL • Fluorinated surfactants • Lipid bilayer insertion

Abbreviations

ANX	annexin B12
DA	distribution analysis
HFTAC and FTAC	hemifluorinated and fluorinated surfactants
FCS	fluorescence correlation spectroscopy
F_{MC}	fraction of membrane-competent form
FRET	Förster resonance energy transfer
IF	interfacial
LUV	large unilamellar lipid vesicles
LysoUB	FRET quencher for NBD fluorescence
MP(s)	membrane protein(s)

A. Kyrychenko (✉)
Institute of Chemistry and School of Chemistry, V. N. Karazin Kharkiv National University, Kharkiv 61022, Ukraine
e-mail: a.v.kyrychenko@karazin.ua

Y.O. Posokhov
Department of Organic Chemistry, Biochemistry and Microbiology, The National Technical University "Kharkiv Polytechnic Institute", Kharkiv 61002, Ukraine

M. Vargas-Uribe
Department of Biochemistry and Microbiology, Universidad Austral de Chile, Valdivia, Chile

C. Ghatak • M.V. Rodnin • A.S. Ladokhin
Department of Biochemistry and Molecular Biology, The University of Kansas Medical Center, Kansas City, KS 66160-7421, USA

C.D. Geddes (ed.), *Reviews in Fluorescence 2016*, Reviews in Fluorescence,
DOI 10.1007/978-3-319-48260-6_10

NBD-PE	fluorescent lipid (1,2-dipalmitoyl-snglycero-3-phosphoethanolamine-N-(7-nitro-2-1,3-benzoxadiazol-4-yl)
POPC	palmitoyloleoylphosphatidylcholine
POPG	palmitoyloleoylphosphatidylglycerol
POPC:POPG 3:1	mixtures of POPC and POPG that contain a molar ratio of
POPC:POPG 1:3	corresponding lipid specified by the number
TM	transmembrane
T-domain	diphtheria toxin translocation domain

10.1 Introduction

Several classes of proteins insert into biological membranes in a translocon-independent manner after being refolded from initial water-soluble structures. A unique characteristic of these proteins is their ability to move from the polar environment of the aqueous medium to the non-polar milieu of the lipid bilayer in response to a given cellular/chemical signal. Figure 10.1a shows some examples of proteins that refold and insert into membranes in response to a external signal, such as change in pH [1–3], proteolytic cleavage or activation by conformational switching due to interaction with lipid or protein partners [1, 4–8]. For instance, the mechanism of action of some bacterial toxins involves the membrane insertion and refolding of specific domains in response to the acidic environment of the endosome [9], which ensures the toxin's entry into the cell. Others, such as the apoptotic regulator Bid, are activated by proteolytic cleavage, resulting in partitioning into the mitochondrial outer membrane. Other apoptotic regulators of the Bcl-2 family, such as Bax and Bcl-xL partition either in a pH-dependent manner or after being activated by Bid *via* the release of hydrophobic anchoring sequences [10]. While in all of the presented examples the soluble structures have been determined by NMR or X-ray crystallography, the structures of the inserted forms or details of the insertion pathway are more challenging to capture.

One of the main reasons for the lack in our understanding of the physiological action of these proteins is the shortage of appropriate experimental tools to study the process of their bilayer insertion. Because refolding on the membrane interface and subsequent transbilayer insertion produces multiple intermediate states (e.g., Fig. 10.1b), high-resolution methods are difficult to apply and structural information is teased out by various types of the fluorescence spectroscopy [9]. In the following sections, we will show how fluorescence-based approaches can be applied to monitor different aspects of protein membrane insertion, including kinetic pathways of membrane insertion and free energy landscapes, specific conformational changes, structure and topology of membrane-inserted states, as well as elucidating the role of lipids on the membrane protein insertion pathways.

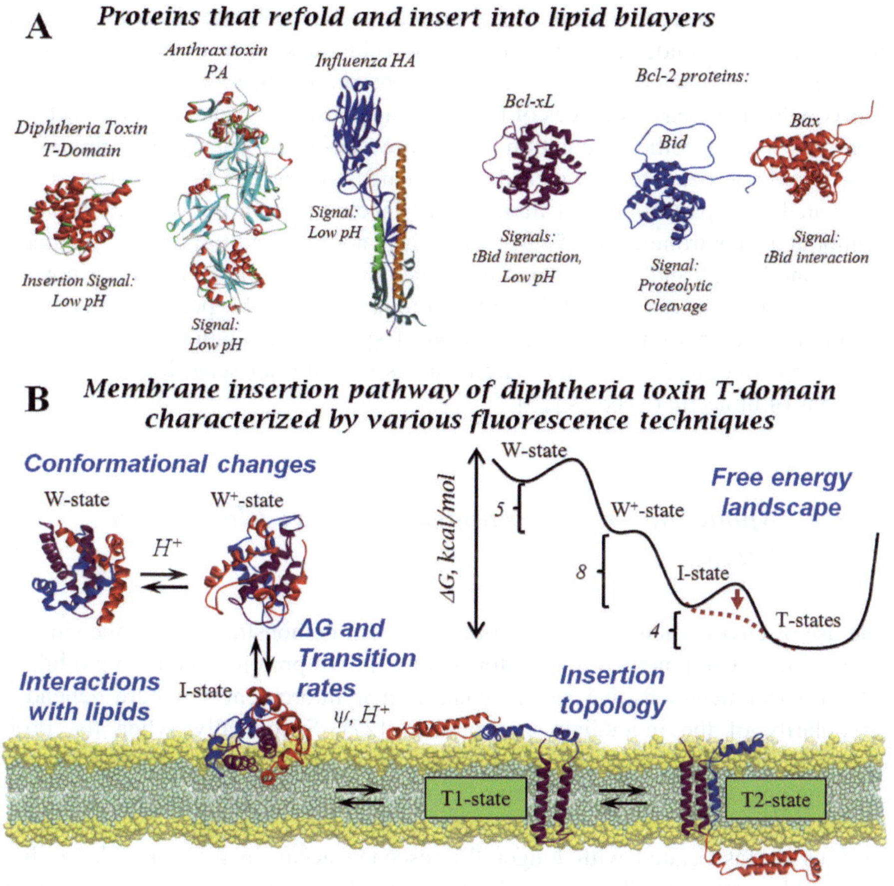

Fig. 10.1 (**a**) Examples of membrane proteins that refold and insert into the membranes posttranslationally in response to a physiological signal, such as acidification or proteolytic cleavage. (**b**) pH-triggered membrane insertion pathway of the diphtheria toxin T-domain [9], responsible for the cellular entry of the toxin. Various fluorescence applications can be used to study different structural and thermodynamic aspects of the membrane insertion process, including refolding, lipid interactions and bilayer insertion topology [6, 11–23]. These spectroscopic approaches are described in Sects. 10.2, 10.3 and 10.4 and their applications to studies of the T domain are summarized in Sect. 10.5

10.2 Methods to Monitor Partitioning into the Membrane

A critical aspect of monitoring of membrane partitioning of proteins is distinguishing the population of protein that is bound to the membrane from that which is free in solution. Classical biochemical techniques, e.g., equilibrium dialysis or size-exchange chromatography, have been used to separate both populations;

the techniques, however, require high protein concentrations, which may lead to aggregation and hinder the analysis. Other techniques, such as surface plasmon resonance and bilayer interferometry, require immobilization of proteins to a matrix, which decreases degrees of freedom and constrains proteins to interact with membranes in specific orientations. Fluorescence-based approaches circumvent these problems because one records the changes in various fluorescence signals associated with partitioning of fluorescence-labeled protein between aqueous and membranous environments [15, 24]. Thus, it is not required to physically separate membrane-bound and free-in-solution populations of protein, use high concentrations of protein nor immobilize the protein. It should be pointed out that not all fluorescence parameters (e.g., position of spectral maximum) are suitable for *quantitative* thermodynamic measurements as they do not scale linearly with the fractions of molecular species [25].

10.2.1 Application of Environment Sensitive Fluorescent Probes

The use of environment-sensitive probes has been demonstrated to be the simplest fluorescence-based method to monitor partitioning of proteins into the lipid bilayer [26]. Environment-sensitive probes change their fluorescent signature depending on polarity of the immediate environment [27]. Specifically, when the probe moves from a polar to a non-polar milieu, e.g., from aqueous to membranous environment, the emission spectrum increases in intensity and the position of maximum emission shifts towards shorter wavelengths (blue-shift) [22, 28]. These changes are associated with longer fluorescence decay in the non-polar milieu. Thus, following these changes in fluorescence allows monitoring the partitioning of a labeled-protein between the aqueous solution and the lipid bilayer [25]. Some examples of environment-sensitive probes include the natural amino acid tryptophan and the commercially available NBD (derivative of 7-nitro-2-1,3-benzoxadiazol-4-amine dye) and Bimane (derivative of 3-(bromomethyl)-2,5,6-trimethyl-1*H*,7*H*-pyrazolo[1,2-*a*]pyrazole-1,7-dione dye) [29].

Intrinsic tryptophan fluorescence is a classical tool to monitor membrane partitioning and the refolding of membrane-active peptides and proteins [25, 30–32]. An important advantage of tryptophan fluorescence is that it does not require site-selective labeling, because the amino acid is incorporated as part of the peptide sequence. However, tryptophan fluorescence has some disadvantages, because proteins usually have multiple tryptophan residues, or the signal-to-noise ratio is often not optimal when proteins have a single tryptophan residue. Additional issues regarding tryptophan fluorescence, such as how to deal with light-scattering effects arising from the presence of vesicles, are discussed elsewhere [25]. These challenges can be bypassed by site-selective labeling of the protein with environment-sensitive dyes, such as NBD, BODIPY or Bimane, which can improve the signal-to-noise ratio [6, 33–35].

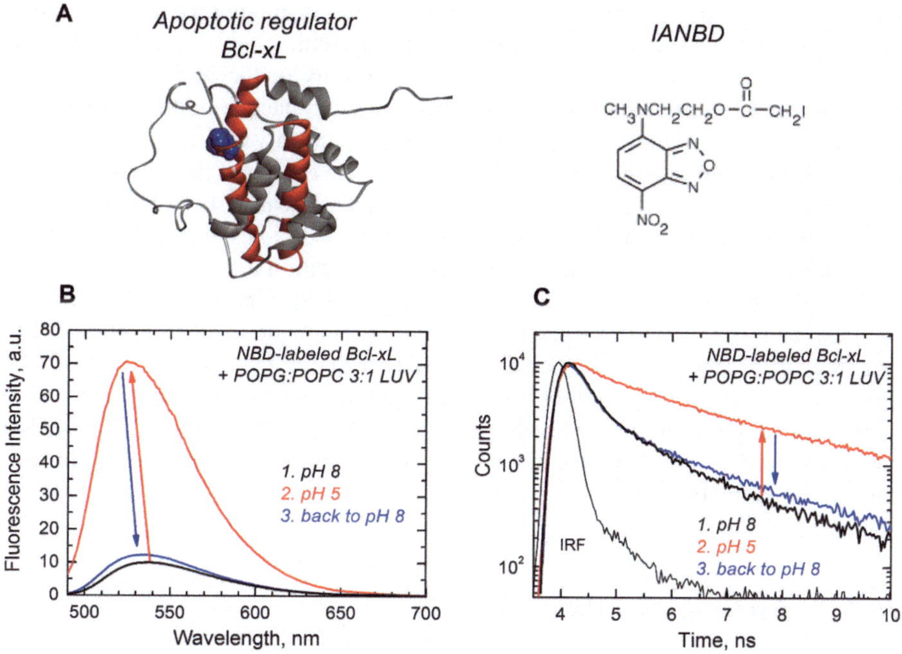

Fig. 10.2 NBD-labeling of Bcl-xL as a tool to monitor reversibility of protein-membrane interactions. (**a**) NMR structure of apoptotic repressor Bcl-xL highlighting transmembrane insertion unit in *red* and NBD-labeling site in blue CPK. Chemical structure of NBD in a thiol-reactive form. (**b**) Steady-state emission spectra of a premixed solution containing NBD-labeled Bcl-xL and LUV under different pH conditions. (**c**) Time-resolved emission traces of the same samples in Panel B. Increase in emission intensity, spectral blue-shift and longer fluorescence decay indicate partitioning into the membrane. The results demonstrate reversibility of the pH-dependent association of Bcl-xL with membranes (Adapted from Ref. [6])

Figure 10.2 shows an example of this approach, where the apoptotic repressor Bcl-xL was selectively-labeled with NBD to monitor pH-dependent interaction with large unilamellar vesicles (LUV) [6]. Bcl-xL is a protein whose cellular action involves interaction with the mitochondrial outer membrane to prevent apoptosis, but this interaction can be triggered by low pH *in vitro*. Figure 10.2a shows the NMR structure of Bcl-xL (*left*) and highlights the residue N175C (*blue CPK*), which was labeled with a thiol-reactive form of NBD (*right*). Because this residue locates in a region suggested to become lipid-embedded (helices α5- α6, red), the pH-triggered membrane partitioning of fluorescently-labeled-Bcl-xL is expected to cause changes in the fluorescence signal of NBD. Indeed, Fig. 10.2b shows that acidification caused a sevenfold increase of NBD emission intensity and 10-nm spectral blue-shift (*red spectrum*) in comparison to the sample at reference pH (*black spectrum*). These changes were associated with longer fluorescence decay (Fig. 10.2c, *red trace*). The transition was reversible because returning to the reference pH caused the recovery of spectral features (Fig. 10.2b, c, *blue*). This fluorescence-based experiment demonstrated that Bcl-xL membrane interactions are reversible, allowing thermodynamic study of the system.

Because these probes sense changes in the polarity of the immediate environment, changes in fluorescence can occur for reasons unrelated to partitioning into the membrane. For example, the increase of emission intensity and spectral blue-shift may be due to shielding the probe from the solvent as a result of protein refolding in solution rather than partitioning into the lipid bilayer. Because this may mislead the researcher, one has to complete control experiments in the absence of lipid vesicles and confirm partitioning into the membrane with additional techniques (discussed below). Nonetheless, this property of environment-sensitive probes can be exploited to study conformational changes occurring in solution or in the membrane. For example, tryptophan fluorescence has served to determine the role of histidine protonation in triggering conformational changes of diphtheria toxin translocation domain in solution and in the membrane [22, 36, 37]; while bimane fluorescence has been used to validate structures of kinetic intermediates modelled through computational methods [19].

10.2.2 FRET-Based Binding Assay

Förster resonance energy transfer (FRET) is a common strategy to detect binding between two fluorescence-labeled particles [38]. In contrast to the use of environment-sensitive probes (discussed above), this approach does not require the probe to become embedded into the lipid bilayer to yield a fluorescence response. Instead, there is a change in signal that can be experimentally monitored because a donor probe transfers energy to an acceptor probe when they are within the Förster's distance [6, 14]. The experiment consists of site-selective labeling the protein and membrane with the donor/acceptor pair, so that the partitioning of the protein causes FRET between the probes.

Figure 10.3 displays a FRET-based experiment applied to the case of the pH-dependent membrane partitioning of Bcl-xL. Figure 10.3a shows the NMR structure of Bcl-xL, with the labeling site previously mutated to cysteine (green CPK); and chemical structures of the donor AlexaFluor-488 as a thiol-reactive probe, and the lipid-attached acceptor (Rhodamine-PE [29]) dye, which was incorporated to the membrane during vesicle preparation at 2 % molar fraction of total lipid. Figure 10.3b, c show the representative results of this type of experiment, where the FRET-based assay was used to confirm the reversibility of the pH-dependent association of Bcl-xL to anionic membranes. Acidification of the sample containing premixed labeled Bcl-xL and labeled vesicles caused a decrease in emission intensity of the donor (Fig. 10.3b, Steps 1 and 2) and a shortening of the fluorescence life-time of the donor (Fig. 10.3c, Steps 1 and 2). Returning to reference pH caused the recovery of the initial spectral features (Fig. 10.3b–c, Step 3). Together, the data indicate that (*i*) there is FRET between the donor/acceptor pair, which occurs under acidic conditions; and (*ii*) FRET vanishes when reversing the pH conditions. These

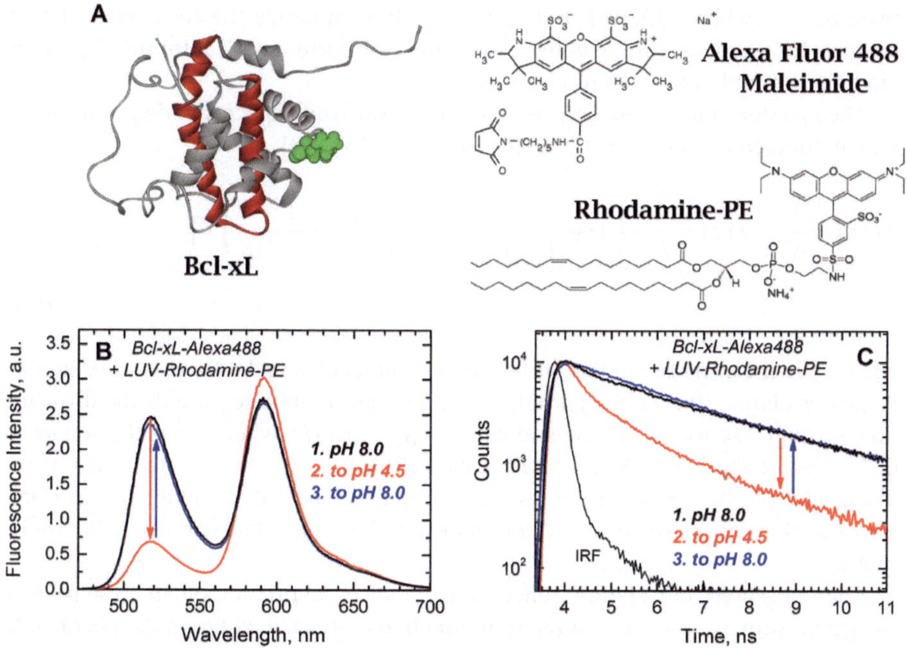

Fig. 10.3 FRET-based assay as a tool to monitor reversibility of protein-membrane interactions. (**a**) NMR structure of apoptotic repressor Bcl-xL highlighting transmembrane insertion unit in red and labeling site for donor dye attachment in green CPK. Chemical structures of thiol-reactive donor and acceptor-labeled phospholipid are shown. (**b**) Steady-state emission spectra of a premixed solution containing Bcl-xL and LUV labeled with donor/acceptor pair under different pH conditions. (**c**) Time-resolved emission traces of the same samples in Panel B. Decrease in emission intensity and shortening of life-time indicate FRET between donor and acceptor. Results demonstrate reversibility of the pH-dependent association of Bcl-xL with membranes (Adapted from Ref. [6])

observations are consistent with a population of labeled Bcl-xL partitioning into the labeled lipid bilayer in a pH-dependent and reversible manner.

10.2.3 FCS-Based Methods to Quantify Membrane Interactions in the Presence of Membrane-Competent and Incompetent Protein Conformations

The FCS technique has been successfully used to study interfacial binding of peptides and proteins [39, 40], as well as to study transmembrane insertion of proteins [15, 39–42]. In recent years, several excellent reviews have been published [43, 44], focusing on the rapid development and applications of FCS for *in cell* studies. In addition, the technical aspects of the FCS method have been intensively reviewed elsewhere [45, 46]. Below we briefly summarize the formalism of FCS analysis for a general case of complex binding equilibrium typical for pH-dependent binding of membrane proteins [15].

The autocorrelation function $G(\tau)$ for single diffusing species undergoing Brownian motion can be described with the Eq. (10.1) [47, 48]:

$$G(\tau) = \frac{1}{N} \times g(\tau) = \frac{1}{N} \cdot \left(1 + \frac{T}{1-T} \cdot e^{-\tau/\tau_{TR}}\right) \cdot \left(\frac{1}{1+\tau/\tau_D}\right) \cdot \left(\frac{1}{1+\tau/S^2\tau_D}\right)^{1/2} \tag{10.1}$$

where N is the average number of fluorescent molecules in the focal volume and τ_{D} is the correlation time of the particles. The correlation time represents the diffusion time through the focus volume and equals $\tau_{\mathrm{D}} = \omega^2/4D$, where ω^2 is the square of the radius of the laser focus and D is the diffusion constant. S is the ratio of the distances from the center of the laser beam focus in the radial and axial directions, respectively. T is the fraction of fluorophores in the triplet state and τ_{TR} is the triplet lifetime.

The measured correlation function $G(\tau)$ of a multicomponent system is a weighted sum of the autocorrelation functions of each component $G_i(\tau)$ with amplitudes A_i [15, 39, 40, 42] as shown by Eq. (10.2).

$$G(\tau) = \sum_{i=1}^{M} q_i^2 N_i^2 G_i(\tau) \Big/ \left[\sum_{i=1}^{M} q_i N_i\right]^2 = \sum_{i=1}^{M} q_i^2 N_i g_i(\tau) \Big/ \left[\sum_{i=1}^{M} q_i N_i\right]^2 = \sum_{i=1}^{M} A_i \cdot g_i(\tau) \tag{10.2}$$

where N_i is the mean particle number and q_i is the ratio of the fluorescence yield of the i_{th} component to that of the first component. $q_i = 1$ if binding of dye-labeled membrane protein to vesicles does not change fluorescence intensity of fluorescent labels. For binding of membrane protein to vesicles only two diffusing species are considered: the fluorescently labeled proteins (index P) and LUV with bound fluorescently labeled proteins (index V):

$$G(\tau) = A_P \cdot g_P(\tau) + A_V \cdot g_V(\tau) \tag{10.3}$$

If the quantum yield of a fluorophore is not environment-sensitive and does not change upon protein association with the membrane (which is often the case with AlexaFluor series of dyes), the amplitudes in Eq. (10.3) will depend only on the numbers of fast- and slow-moving particles in the focal volume (N_P and N_V, respectively)

$$A_P = \frac{N_P}{(N_V + N_P)^2} \quad A_V = \frac{N_V}{(N_V + N_P)^2} \tag{10.4}$$

Under conditions of infinite dilution regime, the fraction of bound protein is equal [15, 42]:

$$f_B = \frac{A_V}{A_V + A_P} \tag{10.5}$$

The resulting titration isotherm is fitted to a mole fraction partitioning equation [25, 49], modified to account for the presence of membrane-incompetent protein species [14, 15, 42]:

$$f_B = F_{MC} \cdot \frac{K_x \cdot [L]}{[W] + K_x \cdot [L]} \tag{10.6}$$

$$\Delta G = -RT \ln K_x \tag{10.7}$$

where F_{MC} is a fraction of the membrane-competent form (determined at lipid saturation), $[L]$ is lipid concentration, $[W]$ is water concentration (55.3 M), and K_x is a mole fraction partitioning coefficient which is used to determine the free energy ΔG of binding.

10.3 Fluorescence and Structural Studies of MPs

10.3.1 Determination of Interfacial and Transmembrane Topology of MPs

Determining the topology of a particular site on a membrane protein (i.e., its position with respect to the two leaflets of the lipid bilayer) is an important step in elucidating its structure. And even if the structure of a membrane protein is known, topology measurements are useful for verifying its correct fold and orientation after reconstitution into model membrane systems [50]. Measurements of topology have been especially important in the determination of membrane insertion pathways for a variety of spontaneously inserting non-constitutive proteins, such as bacterial toxins [9, 51, 52], colicins [53, 54], some annexins [42, 55] and numerous apoptotic regulators of the Bcl-2 family [5–8, 10].

The topology of a membrane protein can be determined by combining site-selective labeling of the protein with a fluorescent probe, and symmetric and asymmetric labeling of the membrane with a quencher [28, 56–59]. The idea is that topology of the membrane protein can be deduced by comparing the fluorescent response of the fluorophore upon interaction of the labeled-protein with symmetric-

or asymmetric-labeled membranes. Examples of these quenchers include LysoMC, a methylcoumarin derivative capable of quenching tryptophan fluorescence [58, 59], and LysoUB (Fig. 10.4a), a UniBlue chromophore derivative capable of quenching NBD fluorescence [60, 61]. In both cases, the quenching occurs through non-radiative Förster type energy transfer (FRET). Because single-chain *lyso*-lipids have significant water solubility and form micelles, they can be incorporated easily into the outer leaflet when added to preformed bilayer vesicles. The bulky charged quencher headgroup and single-acyl chain act in concert to slow lipid flip-flop, resulting in asymmetric distribution, which is stable for topology measurements under equilibrium conditions.

While LysoMC has been useful for determining the topology of relatively short tryptophan-containing peptides [58, 59], several problems arise when it is applied to proteins (e.g., weak tryptophan fluorescence, poor spectral selectivity, parasitic LysoMC fluorescence). These problems can be bypassed by using LysoUB as quencher of NBD fluorescence [60]. Another advantage of LysoUB is that the Förster radius from NBD to LysoUB, which ranges from 30 to 35 Å depending on donor's quantum yield and spectral position, is about the same as the thickness of the bilayer hydrocarbon core. Thus, FRET efficiency will be highest when the quencher and fluorophore are in the same leaflet and lowest when they are separated by the bilayer. This can be measured by monitoring changes in donor intensity or lifetime [61], where the latter is preferred because it is less susceptible to artifacts associated with partial loss of sample.

In the example shown in Fig. 10.4, quenching of NBD by LysoUB-labeled membranes was used to determine topology of annexin B12 (ANX) in the lipid bilayer. Because LysoUB quenching is stronger for NBD located in the same leaflet, quenching is different for the two hypothetical cases shown in the figure (Fig. 10.4b, c, d). The ratio of the longest decay time in the absence and presence of 2 % LysoUB was used to assign the topology of the NBD-labeled site [42]. The results indicate that residue 134 remains on the *cis* side at all times, while residue 162 is rapidly translocated in LUV with low content of anionic lipids (POPC:POPG 3:1). The use of LUV with an increased content of anionic lipids hinders the translocation by stabilizing the interfacial kinetic intermediate, presumably *via* electrostatic attraction of lipid headgroups and R149, located in the middle of the segment [42, 60].

10.3.2 Depth-Dependent Fluorescence Quenching by Lipid-Attached Quenchers

Accurate determination of the penetration depth of proteins and peptides into lipid bilayers is an important step in the study of binding, topology and folding of membrane proteins. However, extracting of quantitative information about the membrane location of an intrinsic fluorophore or a site-selectively attached external probe from

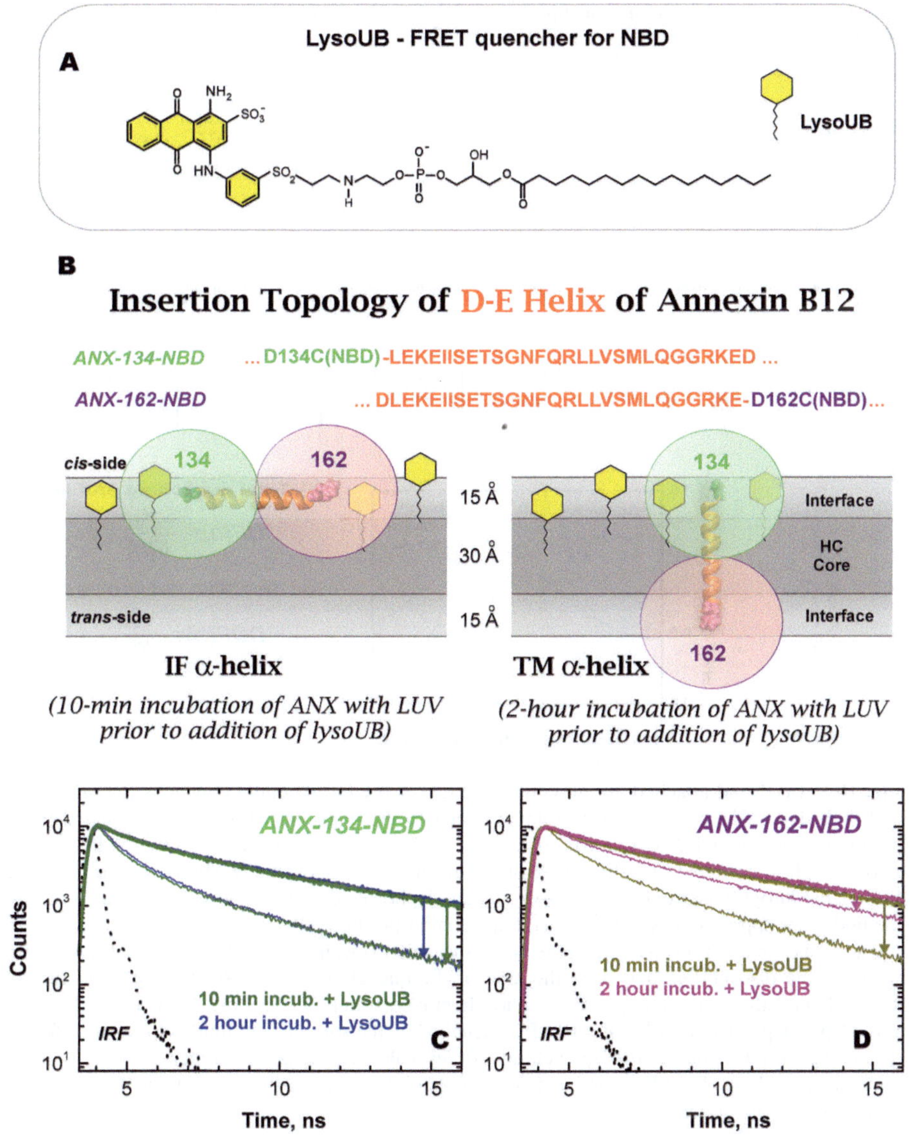

Fig. 10.4 (**a**) Chemical structure of LysoUB, a compound consisting of UniBlue chromophore attached to a *lyso*-lipid. (**b**) Schematic representation of insertion pathway of D-E α-helix of ANX studied with LysoUB/NBD quenching method. Shaded circles illustrate the range of efficient quenching from an NBD-labeling site. (**c–d**) Lifetime experimental scheme for determining membrane topology of the NBD-labeled protein. Typical fluorescence decay experiment used to determine the topology of the NBD probe attached to ANX residue 162 (**c**) or residue 134 (**d**) in POPC:POPG 1:3 LUV. Arrows illustrate the amount of quenching caused by addition of 2 % LysoUB after 10 min or 2 h of the initiation of protein insertion. Strong quenching indicates the *cis* location of the probe (e.g., ANX-134-NBD or ANX-162-NBD after 10 min incubation with POPC:POPG 1:3 LUV). Translocation of the protein-attached NBD across the bilayer results in reduced quenching (e.g., ANX-162-NBD after 2 h incubation)

A

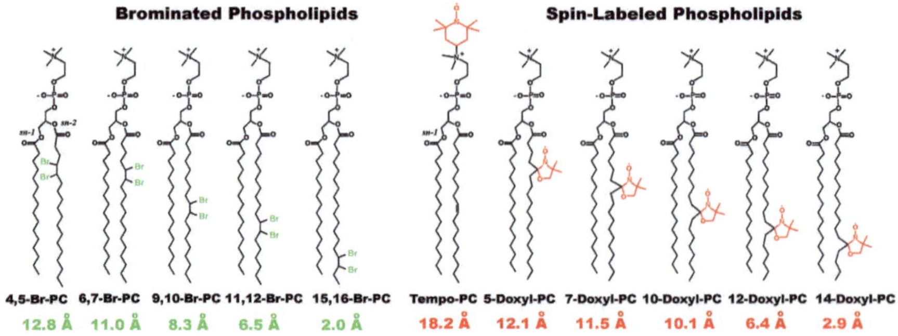

B

$$QP(h) = G(h) + G(-h) = \frac{S}{\sigma\sqrt{2\pi}}\exp\left[-\frac{(h-h_m)^2}{2\sigma^2}\right] + \frac{S}{\sigma\sqrt{2\pi}}\exp\left[-\frac{(h+h_m)^2}{2\sigma^2}\right]$$

C

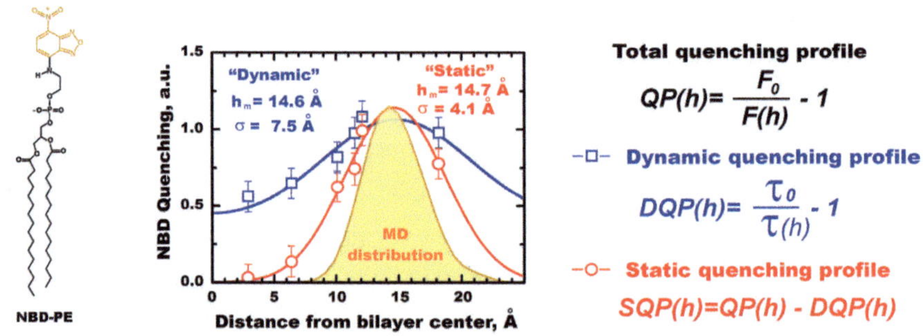

Fig. 10.5 (**a**) Lipid-attached fluorescence quenchers used for depth-dependent measurements and their depth (the distances from bilayer center to the quencher) [66, 67]. Brominated phospholipids: 1-palmitoyl-2-(n,n-dibromo)stearoyl-*sn*-glycero-3-phosphocholine (*n*,*n*-Br-PC). Spin-Labeled phospholipids: 1-palmitoyl-2-oleoyl-*sn*-glycero-3-phospho(TEMPO)choline (Tempo-PC), 1-palmitoyl-2-stearoyl-(*n*-doxyl)-*sn*-glycero-3-phosphocholine (*n*-doxyl-PC). (**b**) The DA methodology approximates the transverse quenching profile ($QP(h)$) with a sum of two mirror-image Gaussian functions ($G(h)$), where h_m = the center (mean) of the quenching profile, σ = the width of the distribution, and S = the area of the quenching profile [62–65]. (**c**) Structure of NBD-PE (1,2-dipalmitoyl-*sn*-glycero-3-phosphoethanolamine-N-(7-nitro-2-1,3-benzoxadiazol-4-yl)). Depth-dependent profiles of NBD-PE quenched by the spin-labeled lipids plotted for lifetime quenching (*squares*) and "*static*" quenching (*circles*) calculated as the difference between total and dynamic quenching. The MD-simulated depth probability profile of the NBD moiety (*yellow area*) is shown for comparison

fluorescence studies is typically complicated by broad distributions of the transverse positions of both quenchers and fluorophores. Therefore, depth-dependent fluorescence quenching with a series of lipid-attached quenchers, such as bromine atoms or paramagnetic Tempo and Doxyl groups (Fig. 10.5a), has become an important spectroscopic tool for calculating the precise depth of a fluorescent moiety [62–65]. While bromolipids can be utilized as quenchers for tryptophan and pyrene fluorescence, the lipid-attached spin moieties, in addition to those, can also quench other organic fluorophores, including NBD, rhodamine, bimane and antracene. The transverse position of bromine atoms has been determined by X-ray diffraction [66], while the position of spin labels has been recently refined by MD simulations [67].

The main assumption of the depth-dependent quenching methodology is that the quenching between a fluorescence dye and a quencher will be greatest when they are located at the same depth within the bilayer. To quantitate membrane penetration the quenching profiles are measured and fitted using the Distribution Analysis (DA) method [62, 65]. DA reconstructs a depth-dependent quenching profile by using the quenching data collected from a series of quenchers located at known depths (h) within the membrane (Fig. 10.5). The DA methodology approximates the transverse quenching profile (QP) of a fluorophore with a symmetrical twin Gaussian function (Fig. 10.5b), which has three fitting parameters: h_m = the center (mean) of the quenching profile, σ = the width of the distribution, and S = the area of the quenching profile. These parameters correspond to the most probable depth of penetration, fluctuations in the transverse position, and overall accessibility to quenching (i.e., quenching efficiency), respectively [63]. Unlike in the previously described LysoUB quenching method (Sect. 10.3.1), the quenchers are distributed in both leaflets. In order to account for possible trans-leaflet quenching of deeply penetrating fluorophores, the mirror-image $G(-h)$ component is added to the main Gaussian component $G(h)$ [63].

The accuracy of membrane depth calculations is still crucially dependent on the precise positioning of the lipid-attached spin-quenchers within the bilayer. Therefore, the average positions of the spin quencher calculated from the center of the lipid bilayer have recently been determined by MD simulations of a series of spin-labeled lipids in the model membrane [67]. A combination of molecular dynamics (MD) simulations and depth-dependent fluorescence quenching was used to calibrate the methodology for extracting quantitative information on membrane penetration of fluorescent-labeled molecules [68].

In the model system NBD-PE, the NBD moiety attached to the lipid headgroup (Fig. 10.5c) was studied in a POPC bilayer. The immersion depth of NBD was estimated by measuring steady-state and time-resolved fluorescence quenching with six different spin-labeled lipids co-incorporated into lipid vesicles (Fig. 10.5c). The overall width of the quenching profile, which is determined by physical sizes of the fluorophore and quencher, and widths of their thermal envelopes are quite large (Fig. 10.5c, *squares*). Therefore, a deconvolution procedure of the total quenching profile into dynamic and static components was suggested [69]. A static profile (Fig. 10.5c, *circles*) can be calculated as the difference between total quenching efficiency and the dynamic component, defined as a quenching efficiency observed in a lifetime quenching experiment. Figure 10.5c shows that although the average depth position estimated by both methods is the same, the static distribution is much narrower than the dynamic profile. Using the methodology of DA, the immersion depth from the bilayer center to the NBD moiety was estimated to be 14.7 Å, which was independently validated by using atomistic MD simulations of NBD-PE lipids in a POPC (14.4 Å) [68, 69].

10.4 Thermodynamic Measurements of Membrane Partitioning.

10.4.1 pH-Dependent Lipid-Protein Interactions

For pH-triggered membrane protein insertion, two thermodynamic transitions should be considered (Fig. 10.1) [14, 15, 42]: (*a*) pH-dependent formation of the membrane-competent state (F_{MC}) and (*b*) its association with the membrane. While both of these processes can be studied by FCS, they require different experimental schemes [15]: (*a*) pH titration in the presence of the large saturating concentration of LUV and (*b*) lipid titration at constant pH.

In the first type of FCS measurement, the association of the protein under lipid-saturating conditions should be followed in order to determine the pH-dependent formation of the membrane-competent form, F_{MC}. According to Eq. (10.6), f_B is equal to F_{MC} at high lipid concentrations. An infinitely large concentration of lipid vesicles over that of the protein ensures that the equilibrium is shifted toward the membrane-bound form [14, 15, 42]. In other words, the fraction of protein bound to the vesicles will be equal to the fraction of membrane-competent form because the excess of membranes ensures that the protein reaching the membrane-competent form will partition in a monomeric form into the bilayer.

In the example shown in Fig. 10.6, where pH-dependent membrane interactions of diphtheria toxin T-domain is studied [23], acidification of the solution results in a progressive shift of mobility from that of the free T-domain to that of the vesicle-bound T-domain (the correlation times are much higher for the vesicle-bound species of the T-domain).

These results confirm that the formation of the membrane-competent form of T-domain occurs in the pH range of 7.0–5.6 [14, 19]. The FCS results of the pH-titration for annexin B12 (ANX) showed that the formation of the membrane-competent form occurs in the pH range of 7.0–5.5 in the case of binding to membranes with high content of anionic lipids and 6.5–4.5 in the case of membranes with low content of anionic lipids [14]. The obtained pH dependencies of F_{MC} were fitted to the Eq. (10.8):

$$F_{MC} = \frac{1}{1 + 10^{n(pH - pKa)}} \tag{10.8}$$

where pK_a is a negative logarithm of the dissociation constant and n is the Hill coefficient. The example of the fit of the pH-titration profile is shown in Fig. 10.6.

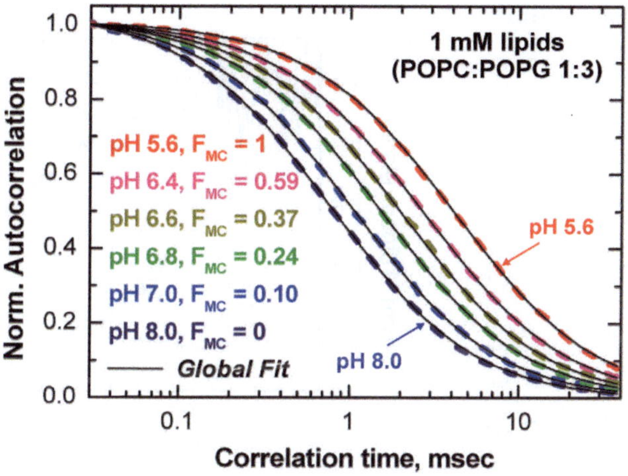

Fig. 10.6 Example of the binding measurement using FCS performed under conditions of the "infinite dilution regime" satisfied at an extremely high lipid excess over fluorescently labeled protein. Acidification results in a progressive shift of mobility from that of the free T-domain to that of an LUV-bound T-domain. Quantitative determination of the membrane-competent fraction, F_{MC}, is achieved by a linked analysis (*solid lines*) of fluorescence autocorrelation curves (*dashed lines*) that links the two correlation times for all curves and allows free fitting of the pre-exponential amplitudes in Eq. (10.3) [14, 15, 42]. For the purpose of better visual representation, all data are normalized to the same number of fluorescent particles in the focal volume. Note that the absolute values of amplitudes are not important under these conditions, just their relative contributions (Eq. 10.5) (Modified from Ref. [14])

10.4.2 Measurements of Free Energy of Insertion

10.4.2.1 Fluorinated Surfactants as New Tools for Chaperoning Thermodynamic Measurements of MP Insertion

Experimental determination of the free energy (ΔG) stabilizing the structure of membrane proteins (MPs) in their native environment has been hampered by MP's aggregation and precipitation outside the lipid bilayer. It has been demonstrated that the latter process can be prevented by the use of fluorinated surfactants, FTACs (Fig. 10.7a), which act as chaperones for MP insertion without partitioning into the membrane themselves [13]. The most important feature that distinguishes FTACs from detergents is their lack of interaction with membranes, caused by poor miscibility of fluorinated and hydrogenated carbons. As a result FTACs do not disrupt or partition into LUV even at concentrations above their critical micelle concentration and at overwhelming excess over lipid [13]. Thus, unlike detergents, they can be used for equilibrium thermodynamic measurements, which require MPs to be distributed between lipid bilayer and aqueous phase, where they are

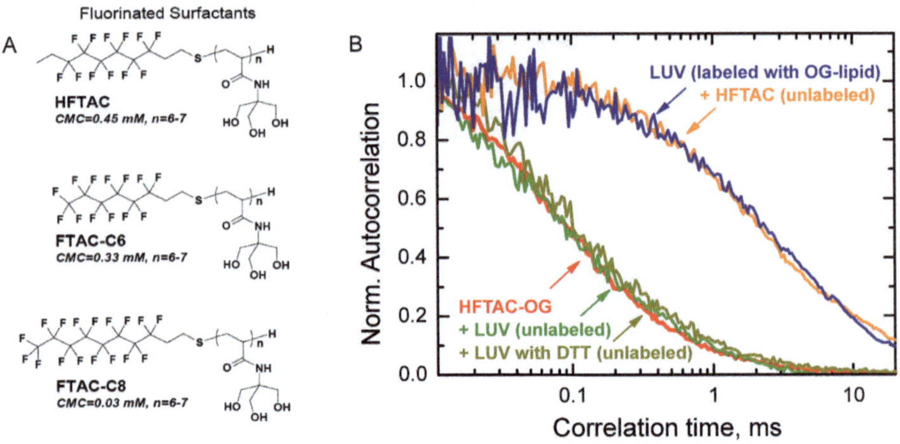

Fig. 10.7 (**a**) The chemical structure of the fluorinated surfactants. Hemifluorinated surfactant (HFTAC) has an unfluorinated tip on its hydrophobic tail, whereas FTAC-C6 and FTAC-C8 contain completely fluorinated hydrophobic tails consisting of six and eight carbons, respectively. The average degree of polymerization of the polar head for the batches is n = 6–7 [70]. (**b**) Comparison of FCS curves of Oregon-Green-labeled surfactant HFTAC-OG and vesicles containing OG-labeled lipid. The fast mobility of the HFTAC-OG (*red line*) is not affected by additions of either LUV alone (*green line*) or in the presence of the T-domain (*gold lines*), which indicates the absence of interaction of the surfactant with the vesicles

chaperoned by the surfactant [12]. This is demonstrated by the comparison of FCS autocorrelation curves for the surfactant and LUV, which were labeled with Oregon-Green (OG) fluorescent dye (Fig. 10.7b). After mixing of 3 nM OG-labeled surfactant with 50 mM of an unlabeled LUV, the FCS decay time is fast and is not affected by the addition of pure lipid LUV, indicating that the surfactant does not associate with the membrane. In the case of surfactant interaction with LUV, the autocorrelation curve would have moved toward that for LUV labeled with OG-attached lipid (Fig. 10.7b, *purple line*). Moreover, addition of the 20-fold molar excess of unlabeled HFTAC did not change vesicle mobility (Fig. 10.7b, *red line*).

10.4.2.2 Establishing Insertion Reversibility as the First Step in Thermodynamic Characterization.

Establishing thermodynamic reversibility is a prerequisite for an equilibrium thermodynamic study of protein-membrane interactions [15]. To establish the insertion reversibility of MPs, the advantages of the chaperone-like ability of FTACs combined with the sensitivity of FCS measurements can be utilized. Figure 10.8 shows an example of the insertion reversibility of annexin B12 (ANX) labeled with AlexaFluor-647 dye. The mobility of ANX in solution is much faster than that of the vesicle; thus, membrane association results in a shift of the autocorrelation curve

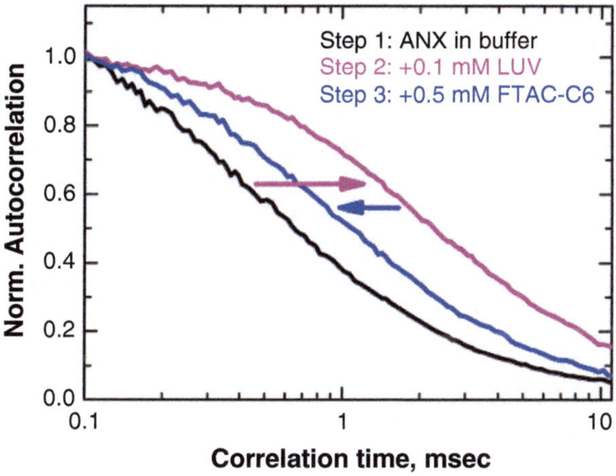

Fig. 10.8 Reversibility of membrane insertion of annex in B12 (ANX) probed by FCS. Partitioning of AlexaFluor-647-labeled ANX into the LUV results in increased correlation time (*magenta arrow*), whereas the subsequent addition of FTAC reverses the binding (*blue arrow*). Curves are normalized for visual comparison (Adopted from Ref. [41])

toward longer times (Fig. 10.8, *magenta arrow*). The addition of FTAC to the same sample results in the redistribution of ANX between LUV and fluorinated surfactant, causing a reversed shift of the FCS autocorrelation curve (Fig. 10.8, *blue arrow*).

Once the reversibility of insertion is established, FCS measurements can be applied to determine ΔG of bilayer insertion of model MPs in the presence of surfactants [41]. Estimation of the ΔG of membrane partitioning in the presence of surfactants was performed by measuring shifts of autocorrelation curves as described in detail in our previous study [42]. The resulting binding data was fitted to a partitioning isotherm and the free energy of partitioning was calculated from the fit [42]. Our designation of partitioning covers both transmembrane insertion and interfacial association. The latter cannot be distinguished by FCS but is established in a fluorescence quenching topology experiment as described in Sect. 10.3 and [42, 60].

10.4.2.3 Measuring Free Energy of Insertion of a Single Transmembrane Helix of WALP Peptides

Peptides of the WALP family have been used in numerous studies and are perhaps the best understood helical transmembrane (TM) peptides [71]. They are composed of a hydrophobic core of variable length, made of alternating Leu and Ala residues, flanked by a pair of Trp residues needed for proper positioning of the termini at each of the bilayer interfaces. As a result of this design, WALP peptides have a unique conformation in the lipid bilayer, namely, a monomeric TM helix, which makes them convenient models for studies of lipid–protein interactions and for development and calibration of new methods.

Accurate determination of the free energy of transfer of a helical segment from an aqueous into a TM conformation is essential for understanding and predicting the folding and stability of membrane proteins [72, 73]. Until recently, direct thermodynamically sound measurements of free energy of insertion of hydrophobic TM peptides were significantly complicated by the protein aggregation outside of the lipid bilayer. The problem can be solved by application of fluorinated surfactants that are capable of preventing aggregation but, unlike detergents, do not themselves interact with the bilayer. The FCS methodology [15, 41] has been applied to study surfactant-chaperoned insertion into preformed POPC vesicles of the two well-studied dye-labeled TM peptides of different lengths: WALP23 and WALP27 [74] (Fig. 10.9). The results of the application of FCS to interactions of WALP peptides with POPC LUV are presented in Fig. 10.9a, b.

The true values of ΔG for bilayer insertion are obtained from linear extrapolation of the apparent ΔG, measured at three different FTAC concentrations; these are equal to -9.0 ± 0.1 and -10.0 ± 0.1 kcal/mol for WALP23 and WALP27, respectively (Fig. 10.9c). Circular dichroism measurements confirmed helical structure of peptides in lipid bilayer, in the presence of surfactants, and in aqueous mixtures of organic solvents. To summarize the results and to provide further analysis the scheme in Fig. 10.9d is used. Left panels represent the free energy of insertion of fluorescently labeled WALP peptides. While the absolute values of ΔG (−9–−10 kcal/mol) provide a useful reference point, the differential value of $\Delta G(LALA) = \Delta G(WALP27) - \Delta G(WALP23) = -1.0$ kcal/mol, which corresponds to partitioning of the Leu–Ala–Leu–Ala region (Fig. 10.9d, *center*), is even more useful for the estimation of the per-residue cost of transfer of helical backbone from water to the hydrocarbon core of the lipid bilayer, i.e. $\Delta G(BB)$ [73]. The latter can be estimated from the measurements by accounting for the free energies of side chains according to the following: $\Delta G(BB)=(\Delta G(LALA)-2\Delta G(Leu)-2\Delta G(Ala))/4$. To estimate $\Delta G(Leu)$ and $\Delta G(Ala)$, an experimental hydrophobicity scale is used: The Wimley–White scale derived from octanol partitioning of model pentapeptides [72]. The Wimley–White scale is a whole-residue scale and contains the contribution of the side chain and that of the peptide bond. The latter is equal to +2.0 kcal/mol and must be subtracted from the whole-residue ΔG value for each amino acid, as illustrated by White and Wimley [75]. The resulting side-chain values are $\Delta G(Leu)=-3.25 \pm 0.11$ kcal/mol and $\Delta G(Ala)= -1.50 \pm 0.12$ kcal/mol. These values combined with our results lead to this value for the helical backbone contribution of $\Delta G(BB)=+2.13 \pm 0.17$ kcal/mol (Fig. 10.9d, *right*).

The free-energy estimates for TM insertion presented in Fig. 10.9 add an important quantitative aspect to our understanding of the thermodynamic laws that govern membrane protein insertion and assist in the further deciphering of the complex nature of membrane protein folding and stability in the native lipid environment.

WALP23: **Acetyl-CGWWLALALALALALALALALWWA-Amide**
WALP27: **Acetyl-CGWWLALALALALALALALALALALWWA-Amide**

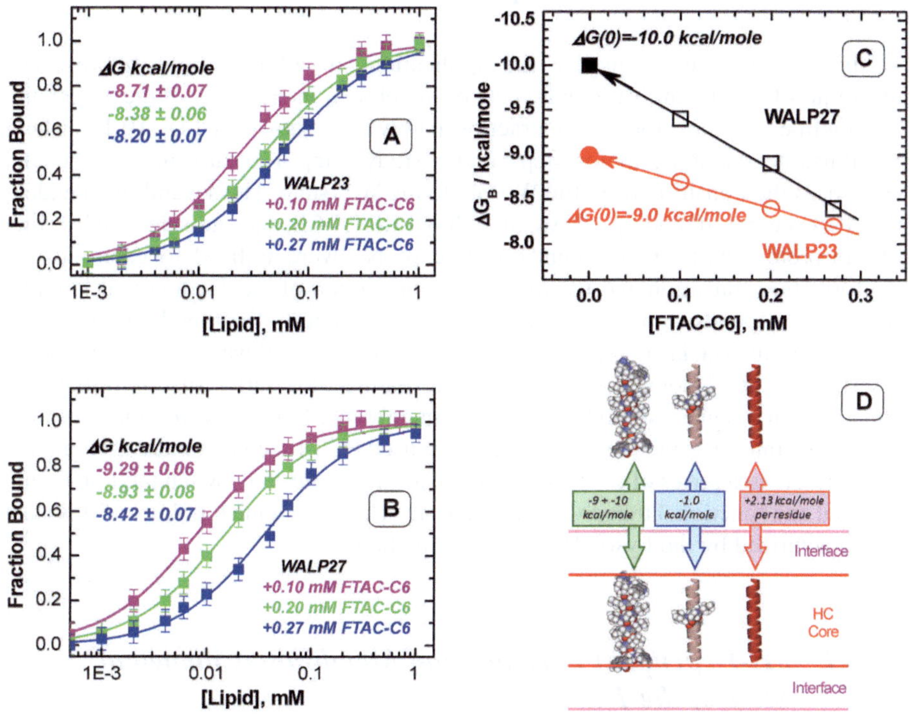

Fig. 10.9 Thermodynamic analysis of the bilayer insertion of model TM WALP peptides chaperoned by fluorinated surfactant FTAC-C6. Lipid titration isotherms for WALP23 (**a**) and WALP27 (**b**) measured by FCS in the presence of 0.1, 0.2, and 0.27 mM FTAC-C6 (*red, green, and blue symbols*, respectively). LUVs made of POPC were used in the titration. (**c**) Free energies of membrane insertion in the absence of surfactants (*filled symbols*) are estimated by the extrapolation (*arrows*) of the apparent ΔG measurements in the presence of surfactants. (**d**) Thermodynamic analysis of WALP insertion in the lipid bilayer. Interpolation of the apparent free-energy values measured in the presence of surfactants to a zero surfactant concentration yielded free-energy values of −9.0 ± 0.1 and −10.0 ± 0.1 kcal/mol for insertion of WALP23 and WALP27, respectively (*left panels*). From a combination of thermodynamic and conformational measurements, one can conclude that the partitioning of a four-residue L–A–L–A segment in the context of a continuous helical conformation from aqueous environment into the hydrocarbon core of the membrane has a favorable free energy of 1 kcal/mol (*center panels*). This estimate combined with the predictions of the hydrophobicity scale of Wimley and White indicates that the per-residue cost of transfer of helical backbone from water to the hydrocarbon core of the lipid bilayer is unfavorable and is equal to + 2.13 ± 0.17 kcal/mol (*right panels*) (Modified from Ref. [74])

10.5 All Together Now: Application of the Battery of Fluorescence Techniques to Characterize Membrane Interactions of the Diphtheria Toxin T Domain

In this section, we summarize the applications of all the above fluorescence techniques for studying various aspects of membrane interactions of the T-domain. A simplified scheme for this interaction pathway is shown in Fig. 10.1b, where we illustrate the three major steps under study. Step 1 occurs in solution and comprises the transition from the W-state to the W^+-state. The several independent fluorescence-based binding assays, including FRET, FCS, and site-selective labeling of the protein with environment-sensitive probe, were utilized to examine this conformation transition and the membrane interaction of the T-domain at low pH. Step 2 corresponds to the partitioning of the W^+-state to the membrane interface to form the interfacial I-state (Fig. 10.1b). The kinetic pathway for the membrane insertion of the T-domain is considered revealing of a series of conformational changes that occur in solution and in the membrane. For the partitioning into the membrane interface and the TM insertion we used FCS-based methodology. Finally, step 3 refers to a series of conformational rearrangements that switch the I-state into the final and functional transmembrane T-state. The membrane topology of this state is characterized by the LysoUB quenching method.

10.5.1 pH-Dependent Formation of Membrane-Competent State of the T-domain

Formation of the membrane-competent form (W^+-state) of the T-domain is the first step along a complex pathway, leading from a soluble conformation with a known crystallographic structure [76] (W-state) ultimately to membrane-inserted states (Fig. 10.1b) [9, 12, 14, 16, 23]. Protonation of T-domain histidine side-chains has been recognized to play an important role in various stages of the membrane insertion process, so that H257 and H223 were implicated to act as a molecular switch that triggers disruption of the native structure of W-state of T-domain in low-pH solution [9, 16, 17, 19]. The pH dependent conformational transition W-to-W^+ is normally identified in a membrane-binding assay conducted under conditions of LUV saturation [9, 23]. The application of three independent fluorescence methods for the membrane binding to the T-domain's interaction with LUV are summarized in Fig. 10.10: (*i*) diffusion measurements by FCS technique (Fig. 10.10, *diamonds*), (*ii*) association of donor-labeled T-domain and acceptor-labeled vesicles by FRET (Fig. 10.10, *circles*) [14], and (*iii*) changes in polarity of a specifically attached environment-sensitive bimane probe by increase in fluorescence intensity (Fig. 10.10, *triangles*) [19]. All combined datasets were fitted with Eq. 10.8 and could be adequately described by a single fitting curve with a pK_a of 6.2 and a Hill coefficient of about 2 (Fig. 10.10, *line*). Thus, these measurements confirmed

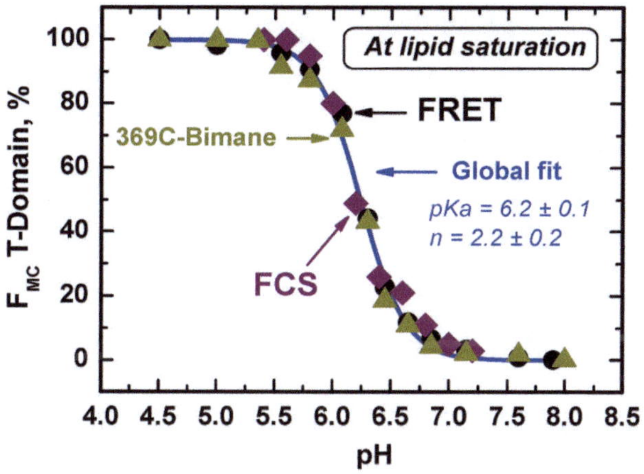

Fig. 10.10 pH-dependent conversion of the T-domain from the soluble W-state into the membrane-competent W^+-state. Identification of W-to-W^+ transition through membrane binding at lipid saturation. Mobility measurements based on FCS (*diamonds*) and measurements of FRET between donor-labeled T-domain and acceptor-labeled LUVs (*circles*) were performed as described in [14]. Triangles represent the increase in fluorescence intensity of environment-sensitive probe bimane attached to a single cysteine in position 369 (Fig. 10.11a) (Modified from Ref. [19])

that the W^+-state is an experimentally observable state of the protein with a conformation different from that of the W-state. The structure and the dynamics of pH-dependent conformation switching of the T-domain have been the subject of MD simulations reported elsewhere [19, 21, 22, 77].

10.5.2 Lipid-Dependent Modulation of the Kinetics of Membrane Association and Insertion of TH8-9 Helical Hairpin of the T-Domain

To test the insertion topology of the diphtheria toxin T-domain, the NBD probe was attached to the very tip of the insertion domain formed by helices TH8 and TH9 using the L350C mutant (Fig. 10.11a) [14]. If these helices adopt the TM conformation upon insertion, the probe will be translocated across the bilayer and become inaccessible to the externally added LysoUB quencher. If the helices remain interfacial, the quenching will be substantial. LysoUB quenching experiments were performed after T-domain insertion had come to equilibrium at 1 *h* after mixing the T-domain and LUV at low pH. In order to describe LysoUB quenching, and the resulting topology, in a quantitative way, the fluorescence decay for all samples was analyzed and the ratio of the average lifetime upon addition of the quencher calculated, as described previously [61]. We found that the degree of quenching depends on the lipid composition. The lifetime ratio τ_α/τ_0 of about 1.1 observed in

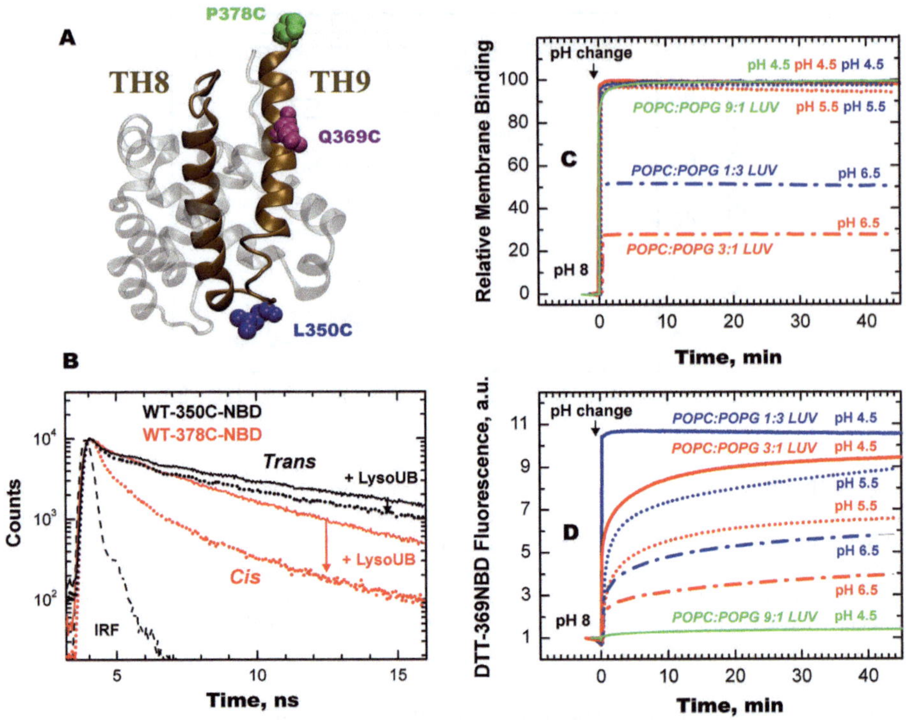

Fig. 10.11 (**a**) Crystallographic structure of the diphtheria toxin T-domain in soluble form at neutral pH [76]. The consensus insertion hairpin consisting of helices TH8 and TH9 is shown solid, while the rest of the structure is shown semitransparent. Three residues, L350, Q369, and P378 (highlighted in CPK), along TH9 were replaced with cysteines (one at a time), providing unique labeling sites for fluorescence dye NBD used in steady-state and kinetic measurements of insertion. (**b**). Application of the Lyso-UB quenching to determine the topology of TH9 in POPC:POPG 1:3 LUV. The difference in quenching of NBD attached at position 350 (*black*) and 378 (*red*) indicates TM insertion of the hairpin and translocation of the loop between the TH8 and 9 (Note that this segment remains in the IF orientation when T-domain binds LUV with a low content of anionic lipids [14]). (**c–d**). Membrane binding and insertion kinetics of the T-domain measured with LUV of specified lipid compositions (color-coded) at various pH (line coded). (**c**) Binding kinetics were followed by the change in FRET signal between donor-labeled T-domain and acceptor-labeled LUV. (**d**) Insertion kinetics were followed by changes in fluorescence intensity of the environment-sensitive probe NBD attached to the center region of the membrane-insertion domain (Q369C-NBD). Differences observed in the two types of kinetics reveal the insertion intermediate, which depends on pH and lipid composition (Modified from Ref. [14])

POPC:POPG 3:1 and POPC:POPG 1:3 is indicative of efficient translocation, while the ratio τ_α/τ_0 of 1.7 observed in POPC:POPG 9:1 confirms interfacial topology. In fact, the same quenching ratio was observed in a control experiment using the NBD-labeled P378C mutant, for which no translocation is expected during TH8-TH9 insertion (Fig. 10.11a).

Formation of intermediates and subsequent transbilayer insertion is mediated by a subtle balance of hydrophobic and electrostatic interactions between proteins and the membrane interface. The applications of fluorescence spectroscopy for kinetic measurements of the T-domain insertion along the pathway from aqueous W-state to interfacial intermediate I-state and finally to transbilayer inserted T-state is summarized in Fig. 10.11c, d. To study time-dependent association with the membrane, kinetic measurements using the FRET signal between AlexaFluor488-labeled T-domain and Rhodamine- labeled LUV were used. The validity of this FRET approach was verified by comparing equilibrium titration profiles with those obtained by FCS (Fig. 10.10, *circles* and *diamonds*). In addition, FRET-based binding kinetics (Fig. 10.11c) were compared to those observed with the environmentally sensitive probe NBD attached to a single-Cys residue in the Q369C mutant of the T-domain (Fig. 10.11a, d). This residue is relatively solvent-exposed in the soluble T-domain at neutral pH (W-state) and according to our topology result (Fig. 10.11b) is in the middle of the TM segment in the inserted T-state. Such a transition is expected to result in a substantial increase of the fluorescence of NBD, which was indeed observed, as shown by the kinetics measurements in Fig. 10.11d. In contrast to the fast FRET-based binding kinetics, fluorescence kinetics of NBD was slow and dependent on the lipid composition. This indicates that there is a membrane-bound, but not fully inserted, kinetic intermediate (I-state) that can be kinetically trapped. Remarkably, the insertion kinetics demonstrated that I-to-T transition is already occurring at pH 6.5, at which point the binding W-to-I transition is only half-completed (*dashed-dotted* lines in Fig. 10.11d) [14]. These results suggest that the protonation transitions allowing membrane binding and TM insertion are staggered, which could be related to an additional protonation of the T-domain on the interface due to variations in pK_a in bulk of solvent and the membrane interfacial zone.

10.5.3 Positioning of TH9 Helix in the Membrane by Depth-Dependent Fluorescence Quenching

We illustrate the application of the DA methodology by examining steady-state and time-resolved depth-dependent quenching of NBD-labeled mutants N366C and P378C. According to our LysoUB topology measurements (Fig. 10.11b and [14]) this helix adopts a transmembrane conformation in the inserted form, which is expected to place the probe attached at position 366 close to the bilayer center, and the one attached at position 378 close to the interface. The membrane depth of the NBD probe attached at positions 366 and 378 was examined by depth-dependent fluorescence quenching. The steady-state and time-resolved quenching were measured for the samples containing 1 μM of the labeled protein and 1 mM of LUV composed of POPC:POPG 3:1 + 30 mol % of spin-labeled lipids after

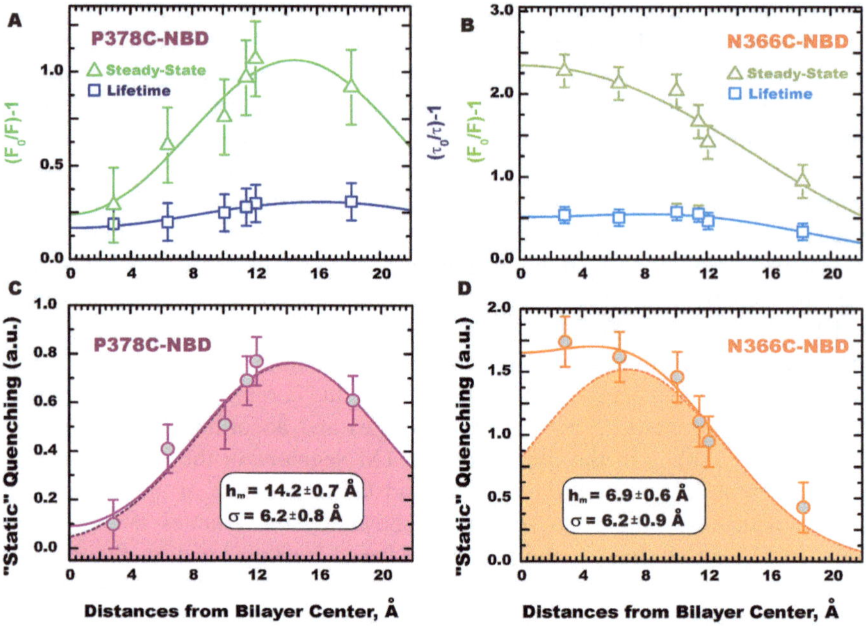

Fig. 10.12 Fluorescence quenching of NBD-labeled T-domain. DA of depth-dependent fluorescence quenching profiles of P378C-NBD and N366C-NBD mutants of the T-domain obtained with a series of the spin-labeled lipids (Tempo-PC and *n*-Doxyl–PCs with n = 5, 7, 10, 12, and 14). Panels **A–B** show application of the DA methodology to steady-state ("total" quenching, *open triangles*) and lifetime ("dynamic" quenching, *open squares*) fluorescence quenching and "static" quenching. (**C–D**) A deconvolution of the "static" fluorescence quenching (*filled circles*) was calculated as the difference between total quenching and dynamic profiles [69]. The single Gaussian depth distributions are shown as shaded profiles

2 h incubation in 50 mM phosphate buffer at pH 4.5. To determine the transverse position of the NBD probe in the lipid bilayer, we constructed depth-dependent quenching profiles, where steady-state and lifetime quenching efficiency $(F_0/F(h))$-1 and $(\tau_0/\tau(h))$-1 are plotted against the depth of the spin quenchers (Fig. 10.12a–b). Quantitative information on membrane penetration of NBD was extracted from these depth-dependent quenching data using the distribution analysis (DA) methodology. Solid lines in Fig. 10.12a–b show that the steady-state and lifetime quenching could be well fitted by a pair of mirror-symmetric Gaussian functions (one for each leaflet, equation in Fig. 10.5b) to take into account transverse distributions of lipid and protein moieties originating from the thermal motion in the bilayer. In addition, to reduce the quenching contribution from the transverse diffusion of a probe occurring during the excited-state lifetime, we calculate the "static" quenching profile by subtracting the "dynamics" lifetime quenching component from the total steady-state quenching [69]. Figure 10.12c–d show that this procedure

leads to narrower, better-defined quenching profiles for the NBD probe compared with those obtained by traditional steady-state quenching experiments. The NBD fluorophores attached to positions 378 and 366 were found to be at depths of about 14.2 Å and 6.9 Å from the bilayer center, which indeed corresponds to the expectation for the transmembrane conformation of TH9. The exact positioning of the helix will allow us to generate realistic starting positions for future studies of the membrane-inserted T-domain by MD computer simulations.

10.5.4 Positioning of TH5 Helix in WT and H223Q Mutant T-Domain

Recent results indicate that H322 is critical for proper insertion of the N-terminal part of the T-domain and formation of the functional open channel state [17, 20]. Replacements with Q, R or K, while not affecting folding in solution or proper insertion of TH8-9, lead to misfolding of the remaining structure on the membrane interface. This lack of proper insertion can be monitored by red-shifted tryptophan fluorescence, associated with changes in insertion of W281 located in TH5. The spectroscopic changes become even more pronounced in a single-Trp mutant W206F (Fig. 10.13 Ghatak, Rodnin, Kyrychenko, Ladokhin in preparation). This

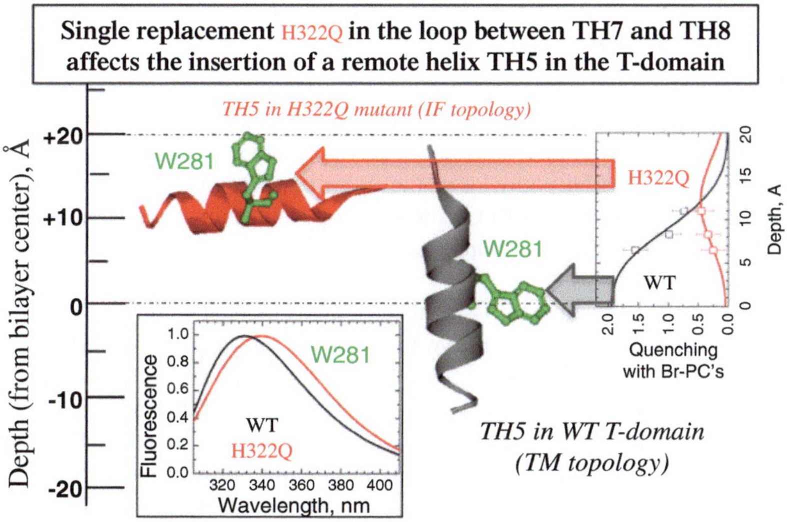

Fig. 10.13 A single replacement of H322 results in repositioning of the TH5 helix from TM state (WT, *black*) to IF state (H322Q mutant, *red*), associated with loss of activity due to uncoupling of TH8-9 insertion and N-terminus translocation [17, 20]. This repositioning can be followed by changes in spectral position of W281 (*lower panel*) and it's quenching with brominated lipids (*right panel*)

variation in the insertion mode is confirmed by the results of independent depth-dependent quenching experiments using three commercially available bromolipids with different depth, h: 6-7BrPC (h=11 Å), 9-10BrPC (h=8.3 Å) and 11-12BrPC (h=6.5 Å) (Fig. 10.5a). Each of the bromolipids was mixed with anionic POPG in 1:1 molar ratio. The intensity of tryptophan fluorescence of the T-domain was measured in the presence of quenching vesicles, $F(h)$ and in quencher-free POPC-POPG LUV, F_0. The "total" quenching profile was generated as $QP(h)=F_0/F(h)-1$ and analyzed using DA method [62, 65] by fitting to the twin Gaussian function (equation in Fig. 10.5b–c). Because the quenching profile for the mutant was poorly defined, the dispersion parameter σ representing the width of the distribution was fixed at 5, 6 or 7 Å. Regardless of the chosen value, the results indicate that W281 in the WT is located close to the center of the lipid bilayer, and remains close to the interface in H223Q mutant (Fig. 10.13). Both spectroscopic experiments presented in Fig. 10.13 support the hypothesis that H322Q mutant is trapped in T1-state and does not form a fully inserted T2-state with a translocated N-terminus (Fig. 10.1b) [17, 20]. While the exact mechanism of how the replacement of H322 influences the topology of the segment some 40 residues away remains a mystery, the result confirms that fluorescence measurements constitute an important tool for studies of membrane protein folding.

10.5.5 FCS Measurements of ΔG of Membrane Insertion of the T Domain, Chaperoned by Fluorinated Surfactants

The discussed FCS-method of the free energy determination of membrane protein lipid bilayer partitioning based on usage of fluorinated surfactants was utilized to characterize the insertion pathway of the T-domain from the thermodynamic perspective [23]. A combination of FCS methodology and chaperoned membrane insertion protocol utilizing the fluorinated surfactant FTAC-C6 was used, as described in Ref. [41].

As mentioned above, the multi-step insertion process starts with a conformational change in solution, in which the water-soluble W-state is converted into the membrane-competent W^+-state as a result of protonation of key histidine residues (Fig. 10.1b) [14, 16, 17, 37]. Upon addition of membranes, the T-domain cascades into multiple conformations in the bilayer, including the interfacial I-state and a number of transmembrane T-states. Figure 10.14a shows that increase in lipid concentration results in a progressive shift of mobility of the T-domain from that of free form to that of vesicle-bound form. Quantitative determination of membrane-bound fraction, f_B, is achieved by the linked analysis (*black lines*) of fluorescence autocorrelation curves (*color coded lines*), which links the correlation times for all curves and allows free fitting of the pre-exponential amplitudes in Eq. (10.1).

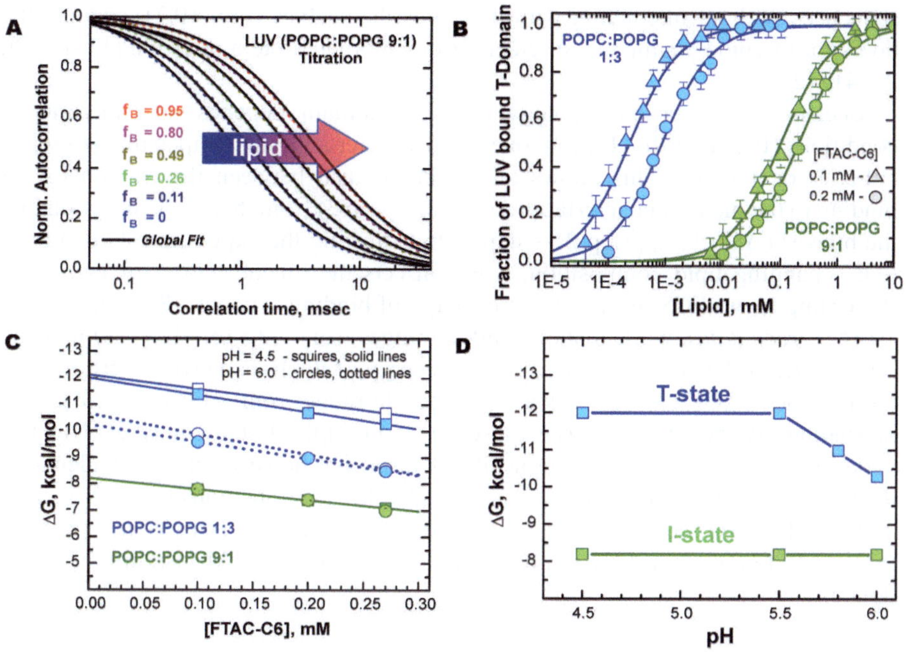

Fig. 10.14 Determination of the free energy of bilayer insertion chaperoned by FTAC-C6. (**a**) Example of LUV binding measurement using FCS performed upon addition of LUV (up to 4 mM total lipid concentration) to fluorescently labeled T-domain (nanomolar concentrations) in the presence of 0.1 mM FTAC-C6. (**b**) Lipid titration isotherms for the T-domain in the presence of different concentrations of the fluorinated surfactant FTAC-C6 at pH 4.5. (**c**) Thermodynamic analysis of the bilayer insertion chaperoned by FTAC-C6. Addition of the surfactant results in gradual linear decrease in ΔG. A true value of ΔG of insertion is estimated from the linear extrapolation of the data to zero concentration of the surfactant. The lipid compositions are color-coded, and concentration of FTAC-C6 and pH values is coded by symbol shape. (**d**) pH dependence of the free energy of interfacial membrane binding (POPC:POPG 9:1, *olive*) and transmembrane insertion (POPC:POPG 3:1, *red*; POPC:POPG 1:3, *blue*) for the T-domain. The free energy of binding to lipid vesicles with formation of trapped interfacial intermediate state is close to −8 kcal/mol, and is pH-independent. The free energy difference between the transmembrane T-state and the interfacial I-state ranges from −1.5 to −4 kcal/mol depending on membrane composition and the pH (Modified from Ref. [23])

While these experiments appear to be similar to those collected at different pH (e.g., Fig. 10.6), they differ in several important aspects. Specifically, the measurements of free energy require that not all the membrane-competent form is bound to the membrane. This is different from measurements done at lipid saturation as a function of pH, as in those it is a membrane-incompetent form that remains in solution. In order to prevent nonproductive aggregation of membrane-competent form in solution, we use fluorinated surfactant FTAC-C6. Unlike regular detergents, fluorinated surfactants can chaperone membrane interactions of hydrophobic sequences without interacting with the lipids themselves (Fig. 10.7), opening the doors for thermodynamic measurements of membrane protein insertion [13–15, 18, 23, 41, 42, 74].

Because the insertion topology of the T-domain depends on experimental conditions (Fig. 10.14b) [14, 23], one can use different lipid compositions to select either I- or T-states. This allows one to distinguish between the free energy of binding to the membrane interface ($\Delta G_{W}{}^{+}{}_{\text{-}to\text{-}I}$, step 2) from that of the insertion into the bilayer ($\Delta G_{W}{}^{+}{}_{\text{-}to\text{-}T}$, step 2 + step 3) by adjusting the experimental conditions (e.g., changing lipid composition, which affects the kinetic barrier of insertion). According to our FCS results, the free energy of binding is about −8 kcal/mol (Fig. 10.14c, *green data*), while that of insertion ranges from −9.5 to −12 kcal/mol (Fig. 10.14c, *red and blue data*). Thus, the free energy of the transbilayer insertion from the interface ($\Delta G_{I\text{-}to\text{-}T}$, step 3) was estimated to range from −1.5 to −4 kcal/mol, depending on the pH [23]. The presence of multiple T-states may explain this variation of the free energy of transbilayer insertion; however, it is a challenge to determine experimentally the free energy value for the transitions between various T-states. The ΔG values for TM insertion of the isolated helix in WALP peptides (Fig. 10.9, [74]) fall in between the ΔG values for IF and TM insertion of the T-domain.

10.6 Conclusions

Fluorescence spectroscopy is an important tool for studies of insertion of proteins into lipid bilayers. As demonstrated with the example of diphtheria toxin translocation domain, various fluorescence approaches can be applied to obtain critical structural and thermodynamic information enabling the deciphering of complex insertion/refolding pathways. Measurements of the insertion topology and bilayer penetration of protein sites, selectively labeled with appropriate probes, have the potential to fill an important knowledge gap in the structure-function relationship of a vide variety of proteins interacting with lipid bilayers. The methods described here are opening the doors for addressing fundamental questions related to lipid-dependent regulation of protein functioning *via* conformational switching and signaling on membrane interfaces.

Acknowledgments A.K acknowledges Grant 0116U000835 of the Ministry of Education and Science of Ukraine. Supported in part by NIH GM-069783 (A.S.L) and FONDECYT 3160734 (M.V.U).

References

1. Young JAT, Collier RJ (2007) Anthrax toxin: receptor binding, internalization, pore formation, and translocation. *Annu Rev Biochem* 76:243–265
2. Murphy JR (2011) Mechanism of diphtheria toxin catalytic domain delivery to the eukaryotic cell cytosol and the cellular factors that directly participate in the process. *Toxicology* 3:294
3. Skehel JJ, Wiley DC (2000) Receptor binding and membrane fusion in virus entry: the influenza hemagglutinin. *Annu Rev Biochem* 69:531–569
4. Shamas-Din A, Kale J, Leber B, Andrews DW (2013) Mechanisms of action of Bcl-2 family proteins. Cold Spring Harb Perspect Biol 5:a008714, doi: 008710.001101/cshperspect.a008714
5. Bogner C, Leber B, Andrews DW (2010) Apoptosis: embedded in membranes. *Curr Opin Cell Biol* 22:845–851
6. Vargas-Uribe M, Rodnin MV, Ladokhin AS (2013) Comparison of membrane insertion pathways of the apoptotic regulator Bcl-xL and the diphtheria toxin translocation domain. *Biochemistry* 52:7901–7909
7. Moldoveanu T, Follis AV, Kriwacki RW, Green DR (2014) Many players in Bcl-2 family affairs. *Trends Biochem Sci* 39:101–111
8. Luna-Vargas MPA, Chipuk JE (2016) The deadly landscape of pro-apoptotic Bcl-2 proteins in the outer mitochondrial membrane. *FEBS Journal*. doi:10.1111/febs.13683
9. Ladokhin A (2013) pH-Triggered conformational switching along the membrane insertion pathway of the diphtheria toxin T-domain. *Toxicology* 5:1362–1380
10. Youle RJ, Strasser A (2008) The Bcl-2 protein family: opposing activities that mediate cell death. *Nat Rev Mol Cell Biol* 9:47–59
11. Ladokhin AS, Legmann R, Collier RJ, White SH (2004) Reversible refolding of the diphtheria toxin T-domain on lipid membranes. *Biochemistry* 43:7451–7458
12. Palchevskyy SS, Posokhov YO, Olivier B, Popot J-L, Pucci B, Ladokhin AS (2006) Chaperoning of insertion of membrane proteins into lipid bilayers by hemifluorinated surfactants: application to diphtheria toxin. *Biochemistry* 45:2629–2635
13. Rodnin MV, Posokhov YO, Contino-Pepin C, Brettmann J, Kyrychenko A, Palchevskyy SS, Pucci B, Ladokhin AS (2008) Interactions of fluorinated surfactants with diphtheria toxin T-domain: testing new media for studies of membrane proteins. *Biophys J* 94:4348–4357
14. Kyrychenko A, Posokhov YO, Rodnin MV, Ladokhin AS (2009) Kinetic intermediate reveals staggered pH-dependent transitions along the membrane insertion pathway of the diphtheria toxin T-domain. *Biochemistry* 48:7584–7594
15. Ladokhin AS (2009) Fluorescence spectroscopy in thermodynamic and kinetic analysis of pH-dependent membrane protein insertion. Methods in enzymology, vol 466. Academic Press, pp 19-42
16. Rodnin MV, Kyrychenko A, Kienker P, Sharma O, Posokhov YO, Collier RJ, Finkelstein A, Ladokhin AS (2010) Conformational switching of the diphtheria toxin T domain. *J Mol Biol* 402:1–7
17. Rodnin Mykola V, Kyrychenko A, Kienker P, Sharma O, Vargas-Uribe M, Collier RJ, Finkelstein A, Ladokhin Alexey S (2011) Replacement of C-terminal histidines uncouples membrane insertion and translocation in diphtheria toxin T-domain. *Biophys J* 101:L41–L43
18. Kyrychenko A, Rodnin MV, Vargas-Uribe M, Sharma SK, Durand G, Pucci B, Popot J-L, Ladokhin AS (2012) Folding of diphtheria toxin T-domain in the presence of amphipols and fluorinated surfactants: toward thermodynamic measurements of membrane protein folding. *Biochim Biophys Acta* 1818:1006–1012
19. Kurnikov IV, Kyrychenko A, Flores-Canales JC, Rodnin MV, Simakov N, Vargas-Uribe M, Posokhov YO, Kurnikova M, Ladokhin AS (2013) pH-triggered conformational switching of the diphtheria toxin T-domain: the roles of N-terminal histidines. *J Mol Biol* 425:2752–2764

20. Vargas-Uribe M, Rodnin MV, Kienker P, Finkelstein A, Ladokhin AS (2013) Crucial role of H322 in folding of the diphtheria toxin T-domain into the open-channel state. *Biochemistry* 52:3457–3463
21. Flores-Canales JC, Vargas-Uribe M, Ladokhin AS, Kurnikova M (2015) Membrane association of the diphtheria toxin translocation domain studied by coarse-grained simulations and experiment. *J Membr Biol* 248:529–543
22. Ghatak C, Rodnin M, Vargas-Uribe M, McCluskey A, Flores-Canales J, Kurnikova M, Ladokhin A (2015) Role of acidic residues in helices TH8–TH9 in membrane interactions of the diphtheria toxin T domain. *Toxicology* 7:1303–1323
23. Vargas-Uribe M, Rodnin MV, Öjemalm K, Holgado A, Kyrychenko A, Nilsson I, Posokhov YO, Makhatadze G, von Heijne G, Ladokhin AS (2015) Thermodynamics of membrane insertion and refolding of the diphtheria toxin T-domain. *J Membr Biol* 248:383–394
24. Walla PJ (2014) Optical properties of biomolecules. Modern biophysical chemistry. Wiley-VCH Verlag GmbH & Co. KGaA, pp 41-60
25. Ladokhin AS, Jayasinghe S, White SH (2000) How to measure and analyze tryptophan fluorescence in membranes properly, and why bother? *Anal Biochem* 285:235–245
26. Demchenko AP, Duportail G, Oncul S, Klymchenko AS, Mély Y (2015) Introduction to fluorescence probing of biological membranes. In: Owen DM (ed) Methods in Membrane Lipids. Methods in Molecular Biology, vol 1232. Springer, New York, pp 19–43
27. Demchenko AP, Mély Y, Duportail G, Klymchenko AS (2009) Monitoring biophysical properties of lipid membranes by environment-sensitive fluorescent probes. *Biophys J* 96:3461–3470
28. Ladokhin AS (2000) Fluorescence spectroscopy in peptide and protein analysis. In: Meyers RA (ed) Encyclopedia of analytical chemistry: applications of Instrumental Methods. Wiley, Chichester, pp 5762–5779
29. Kyrychenko A (2015) Using fluorescence for studies of biological membranes: a review. *Methods Appl Fluoresc* 3:042003/042001–042003/042019
30. Reshetnyak YK, Segala M, Andreev OA, Engelman DM (2007) A monomeric membrane peptide that lives in three worlds: in solution, attached to, and inserted across lipid bilayers. *Biophys J* 93:2363–2372
31. Kyrychenko A, Freites JA, He J, Tobias DJ, Wimley WC, Ladokhin AS (2014) Structural plasticity in the topology of the membrane-interacting domain of HIV-1 gp41. *Biophys J* 106:610–620
32. Kyrychenko A, Vasquez-Montes V, Ulmschneider MB, Ladokhin AS (2015) Lipid headgroups modulate membrane insertion of pHLIP peptide. *Biophys J* 108:791–794
33. Kaback HR, Wu J (1999) What to do while awaiting crystals of a membrane transport protein and thereafter. *Acc Chem Res* 32:805–813
34. Wang J, Rosconi MP, London E (2006) Topography of the hydrophilic helices of membrane-inserted diphtheria toxin T domain: TH1—TH3 as a hydrophilic tether. *Biochemistry* 45:8124–8134
35. Rosconi MP, London E (2002) Topography of helices 5–7 in membrane-inserted diphtheria toxin T domain: identification and insertion boundaries of two hydrophobic sequences that do not form a stable transmembrane hairpin. *J Biolumin Chemilumin* 277:16517–16527
36. Montagner C, Perier A, Pichard S, Vernier G, Ménez A, Gillet D, Forge V, Chenal A (2007) Behavior of the N-terminal helices of the diphtheria toxin T domain during the successive steps of membrane interaction. *Biochemistry* 46:1878–1887
37. Perier A, Chassaing A, Raffestin S, Pichard S, Masella M, Ménez A, Forge V, Chenal A, Gillet D (2007) Concerted protonation of key histidines triggers membrane interaction of the diphtheria toxin T domain. *J Biolumin Chemilumin* 282:24239–24245
38. May V, Kühn O (2011) Excitation energy transfer. Charge and Energy Transfer Dynamics in Molecular Systems. Wiley-VCH Verlag GmbH & Co. KGaA, Weinheim, Germany, pp 467–558
39. Rusu L, Gambhir A, McLaughlin S, Rädler J (2004) Fluorescence correlation spectroscopy studies of peptide and protein binding to phospholipid vesicles. *Biophys J* 87:1044–1053

40. Rhoades E, Ramlall TF, Webb WW, Eliezer D (2006) Quantification of α-synuclein binding to lipid vesicles using fluorescence correlation spectroscopy. *Biophys J* 90:4692–4700
41. Posokhov YO, Rodnin MV, Das SK, Pucci B, Ladokhin AS (2008) FCS study of the thermodynamics of membrane protein insertion into the lipid bilayer chaperoned by fluorinated surfactants. *Biophys J* 95:L54–L56
42. Posokhov YO, Rodnin MV, Lu L, Ladokhin AS (2008) Membrane insertion pathway of annexin B12: thermodynamic and kinetic characterization by fluorescence correlation spectroscopy and fuorescence quenching. *Biochemistry* 47:5078–5087
43. Melo AM, Prieto M, Coutinho A (2014) Quantifying lipid-protein interaction by fluorescence correlation spectroscopy (FCS). In: Engelborghs Y, Visser AJWG (eds) Fluorescence spectroscopy and microscopy. Methods in Molecular Biology, vol 1076. Humana Press, New York/Heidelberg/Dordrecht/London, pp 575–595
44. Nguyen TT, Swift JL, DT C (2012) Fluorescence correlation spectroscopy: the measurement of molecular binding. In: CD G (ed) Reviews in fluorescence 2010. Reviews in fluorescence, vol 2010. Springer, New York, pp 45–66
45. Schwille P, Korlach J, Webb WW (1999) Fluorescence correlation spectroscopy with single-molecule sensitivity on cell and model membranes. *Cytokine+* 36:176–182
46. Hink MA (2015) Fluorescence correlation spectroscopy. In: PJ V (ed) Advanced fluorescence microscopy. Methods in molecular biology, vol 1251. Springer, New York, pp 135–150
47. Hess ST, Huang S, Heikal AA, Webb WW (2002) Biological and chemical applications of fluorescence correlation spectroscopy: a review. *Biochemistry* 41:697–705
48. Haustein E, Schwille P (2007) Fluorescence correlation spectroscopy: novel variations of an established technique. *Annu Rev Biophys Biomol Struct* 36:151–169
49. White SH, Wimley WC, Ladokhin AS, Hristova K (1998) Protein folding in membranes: determining the energetics of peptide-bilayer interactions. *Methods Enzymol* 295:62–87
50. Lee H, Kim H (2014) Membrane topology of transmembrane proteins: determinants and experimental tools. *Biochem Biophys Res Commun* 453:268–276
51. Heuck AP, Tweten RK, Johnson AE (2001) β-Barrel pore-forming toxins: intriguing dimorphic proteins. *Biochemistry* 40:9065–9073
52. Collier RJ, Young JA (2003) Anthrax toxin. *Annu Rev Cell Dev Biol* 19:45–70
53. Zakharov SD, Cramer WA (2002) Colicin crystal structures: pathways and mechanisms for colicin insertion into membranes. *Biochim Biophys Acta* 1565:333–346
54. Zakharov SD, Cramer WA (2004) On the mechanism and pathway of colicin import across the *e. coli* outer membrane. *Front Biosci* 9:1311–1317
55. Isas JM, Cartailler J-P, Sokolov Y, Patel DR, Langen R, Luecke H, Hall JE, Haigler HT (2000) Annexins V and XII insert into bilayers at mildly acidic pH and form ion channels. *Biochemistry* 39:3015–3022
56. Matsuzaki K, Murase O, Fujii N, Miyajima K (1995) Translocation of a channel-forming antimicrobial peptide, Magainin 2, across lipid bilayers by forming a pore. *Biochemistry* 34:6521–6526
57. Everett J, Zlotnick A, Tennyson J, Holloway PW (1986) Fluorescence quenching of cytochrome B5 in vesicles with an asymmetric transbilayer distribution of brominated phosphatidylcholine. *J Biolumin Chemilumin* 261:6725–6729
58. Wimley WC, White SH (2000) Determining the membrane topology of peptides by fluorescence quenching. *Biochemistry* 39:161–170
59. Ladokhin AS, White SH (2004) Interfacial folding and membrane insertion of a designed helical peptide. *Biochemistry* 43:5782–5791
60. Ladokhin AS, Isas JM, Haigler HT, White SH (2002) Determining the membrane topology of proteins: insertion pathway of a transmembrane helix of Annexin 12. *Biochemistry* 41:13617–13626
61. Posokhov YO, Ladokhin AS (2006) Lifetime fluorescence method for determining membrane topology of proteins. *Anal Biochem* 348:87–93

62. Ladokhin AS (1997) Distribution analysis of depth-dependent fluorescence quenching in membranes: a practical guide. Ludwig Brand MLJ (ed) Methods in Enzymology, vol 278. Academic, New York, pp 462-473
63. Ladokhin AS (1999) Analysis of protein and peptide penetration into membranes by depth-dependent fluorescence quenching: theoretical considerations. *Biophys J* 76:946–955
64. London E, Ladokhin AS (2002) Measuring the depth of amino acid residues in membrane-inserted peptides by fluorescence quenching. In: SA S, TJ MI (eds) Peptide-lipid interactions. Current topics in membranes, vol 52. Elsevier, Amsterdam, pp 89–115
65. Ladokhin AS (2014) Measuring membrane penetration with depth-dependent fluorescence quenching: distribution analysis is coming of age. *Biochim Biophys Acta, Bioenerg* 1838:2289–2295
66. McIntosh TO, Holloway PW (1987) Determination of the depth of bromine atoms in bilayers formed from bromolipid probes. *Biochemistry* 26:1783–1788
67. Kyrychenko A, Ladokhin AS (2013) Molecular dynamics simulations of depth distribution of spin-labeled phospholipids within lipid bilayer. *J Phys Chem B* 117:5875–5885
68. Kyrychenko A, Rodnin MV, Ladokhin AS (2015) Calibration of distribution analysis of the depth of membrane penetration using simulations and depth-dependent fluorescence quenching. *J Membr Biol* 248:583–594
69. Kyrychenko A, Ladokhin AS (2014) Refining membrane penetration by a combination of steady-state and time-resolved depth-dependent fluorescence quenching. *Anal Biochem* 446:19–21
70. Barthélémy P, Ameduri B, Chabaud E, Popot J-L, Pucci B (1999) Synthesis and preliminary assessments of ethyl-terminated perfluoroalkyl nonionic surfactants derived from tris(hydroxymethyl)acrylamidomethane. *Org Lett* 1:1689–1692
71. Killian JA (2003) Synthetic peptides as models for intrinsic membrane proteins. *FEBS Lett* 555:134–138
72. Wimley WC, White SH (1996) Experimentally determined hydrophobicity scale for proteins at membrane interfaces. *Nat Struct Biol* 3:842–848
73. White SH, Ladokhin AS, Jayasinghe S, Hristova K (2001) How membranes shape protein structure. *J Biolumin Chemilumin* 276:32395–32398
74. Kyrychenko A, Rodnin MV, Posokhov YO, Holt A, Pucci B, Killian JA, Ladokhin AS (2012) Thermodynamic measurements of bilayer insertion of a single transmembrane helix chaperoned by fluorinated surfactants. *J Mol Biol* 416:328–334
75. White SH, Wimley WC (1999) Membrane protein folding and stability: physical principles. *Annu Rev Biophys Biomol Struct* 28:319–365
76. Bennett MJ, Eisenberg D (1994) Refined structure of monomelic diphtheria toxin at 2.3 Å resolution. *Protein Sci* 3:1464–1475
77. Flores-Canales JC, Kurnikova M (2015) Microsecond simulations of the diphtheria toxin translocation domain in association with anionic lipid bilayers. *J Phys Chem B* 119:12074–12085

Chapter 11
Organic Material Based Fluorescent Sensor for Hg^{2+}: A Brief Review on Recent Development

Muhammad Saleem, Muhammad Rafiq, and Muhammad Hanif

Keywords Mercury • Pollutants • Environmental contamination • Organic material • Fluorescent sensor

11.1 Introduction

The need for public safety and environmental protection spurs the interest of scientific communities toward the detection of harmful environmental analytes [1]. The rapid development of industries and competitive use of fresh water in many parts of worlds, water contaminated by heavy metal ions has become a serious global issue nowadays [2]. The presence of heavy metals in the environment is a major concern because of their toxicity and threat to human life and ecosystem [3]. The emission of toxic heavy metal ions has posed a challenge to the global plan of maintaining a sustainable environment [4]. Among the several metal ions-based pollutants, the mercury ion is a frontier contaminant to the human health and environment. Both natural and anthropogenic activities can generate Hg contaminations. The main routes for mercury entrance in the body are by breathing, skin absorption and by the mouth [5]. Mercury in all of its oxidation states is released by combustion of coal and waste, but also as a consequence of industrial processes such as the chloralkali process and gold mining. Through high bioaccumulation and bio-amplification factors, the mercury compounds end up in the food chain [6]. Its small amount in the body triggered the long-term irreversible damage to the human health by incorporating the unfavorable impact on the vital organs and tissues such as brain, nervous/immune system, kidney, liver

M. Saleem (✉)
Department of Chemistry, University of Sargodha, Sub-campus Mianwali, Sargodha, Pakistan
e-mail: saleemqau@gmail.com

M. Rafiq
Department of Biochemistry and Biotechnology (Baghdad-ul-Jadeed Campus), The Islamia University of Bahawalpur, Bahawalpur 63100, Pakistan

M. Hanif
Department of Chemistry, GC University Faisalabad, Sub Campus Layyah, Faisalabad, Pakistan

C.D. Geddes (ed.), *Reviews in Fluorescence 2016*, Reviews in Fluorescence,
DOI 10.1007/978-3-319-48260-6_11

and induce the cognitive and motion disorders [7–9]. In addition, the inorganic mercury pollutants possessed capability to absorb and transformed into organic ones by bacteria and microbes. After being accumulated through food chains, the concentration of Hg contaminations can be increased to a hazard level [10]. The maximum allowable concentration of Hg^{2+} is 10 nM in drinking water, which is defined by United States Environment Protection Agency [11].

Therefore, it is of great importance to develop a rapid and ecofriendly method to detect Hg(II) with high sensitivity and selectivity. Among several metal detection strategies, the fluorescent and colorimetric methods possess some advantages such as easy readout with the naked eye and potential for high throughput formats. However, still there exists a need toward improvements in some areas [12, 13].

11.2 Fluorescent Sensors for the Trace Mercuric Ion Detection

11.2.1 Rhodamine Based Fluorescent Sensors

Zhang et al. [14] reported the reversible Hg^{2+}-selective fluorescent chemosensor **1** in the nearly pure aqueous media (Fig. 11.1). The molecule **1** showed very weak fluorescence signal which was enhanced up to 90 fold on 4 eq. addition of mercuric ion alongside the colorimetric change in the reaction solution for naked eye metal detection. The 1:2 ligand-metal binding stoichiometry was estimated by the Job's curve which was further confirmed by the mass spectrometric analysis and DFT calculation. The association constant was determined by the Benesi-Hilderbrand equation which comes to be 3.8×10^9 M^{-2}. The sensor behaves reversibly on access addition of sodium sulfide.

Kaewtong et al. [15] reported a multifunctional FRET based sensor **2** constituting with rhodamine and terthiophene for selective Hg^{2+} detection (Fig. 11.2). Without the metallic induction, the sensor molecule stay colorless and fluorescence inactive in the less or more concentration indicating the existence in the spyrolactam

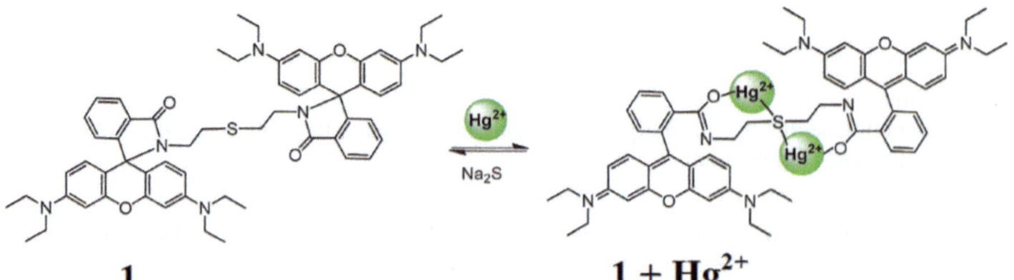

Fig. 11.1 The chemical structure of the ligand **1** and proposed ligand- Hg^{2+} complexation mechanism

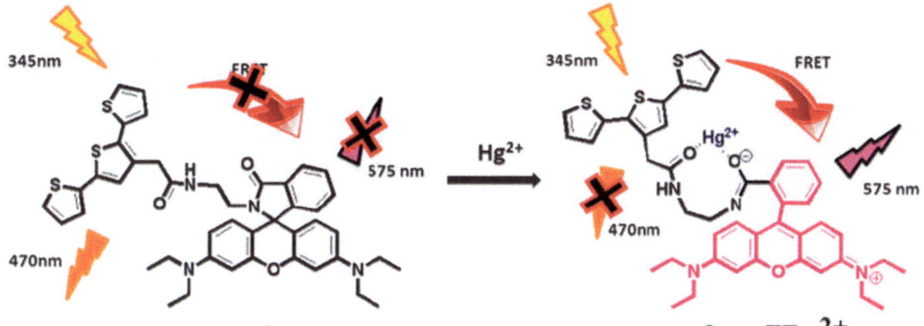

2 2 + Hg^{2+}

Fig. 11.2 The chemical structure of the ligand **2** and proposed complexation mechanism

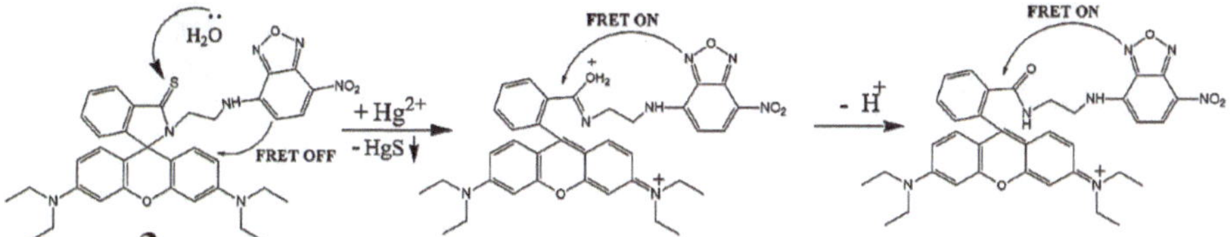

Fig. 11.3 The molecular structure of the receptor **3** and its proposed sensing mechanism

conformation. The mercuric ion promote the significant signal at 578 nm and change in the color of solution indicating the Hg^{2+} sensing ability of the ligand. The 2:1 ligand-metal binding stoichiometry and association constant value of 8.46 was assessed for the sensor. The minimum detection ability of this FRET based sensor was determined by the 3 σ slope^{-1} method which comes to be 1.34×10^{-6} M.

Interestingly, the FRET based sensor **3** [16] having the rhodamine B as acceptor and 4-chloro-7-nitro-2,1,3-benzoxadiazole (Fig. 11.3) as donor unit considerably increased the limit of detection up to 120 nm, indicating the appreciable prestige of molecule **3** over the molecule **2** (LOD = 1.34×10^{-6} M) where in **3**, there exist the terthiophene moiety as the donor unit (Fig. 11.2).

Shen et al. [17] and Chen et al. [18] reported the rhodamine functionalized iron nanocomposite **4** and **5** for selective mercuric ion detection. The prior used the rhodamine 6G, core–shell structured composite with superparamagnetic ferric oxide as the inner core and silica molecular sieve as the outer shell while the later used the core–shell structured Fe_3O_4 nanoparticles functionalized with the rhodamine B (Fig. 11.4). The nano particle based sensor exhibited advantageous features over the solution phase sensor that their suspension can be made and studied in the pure aqueous media without involvement of organic solvent. Hence seems more practical for direct environmental mercuric ion detection in the river, pond or see water. Furthermore, the nanocomposite with the rhodamine 6G exhibited the detection limit of 0.8×10^{-7} M while that with the rhodamine B showed the LOD = 1.5×10^{-8} M.

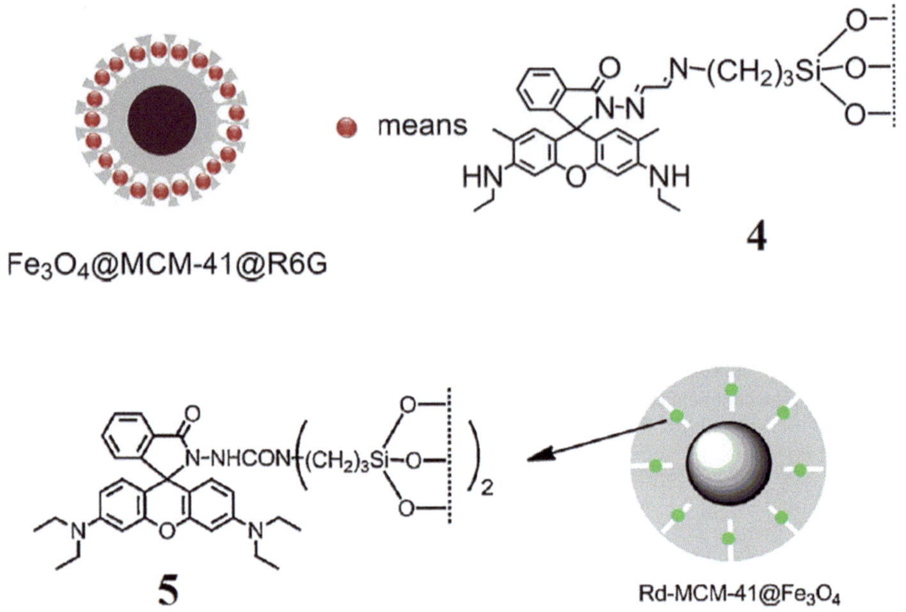

Fig. 11.4 The construction routs for the nanocomposite **4** and **5** having rhodamine 6G and rhodamine B

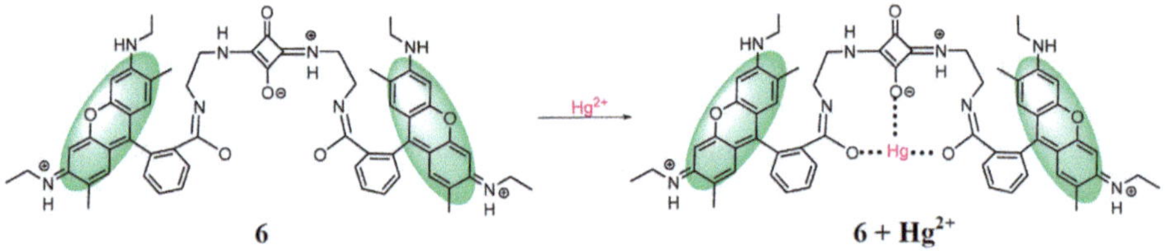

Fig. 11.5 The molecular structure of the receptor **6** and proposed Hg^{2+}-promoted ring opening of spirolactum

From these results, it can be concluded that the nanocomposite using rhodamine B are more fruitful for sensing purposes in comparison with that of rhodamine 6G.

However, some of the fluorescent sensor made up of rhodamine 6G exhibited low detection limit as well like squaraine–bis(rhodamine-6G) **6** (Fig. 11.5) [19]. The ligand **6** undergoes "Off-On" fluorescent change when it was treated with the trace mercuric ions. The 200 fold enhancement in the fluorescent emission signal intensity at 553 nm as well as UV-visible absorption signal enhancement centered at 527 nm showed the susceptibility of the reported ligand toward the mercuric ions. The detection limits and association constant Ka of the complex was reported as 1.69×10^{-8} M and 9×10^{8} M^{-1}, respectively. The 1:1 ligand-metal binding stoichiometry was found by the Job's plot which was then confirmed by the mass spectrometric analysis and DFT calculation.

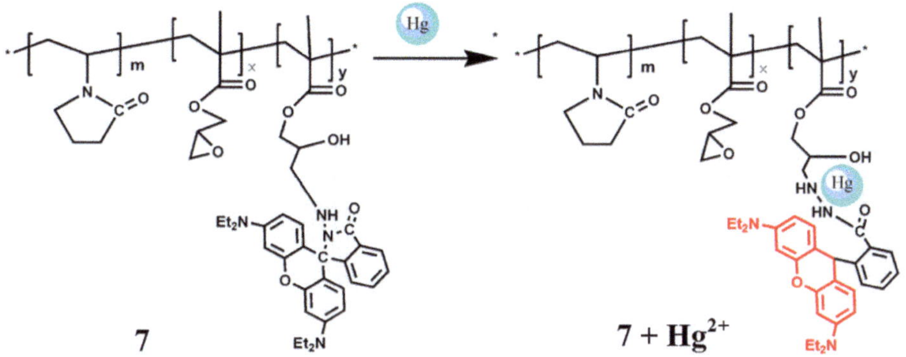

Fig. 11.6 The molecular structure of the receptor **7** and possible Hg^{2+}-induces ring opening of spirolactum

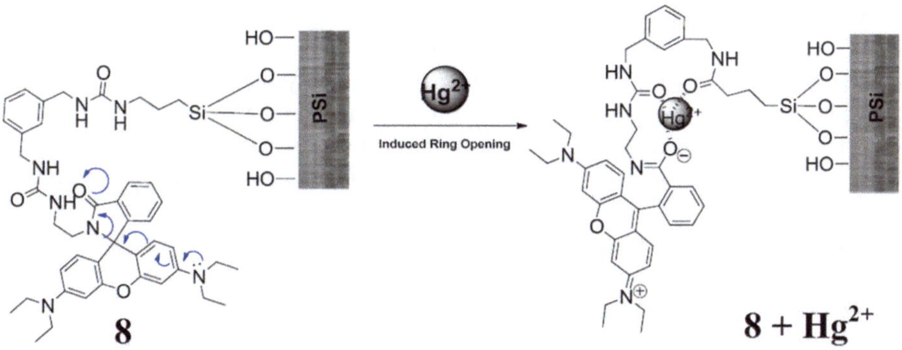

Fig. 11.7 Proposed mechanism of the ligand-metal coordination mode

One of the major drawbacks of the fluorescent chemosensors is the water insolubility due to their organic skeleton. They need organic media for the solubility which limits their direct detection applicability of the metal contamination from water sample and needs the toxic organic solvents for their working. One way to increase the water susceptibility of the organic sensor is to make their polymeric structure which induces their solubility in the pure aqueous media. One of such kind of chemosensor **7** (Fig. 11.6) was reported by the Luo et al. [20] containing rhodamine as switching moiety and copolymer for water solubility. The sensor exhibited the turn on fluorescent behavior alongside the colorimetric change in the reaction mixture for the naked eye metal detection. The limit of detection for this sensor was reported as 0.05 μM in the pure aqueous solution.

Cruz-Guzman [21] reported a turn-on fluorescent solid sensor **8** (Fig. 11.7) for detection and quantification of mercuric ion. The organosilane derivative was synthesized by the one pot synthesis which was then covalently immobilized on the mesoporous silicon surface in order to get the organic-inorganic hybrid sensing material for the environmental metal detection. The solid phase sensors

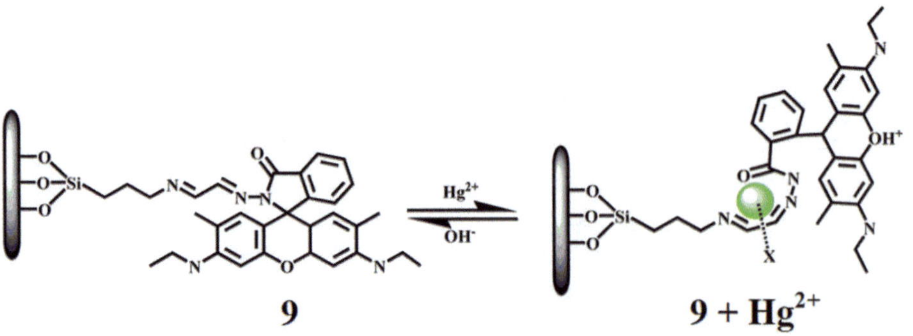

Fig. 11.8 The structural representation of the solid phase sensor **9** and the plausible Hg^{2+} complexation mechanism

have advantageous features of facile utility in the field without involvement of complicated instruments. Furthermore, the solid phase sensor provides the easy removals of the sensor material from the sample just by picking out which can be recycled hence, opening the doors toward economic and feasible strategies. The author explored the applicability of the developed sensor for the solution as well as solid phase. In both the cases the materials showed the clear-cut fluorescent turn-on response on treatment with the heavy metal ions.

Similar type of solid phase sensor **9** (Fig. 11.8) have been reported by Huang et al. [22] by utilizing the rhodamine as indicating unit and silica gel as solid support. The developed sensor can be used for the detection of metal ions in the mixed aqueous-organic media, in the pure aqueous media as well as in the pure organic media because it can be employed for the fluorescent studies by making the suspension. Therefore by using these kinds of solid detection strategies, the sensing experiment can efficiently be conducted in the pure aqueous or spiked environmental samples rather than mixed aqueous-organic media. Moreover, these sensors exhibited the reproducibility and facile removal from the samples.

11.2.2 BODIPY Based Fluorescent Sensors for Mercuric Ion Recognition

The boron-dipyrromethene (BODIPYs) dyes exhibited the interesting optical properties *e.g.,* high molar absorption co-efficient, considerable stability toward the environmental condition, narrow emission bands at low energy and tunable structural properties attracted its interest of the scientific community to utilize as a sensor for analytical detection. Herein, we summarized few interesting sensing materials for trace mercuric ion detection based on BODIPY exhibiting the low detection limits below than the mercury intake upper limit.

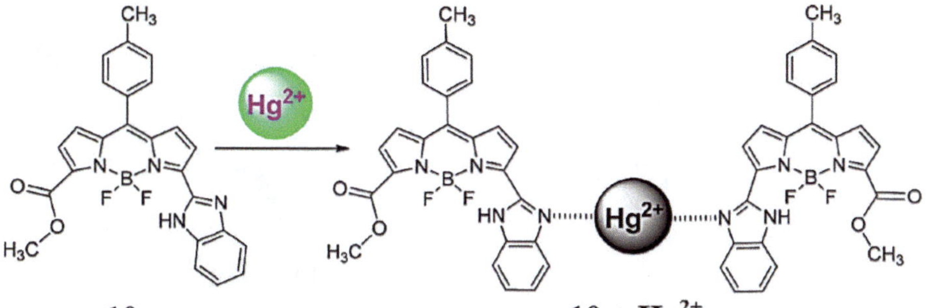

Fig. 11.9 The molecular structure of the receptor **10** and proposed binding mechanism between mercuric ion and BODIPY based receptor

Madhu et al. [23] reported the benzimidazole substituted BODIPY based sensor **10** (Fig. 11.9) for trace mercuric ion detection in the aqueous solution and living cells. The receptor **10** showed the absorption band at 577 nm along with the shoulder signal at 545 nm and higher wavelength narrow emission band at 603 nm. The Hg^{2+} addition induce the strong colorimetric as well as fluorimetric change in the receptor solution indicating the binding efficacy of the receptor toward the mercuric ions. From the fluorescent titration experiment, the binding constant for the probe toward the mercuric ions was estimated to be 6.18×10^6 M^{-1} by using the Benesi-Hilderbrand plot. The submicromolar sensitivity of the probe was estimated by using the linear dynamic response with the LOD = 0.77 μM. The 2:1 ligand-metal binding stoichiometry was estimated by the Job's curve which was further confirmed by the NMR titration experiment and mass spectrometric analysis. The probe was further employed for the bioimaging experiment using MDA-MB-231 cells under confocal fluorescent microscope and the appearance of the bright red fluorescent from the live cells upon treatment with the mercuric ion and receptor without any deformation in the cells indicate the appreciable cell permeability and low toxicity of the receptor for probing the intracellular mercury contamination.

Near infrared probe possessed several advantageous features for trace metal detection especially in the biological specimen due to low excitation energy which preserves the cell culture from photo-damage. Such similar type of near infrared probe containing BODIPY as core skeleton and 6-hydroxy indole as side group **11** (Fig. 11.10) was introduces by the Zhao and co-workers [24] which showed turn-on fluorescence response on treatment with the mercuric ions. The probe exhibited the well-defined absorption band at 548 nm in the absence of mercuric ion while after mercuric ion induction; there was constant decrease in the intensity of absorption band at 548 nm with the concomitant appearance and increment in the new absorption band centered at 591 nm. Meanwhile, the appearance of isosbestic point at 563 nm suggests the formation of new species *i.e.*, probe-metal complex. Similarly, there was enhancement in the fluorescence emission signal intensity at 637 nm on treatment with mercuric ion when it was excited at 563 nm (isosbestic

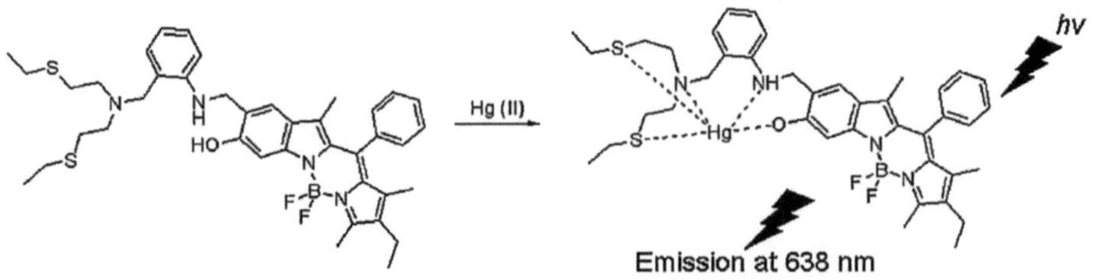

11 **11 + Hg^{2+}**

Fig. 11.10 The molecular structure of the receptor **11** and proposed ligation mechanism with the mercuric ions

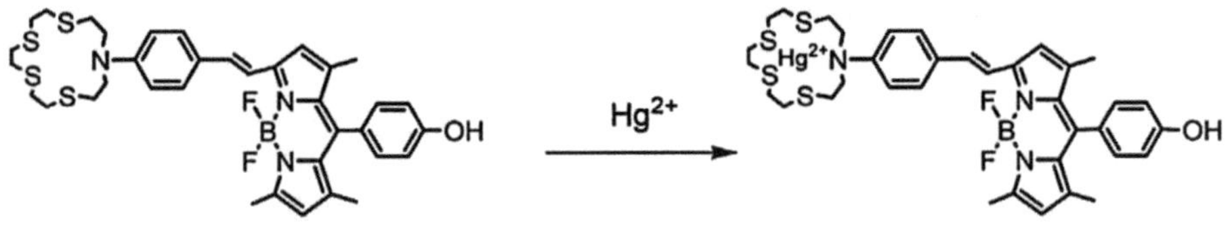

12 **12 + Hg^{2+}**

Fig. 11.11 The molecular structure of the receptor **12** and proposed ligand-metal complexation mechanism

point position) or at maxima of new coming band. The enhancement in the emission signal at 637 nm is due to inhibition of PET by the phenol moiety due to Hg^{2+} induced deprotonation. The 1:1 ligand-metal binding stoichiometry was estimated by using the optical titration graph. The minimal detection ability of the ligand was determined using standard deviation of 3-blank reading which comes to be 5×10^{-7} M. Meanwhile, successful bioimaging experiment was conducted by using HeLa cells under confocal fluorescence microscope.

Zhang et al. [25] described the ICT based fluorescent sensor for trace mercuric ion detection composed of BODIPY fluorophore and a thia aza crown ether receptor **12** (Fig. 11.11). The receptor exhibited the absorption band at 594 nm which are about 90 nm red shifted in comparison to the standard BODIPY skeleton and this red shift is attributed to the intermolecular charge transfer from the thiaza crown ether to the BODIPY nucleus. However the mercuric ion addition leads to the reduction of signal at 594 nm with the concomitant appearance of new signal at 564 nm. These signal variations are due to the inhibition of ICT by the engagement of nitrogen lone pair of thiazacrown ether in the complexation with the mercuric ion. The decrease in the ICT effect caused the blue shift in the spectrum alongside the colorimetric change visible to the naked eye. The ligand-metal binding constant and the limit of detection of the presented sensor were calculated which comes to be 3.19×10^6 M^{-1} and 4.36 nm, respectively. Due to low energy emission band of the sensor, the receptor **12** is more useful for the bioimaging experiment which was successfully employed for the bioimaging experiment using the HeLa cell lines.

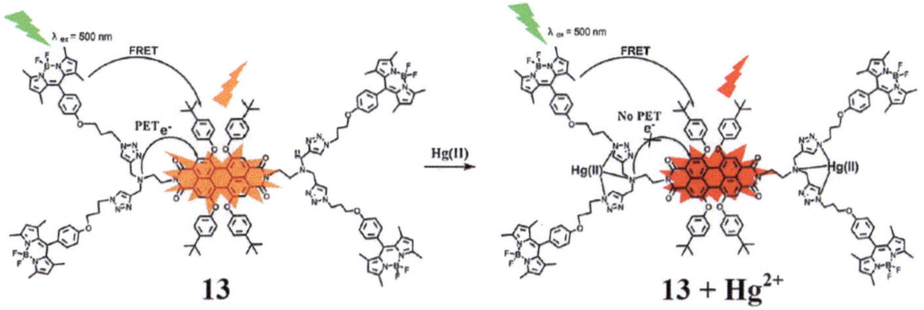

Fig. 11.12 The molecular structure of the receptor **13** and proposed ligand-mercury complexation mechanism

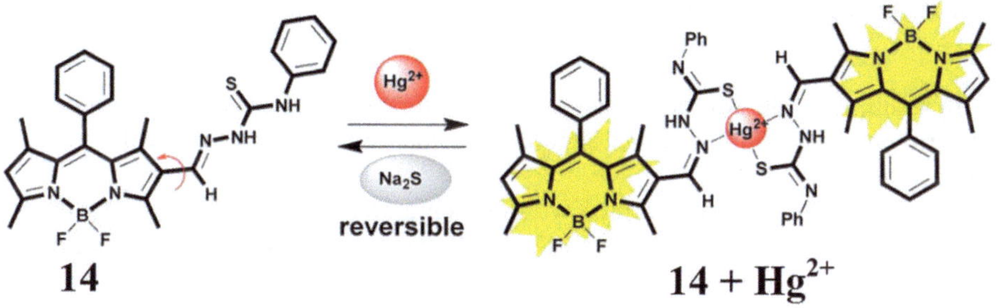

Fig. 11.13 The molecular structure of the receptor **14** and proposed receptor-mercury complexation mechanism

Cheng et al. [26] described the FRET and PET based fluorescence probe for mercuric ion detection made up of perylene bisimide as central unit which act as acceptor and four BODIPY side group acting as donor unit. The side groups are linked via click reaction making the dendrimeric type receptor **13** (Fig. 11.12). The ligand exhibited the prominent optical variation on binding with the mercuric ion. The 1:2 ligand-metal binding stoichiometry was calculated by the Job's curve. The dissociation constant of $6.67 \times 10^{12}\ M^{-1}$ was calculated by the non-linear curve fitting of the fluorescence experiment and the detection limit of 2×10^{-8} M was estimated from the titration results. The resulting probe was further applied for the bioimaging experiment by using the HeLa cells under confocal fluorescent microscope and harvested the significantly useful results.

As the sulfur atom showed very good affinity toward the mercury recognition, on this basis, there was low detection limit probe presented by the Canturk and co-workers [27] containing BODIPY as fluorescent unit and thiosemicarbazone as metal capturing motif **14** (Fig. 11.13). The mercury binding prevent the electron transfer process by the blockage of the C=N isomerism in the thiosemicarbazone skeleton. The detection limit of this turn-on fluorescent sensor was found to be 160 nM. The binding constant was determined to be $4.2 \times 10^{4}\ M^{-2}$ by using optical titration results. The reversibility of the receptor was found on the treatment

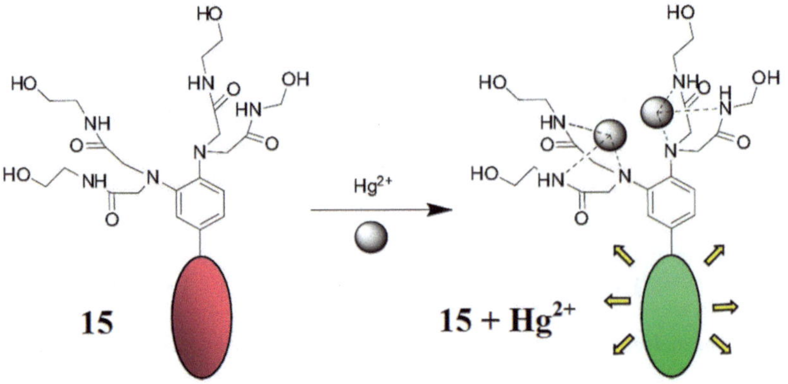

Fig. 11.14 The molecular structure and the proposed metal complexation mode of the receptor **15**

with sodium sulfide. Moreover, the ligand was further employed for the bioimaging experiment using human lungs adenocarcinoma cells under confocal fluorescent microscope.

The BODIPY derivative armed with the tetrapode receptor **15** (Fig. 11.14) was presented by the Culzoni and co-worker [28] for the trace recognition of mercuric ion based on PET mechanism. The fluorescent silent ligand undergoes turn-on response with 40 fold fluorescent increment by the blockage of PET on treatment with the mercuric ion with the selectivity over other competing metal ions. One more plus point of the receptor was the recognition of metal contamination in the environmental water sample. The 1:2 ligand-metal binding stoichiometry was determined by the Job's curve which was confirmed by the other spectroscopic techniques. The minimum mercury detection ability of the receptor was found to be 2.7 ng/mL.

11.2.3 Thiourea Based Fluorescent Sensors for Mercuric Ion Recognition

Liu et al. [29] reported the fluorescent sensor **16** (Fig. 11.15) for selective mercuric ion detection based on the thiosemicarbazide as binding site and nitrophenyl moiety as signaling group. On increasing concentration of the mercuric ion, the absorption signal at 470 nm decreases with the concomitant appearance of new signal at 401 nm alongside the colorimetric change in the reaction solution from brown to colorless. The appearance of two isosbestic points at 435 and 363 nm indicate the presence to two species *i.e.*, ligand and ligand-metal complex. The 1:1 ligand-metal binding stoichiometry was found by the Job's plot which was further confirmed

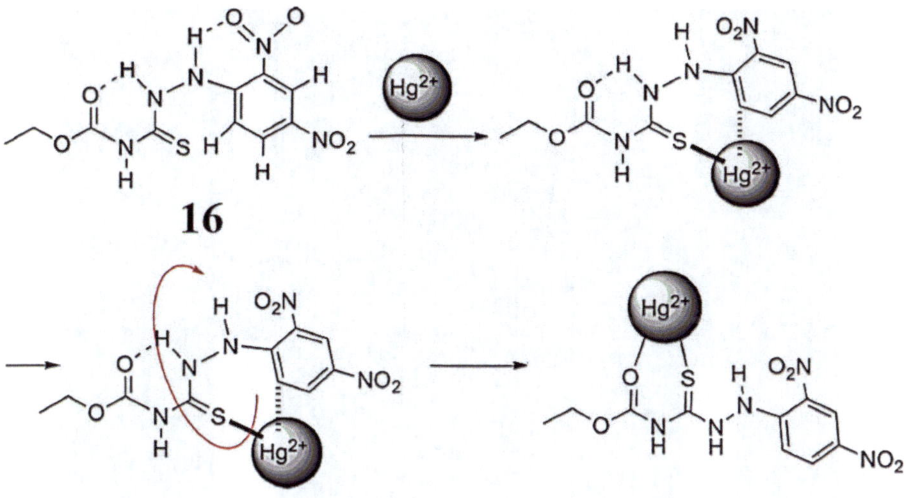

Fig. 11.15 The molecular structure of the receptor **16** and proposed complexation mechanism with the mercuric ions

by the mass spectrometric and NMR titration analysis. The association constant of $6.04 \times 10^4\ M^{-1}$ was found by the non-linear least square fitting analysis. The ligand exhibited the colorimetric detection limit of 5×10^{-6} M while 1×10^{-7} M in solution phase utilizing the UV-vsible experimental data.

Vengaian et al. [30] reported the phenothiazine based fluorescent probe **17** (Fig. 11.16) having thiourea as terminal moiety for selective recognition of mercuric ion. The mercuric ion induction to the probe solution caused the significant variation in the absorption signal intensity as well as position alongside two well defined isosbestic points which represent the clear cut formation of new species *i.e.*, ligand-mercury complex. Meanwhile, the emission band at 557 nm undergo constant downfall in intensity on gradual addition of mercuric ion when the excitation slit was adjusted at 450 nm. The ligand behaves reversibly on treatment with the fluoride ions due to high susceptibility of the mercuric ion to form more stable mercuric fluoride in comparison to the probe-metal complex. The 2:1 ligand-metal binding stoichiometry was assessed by the methods of continuous variation which was further confirmed by the mass spectrometric analysis and NMR titration results. The binding constant was calculated by using fluorescent titration experiment which comes to be $3.4 \times 10^3\ M^{-1}$. The ligand showed the lower detection limit of 1.5×10^{-8} M. The proposed sensor was further employed for the cytotoxic analysis and bioimaging experiment by using HeLa cell lines. The appearance of bright green fluorescence from the live cells on incubation with mercury and ligand without any deformation suggest the appreciable cells permeability of the probe. Meanwhile, the 5 μM safer dose of the ligand was determined by the MTT assay.

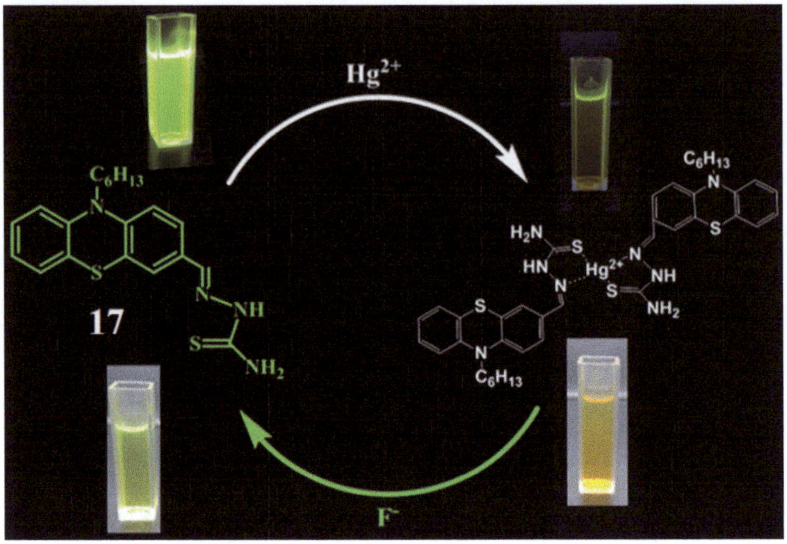

Fig. 11.16 The molecular structure of the receptor **17** and proposed complexation/decomplexation mechanism

11.2.4 Crown-Ether Based Fluorescent Sensors for Mercuric Ion Recognition

Dai et al. [31] presented the intramolecular charge transfer (ICT) based ratiometric fluorescent sensor for trace mercuric ion detection which consist of dansyl moiety as fluorophore and an azathia-crown ether as the receptor **18** (Fig. 11.17). The ligand exhibited the maximum emission at 545 nm when it was excited at 343 nm. Upon successive mercuric ion induction, the fluorescent intensity of the receptor was gradually decreased and this fluorescent quenching was due to the inhibition of intermolecular charge transfer by the involvement of nitrogen lone pair with the mercury binding. The Job's curve showed the 1:1 ligand-metal binding stoichiometry. The detection limit and association constant of the receptor were found to be 1.0×10^{-7} M and 6.5×10^{6} M^{-1}, respectively.

Nearly similar crown-ether based fluorescent probe **19** (Fig. 11.18) was designed and presented by Yan et al [32] with the improved detection limit in comparison to the receptor **18** (Fig. 11.17). The receptor was synthesized by combining the crown ether ring, azo and carboxyl group which exhibited the ratiometric mercury detection with the lower detection limit of 2.9×10^{-8} M. The ligand showed the two absorption signals at 250 and 354 nm in the absence of metallic induction while metallic addition caused the red shift in the absorption signal centered at 354 nm without considerable change in the high energy signal at 250 nm. The red shifts in the receptor absorption signal turn the colorimetric change in the solution

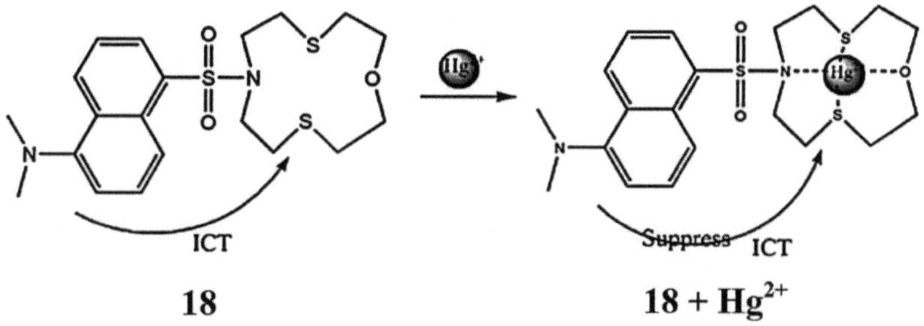

Fig. 11.17 The molecular structure of the receptor **18** and proposed complexation mechanism

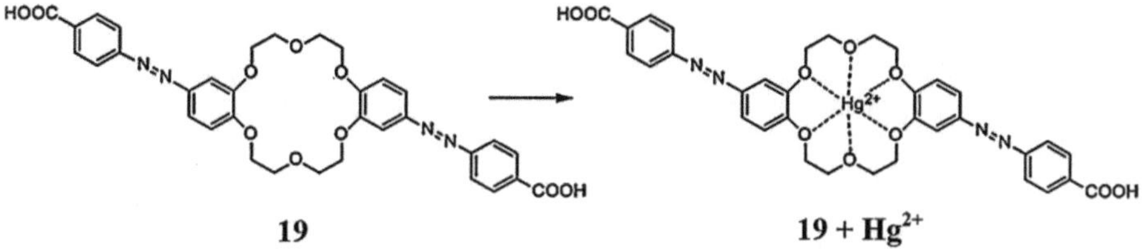

Fig. 11.18 The molecular structure of the receptor **19** and proposed complexation mechanism with the mercuric ions

which enables the naked eye mercuric ion detection. The 1:1 ligand-metal binding stoichiometry was found by the Job's curve.

11.2.5 Coumarine Based Fluorescent Sensors for Mercuric Ion Recognition

El-Shekheby et al. [33] reported the fluorescent chemosensor **20** (Fig. 11.19) based on coumarine dithioate derivative which can detect the mercuric ion following the chelation enhances fluorescence mechanism. The probe exhibited the off-on fluorescent response on mercuric ion induction which was used as a tool to sense the mercuric ion by this sensor material. The 1:1 ligand-metal binding stoichiometry was calculated by the Job's curve and the detection limit was calculated to be 0.029×10^{-5} M. The binding constant was determined from the ratio of the intercept and the slope of the Benesi-Hilderbrand plot of fluorescence emission intensities which comes to be $2.25 \times 10^4\ M^{-1}$.

The chemodosimetric sensors are the materials which can detect the analyte in the irreversible manners. The chemodosimeteric sensor attracted the immense interest of the scientific community due to their easily recognizable mechanism of action by the simple spectroscopic techniques. Moreover, these provide the low detection limit but having disadvantages of single use. One such type of coumarin based sensor **21** (Fig. 11.20) was reported by the Gao et al. [34] which

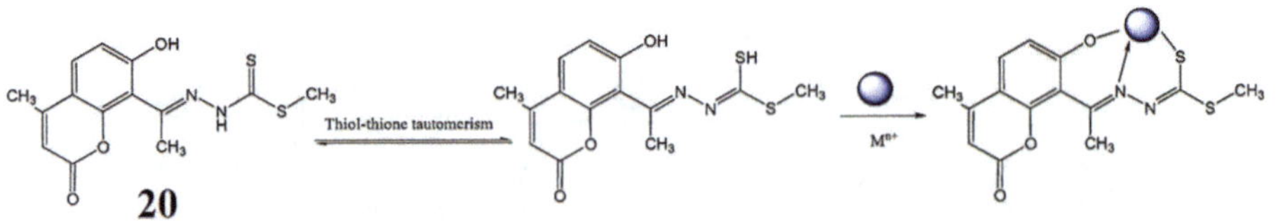

20

Fig. 11.19 The molecular structure of the ligand **20** and proposed ligand-metal complexation mechanism

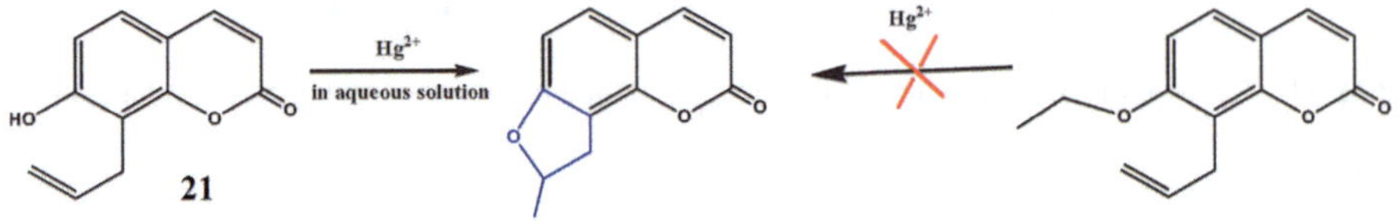

21

Fig. 11.20 The molecular structure of the sensor **21** and the proposed chemodosimetric reaction pathway

behave irreversibly on treatment with the mercuric ions. Meanwhile, the presented chemodosimeter exhibited the advantageous features of aqueous system utility. The detection limit of this turn-on fluorimetric as well as colorimetric sensor was calculated to be 5.12×10^{-7} M.

11.2.6 Squaraines Based Fluorescent Sensors for Mercuric Ion Recognition

Squaraines contains the extensively conjugated skeleton with the resonance stabilized zwitterionic structures exhibiting the strong absorption bands in the visible to near infrared regions. These absorption bands arise from donor–acceptor–donor type of charge transfer process. The sophisticated optical properties, redox activity, sensitivity toward the temperature fluctuation and solvent polarity make the squarains more favorable candidate for the design and development of fluorescent chemosensors for trace analyte detection.

Fan et al. [35] introduced the squaraines based fluorescent chemosensors **22** (Fig. 11.21) for trace mercuric ion detection. The addition of the mercuric ions caused the shifting in the absorption signal position and intensities alongside the colorimetric change in the reaction mixture which was useful for the naked eye metal detection. In case of fluorescence, the gradual addition of Hg^{2+} caused the progressive quenching of the fluorescence emission signal intensity. The formation of 1:1 ligand-metal complex was confirmed by the Job's curve and mass spectrometric analysis. The association constant and the detection limit of the sensor were determined to be 3.05×10^3 M^{-1} and 1.3×10^{-7} M, respectively. The ligand exhibited the reversibility on EDTA and thiol containing amino acid like cysteine and homocysteins addition, so, might be useful for multi-sensation as well.

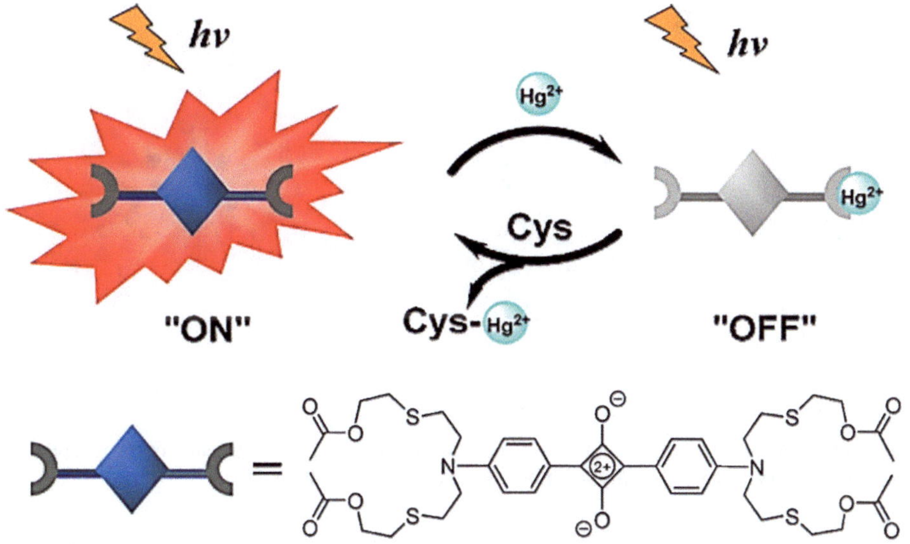

22

Fig. 11.21 The molecular structure of the receptor **22**, proposed complexation mechanism and reversibility

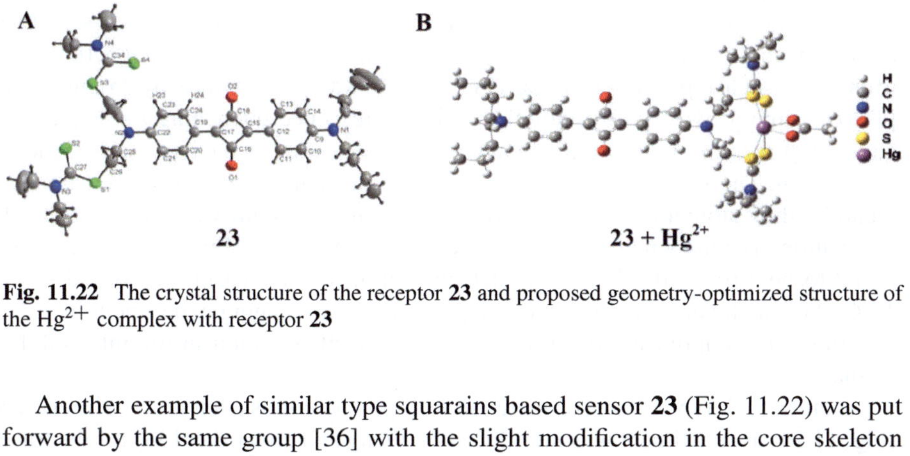

23 | **23 + Hg^{2+}**

Fig. 11.22 The crystal structure of the receptor **23** and proposed geometry-optimized structure of the Hg^{2+} complex with receptor **23**

Another example of similar type squarains based sensor **23** (Fig. 11.22) was put forward by the same group [36] with the slight modification in the core skeleton which improved the detection limit of the sensor in comparison to **22** (Fig. 11.21). This sensor material caused the colorimetric change in the reaction solution for naked eye metal detection. The addition of mercuric ion caused the successive increment in the fluorescence emission signal intensities seems useful for the ratiometric metal detection. The limit of detection for this sensor was calculated which gave the results of 4.5×10^{-8} M. The 1:1 ligand-metal binding stoichiometry was calculated by the Job's curve which harvested the maximum intensity on 0.5 mole fraction. The further mechanistic insight was got by the NMR titration experiment, mass spectrometric analysis and DFT calculations.

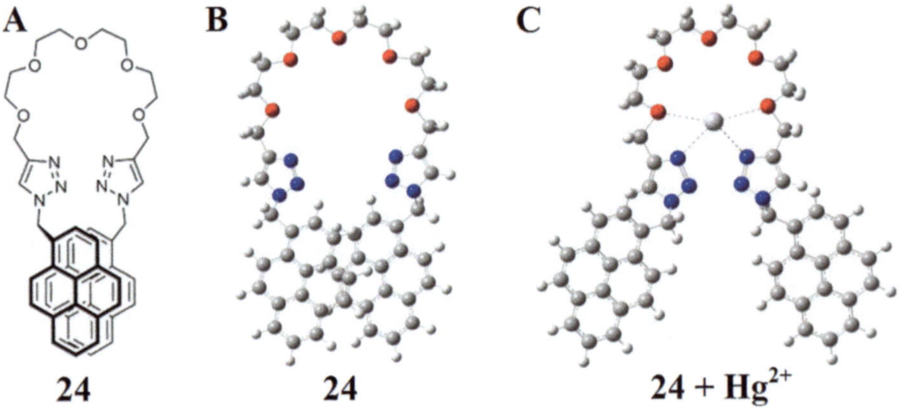

Fig. 11.23 Molecular structure of the receptor **24** (**a**), DFT optimized structure of the receptor (**b**) and its mercury complex (**c**); *Red* atom, O; *blue* atom, N; *gray* atom, Hg

11.2.7 Pyrene Based Fluorescent Sensors for Mercuric Ion Recognition

Another important scaffold frequently used in the construction of fluorescent chemosensors is the pyrene moiety which showed the efficient and sensitive monomer and excimer emission in the visible range of the electromagnetic spectrum. One such type of sensor **24** (Fig. 11.23) was put forward by Wang et al. [37] containing two triazole moieties. The fluorescence quenching on sequential addition of mercury ions into the ligand solution was used as the basis for the metal recognition. The Job's plot and mass spectrometric analysis suggest the 1:1 ligand-metal binding stoichiometry. The further mechanistic insight was got by the DFT calculation. The association constants (Ka) of chemosensor-Hg^{2+} complexes were found to be $1.68 \times 10^3\ M^{-1}$. The detection limit of the sensor was found to be 1.74 μM. The ligand was further employed for the bioimaging experiment in order to figure out its applicability to trace the metal contamination in the intracellular media.

Another such type of sensor material with pyrene core skeleton and diazene **25** (Fig. 11.24) as substituent was introduced by Weng and co-worker [38] for the trace recognition of mercuric ion. The mercury addition caused the appearance of two new absorption bands at 300 and 415 nm with the concomitant disappearance of the absorption signals at 279 and 350 nm, respectively. The 2:1 binding stoichiometry of the complexation species was determined by the Job's plot employing the ligand-mercury titration experimental graphs. The further mechanistic information was obtained by the mass spectrometric and NMR titration experiment. The titration experiment revealed two well defined isosbestic points at 322 and 369 nm which give information about the existence of two species in the mixture. The colorimetric change in the reaction mixture after metallic induction can be used as naked eye

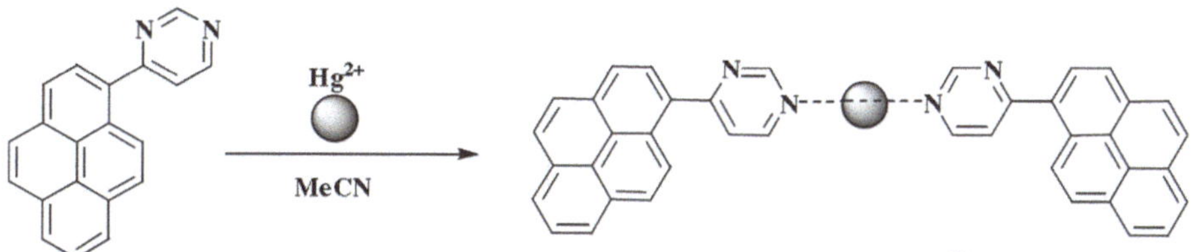

25 **25 + Hg^{2+}**

Fig. 11.24 The molecular structure of the receptor **25** and proposed complexation mechanism

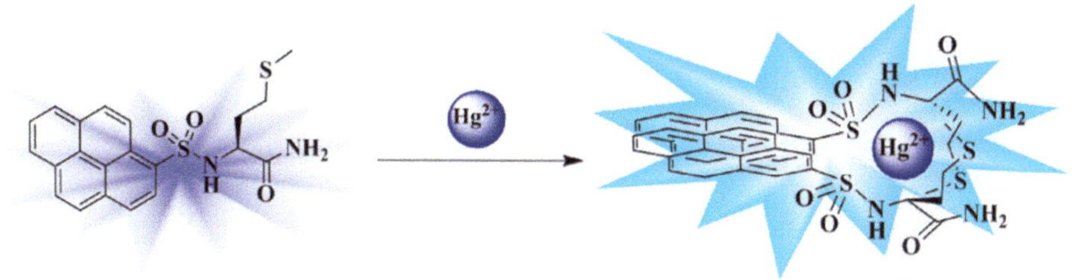

26 **26 + Hg^{2+}**

Fig. 11.25 The molecular structure of the receptor **26** and proposed complexation mechanism

detection. Similarly, in case of fluorescence experiment, the band at 440 nm decrease while the band centered at 545 nm increased gradually upon successive mercury addition. The estimated detection limit from the titration results was 4.69×10^{-6} M.

Yang et al. [39] reported the pyrene based fluorescent sensor for mercuric ion detection linked with amino acid as metal capturing site **26** (Fig. 11.25). The mercury addition caused the incremental appearance of excimer emission at 480 nm with the concomitant decrease in the monomer emission at 383 nm. The variation in the optical spectra as well as colorimetric change in the reaction solution in the presence of mercuric ions was used as a tool for the metal detection. The precisely low detection limit of 57.2 nM was calculated by the reported sensor. The 1:1 ligand-metal complexation mechanism was estimated by the Job's curve which was further supported by the mass spectrometric analysis and NMR titration experiment. The association constant was calculated as 1.68×10^3 M^{-1} which showed the good binding affinity between the metal and the receptor.

11.2.8 Imidazole Based Fluorescent Sensors for Mercuric Ion Recognition

Choi et al. [40] reported the 2-((E)-(3-(1H-imidazol-1-yl)propylimino)methyl)-5-(diethylamino)phenol based fluorescent chemosensor **27** (Fig. 11.26) for mercuric ion detection synthesized by the combination of diethylaminosalicylaldehyde and

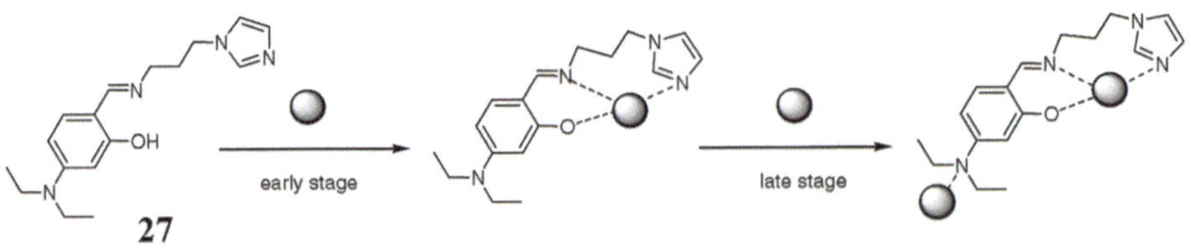

27

Fig. 11.26 The molecular structure of the receptor **27** and proposed two steps complexation mechanism

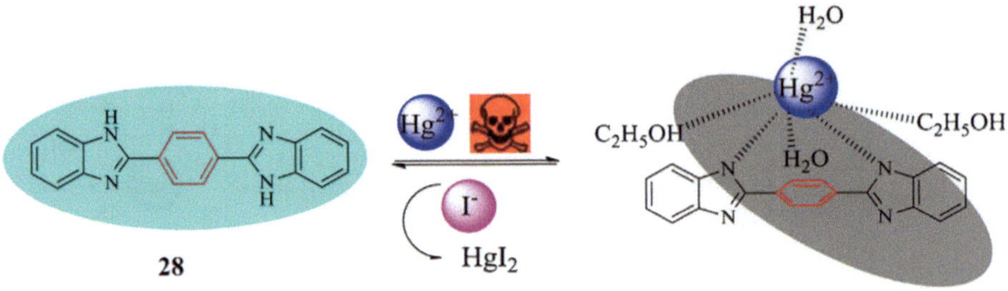

28

Fig. 11.27 The molecular structure of the receptor **28** and proposed complexation mechanism/reversibility

imidazole moieties. The sensor works on the basis of ligand to metal charge transfer process (LMCT). The mercury induction caused the two steps change in the absorption spectra. The first change include the reduction of absorption signal at 345 nm and increment in the absorption signal at 385 nm alongside the concomitant appearance of two isosbestic points at 298 and 357 nm, respectively. The distinctly different pattern was observed in the 2nd change with the appearance of absorption signal at 448 nm along with the new isosbestic points at 317 nm. On the basis of two steps UV-visible process, mass spectrometric analysis and NMR titration, the 1:1 and 1:2 ligand metal binding stoichiometry was estimated. The ligand also caused colorimetric change on metallic induction which might be useful for the naked eye analytical detection.

The 1, 4-bis (2-benzimidazolyl) benzene based fluorescent chemosensor **28** (Fig. 11.27) for mercuric ion detection was presented by Liu and co-workers [41] with the nano-molar detection ability. The receptor works following the fluorescent on-off mechanism. The induction of mercuric ion caused the selected fluorescent quenching and this quenched fluorescent was recovered on iodide ion addition. The 1:1 ligand metal binding stoichiometry was established by the methods of continuous variation and mass spectrometric analysis. The further mechanistic insight was got via NMR titration experiment. The ligand was further employed for the logic behavior and test strips. The 1 nM detection limit of the sensor was calculated by employing the optical results.

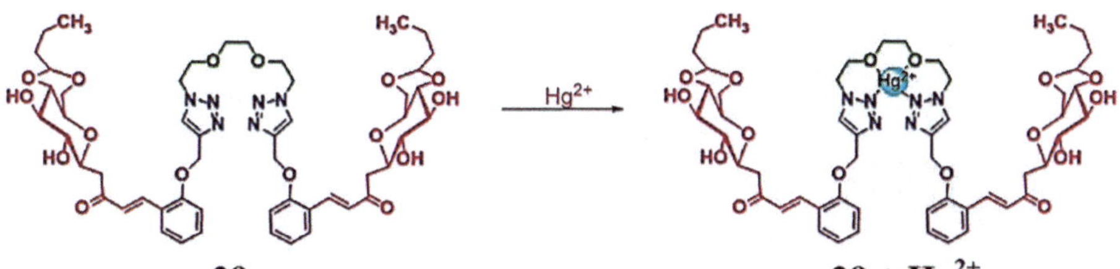

29 **29 + Hg^{2+}**

Fig. 11.28 The molecular structure of the receptor **29** and proposed mercury complexation mechanism

11.2.9 Triazole Based Fluorescent Sensors for Mercuric Ion Recognition

Hemamalini et al. [42] reported the ether-linked-bis-triazole derivative **29** (Fig. 11.28) for the fluorescent recognition of mercuric ion. The receptor exhibited the broad fluorescent emission signal at 460 nm when it was excited at 330 nm. The progressive addition of mercuric ion caused the enhancement in the fluorescence emission signal intensity alongside the significant hypsochromic shift in the position of spectrum which was used as the basis for this sensor material. Similarly, the molecule exhibited the two absorption bands at 284 and 330 nm while the mercury induction cause the downfall in the emission signal intensity at 284 nm with the continuous increment in the intensity of 2nd signal at 330 nm. This behavior was found selective for the mercuric ion. On the basis of optical results, the 1:1 ligand-metal complexation stoichiometry was determined which was further supported by the NMR titration experiment, mass spectrometric results and DFT calculation. However, the binding constant and the detection limit was not reported.

11.2.10 Anthracene Based Fluorescent Sensors for Mercuric Ion Recognition

Praveen et al. [43] reported the anthracene-oxyquinoline dyad **30** (Fig. 11.29) as fluorescent sensor for the trace mercuric ion detection. The absorption spectrum of the receptor exhibited the two electronic transitions at 200–400 nm and other one in the range of 245–280 nm. The mercury addition caused the decrease in the absorption band at 254 nm alongside the appearance of shoulder signal at 273 nm. Meanwhile, the well resolve smaller signals at 366 and 386 nm undergo the blue shift and one new signal appears at 425 nm. The absorption titration experiment caused the emergence of several well defined isosbestic points at 370, 360, 317 and 290 nm gives the indication about the formation of 2nd species in the reaction mixture. Similarly, the ligand exhibited the emission band at 480 nm along with the

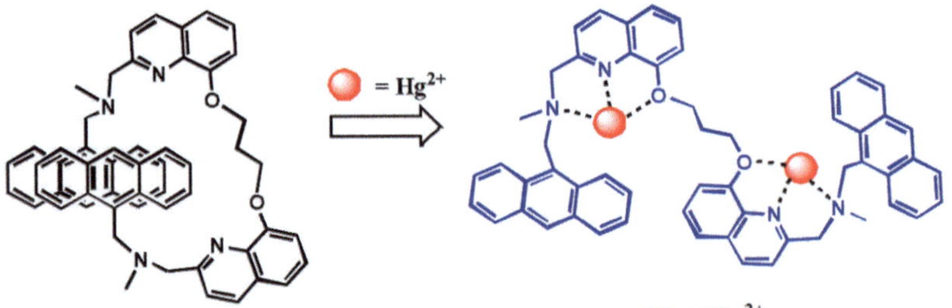

30

30 + Hg^{2+}

Fig. 11.29 The molecular structure of the receptor **30** and proposed mercury complexation mechanism

highly structured signal at 410 nm. The mercury addition gradually shifts the higher wavelength signal at 410 nm alongside the visual color change from bluish green to blue. The 1:1 binding stoichiometry of the complex was determined by using the optical titration results and the detection limit of 3.2×10^{-6} M was calculated by using the standard deviation of three blank reading.

11.2.11 Dansyl Based Fluorescent Sensors for Mercuric Ion Recognition

Lohani et al. [44] reported the dansyl fluorophore based on un-natural amino acid **31** (Fig. 11.30) synthesized by the solid phase protocol as fluorescence sensor for the recognition of mercuric ion. The mercury induction into the receptor solution caused the significant shifting and variation in the absorption and emission spectrum which was used as basis for these sensor materials. One major drawback of the organic fluorescent chemosensor is the water in-solubility which needs the organic solvent for its operation. However, the receptor **31** can work well in the pure aqueous media which is the appreciable point for this sensing system. The 1:1 ligand-metal binding stoichiometry was determined by this sensor by recording the optical spectrum on varying the mole fraction. The further mechanistic insight was got by the NMR titration experiment and mass spectrometric analysis. The ligand exhibited the dissociation constant and the detection limit of 1.2 pM and 9.3 nM, respectively. The ligand behaves reversibly on treatment with the EDTA solution which turns back the ligand-metal complex into the free ligand.

Tharmaraj et al. [45] reported the acyclic dansyl based bipodal fluorescent chemosensor **32** (Fig. 11.31) for the mercuric ion recognition. The sensor exhibited the on-off fluorescent response based on the twisted intramolecular charge transfer mechanism (TICT). The ligand showed the absorption signal at 342 nm which was blue shifted to 330 nm on addition of the mercuric ion. Similarly, the fluorescence emission signal at 496 nm was red shifted to 514 nm on mercuric ion addition

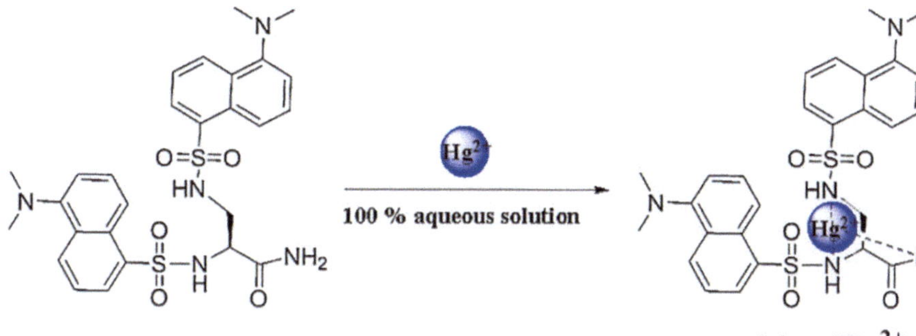

31 **31 + Hg^{2+}**

Fig. 11.30 The molecular structure of the ligand **31** and proposed complexation mode

Fig. 11.31 The molecular structure of the receptor **32** and proposed complexation mechanism

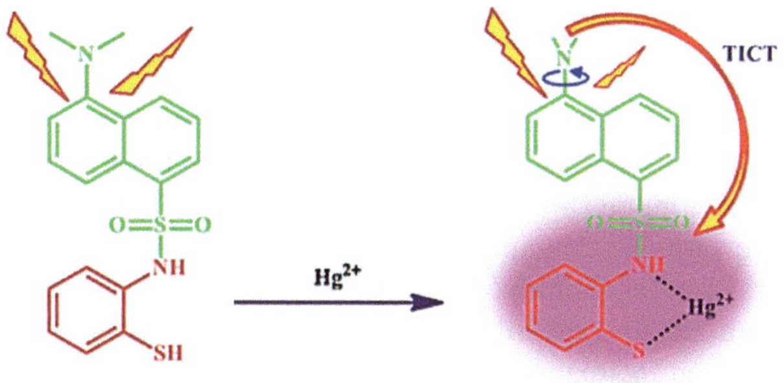

32 **32 + Hg^{2+}**

alongside the remarkable quenching. The 1:1 ligand-metal binding stoichiometry was estimated by the Job's curve which was further established by the NMR titration experiment, mass spectrometric analysis and the DFT calculation. The detection limit of the sensor was found to be 5.0×10^{-10} M. The binding constant for this complexation process was estimated to be 5.0×10^{4} M^{-1}.

11.2.12 Naphthalenedimide/Naphthalene/Naphthalimide-Based Fluorescent Sensors for Mercuric Ion Recognition

Naphthalimide derivative showed the excellent fluorescent properties with high fluorescence quantum yield. On the basis of interesting optical properties, these can be used as fluorophore unit in the design and manufacturing of the organic chemosensor. Li et al. [46] introduced the novel near-infrared "turn-on" fluorescent chemosensor **33** (Fig. 11.32) with selective detection of Hg^{2+} following the twisted intramolecular charge transfer (TICT) process. The fluorescence selectivity can be modulated by changing the solvent polarity. The ligand behaves selectively toward

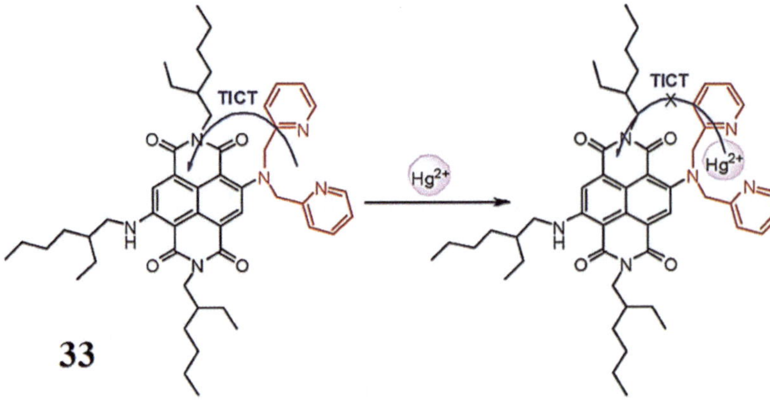

33

Fig. 11.32 The chemical structure of probe **33** and proposed complexation mechanism

mercuric ion induction in comparison to the variety of competing species over broad pH span. The 1:1 ligand-metal binding stoichiometry was estimated by the methods of continuous variation, ^{1}H NMR and ESI-MS spectral analysis. In order to trace the metallic contamination in the living cells and biomedia, the sensor material was further employed for the bioimaging experiment using HeLa cell lines under confocal fluorescent microscope. The appearance of bright fluorescent from the live cells on incubation with the probe and mercuric ion suggest the utility of receptor for the *in vivo* applications.

Wei et al. [47] introduced the fluorescent chemosensor **34** (Fig. 11.33) for the mercuric ion detection based on naphthalene as signaling unit and imine moiety as metal binding site. The receptor exhibited the weak fluorescent emission signal at 470 nm when it was excited at 305 nm in the absence of metallic ions while only the mercuric ion induction caused the rapid and ratiometric increment in the fluorescence emission signal intensity. The 1:1 ligand-metal binding stoichiometry was confirmed by the Job's plot, mass spectrometric analysis, NMR titration experiment and FT-IR spectral analysis. The binding constant of the metal complex was determined as $9.35 \times 10^4\ M^{-1}$. The ligand exhibited the detection limit values of 5.59×10^{-8} M. The ligand was further employed for the preparation of test strips on dipping with the sensor solution for the convenient diagnosis of metallic contamination in the solution.

Li et al. [48] designed and developed the naphthalimide based fluorescent sensor containing hydrophilic hexanoic acid group **35** (Fig. 11.34) for the mercuric ion detection. The ligand alone showed the fluorescence emission signal at 529 nm which is typical emission signal related to the naphthalimide. The gradual induction of mercuric ion caused the significant enhancement in the emission signal intensity at 529 nm, hence used for the ratiometric detection of analytes. Meanwhile the absorption signal at 396 nm underwent continuous decrease on progressive increase in mercuric ion concentration. The methods of continuous variation suggest the

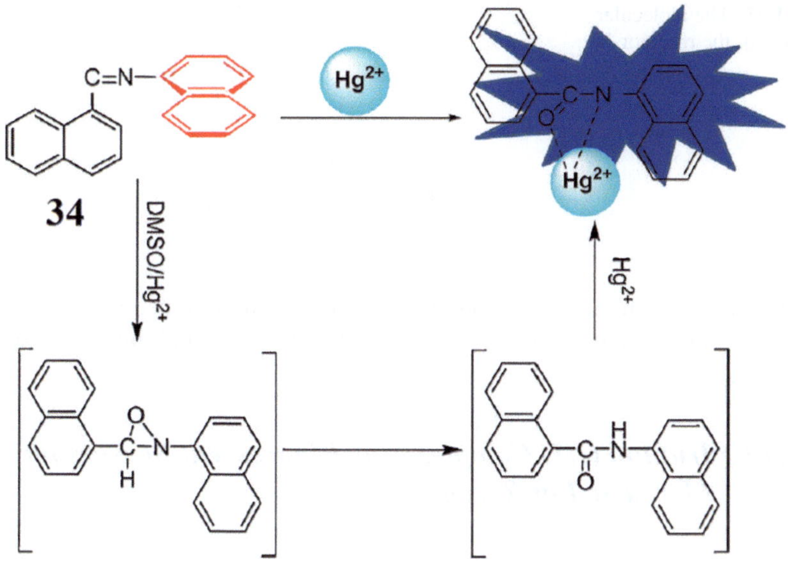

Fig. 11.33 The molecular structure of the receptor **34** and proposed mercury detection mechanism

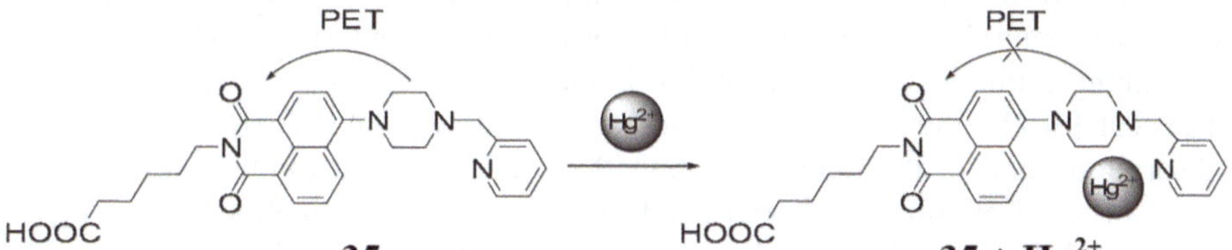

35 **35 + Hg^{2+}**

Fig. 11.34 The molecular structure of the receptor **35** and proposed mercury detection mechanism

1:1 ligand-metal binding stoichiometry. The detection limit of the sensor was determined to be 4.93×10^{-8} M.

11.2.13 Naphthyridine-Based Fluorescent Sensors for Mercuric Ion Recognition

Mahapatra et al. [49] reported the 1,8-naphthyridine based di-olefinic chemosensor **36** (Fig. 11.35) for the fluorescent detection of mercuric ion. The ligand exhibited the absorption signal centered at 441 nm which underwent the gradual decrease on successive increasing the mercuric ion alongside the concomitant appearance of a red shifted peak around 579 nm. The bathochromic shift in the absorption spectrum on mercuric ion addition caused the colorimetric change in the reaction mixture for the naked eye mercury detection. In case of fluorescence emission spectral

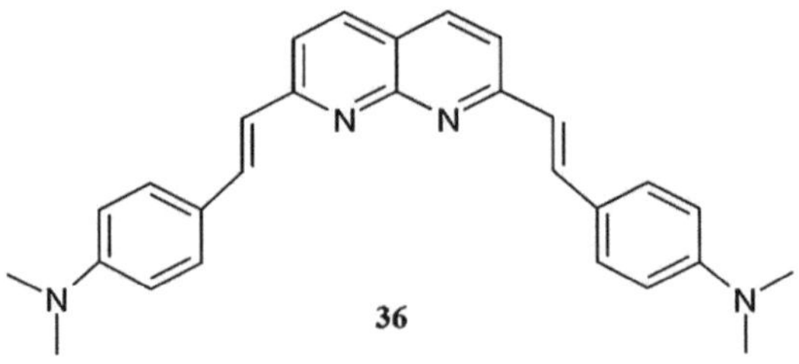

36

Fig. 11.35 The molecular structure of the receptor **36**

measurement, the mercuric ion addition caused the appearance of new emission band at 514 nm with the decrease in the intensity of initial emission peak at 608 nm.

11.2.14 Iridium (III) Complexes as Fluorescent Sensors for Mercuric Ion Recognition

The iridium (III) complexes are the highly efficient phosphorescent dyes due to their short excited state life time, higher photoluminescence efficiency and tunable optical properties from blue to near infrared region by bringing the structural variation. One such neutral iridium (III) complex based fluorescent sensor **37** (Fig. 11.36) containing 2-thiophen-2-yl-benzothiazole metal capturing site was reported by the Yan et al. [50] for the trace detection of mercuric ion. The ligand exhibited the strong absorption band in the range of 250–350 nm and weak absorption signal in the range of 350–550 nm. The shorter bands were assigned to spin allowed inter-ligand transfer while the higher wavelength peaks refers to the mixed singlet and triplet metal-to-ligand charge-transfer (MLCT). However the mercuric ion addition caused the disappearance of the signal at 480 nm and the gradual reduction in the intensity of signal at 405 nm. These spectral variations in the ligand solution caused the colorimetric change in the reaction mixture from orange to yellow. The appearance of well-defined isosbestic points in the absorption titration graph represents the clear cut formation of ligand-mercury complex. Meanwhile the ligand showed the intense emission band at 608 nm which undergo gradual decrease alongside the concomitant appearance of new emission signal at 585 nm on sequential addition of mercuric ion. The estimated detection limit from the titration results for the sensor toward mercuric ion was 2.14×10^{-6} M.

Another similar type of iridium complex **38** (Fig. 11.37) with the slight modification was introduced by the same research group [51] and explored their utility as fluorescent sensor for mercuric ion in the extracellular as well as intracellular media. The ligand exhibited the significant variation in the absorption and emission spectra before and after the treatment with the mercuric ion alongside the colorimetric change in the reaction solution which might be useful for the naked eye mercury detection in the environmental sample. The mechanistic insight was got from the NMR titration experiment, mass spectrometric analysis, DFT calculation and optical

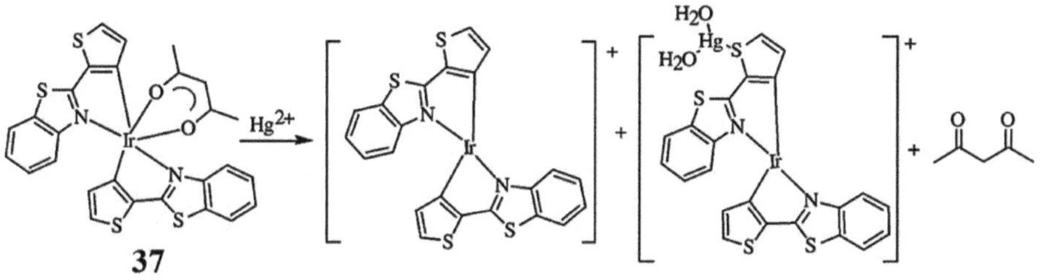

Fig. 11.36 The molecular structure of the receptor **37** and the possible mercury detection mechanism

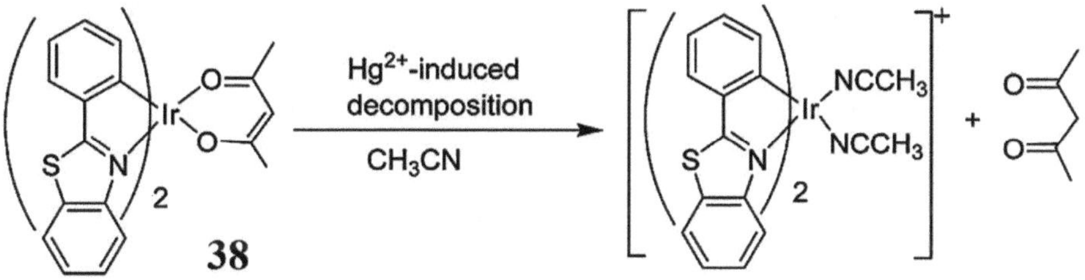

Fig. 11.37 The molecular structure of the receptor **38** and illustration of its mercury sensing mechanism.

studies which suggest the chemodosimetric operational strategy for the sensor. The cytotoxicity of the sensor was determined by the MTT assay utilizing the two different cell lines (KB and HeLa cells) which suggest the 10 μM safer dose for the receptor toward the live cells. Furthermore, the ratiometric mercury detection inside the intracellular media was done by the bioimaging experiment under confocal fluorescence microscope. The appearance of bright fluorescence from the live cells without any deformation represents the appreciable membrane permeability by the receptor molecule. Moreover, the brightness from the cells goes on increasing on incremental treatment with the mercuric ion represent the ratiometric mercury detection by the receptor in the intracellular media.

11.2.15 Polymeric Materials as Fluorescent Sensors for Mercuric Ion Recognition

Geng et al. [52] reported the monomeric 1-acrolyl-2-rhodamine 6G hydrazide material **39** (Fig. 11.38) for metallic detection. The simple rhodamine 6G are water insoluble while on conjunction with the monomeric unit become water soluble and incredibly enhance the applicability of the materials as fluorescence sensor can be used directly in the aqueous media. In the absence of mercuric ion, the sensor exhibited the maximal peak at 545 nm which was greatly enhanced on induction of

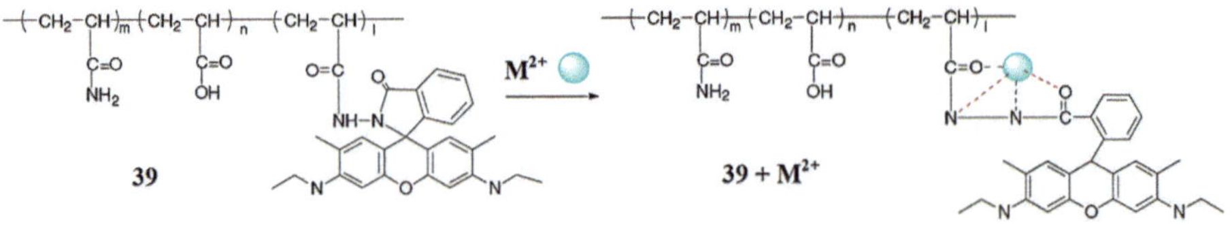

Fig. 11.38 The molecular structure of the receptor **39** and proposed mercury complexation mechanism

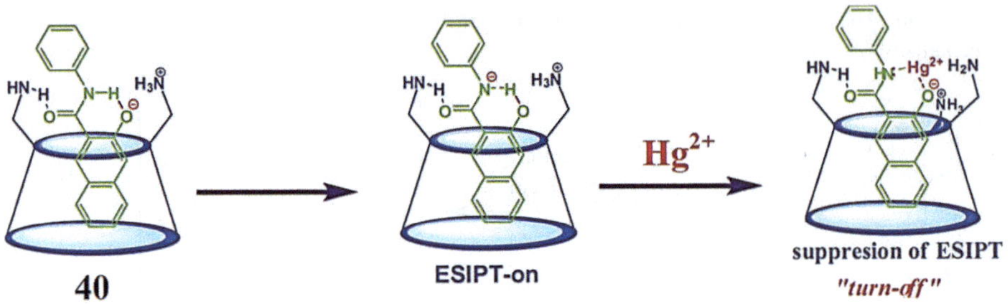

Fig. 11.39 The structure of the sensor **40** and proposed mercury complexation mechanism

mercuric ions. Furthermore there was a sudden color change in the reaction mixture on addition of mercuric ion and this colorimetric change might be useful for the naked eye detection of the environmental contamination. The minimum mercury detection ability of the sensor was found to be 36.1 μmol/L.

11.2.16 Cyclodextrin Based Fluorescent Sensors for Mercuric Ion Recognition

Kanagaraj et al. [53] reported the fluorescent "turn-off" colorimetric sensor **40** (Fig. 11.39) for the mercuric ion detection based on aminocyclodextrin: 3-hydroxy-N-phenyl-2-naphthamide complex. The mercuric ion addition into the ligand solution caused fluorescent quenching which was attributed due to suppression of excited-state intramolecular processes (ESIPT) on complexation with mercuric ion and ligand. The detection limit of the sensor was estimated to be 1 pM. The titration experiment exhibited the well-defined isosbestic points which represent the clear cut formation of the complexation species. Meanwhile the colorimetric change in the reaction mixture upon addition of mercuric ion can be used for the naked eye mercury detection. The binding constant of the receptor for mercury complexation was found to be 2166 M^{-1}.

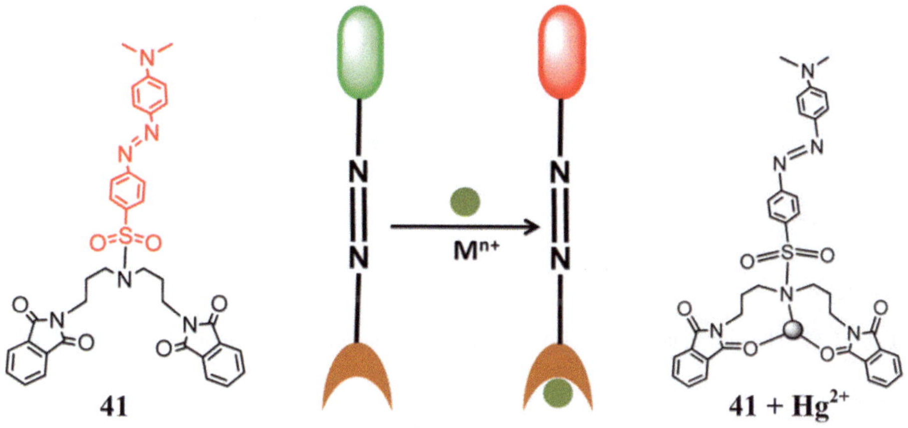

Fig. 11.40 The molecular structures of the receptor **41** and proposed complexation mechanism

11.2.17 Phthalic Anhydride Based Fluorescent Sensors for Mercuric Ion Recognition

Das et al. [54] reported the fluorescent chemosensor **41** (Fig. 11.40) for the selective detection of mercuric ion based on 4-((4 (dimethylamino)phenyl)diazenyl)benzene-1-sulfonyl chloride having dimethylaminophenyldiazenyl fragment as signaling unit and 2,2′-(3,3′-azanediylbis(propane-3,1-diyl))diisoindoline-1,3-dione as the receptor unit. The mercuric ion addition caused the appearance of new absorption band at 509 nm alongside the colorimetric change in the reaction solution for naked eye detection. Meanwhile the appearance of clear cut isosbestic points in the absorption titration experiment indicates the existence of two species in the reaction mixture. Similarly the mercury ion induced the increment in the intensity of fluorescent intensity on progressive addition of mercuric ion. The 1:1 ligand-metal binding stoichiometry was assessed by the titration experiment which was further confirmed by the mass spectrometric analysis and NMR titration experiment. The 10 ppm detection ability of this sensor has been reported by the proposed sensor. Furthermore, the test strips was prepared by dipping the strips in the ligand solution which showed the rapid and facile detection of mercuric ion.

11.2.18 Indole Based Fluorescent Sensors for Mercuric Ion Recognition

Wu et al. [55] reported the indole-based fluorescent chemosensor **42** (Fig. 11.41) for the selective recognition of mercuric ion. The backbone of the sensor was composed of indole as fluorophore and the triaminoethylene as the linker. The ligand exhibited

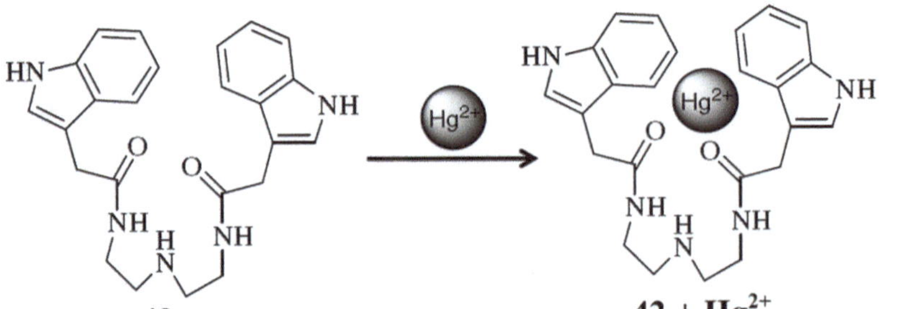

42 **42 + Hg^{2+}**

Fig. 11.41 The molecular structures of the receptor **42** and proposed complexation mechanism

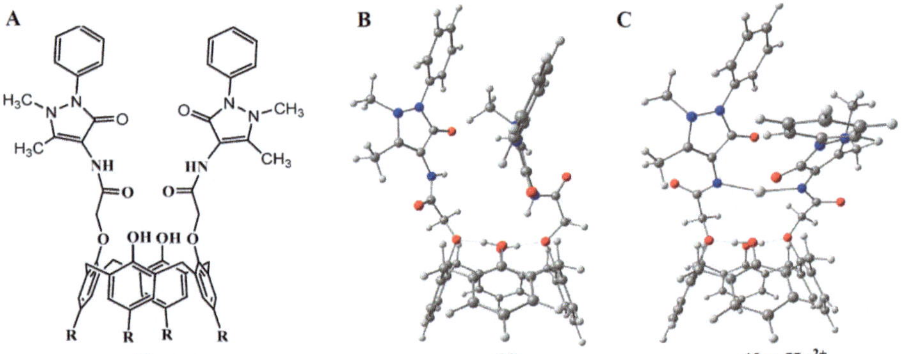

43 43 43 + Hg^{2+}

Fig. 11.42 The molecular structure of the receptor **43** (**a**), DFT optimized geometry of the receptor **43** (**b**) and receptor-mercury complex **43** (**c**)

the fluorescent emission signal at 358 nm which undergo strong quenching in the presence of mercuric ion. The maximum fluorescent change was observed at 0.1 M mole fraction represents the formation of 1:1 complex. The binding mechanism was further confirmed by the mass spectrometric analysis, NMR titration results and DFT calculation. The association constant of this sensor was calculated as 9.57×10^3 M^{-1}. The minimum mercury detection ability of this sensor was found to be 2.25×10^{-5} M.

11.2.19 Calix[4]arene Based Fluorescent Sensors for Mercuric Ion Recognition

Dessingou et al. [56] reported the fluorescent chemosensor **43** (Fig. 11.42) for the selective recognition of mercuric ion. The ligand showed the fluorescence emission signal at 400 nm when it was excited at 280 nm which undergo constant decrease on successive addition of mercuric ion. This titration graph was employed for the calculation of 1:1 ligand-metal binding stoichiometry by the methods of continuous variation. Meanwhile, in case of absorption spectral analysis, the mercuric ion addition caused the decrease in the signal at 280 nm with the concomitant increase in the peak intensity centered at 315 and 256 nm. The appearance of clear isosbestic point at 294 nm further support the ligand-metal complex formation. The further mechanistic studies were done by the mass spectrometric analysis and DFT calculation. The detection limit of the sensor was found to be 1.15 ± 0.1 ppm.

11.2.20 Chromenone Based Fluorescent Sensors for Mercuric Ion Recognition

Kim et al. [57] reported the 7-Amino-4-methyl-2*H*-chromen-2-one as fluorescent chemodosimeter **44** (Fig. 11.43) for the selective detection of mercuric ion based on desulfurization mechanism. The sensor exhibited the absorption maxima at 329 nm while the mercuric ion addition caused the bathochromic shift in the absorption spectrum. Similarly the mercury induction caused the increment in the fluorescent emission signal intensity at 403 nm while decrease in the peak intensity centered at 442 nm hence generating the well-defined isosbestic point which support the formation of the second species in the reaction mixture. The fluorescent change in the reaction mixture can also be observed under UV-lamp by illuminating at higher wavelength. The mechanistic insight for the mercury desulfurization process was got by the NMR titration experiment, mass spectrometric analysis and optical studies. The detection limit of the sensor was reported to be less than 3 nM.

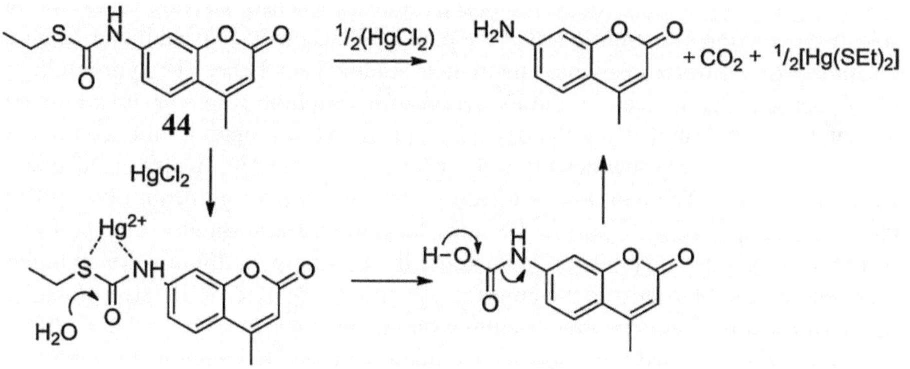

Fig. 11.43 The proposed mechanism of mercury selective recognition by the fluorescent indicator **44**

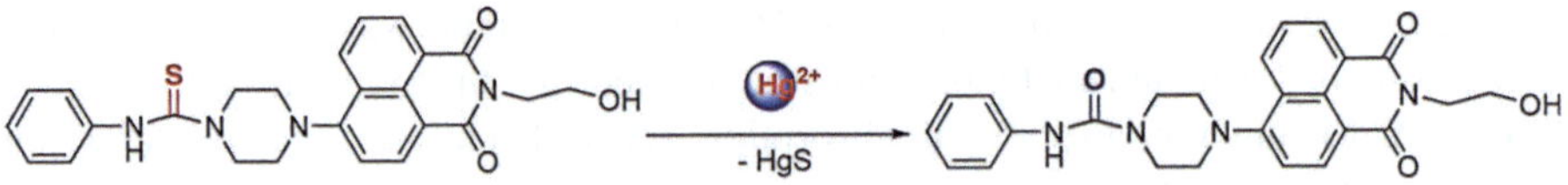

45

Fig. 11.44 The molecular structure of the receptor **45** and proposed desulfurization mechanism

11.2.21 1,8-Naphthalimides Based Fluorescent Sensors for Mercuric Ion Recognition

Another similar type of chemodosimeter **45** (Fig. 11.44) was introduced by the Yan et al. [58] for the detection of mercuric ion based on the desulfurization mechanism. The mercuric ion induction caused the increment in the fluorescence emission signal intensity at 520 nm when it was excited at 382 nm, accompanied by the transformation from weak fluorescence to the strong green fluorescence. The detection limit of this chemodosimetric sensor was found to be 1×10^{-7} M. The operational desulfurization mechanism for this sensor was found by the mass spectrometric analysis and NMR titration experiment. The naked eye mercury detection by the sensor can be performed by keeping the ligand solution under UV-lamp by illuminating at higher wavelength.

11.2.22 Lysine Based Water Soluble Fluorescent Sensors for Mercuric Ion Recognition

Chen et al. [59] introduced the lysine-based fluorescent sensor **46** (Fig. 11.45) for the selective detection of mercuric ion in the pure aqueous media. Normally the organic chemosensor are water insoluble due to their organic back bone. The hydrophilicity of this sensor was induced by the induction of hydrophilic lysine group which act as metal capturing unit while the densyl group act as fluorophore unit. The ligand exhibited the precisely low detection limit for the mercuric ion with the calculated values of 2.0 nM. The addition of mercuric ion caused the significant blue shift in the emission spectra from 540 to 489 nm alongside the colorimetric change in the reaction solution from yellow to green under the UV-lamp on illumination at higher wavelength. The blue shift in the emission spectra corresponds to the deprotonation of the sulfonamide moiety upon binding with the mercuric ion. On further addition, the mercuric ion caused the fluorescence quenching due to heavy metal ion effect along with another blue shift in the spectral position. The 1:1 ligand-metal binding stoichiometry was assessed by the methods of continuous variation and the binding constant of 3.0×10^{5} L mol^{-1} was determined by the Hildebrand-Benesi plots. The further mechanistic insight was got by the NMR titration experiment and mass spectrometric analysis. Because of appreciable optical properties of the receptor, it

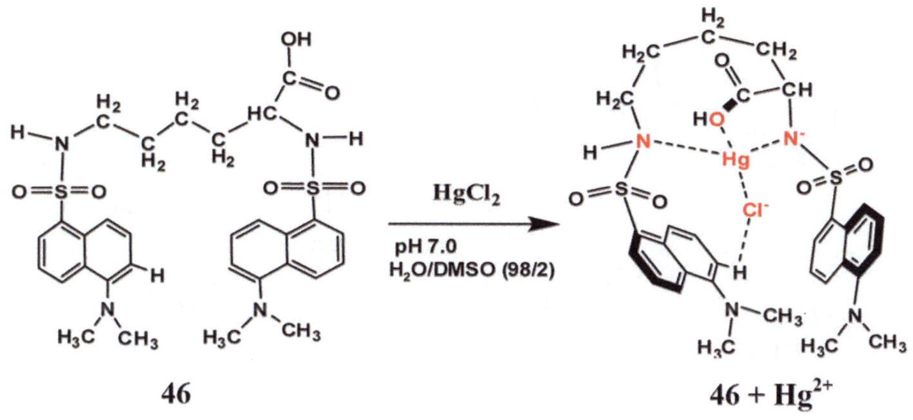

Fig. 11.45 The molecular structure of the receptor **46** and proposed ligand-mercury complexation mechanism

was further utilized for the bioimaging experiment for the monitoring of mercury ions in living cells. The confocal fluorescent microscopic images represent the bright fluorescent from the live cells without any deformation reflects the good membrane permeability of the receptor and low level of cellular toxicity.

11.2.23 Styrylindolium Dye Based Water Soluble Fluorescent Sensors for Mercuric Ion Recognition

Li et al. [60] reported the styrylindolium dye based fluorescent sensor **47** (Fig. 11.46) for the selective detection of mercuric ion. The ligand exhibited the mercury selective discrimination in the pure aqueous media. The sensor showed the absorption and emission signal at higher wavelength due to intramolecular charge transfer process between the nitrogen atoms in the sensor. However the metal binding engages the nitrogen lone pair in the ligation process which inhibits the intramolecular charge transfer process. Ultimately shift the absorption and emission signal toward the shorter wavelength alongside the colorimetric change in the reaction mixture. The ligand showed the characteristic intense absorption signal at 540 nm in water and 525 nm in the mixed water/ethanol solvent system. The 50 equivalent mercury additions drastically vanish the absorption band centered at 540/525 nm alongside the appearance of new absorption band at shorter wavelength *i.e.*, 406/400 nm which keep go on increasing by the incremental induction of mercuric ion. The mercury titration experiment revealed the occurrence of well-defined isosbestic points at 466/455 nm in the variable solvent clearly represent the existence of second species in the reaction mixture *i.e.*, ligand-mercury complex. The 1:1 ligand-metal binding stoichiometry was assessed by the

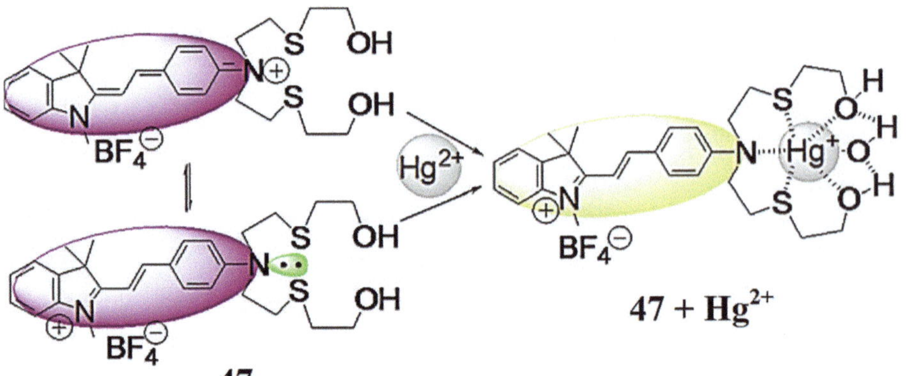

Fig. 11.46 The molecular structure of the receptor **47**, its equilibrium between the benzenoid form and the quinoid form and the proposed mercury complexation mechanism

methods of continuous variation employing the absorption titration experiment. The detection limit of the sensor was estimated to be 1.5×10^{-7} M in water. Similarly the chemosensor showed the intense emission signal at 590 nm when it was excited at 520 nm which was significantly quenched on mercuric ion addition. The fluorescence titration experimental results were utilized to find out the binding constant for the complex which comes to be $3.46 \times 10^{4}\ M^{-1}$ and $1.17 \times 10^{4}\ M^{-1}$ in the pure water and water/ethanol, respectively. These results represents that the sensor showed efficient binding in the pure aqueous media instead of mixed aqueous-organic media.

11.2.24 Phenothiazine Based Fluorescent Sensors for Mercuric Ion Recognition

Zhou et al. [61] reported the phenothiazine based fluorescent sensor **48** (Fig. 11.47) for the trace mercuric ion detection. The ligand showed the absorption maxima at 368 nm which was significantly up hilled on mercuric ion induction. Similarly the intense fluorescence emission signal was observed at 526 nm by the receptor in the absence of metal ions. The mercuric ion addition caused the gradual shorten in the intensity of the emission signal which end up on complete quenching. From the fluorescence titration experiment, the detection limit of the sensor was determined as 3.13×10^{-7} M. The Job's plot indicates the maximum intensity at 0.5 mole fraction which represents the 1:1 complexation stoichiometry. The association constant for the receptor was determined by the Benesi-Hilderbrand plot which comes to be $8.66 \times 10^{6}\ M^{-1}$. Moreover, the filter paper strips was prepared by using the sensor material in order to explore more practical utility of the receptor for mercury recognition in the solution phase. Furthermore, the ligand was used for the

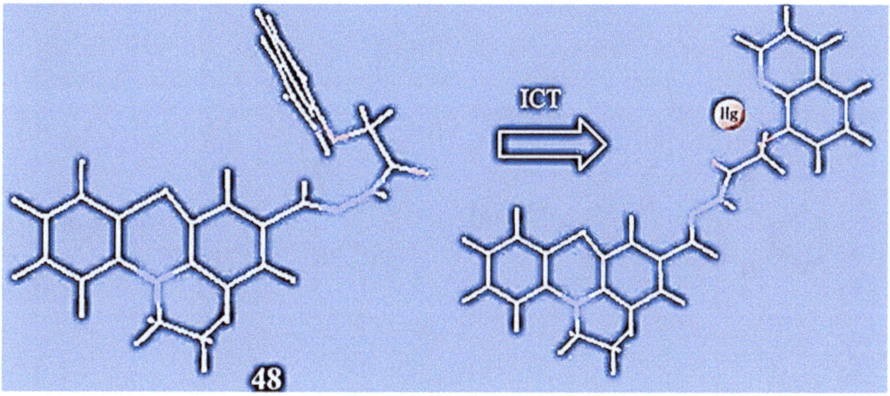

Fig. 11.47 The molecular structure of the receptor **48** and proposed ligand-mercury complexation mechanism

bioimaging experiment by using the human stomach lining cells by incubating with the mercuric ions and ligands and capturing the images under confocal fluorescence microscope. The appearance of bright fluorescence from the live cells represents the applicability of the receptor for the monitoring and understanding the mercury related biological process in the living cells.

11.2.25 Thiocarbonyl Quinacridone-Based Water Soluble Fluorescent Sensors for Mercuric Ion Recognition

Quinacridones are the class of the organic molecule which behaves as organic dyes possessing the strong fluorescent properties and are the exceptional pigments with the long term resistance toward the decomposition from the light, weather and heat. The Qu et al. [62] introduced the thiocarbonyl quinacridone-based fluorescent chemodosimetric sensor **49** (Fig. 11.48) for the ratiometric detection of mercuric ion. The fluorescence emission signal intensity of the receptor at 531 nm in chloroform and 555 nm in aqueous solution was enhanced in the presence of mercuric ion up to 50 fold. These spectral changes further bring the colorimetric change in the receptor solution from dark yellow green to bright yellow. On the basis of optical experimental results, mass spectrometric and NMR titration, the desulfurization mechanism by the receptor was proposed. The detection limit of the sensor was found to be 91 nM in the chloroform and 82 nM in the aqueous solution. Meanwhile, on introduction of the electron donating triphenyl amine (TPA) moiety, the selectivity and sensitivity of the sensor has been increased up to 25 fold in comparison to the sensor without TPA unit. The detection limit of the sensor containing the TPA moiety reached to 4.7 nM.

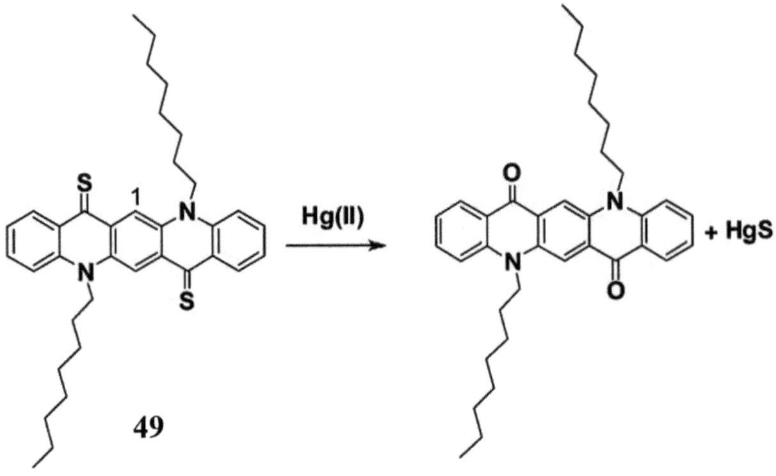

Fig. 11.48 The molecular structure of the receptor **49** and proposed desulfurization mechanism

11.2.26 Oxadiazole-Based Fluorescent Sensors for Mercuric Ion Recognition

Choi et al. [63] reported the oxadizole based fluorescent chemosensor **50** (Fig. 11.49) for the fluorescent detection of mercuric ion. The ligand exhibited the strong absorption band at 360 and 508 nm which was completely disappear in the presence of mercuric ion alongside the concomitant appearance of new absorption band at 430 nm which brings the colorimetric change in the reaction solution from red to yellow for the naked eye analytical detection. Similarly the fluorescence emission signal at 545 nm was greatly enhanced on treatment with the mercuric ion along with the appearance of minor blue shifted signal at 530 nm. The ligand behave reversibly on treatment with the EDTA which transform the ligand metal complex back to ligand and the liberated mercury ion made another more stable complex with the tetradentate ligand EDTA. The detection limit and the association constant of the receptor were determined by using the optical results which comes to be 1.5×10^{-6} M and 4.5×10^{4} M^{-1}, respectively.

Another oxadiazole based probe **51** (Fig. 11.50) was reported by the Shen et al. [64] for the recognition of mercuric ion. The probe exhibited the week fluorescence in the absence of mercuric ion while the mercury addition caused the significant enhancement in the fluorescence emission signal intensity at 572 nm when it was excited at 472 nm. Similarly, the compound showed the week absorption band at 342 nm and strong absorption signal at 472 nm in the absence of mercuric ion while mercuric ion addition caused the bathochromic shift in the both signal with the new position at 362 and 492 nm. The maximum intensity at 0.5 mole fraction in the methods of continuous variation establishes the 1:1 ligand-metal binding stoichiometry.

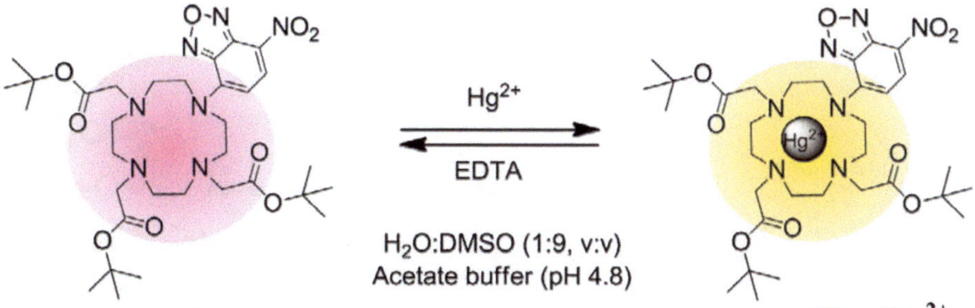

50 **50 + Hg2+**

Fig. 11.49 The molecular structure of the receptor **50** and proposed ligand-mercury complexation mechanism

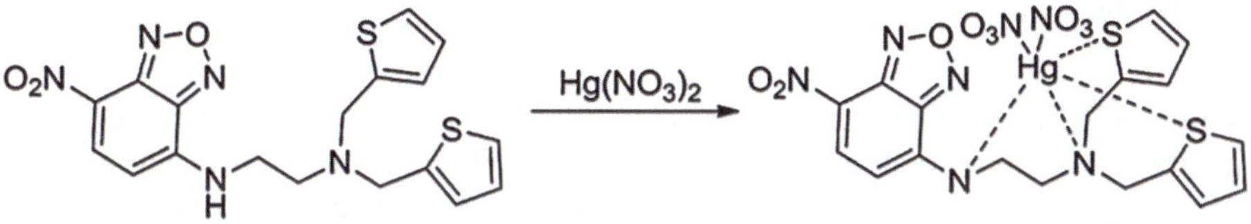

51 **51 + Hg2+**

Fig. 11.50 The molecular structure of the receptor **51** and proposed ligand-mercury complexation mechanism

The further mechanistic studies were done by using the mass spectrometric analysis, NMR titration experiment and DFT calculation. The association constant for this complexation process was estimated to be $7.14 \times 10^3\ M^{-1}$. The detection limit of 30 nM was determined by using the fluorescent titration results. The bioimaging experiment was performed by using the MCF-7 cells under confocal fluorescence microscope. The appearance of bright green fluorescent from the live cells after treatment with the ligand and mercury ion suggests the efficient viability of the receptor for the intracellular mercury detection.

11.2.27 Triphenylamine–Triazines-Based Fluorescent Sensors for Mercuric Ion Recognition

Triphenyl amine has been widely used in the opto and electroactive material due to its high electron donating and electron transporting ability. It has been extensively utilized for the construction of the hole transporting material for the perovskite solar cells. Meanwhile it possesses the good optical response which can open its utility in the construction of organic sensing materials for the analytical detection of ionic species. Zang et al. [65] reported the triphenylamine–triazines based fluorescent

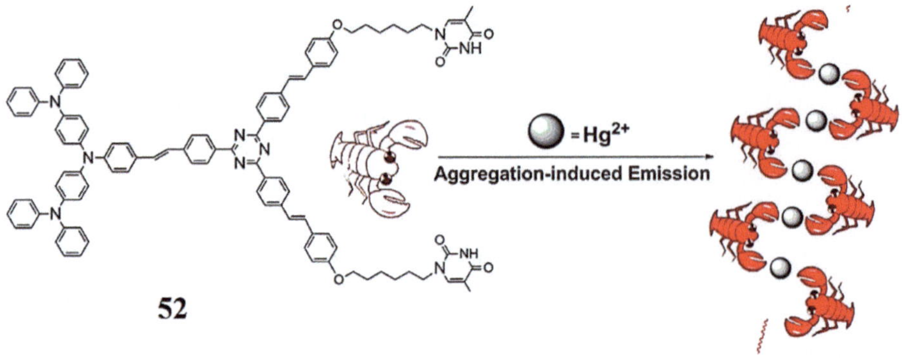

52

Fig. 11.51 The molecular structure of the receptor **52** and proposed ligand-mercury complexation mechanism

sensor **52** (Fig. 11.51) for selective recognition of the mercuric ion. The sensor exhibited the very weak emission when it was excited at 400 nm in the absence of metallic ion while the mercury addition triggered an intense emission signal at 610 nm whose intensity goes on increasing on the incremental addition of mercuric ion. The ligand can behave as naked eye mercury detection by keeping the solution under UV-lamp on illumination at longer wavelength. The detection limit of the sensor was found by the fluorescence titration graph which comes to be 6.6×10^{-8} M. Similarly, the mercury ion addition caused the decrease in the absorption signal intensity at 373 nm while increase in the absorption peak intensity in the range of 400–600 nm. The mechanistic insight was got by the NMR and mass spectrometric analysis.

11.2.28 Tetraphenylethene-Based Fluorescent Sensors for Mercuric Ion Recognition

Ozturk et al. [66] introduced the turn on fluorescent chemodosimeter **53** (Fig. 11.52) for the trace mercuric ion detection. The turn on behavior in the sensor was due to aggregation induced emission phenomena in the tetraphenylethene moiety. The mercury induction caused the elimination of polar moiety from the ligand leaving behind the formation of non-polar fluorescent aggregates. The mercury ion induction turn the non-emissive ligand solution into high fluorescent up to 410 fold enhancement in the emission intensity with the emission band centered at 474 nm. The incremental addition of the mercuric ion increased the formation of fluorescent aggregates species which is associated for the increment in the emission intensity. The 1:1 ligand metal binding stoichiometry was determined by the methods of continuous variation by employing the fluorescence titration graph. For practical applicability of the receptor to trace the mercury contamination intracellulary, the

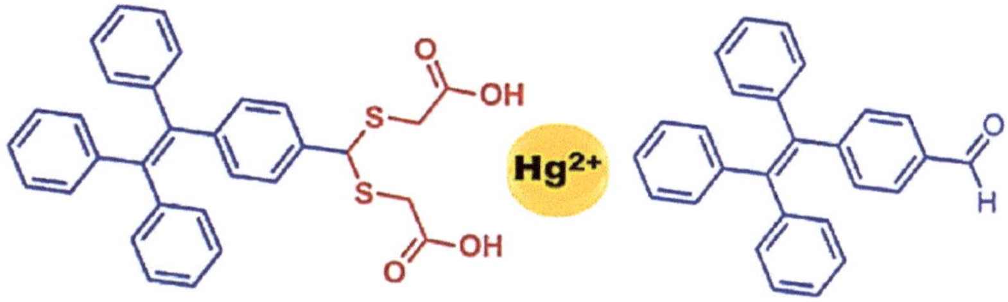

53

Fig. 11.52 The molecular structure of the receptor **53** and proposed chemodosimetric mercury detection mechanism

bioimaging experiment was conducted by using the HeLa cells and images were taken under confocal fluorescent microscope. The appearance of bright fluorescent from the live cells without any deformation inside them represents the efficient cells permeability and viability of the receptor for tracking the mercuric ion in the intracellular media. The detection limit of the sensor was determined to be 0.1 μM.

11.2.29 Peptidyl-Based Fluorescent Sensors for Mercuric Ion Recognition

Peptide based fluorescent sensor have advantageous features of their water solubility. On that basis they can be employed directly for the recognition of metallic contamination in the environmental sample directly without the consumption of any organic solvent. Moreover they have advantageous features of structural diversity, solubility in the aqueous media and compatibility toward the biological and environmental related analytes. Meanwhile the pyrene used as fluorophore in the present sensor also showed the interesting photophysical properties, high fluorescent quantum yield, dual fluorescent emission due to monomeric and excimeric unit as well as appreciable chemical stability. Thirupathiet al. [67] reported one such peptidyl based fluorescent sensor **54** (Fig. 11.53) by the incorporation of two pyrene moieties as fluorophores and tetrapeptide using sulfonamide group as the metal capturing site. In the aqueous buffer solution, the ligand showed the strong excimer emission at 494 nm originated in the molecule due to π–π-stacking in-between two pyrene units and weak monomer emission at 381 and 400 nm. However, the mercuric ion addition caused the significant decrease in the both monomer and excimer emission which was selective in comparison to the other environmentally relevant metallic species. The maximum intensity at 0.5 mole fraction in the Job's plot represents the 1:1 ligand-metal binding stoichiometry which was further confirmed by the other spectroscopic techniques. The detection

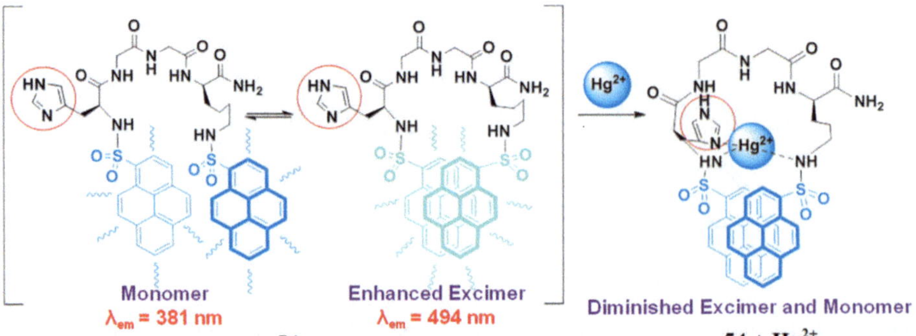

54

54 + Hg^{2+}

Fig. 11.53 The molecular structure of the receptor **54** and proposed mercury complexation mechanism

limit by the sensor for the mercuric ion was calculated to be 31.2 nM by using the standard deviation of three blank readings. The dissociation constant of the complex was determined to be 120 nM. The mercury ion induction also brings the colorimetric change in the reaction solution when observed under UV-lamp by illuminating at higher wavelength. The ligand was further employed for the bioimaging experiment utilizing the confocal fluorescent microscopic experiments and harvested the excellent output.

11.2.30 Semicarbazone-Based Fluorescent Sensors for Mercuric Ion Recognition

Qu et al. [68] introduces the (E)-1-((5-(4-nitrophenyl)furan-2-yl)methylene) semicarbazone based fluorescent sensor **55** (Fig. 11.54) for the selective detection of mercuric ion. The addition of the mercuric ion into the ligand solution brings the red shift in the absorption spectrum from 400 to 414 nm. In case of fluorescent spectral measurement, the sensor exhibited the strong emission band at 568 nm when it was excited at 417 nm and this emission was significantly quenched upon addition of the mercuric ion. The colorimetric change in the receptor solution from bright red to dim red can be observed under UV-lamp upon higher wavelength illumination. The mechanistic insight was got by the spectrophotometric results, NMR titration experiment, mass spectrometric, FT-IR and XRD analysis. The ligand can be used for the direct detection of the environmental metallic contamination by using its test strips. The detection limit of the sensor was found to be 2.08×10^{-9} M. The ligand behaves reversibly with the iodide ion addition which might be due to the formation of more stable mercuric iodide in comparison to the ligand-mercury complex.

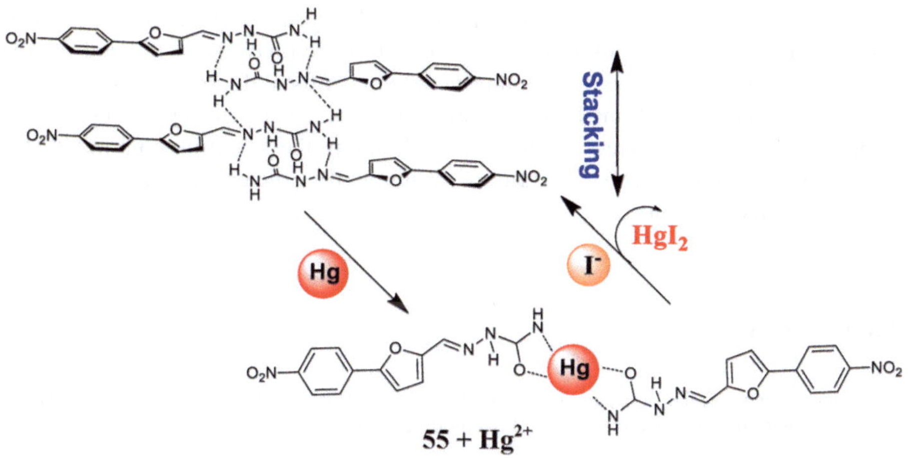

Fig. 11.54 The molecular structure of the receptor **55**, proposed mercury complexation mechanism and reversibility

11.3 Conclusion

There is an ever growing research on the design, synthesis and improvement in the sensitivity of the metal recognition materials in the environmental and biological system. The fluorescent detection methodologies are the non-invasive means for the detection and quantification of the multichannel analytes. Transition metal ions have particular interest due to involvement in the daily life of human being including automobiles, electric cables, storage batteries, electrode development, utensils/domestic usage, doors, windows, protective sheets, roofing, construction, heavy machineries developments, military usage, explosive, electronic products, tower construction, turbines, motors, laboratory usage, surgical instruments, dental filling and railway tracks. The heavy metal ions released from both the natural and industrial sources induct the adverse effects on human health and ecosystem. Among the several metal ions, the Hg^{2+} is extensively distributed in the atmosphere, air, water, soil, cosmetic products and anthropogenic materials which have ability to incorporate into the living system via skins, respiration, water and affected food products. It would consider as one of the major threatening hazardous pollutants and highly dangerous elements for the biological media due to its high accumulative and toxic character in the *in-vivo* and *in-vitro* systems. Furthermore, the mercury amalgam is frequently used in the dental filling which open the ways for the mercury induction into the stomach.

Herein, we summarized the several organic based fluorescent chemosensor/chemodosimeter for the trace mercuric ion detection. The complexation binding stoichiometry, detection limit and the binding/association/dissociation constant for the complexation process were specially highlighted during the representation of each sensing materials. Furthermore, the sensors which have been utilized for the detection of analytes inside the cellular media via confocal fluorescent microscopic means have been mentioned in the description of each sensing strategies. We hope that the present review might be useful among the scientific community for understanding the mechanism and future designing the efficient sensor for mercuric ion analytical detection.

References

1. Zhang J, Zhu Z, Di J, Long Y, Li W, Tu Y (2015) A sensitive sensor for trace Hg^{2+} determination based on ultrathin g-C 3 N4 modified glassy carbon electrode. Electrochim Acta 186:192–200
2. Zhan S, Xu H, Zhang D, Xia B, Zhan X, Wang L, Lv J, Zhou P (2015) Fluorescentdetectionof Hg^{2+} and Pb^{2+} using Gene Finder™ and an integratedfunctionalnucleicacid. Biosens Bioelectron 72:95–99
3. Xing H, Xu J, Zhu X, Duan X, Lu L, Wang W, Zhang Y, Yang Y (2016) Highly sensitive simultaneous determination of cadmium (II), lead (II), copper (II), and mercury (II) ions on N-doped graphene modified electrode. J Electroanal Chem 760:52–58
4. Mei C, Lin D, Fan C, Liu A, Wang S, Wang J (2016) Highly sensitive and selective electrochemical detection of Hg^{2+} through surface-initiated enzym atic polymerization. Biosens Bioelectron 80:105–110
5. Lin SM, Geng S, Li N, Li NB, Luo HQ (2016) D-penicillamine-templated copper nanoparticles via ascorbic acid reduction as a mercury ion sensor. Talanta 151:106–113
6. Petzoldt M, Eschenbaum C, Schwaebel ST, Broedner K, Lemmer U, Hamburger M, Bunz UHF (2015) A biphasic mercury-ion sensor: exploiting microfluidics to make simple anilines competitive ligands. Chem Eur J 21:14297–14300
7. Wen G, Wen X, Choi MMF, Shuang S (2015) Photoelectrochemical sensor for detecting Hg^{2+} based on exciton trapping. Sensors Actuators B 221:1449–1454
8. Li J, Lu L, Kang T, Cheng S (2016) Intense charge transfer surface based on graphene and thymine–Hg (II)–thymine base pairs for detection of Hg^{2+}. Biosens Bioelectron 77:740–745
9. Li Z, Miao X, Xing K, Peng X, Zhu A, Ling L (2016) Ultrasensitive electrochemical sensor for Hg^{2+} by using hybridization chain reaction coupled with Ag @ Aucore–shell nanoparticles. Biosens Bioelectron 80:339–343
10. Khodaveisi J, Shabani AMH, Dadfarnia S, Moghadam MR, Hormozi-Nezhad MR (2016) Development of a novel method for determination of mercury based on its inhibitory effect on horseradish peroxidase activity followed by monitoring the surface plasmon resonance peak of gold nanoparticles. Spectrochim Acta A Mol Biomol Spectrosc 153:709–713
11. Jiqu H, Qixia Y (2015) A core–shell structured inorganic–organic hybrid nanocompositefor Hg(II) sensing and removal. Spectrochim Acta A Mol Biomol Spectrosc 149:487–493
12. Huang YL, Gao ZF, Jia J, Luo HQ, Li NB (2016) A label-free electrochemical sensor for detection of mercury(II) ionsbased on the direct growth of guanine nanowire. J Hazard Mater 308:173–178
13. Wang Y-W, Tang S, Yang H-H, Song H (2016) A novel colorimetric assay for rapid detection of cysteine and Hg^{2+} based on gold clusters. Talanta 146:71–74
14. Zhang X, Huang XJ, Zhu ZJ (2013) A reversible Hg(II)-selective fluorescent chemosensor-based on a thioether linked bis-rhodamine. RSC Adv 3:24891–24895
15. Kaewtong C, Niamsa N, Wanno B, Morakot N, Pulpoka B, Tuntulani T (2014) Optical chemosensors for Hg^{2+} from terthiophene appended rhodamine derivatives: FRET based molecular and in situ hybrid gold nanoparticle sensors. New J Chem 38:3831–3839

16. Chen J, Li Y, Zhong W, Wang H, Zhang P, Jiang J (2016) A highly selective fluorescent and colorimetric chemosensor for Hg^{2+} based on a new rhodamine derivative. doi: 10.1039/c5ay03281d
17. Shen L, Wu Y, Ma W (2015) Rhodamine functionalized magnetic core–shell nanocomposite: an emission "Off–On" sensing system for mercury ion detection and extraction. Spectrochim Acta A Mol Biomol Spectrosc 138:348–356
18. Chen Y, Mu S (2014) Core–shell structured Fe_3O_4 nanoparticles functionalized with rhodamine derived probe for the detection, adsorption and removal of Hg(II): a sensing system with 'warning' signal. Sensors Actuators B 192:275–282
19. So HS, Rao BA, Hwang J, Yesudas K, Son Y (2014) Synthesis of novel squaraine–bis(rhodamine-6G): a fluorescent chemosensor for the selective detection of Hg^{2+}. Sensors Actuators B 202:779–787
20. Luo J, Jiang S, Qin S, Wu H, Wang Y, Jiang J, Liu X (2011) Highly sensitive and selective turn-on fluorescent chemosensor for Hg^{2+} in pure water based on a rhodamine containing water-soluble copolymer. Sensors Actuators B 160:1191–1197
21. Cruz-Guzman MD, Aguilar-Aguilar A, Hernandez-Adame L, Bañuelos-Frias A, Medellín-Rodríguez FJ, Palestino G (2014) A turn-on fluorescent solid-sensor for Hg(II) detection. Nanoscale Res Lett 9:431–440
22. Huang W, Wu D, Wu G, Wang Z (2012) Dual functional rhodamine-immobilized silica toward sensing and extracting mercury ions in natural water samples. Dalton Trans 41:2620–2625
23. Madhu S, Sharma DK, Basu SK, Jadhav S, Chowdhury A, Ravikanth M (2013) Sensing Hg(II) in vitro and in vivo using a benzimidazole substituted BODIPY. Inorg Chem 52:11136–11145
24. Zhao C, Li K, Xie N, Zhao M, Peng S (2014) Deprotonation and chelation synergically triggered near infrared fluorescence for selective detection of Hg(II). J Photochem Photobiol A Chem 290:72–76
25. Zhang T, She G, Qi X, Mu L (2013) A BODIPY-based sensor for Hg^{2+} in living cells. Tetrahedron 69:7102–7106
26. Cheng H, Qian Y (2015) Intramolecular fluorescence resonance energy transfer in a novelPDI–BODIPY dendritic structure: synthesis, Hg^{2+} sensor and livingcell imaging. Sensors Actuators B 219:57–64
27. Canturk C, Uçuncu M, Emrullahoglu M (2015) A BODIPY-based fluorescent probe for the differential recognition of Hg(II) and Au(III) ions. RSC Adv 5:30522–30525
28. Culzoni MJ, Peña AM, Machuca A, Goicoechea HC, Brasca R, Babiano R (2013) Photoinduced electron transfer fluorometric Hg(II) chemosensor based on a BODIPY armed with a tetrapod receptor. Talanta 117:288–296
29. Liu J, Yua M, Wang XC, Zhang Z (2012) A highly selective colorimetric sensor for Hg^{2+} based on nitrophenyl-aminothiourea. Spectrochim Acta A 93:245–249
30. Vengaian KM, Britto CD, Sivaraman G, Sekar K, Singaravadivel S (2015) Phenothiazine based sensor for naked-eye detection and bioimaging of Hg(II) and F^- ions. RSC Adv 5:94903–94908
31. Dai H, Liu F, Gao Q, Tianhua F, Kou X (2011) A highly selective fluorescent sensor for mercury ion (II) based on azathia-crown ether possessing a dansyl moiety. Luminescence 26:523–530
32. Yan Z, Hu L, Nie L, Lv H (2011) Preparation of 4,4-bis-(carboxyl phenylazo)-dibenzo-18-crown-6 dye and its application on ratiometric colorimetric recognition to Hg^{2+}. Spectrochim Acta A 79:661–665
33. El-Shekheby HA, Mangood AH, Hamza SM, Al-Kady AS, Ebeid EM (2014) A highly efficient and selective turn on fluorescent sensor for Hg^{2+}, Ag^+ and Ag nanoparticles based on a coumarin dithioate derivative. Luminescence 29:158–167
34. Gao B, Gong WT, Zhang QL, Ye JW, Ning GL (2012) A selective "turn-on" fluorescent sensor for Hg^{2+} based on "reactive" 7-hydroxycoumarin compound. Sensors Actuators B 162:391–395
35. Fan J, Chen C, Lin Q, Fu N (2012) A fluorescent probe for the dual-channel detection of Hg^{2+}/Ag^+ and its Hg^{2+}-based complex for detection of mercapto biomolecules with a tunable measuring range. Sensors Actuators B 173:874–881

36. Chen C, Dong H, Chen Y, Guo L, Wang Z, Sun JJ, Fu N (2011) Dual-mode unsymmetrical squaraine-based sensor for selective detection of Hg^{2+} in aqueous media. Org Biomol Chem 9:8195–8201
37. Wang HF, Wu SP (2013) Highly selective fluorescent sensors for mercury(II) ions and their applications in living cell imaging. Tetrahedron 69:1965–1969
38. Weng J, Mei Q, Ling Q, Fan Q, Huang W (2012) A new colorimetric and fluorescent ratiometric sensor for Hg^{2+} based on 4-pyren-1-yl-pyrimidine. Tetrahedron 68:3129–3134
39. Yang MH, Thirupathi P, Lee KH (2011) Selective and sensitive ratiometric detection of Hg(II) ions using a simple amino acid based sensor. Org Lett 13:5028–5031
40. Choi YW, You GR, Lee MM, Kim J, Jung KD, Kim C (2014) Highly selective recognition of mercury ions through the "naked-eye". Inorg Chem Commun 46:43–46
41. Liu J, Lin Q, Zhang YM, Wei TB (2014) A reversible and highly selective fluorescent probe for monitoring Hg^{2+} and iodide in aqueous solution. Sensors Actuators B 196:619–623
42. Hemamalini A, Mudedl SK, Subramanian V, Das TM (2015) Design, synthesis and metal sensing studies of ether-linked bis-triazole derivatives. New J Chem 39:3777–3784
43. Praveen L, Babu J, Reddy MLP, Varma RL (2012) Unfolding with mercury: anthracene-oxyquinoline dyad as a fluorescent indicator for Hg(II). Tetrahedron Lett 53:3951–3954
44. Lohani CR, Neupane LN, Kim JM, Lee KH (2012) Selectively and sensitively monitoring Hg^{2+} in aqueous buffer solutions with fluorescent sensors based on unnatural amino acids. Sensors Actuators B 161:1088–1096
45. Tharmaraj V, Pitchumani K (2012) An acyclic, dansyl based colorimetric and fluorescent chemosensor for Hg(II) via twisted intramolecular charge transfer (TICT). Anal Chim Acta 751:171–175
46. Li Q, Peng M, Li H, Zhong C, Zhang L, Cheng X, Peng X, Wang Q, Qin J, Li Z (2012) A new "Turn-on" naphthalenedimide-based chemosensor for mercury ions with high selectivity: successful utilization of the mechanism of twisted intramolecular charge transfer, near-IR fluorescence, and cell images. Org Lett 14:2094–2097
47. Wei TB, Gao GY, Qu WJ, Shi BB, Lin Q, Yao H, Zhang YM (2014) Selective fluorescent sensor for mercury(II) ion based on an easy toprepare double naphthalene Schiff base. Sensors Actuators B 199:142–147
48. Li CY, Xu F, Li YF, Zhou K, Zhou Y (2012) A fluorescent chemosensor for Hg^{2+} based on naphthalimide derivative by fluorescence enhancement in aqueous solution. Anal Chim Acta 717:122–126
49. Mahapatra AK, Hazra G, Das NK, Sahoo P, Goswami S, Fun HK (2011) A highly sensitive and selective ratiometric fluorescent probe based on conjugated donor–acceptor–donor constitution of 1,8-naphthyridine for Hg^{2+}. J Photochem Photobiol A Chem 222:47–51
50. Yan F, Mei Q, Wang L, Tong B, Xu Z, Weng J, Wang L, Huang W (2012) A highly selective and ratiometric sensor for Hg^{2+} based on a phosphorescent iridium (III) complex. Inorg Chem Commun 22:178–181
51. Wu Y, Jing H, Dong Z, Zhao Q, Wu H, Li F (2011) Ratiometric phosphorescence imaging of Hg(II) in living cells based on a neutral Iridium(III) complex. Inorg Chem 50:7412–7420
52. Geng TM, Wu DY, Huang W, Huang RY, Wu GH (2014) Fluorogenic detection of Hg^{2+}, Cd^{2+}, Fe^{2+}, Pb^{2+} cations in aqueous media by means of an acrylamide-acrylic acid copolymer chemosensor with pendant rhodamine-based dyes. J Polym Res 21:354–362
53. Kanagaraj K, Bavanidevi K, Chow TJ, Pitchumani K (2014) Selective "turn-off" fluorescent sensing of mercury ions using aminocyclodextrin: 3-hydroxy-Nphenyl-2-naphthamide complex in aqueous Solution. RSC Adv 4:11714–11722
54. Das P, Ghosh A, Bhatt H, Das A (2012) A highly selective and dual responsive test paper sensor of Hg^{2+}/Cr^{3+} for naked eye detection in neutral water. RSC Adv 2:3714–3721
55. Wu HH, Sun YL, Wan CF, Yang ST, Chen SJ, Hu CH, Wu AT (2012) Highly selective and sensitive fluorescent chemosensor for Hg^{2+} in aqueous solution. Tetrahedron Lett 53:1169–1172

56. Dessingou J, Tabbasum K, Mitra A, Hinge VK, Rao CP (2012) Lower rim 1,3-Di{4-antipyrine}amide conjugate of Calix[4]arene: synthesis, characterization, and selective recognition of Hg^{2+} and its sensitivity toward pyrimidine bases. J Org Chem 77:1406–1413
57. Kim SB, Cho DG (2012) HgII-selective fluorescent indicator: one-step synthesis. Eur J Org Chem 2012:2495–2498
58. Yan Y, Zhang Y, Xu H (2013) A selective "turn-on" fluorescent probe for recognition of mercury(II) ions in aqueous solution based on a desulfurization reaction. Chem Plus Chem 78:1–14. doi:10.1002/cplu.201300150
59. Chen L, Yang L, Li H, Gao Y, Deng D, Wu Y, Ma LJ (2011) Tridentate lysine-based fluorescent sensor for Hg(II) in aqueous solution. Inorg Chem 50:10028–10032
60. Li Y, Wei F, Lu Y, He S, Zhao L, Zeng X (2013) Novel mercury sensor based on water soluble styrylindolium dye. Dyes Pigments 96:424–429
61. Zhou C, Song Y, Li Y (2014) A novel highly sensitive and selective fluorescent sensor for imaging mercury(II) in living cells. RSC Adv 4:33614–33618
62. Qu Y, Yang J, Hua J, Zou L (2012) Thiocarbonyl quinacridone-based "turn on" fluorescent chemodosimeters for highly sensitive and selective detection of Hg(II). Sensors Actuators B 161:661–668
63. Choi J, Lee SK, Bae J, Chang SK (2014) Colorimetric signaling of Hg^{2+} ions by a nitrobenzoxadiazole-appended cyclen-triester. Tetrahedron Lett 55:5294–5297
64. Shen Y, Zhang Y, Zhang X, Zhang C, Zhang L, Jin J, Li H, Yao S (2014) A new turn-on fluorescent sensor based on NBD for highly selective detection of Hg^{2+} in aqueous media and imaging in live cells. Anal Methods 6:4797–4802
65. Zhang H, Qu Y, Gao Y, Hua J, Li J, Li B (2013) A red fluorescent 'turn-on' chemosensor for Hg^{2+} based on triphenylamine–triazines derivatives with aggregation-induced emission characteristics. Tetrahedron Lett 54:909–912
66. Ozturk S, Atilgan S (2014) A tetraphenylethene based polarity dependent turn-on fluorescence strategy for selective and sensitive detection of Hg^{2+} in aqueous medium and in living cells. Tetrahedron Lett 55:70–73
67. Thirupathi P, Lee KH (2013) A new peptidyl fluorescent chemosensors for the selective detection of mercury ions based on tetrapeptide. Bioorg Med Chem 21:7964–7970
68. Qu WJ, Gao GY, Shi BB, Wei TB, Zhang YM, Lin Q, Yao H (2014) A highly selective and sensitive fluorescent chemosensor for mercuryions based on the mechanism of supramolecular self-assembly. Sensors Actuators B 204:368–374

Chapter 12
Fluorescence Studies for Biomolecular Structure and Dynamics

Vishvanath Tiwari and Vandana Solanki

Keywords Tryptophan fluorescence • DNA Triplex • Fluorescence spectroscopy • Fluorescence resonance energy transfer • Biomolecular structure and dynamics

12.1 Introduction

Fluorescence is the emitted light which have wavelength higher than the absorbed light. Emitted light can reveal some properties of biological molecules which are not obtained from simple absorption spectroscopy. Molecules which have delocalized π-electrons in their valence shells like polyaromatic or heterocyclic hydrocarbons generally absorb a photon, which is followed by the emission of a longer wavelength. The compound which gives fluorescence is called "fluorophore". The process of fluorescence is illustrated by the simple electronic-state diagram, Jablonski diagram in Fig. 12.1.

At room temperature, most of the molecules occupy the lowest vibrational level of the ground electronic state, and on absorption of light they move to their excited states. The simplified diagram above shows absorption by molecules to produce either the first, S1, or second S2, excited state (Fig. 12.1). Excitation can cause the molecule to reach any of the vibrational sub-levels associated with each electronic state. Since the energy is absorbed as discrete quanta, this should result in a series of distinct absorption bands. However, the simple diagram above (Fig. 12.1) neglects the rotational levels associated with each vibrational level and which normally increase the number of possible absorption bands to such an extent that it becomes impossible to resolve individual transitions. Therefore, most compounds have broad absorption spectra except for those where rotational levels are restricted (for example, planar, aromatic compounds). Having absorbed energy and reached one of the higher vibrational levels of an excited state (known as excitation lifetime mostly 1–10 nanoseconds(10^{-9}), the molecule rapidly loses its excess of vibrational

V. Tiwari (✉) • V. Solanki
Department of Biochemistry, Central University of Rajasthan, Ajmer 305817, Rajasthan, India
e-mail: vishvanath7@yahoo.co.in

C.D. Geddes (ed.), *Reviews in Fluorescence 2016*, Reviews in Fluorescence,
DOI 10.1007/978-3-319-48260-6_12

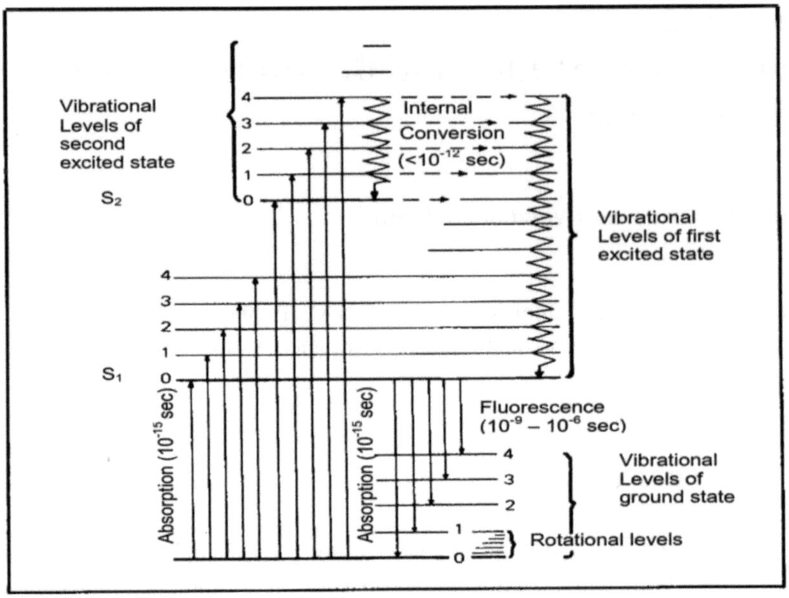

Fig. 12.1 Jablonski diagram: Transitions giving rise to absorption and fluorescence emission spectra [1]

energy by collision and falls to the lowest vibrational level of the excited state. In addition, almost all molecules occupying an electronic state higher than the second undergo internal conversion and pass from the lowest vibrational level of the upper state to a higher vibrational level of a lower excited state which has the same energy. From there the molecules again lose energy until the lowest vibrational level of the first excited state is reached. From this level, the molecule can return to any of the vibrational levels of the ground state, emitting its energy in the form of fluorescence. If this process takes place for all the molecules that absorbed light, then the quantum efficiency of the solution will be a maximum, unity. If, however, any other route is followed, the quantum efficiency will be less than one and may even be almost zero. The transition from the lowest vibrational level in the ground electronic state to the lowest vibrational level in the first excited state, the 0-0 transition, is common to both the absorption and emission phenomena, whereas all other absorption transitions require more energy than any transition in the fluorescence emission. We can therefore expect the emission spectrum to overlap the absorption spectrum at the wavelength corresponding to the 0-0 transition and the rest of the emission spectrum to be of lower energy, or longer wavelength (Fig. 12.2).

In practice, the 0-0 transitions in the absorption and emission spectra rarely coincide exactly, the difference representing a small loss of energy by interaction of the absorbing molecule with surrounding solvent molecules. The absorption of energy to produce the first excited state does not perturb the shape of the molecule greatly and this means that the distribution of vibrational levels is very similar in

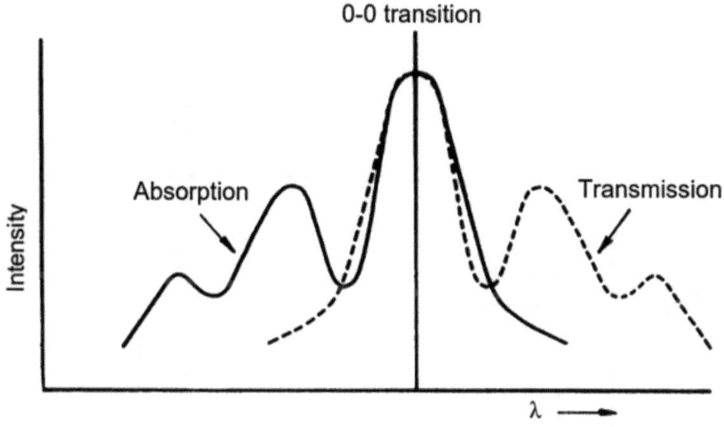

Fig. 12.2 Idealised absorption and emission spectra [1]

both the ground and first excited states. The energy differences between the bands in the emission spectrum will be similar to those in the absorption spectrum and frequently the emission spectrum will be approximate to a mirror image of the absorption spectrum. Since the emission of fluorescence always takes place from the lowest vibrational level of the first excited state, the shape of the emission spectrum is always the same, despite changing the wavelength of exciting light. A plot of emission against wavelength for any given excitation wavelength is known as the emission spectrum. If the wavelength of the exciting light is changed and the emission from the sample plotted against the wavelength of exciting light, the result is known as the excitation spectrum. Furthermore, if the intensity of exciting light is kept constant as its wavelength is changed, the plot of emission against exciting wavelength is known as the corrected excitation spectrum. The quantum efficiency of most complex molecules is independent of the wavelength of exciting light and the emission will be directly related to the molecular extinction coefficient of the compound; in other words, the corrected excitation spectrum of a substance will be the same as its absorption spectrum.

Phosphorescence is the production of excited states by promotion of an electron into a higher orbital, with the direction of spin of the electron being preserved. Since most molecules have an even number of electrons and these are normally arranged in pairs of opposite spin, the promotion of an electron does not disturb this parity. However, it is possible for the spin of the promoted electron to be reversed so that it is no longer paired and the molecule has two independent electrons of the same spin in different orbitals [1]. Quantum theory predicts that such a molecule can exist in three forms of very slightly differing, but normally indistinguishable energy, and the molecule is said to exist in a triplet state. The indirect process of conversion from the excited state produced by absorption of energy, the singlet state, to a triplet state, is known as intersystem crossing (Fig. 12.3) and can occur in many substances when the lowest vibrational level of the excited singlet state, S1, has the same energy

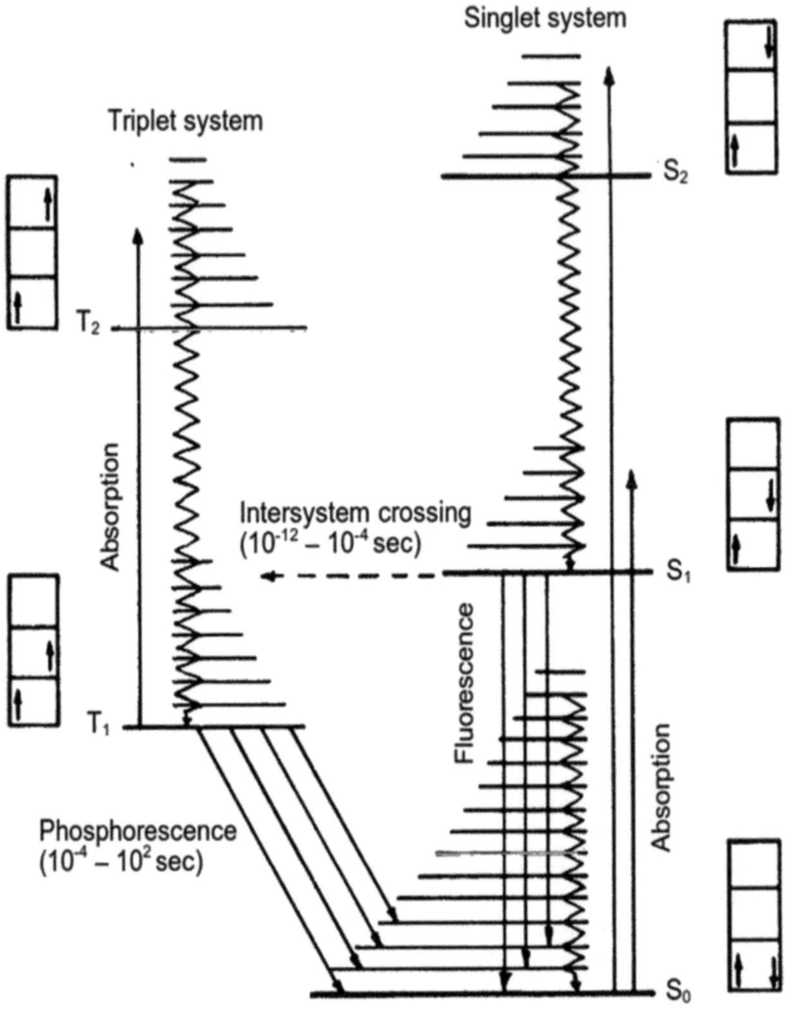

Fig. 12.3 Transition from the excited singlet state (S1) to the triplet state (intersystem crossing) [1]

level as an upper vibrational level of the triplet state. Direct transition from the ground state, usually a singlet state, for a molecule with an even number of electrons, to an excited triplet state is theoretically forbidden, which means that the reverse transition from triplet to ground state will be difficult. Thus, while the transition from an excited singlet state, for example, S1, to the ground state with the emission of fluorescence can take place easily and within 10^{-9}–10^{-6} s, the transition from an excited triplet state to the ground state with the emission of phosphorescence requires at least 10^{-4} s and may take as long as 10^{2} s. This delay was once used as the characterisation of phosphorescence, but a more precise definition requires

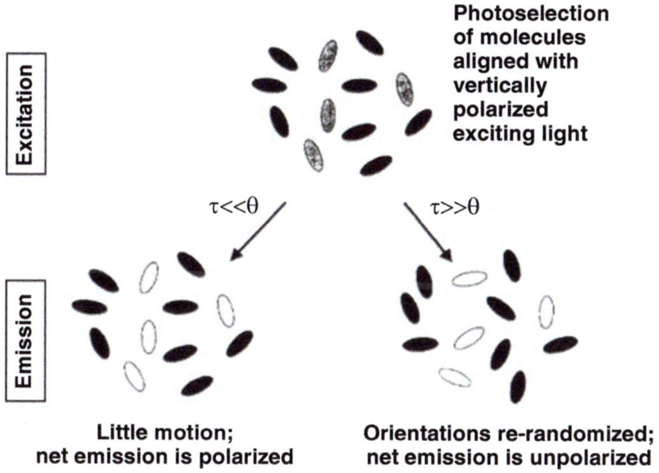

Fig. 12.4 Relationship between Fluorescence lifetime T and rotational correlation time θ

Fig. 12.5 Measurement of fluorescence polarization

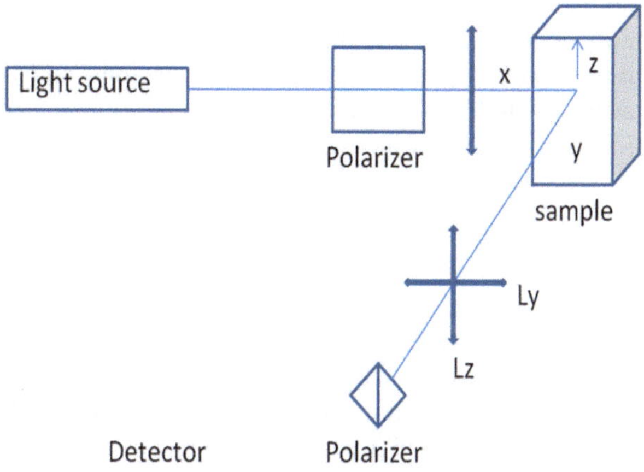

that phosphorescence be derived from transitions directly from the triplet state to the ground state (Figs. 12.4 and 12.5).

According to the Beer-Lambert Law, fraction of a parallel beam of light absorbed by a sample is independent of the intensity of the incident beam and is related to the concentration of the absorbing species by the familiar Beer-Lambert Law:-

$$\frac{Io}{I} = e^{-EcL} \tag{12.1}$$

Normally written as:

$$log_{10}\frac{Io}{I} = EcL$$

where,

I = intensity of transmitted light, Io = intensity of incident light, E = molecular extinction coefficient, c = concentration in gm moles/litre, L = pathlength of sample

The quantity $\log_{10}$ Io/I is known as the absorbance or optical density of the sample. Absorption spectra are simply a plot of absorbance against wavelength for any sample. The intensity of fluorescence emission of weakly absorbing, dilute samples is related to the molecular extinction coefficient by the relation:

$$FI = Io\ (2.303\ \mathrm{EcL})\ \Phi_F \tag{12.2}$$

where,

FI = intensity of fluorescence emission, Φ_F = fluorescence yield.

Hence, the intensity of fluorescence emission is directly proportional to the intensity of the incident radiation. These two equations explain the difference in sensitivity between absorption and fluorimetric procedures. Sensitivity in absorption measurements is limited by the ability of the instrument to discriminate between the two nearly equal signals due to I and Io, and the detection limit for even the most favourable cases rarely exceeds 10. Conversely, fluorimetric instruments are limited only by the intensity of exciting light and the ability to detect low light levels, so under ideal conditions concentrations of 10^{-12} moles can be measured. Equation 12.2 only holds for dilute samples and it is possible to calculate the concentration level (Cmax) above which non-linearity is obtained from:

$$Cmax = \frac{0.05}{\mathrm{EL}} \tag{12.3}$$

It is most important to establish the concentration at which a plot of fluorescence emission against concentration becomes non-linear either by calculation or by measurement of a suitable range of standards. Direct interpolation of concentrations outside this limit will give inaccurate results.

Fluorescence Polarization (FP) and Anisotropy: Fluorescence polarization (FP) measurements are based on the assessment of the rotational motion of species. FP can be considered a competition between the molecular motion and the lifetime of fluorophores in solution. If linear polarized light is used to excite an ensemble of fluorophores; only those fluorophores that are aligned with the plane of polarization will be excited. There are two scenarios for the emission, provided the fluorescence lifetime of the excited fluorescent probe is much longer than the rotational correlation time θ of the molecule it is bound to ($\tau >> \theta$) (θ is a parameter that describes how fast a molecule tumbles in solution), the molecules will randomize in solution during the process of emission, and, as a result, the emitted light of the fluorescent probe will be depolarized (Fig. 12.6). If the fluorescence lifetime of the fluorophore is much shorter than the rotational correlation time θ ($\tau << \theta$) the excited molecules will stay aligned during the process of emission and as a result the emission will be polarized (Fig. 12.6) [2]. The fluorescence polarization

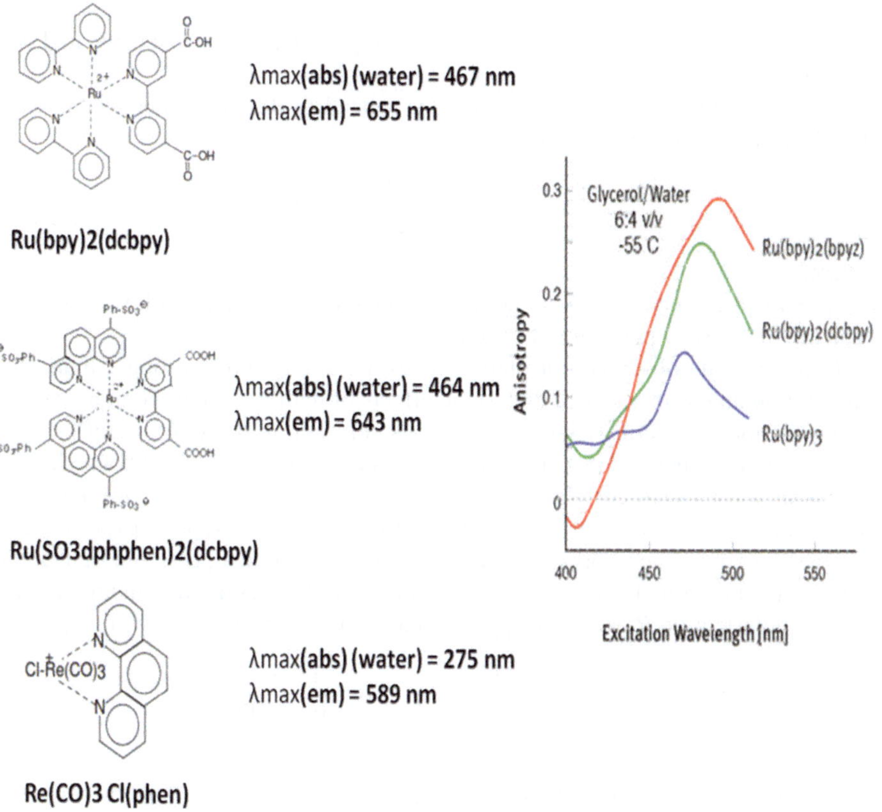

Fig. 12.6 Excitation polarization spectra of Ru metal ligand complexes in solution at −55 °C

(*P*) of a labeled macromolecule depends on the fluorescence lifetime (τ) and the rotational correlation time (θ) is the polarization observed in the absence of rotational diffusion. The effect of the molecular weight on the polarization values can be seen from an alternative form of the above Eq. (12.3):
where

$$\left(\frac{1}{P} - \frac{1}{3}\right) = \left(\frac{1}{Po} + \frac{1}{3}\right) \left(1 + \frac{\tau}{\theta}\right) \tag{12.4}$$

where,

Po = polarization observed in the absence of rotational diffusion, P = fluorescence polarization

The effect of the molecular weight on the polarization values can be seen from an alternative form of the above equation:

$$\left(\frac{1}{P} - \frac{1}{3}\right) = \left(\frac{1}{Po} + \frac{1}{3}\right) \left(1 + \frac{kT}{\eta V}\tau\right) \tag{12.5}$$

where,

k= Boltzman constant, T = absolute temperature, η = viscosity and V = molecular volume [3]

The molecular volume of the protein is related to the molecular weight (MW) and the rotational correlation time as given by

$$\theta = \frac{\eta V}{kT} = \eta \frac{MW}{RT} (\overline{\upsilon} + h) \quad (12.6)$$

where,

R = ideal gas constant, $\overline{\upsilon}$ = specific volume of the protein and h = hydration, typically 0.2 g H_2O per gram of protein

Generally, the observed correlation times are about twofold longer than calculated for an anhydrous sphere due to the effects of hydration and the non- spherical shapes of most proteins. Hence, in aqueous solution at 20 °C (η =LcP) one can expect a protein such as HSA (human serum albumin) (MW ~ 65,000, with h = 1.9) to display a rotational correlation time θ~50 *ns*. The measurement of fluorescence polarization is relatively straight-forward (Fig. 12.5). In a typical experiment the sample containing the fluorescent probe is excited with linear polarized light and the vertical and horizontal components of the intensity of the emitted light are measured and the polarization *P* or anisotropy *r* are calculated using the following equations:

$$P = \frac{\mathrm{Iv} - \mathrm{Ih}}{\mathrm{Iv} + \mathrm{Ih}} \qquad r = \frac{\mathrm{Iv} - \mathrm{Ih}}{\mathrm{Iv} + 2\mathrm{Ih}} \quad (12.7)$$

where,

Iv = intensity parallel to the excitation plane, Ih = emission perpendicular to the excitation plane

They are interchangeable quantities and only differ in their normalization. Polarization, *P* ranges from 0.33 to +0.5 while the range for anisotropy, *r* is −0.25 to +0.4. When the polarization is known, one can calculate the anisotropy, and vice-versa, using the following relations:

$$P = \frac{3r}{2 + r} \qquad r = \frac{2P}{3 - P} \quad (12.8)$$

Figure 12.5 shows the schematics of a typical instrument for the measurement of the polarization. The light from the source goes through a polarizer before impinging on the sample.

The fluorescence is collected at 90° through a movable polarizer located in the emission channel of the instrument (Fig. 12.5). Typical fluorophores used in FP are fluorescein and rhodamines. BODIPY dyes have longer excited-state lifetimes than fluorescein and rhodamine dyes, making their fluorescence polarization sensitive to binding interactions over a larger molecular weight range [4]. A limitation of current fluorescence polarization immunoassays (FPIs) is that they are useful only for measurement of low molecular weight antigens. This limitation is the result of the use of fluorophores, such as fluorescein, which display lifetime of about 4 ns. An FPI requires that the emission from the unbound labeled antigen be depolarized, so that an increase in polarization may be observed upon binding to antibody. For depolarization to occur the antigen must display a rotational correlation time much shorter than 4 ns, which limits the FPI to antigens with molecular weight less than several thousand Daltons. A class of dyes that have been shown to combine long lifetime and high polarization are the metal-ligand complexes (MLC) of Ru, Os and Re (Fig. 12.6) [5]. These labels have lifetimes in the range of a few hundred ns to microseconds and would therefore allow measurement of higher molecular weight antigens but the strong propeller effect of the MLC when labeled to proteins other than HSA has limited their use as labels for high molecular weight analytes [3].

Fluorescence polarization measurements have been used in analytical and clinical chemistry [6] and as a biophysical research tool for studying membrane lipid mobility [7], domain motions in proteins, and interactions at the molecular level [8]. Fluorescence polarization based immunoassays are also extensively utilized for clinical diagnostics [Y and Potter JM.1992]. FP has the advantage that it requires only one labeled species for the assay (unlike energy-transfer based read outs that require two labeled species) and thus FP has become a very popular read out format for HTS [9, 10]. Many of these assays are based on the use of antibodies that provide the specificity needed to selectively detect a wide variety of antigens. An example for a homogeneous binding assay based on FP is shown in Fig. 12.7. Any material that enables a mass increase of the labeled species can replace antibodies. In the IMAP assay™ (Molecular Devices) the high affinity of trivalent metal-ions to phosphate is utilized to generate the FP read-out [11].

12.2 Fluorophore and Their Types and Examples

Fluorescent label or probe is a molecule that is attached chemically to aid in the labelling and detection of biomolecules such as protein, antibody, or amino acid. Generally, fluorescent tagging, or labelling, uses a reactive derivative of a fluorescent molecule known as a fluorophore. The fluorophore selectively binds to a specific region or functional group on the target molecule and can be attached chemically or biologically. Various labelling techniques such as enzymatic labelling, protein labelling and genetic labelling are widely utilized. In this protein are mostly derivatives of maleimide, alkyl halide, idoacetamide, sulfonyl chloride, succinimidyl ester and isothiocyanate (Fig. 12.8). Two types of fluors are used in fluorescence analysis of macromolecules-intrinsic fluors (contained in the macromolecules themselves) and extrinsic fluors (added to the system, usually binding to one of the components) (Figs. 12.9 and 12.10).

12.2.1 Intrinsic Fluors

There are only three types of intrinsic fluors namely, tryptophan, tyrosine and phenylalanine (listed in order of quantum yield, from larger to smaller). The

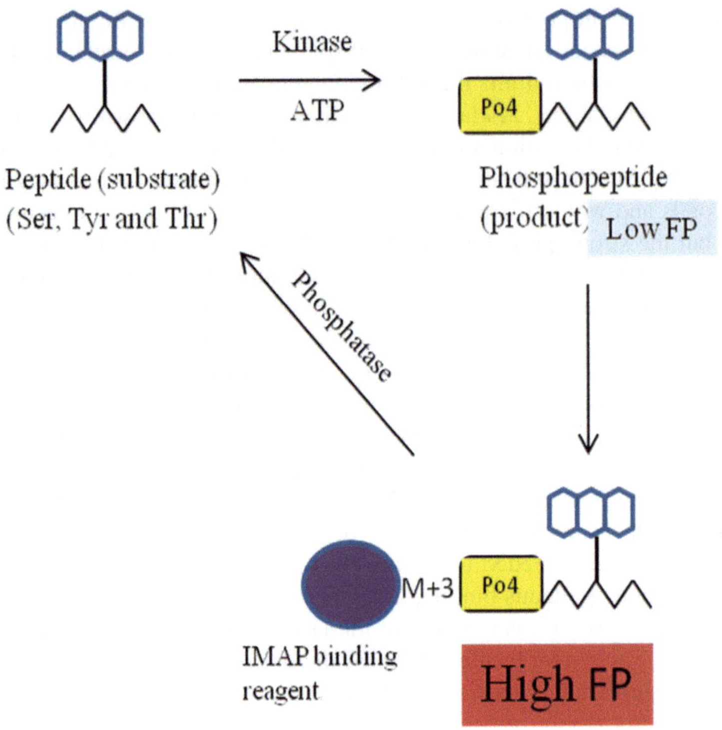

Fig. 12.7 IMAPTM homogeneous binding assay for kinases and phosphatases

fluorescence of each of them can be distinguished by exciting with and observing at the appropriate wavelength. In practice, tryptophan fluorescence is most commonly studied because phenylalanine has a very low Q and tyrosine fluorescence is frequently very weak due to quenching. The fluorescence of tyrosine is almost totally quenched if it is ionized, or near an amino group, a carboxyl group, or a Tryptophan. In special situations, however, it can be detected by excitation at 280 nm. The principal reason for studying the intrinsic fluorescence of proteins is to obtain information about conformation. This is possible because the fluorescence of both tryptophan and tyrosine depends significantly on their environment (solvent, pH and presence of quencher, small molecules, or a neighbouring group in the protein). Enzyme cofactors NADH (oxido-reductases) Ex/Em 340/460 nm and FAD (metabolic enzymes) Ex/Em 450 nm/540, Porphyrins Ex/Em 550 nm/620 nm, (Heme) Fe^{2+}, myoglobin, haemoglobin, Mg^{2+} chlorophylls cytochromes b and c, cytochrome P450 and cytochrome oxidase are also known as natural fluorphores.

The use of measurement of intrinsic fluorescence in protein is based on empirical "rules" obtained from studies of model compounds whose structure and conformation are well known. The rules are mentioned below [12]:

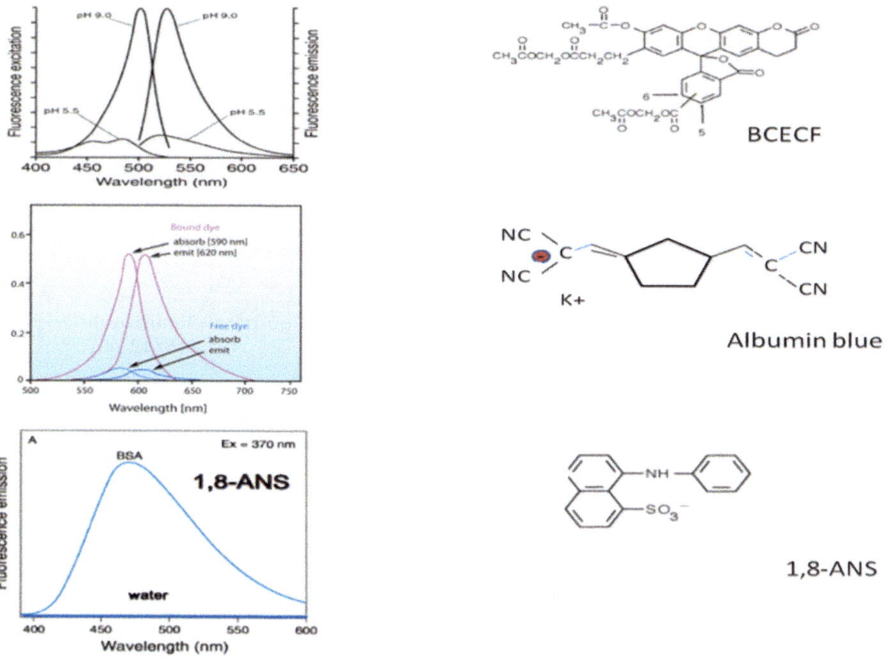

Fig. 12.8 Examples of fluorescence probes with their emission spectras

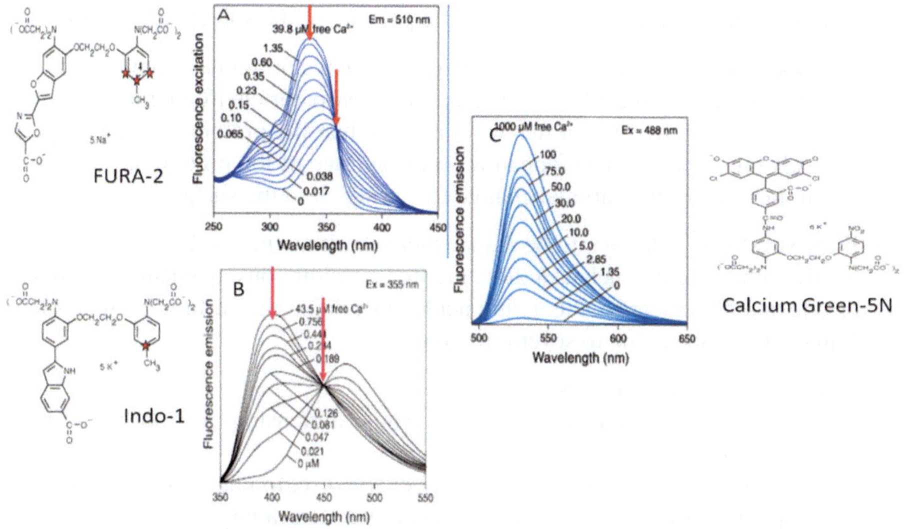

Fig. 12.9 Calcium probes: (**a**) FURA-2, (**b**) Indo-1 ratiometric fluorescence emission spectra and (**c**) calcium Green-5 N non-ratiometric fluorescence emission spectra

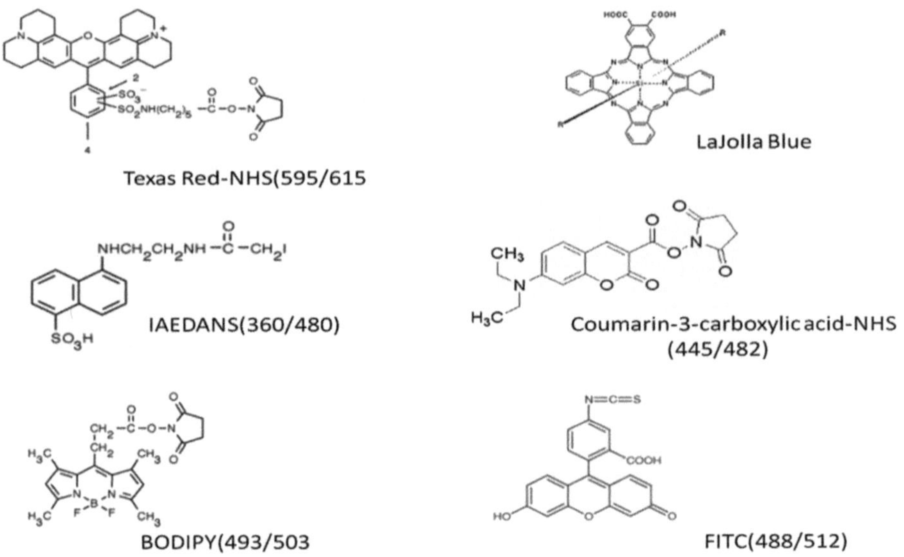

Fig. 12.10 Examples of Extrinsic labels

(a) All fluorescence of a protein is due to tryptophan, tyrosine and phenylalanine unless the protein is known to contain another fluorescent component.

(b) The λ_{max} of the tryptophan fluorescence spectrum shifts to shorter wavelengths and the intensity of λ_{max} increases as the polarity of the solvent decreases.

   (i) If λ_{max} is shifted to shorter wavelength when the protein is in a polar solvent, the tryptophan must be internal and in a non-polar environment.*

   (ii) If λ_{max} is shifted to shorter wavelength when the protein is in a non-polar solvent, either the tryptophan is on the surface of the protein or the solvent induces a conformational change that brings it to the surface.*

(c) If the substance is known to be a quencher (i.e., it quenches the fluorescence of the free amino acid), such as iodide, nitrate or cesium ions, quenches tryptophan or tyrosine fluorescence, the amino acids must be on the surface of the protein. It may fail to do so due to several reasons:

   (i) The amino acid may be internal.

   (ii) The amino acid may be in a crevice whose dimensions are too small for the quencher to enter it.

   (iii) The amino acid may be in a highly charged region and the charges might repel the quencher. For example, the iodide ion (a negative quencher) fails to quench tryptophan fluorescence if the tryptophan is in a negative region; the Cs^+ ion is ineffective if the fluor is in a positive region. The neutral quencher, acrylamide, disregards the charges.

(d) If a substance that does not affect the quantum yield of the free amino acid but affects the fluorescence of a protein, it must be due to production of a conformational change in the protein.
(e) If tryptophan and tyrosine are in the polar environment, their quantum yield, Q decreases with increasing temperature, T, whereas, in a nonpolar environment, there is little change. Hence, deviation from the monotonic decrease in Q with increasing T indicates that heating induced conformational changes as the polarity of the region to which the tryptophans are being exposed must be changing. An increases with T of the temperature dependence of Q when the protein is in a polar solvents such as water indicates that more tryptophan molecules are being exposed to the solvents- that is, the protein is unfolding.
(f) The Q for the tryptophan and tyrosine are decreased if α-carboxyl groups of these amino acids is protonated. In order to be protonated during the titration, the amino acid must be accessible to the solvent, that is, it must be on the surface or in a crevice.
(g) Tryptophan fluorescence is quenched by neighbouring protonated acidic groups. Hence, if the pK measured by monitoring tryptophan fluorescence is the same as the pK for a known ionisable group (e.g., the imidazole of histidine or the SH bond of cysteine), then that group must be very near to a tryptophan. This rule applies only if it can be shown independently that the pH changes do not introduce any conformational change.
(h) If a substance binds to a protein and tryptophan fluorescence is quenched, either there is a gross conformational change as a result of binding or some tryptophan is in or very near the binding site. Furthermore, because a decrease in the polarity of the solvent cause shifts in λ_{max} to shorter wavelength, such a shift associated with binding indicates that water is excluded in the complex.
(i) If the absorption spectrum of a small molecules overlaps the emission spectrum of tryptophan and the distance is small, there quenching. Hence if the binding of molecules to a protein quenches tryptophan fluorescence, tryptophan must be in or near the binding site.

NOTE: Two important points must be made about these rules: (1) fluorescence is so sensitive to environmental factors that other interpretations must always be sought for the results described in these rules; (2) if a protein contains several tryptophans, as is usually the case, each may have a different quantum yield. Therefore, the absolute magnitude of changes cannot be used to determine the fraction in a given environment-for example, internal versus external.

*A shift of λ_{max} in macromolecules is, in standard terminology, in relation to λ_{max} of the free amino acid in water.

12.2.2 Extrinsic Fluor

Nature does not always supply the investigator with a fluorescent group in the appropriate place in macromolecules. However, in many cases, a fluor can be introduced into the molecules to be studied either by chemical coupling or by simple binding (as in the use of reporter groups in absorption spectroscopy) [13]. The use of such added molecules in fluorescence analysis is called the method of extrinsic fluorescence. Several requirements of the fluor must be met for using extrinsic fluorescence: (i) The fluor must be tightly bound at a unique location, (ii) Its fluorescence should be sensitive to environmental condition and (iii) It should not itself affect the features of the macromolecules being investigated. These three criteria must always be verified.

For protein, (Table 12.1) the most common extrinsic fluors are 1-anilino-8-naphthalene sulfonate (ANS); 1-dimethylaminonaphthalene-5-sulfonate (DNS) and its chlorinated derivative, dansyl chloride; 2-p-toluidylnaphthalene-6-sulfonate (TNS); rhodamine; fluorescein; and the isothiocyanates of rhodamine and fluorescein. For nucleic acid, various acridines (acridine orange, proflavin, acriflavin), SYBR green II, Hoechst 33,258 and ethidium bromide are used. ANS, DNS and TNS have the valuable property that in aqueous solution they fluoresce very weakly. However, in a polar environment, Q increases markedly and the spectrum shifts towards shorter wavelength, both effects increasing as polarity decreases. Studies with model compounds have provided a rough relation between Q and the degree of nonpolarity. Dansyl chloride and the isothiocyanates have the useful property of reacting with specific amino acids in proteins. Hence, these substances are widely used to detect nonpolar region in protein because, on binding to such a region, their fluorescence increases [12].

12.2.3 Classes of Probes, Dyes and Labels

Extrinsic fluor has different types of probes, dyes and labels. Fluorescent probes interact with molecules through non covalent interaction within a specific region whereas fluorescent labels interact covalently with molecules and emits fluorescence. Different classes of extrinsic fluors are: Metal-Ligand Complexes, Organic Dyes, Quantum Dots or Nanoparticles, Fluorescent Beads or Polymers and Fluorescent proteins. 1, 8-ANS strongly enhances its barely colour fluorescence in hydrophobic environments. It is one of the valuable probes used for studying membranes and protein. Albumin blue is used for quantitative determination of albumin levels in biological samples including serum and urine. BCECF (2′,7′-Bis-(2-Carboxyethyl)-5-(and-6-)-carboxyfluorescein, is widely used as membrane permeant fluorescent indicator for intracellular pH. Fluorescence measurement using chelators and ion probes provide superior sensitivity and is well established for the determination of the extra- and intracellular concentration of ions. Although, the tracking of calcium ions has a long tradition but interest in other ions has also increased, especially, zinc ions have seeked more attention. Fluorescence ion probe have been developed for a wide range for the ions: cations: H^+, Pb^{2+}, Zn^{2+}, Mg^{+2}, K^+, Na^+, Li^{2+}, Ca^{+2} and others, for anions Cl^-, PO_4^{-2}, citrate and ATP. There are so many probes used for measuring calcium ion concentration: For

Table 12.1 Overview of Extrinsic fluorescence dyes applied for protein characterization [14]

Dye	Excitation (nm)	Emission maximum (nm)	Extinction coefficient (m^{-1} cm^{-1})	Application
ANS	350–380	505[a]	5000 (350 nm, water)	Surtace hydrophobicity unfolding/folding aggregation conformation
			4950 (350 nm, water)	
Bis-ANS	385–400	515[a]	16,790 (350 nm, water)	Surtace hydro-phobicity unfolding/folding aggregation conformation
Nile Red	540–580	660[a]	19,600 (552 nm, DMSO)	Surtace hydrophobicity unfolding/folding aggregation conformation
Thioflavin T	450	480–490[b]	36,000 (412 nm, Water)	Fibrillation
DCVJ	450	480–505	659,000 (453 nm, ethanol)	Micro-viscosity of protein environment and rigidity
Congo Red	550		45,000 (498 nm, water)	Fibrillation
			59,300 (505 nm, ethanol)	

[a]In water; blue shift in the hydrophobic environment
[b]In presence of the amyloids

ratiometric methods which are based on the use of a ratio between two fluorescence intensities; correction of artifacts due to bleaching, changes in focus, variation in laser intensity, etc. can be performed. The Ca+2 probes which are most commonly used in ratiometric methods include FURA (Fura-2, Fura-4F, Fura-5F, Fura-6F, Fura-FF) and INDO (Indo-1, indo-5F). In case of non ratiometric analysis, when a molecule binds to calcium, indicators show a shift in their fluorescence intensity (the free indicator has usually a very weak fluorescence). Here, cytosolic free calcium concentration is related with fluorescence intensity and for this FLUO(Fluo-3, Fluo-4, Fluo-5F, Fluo-5 N, Fluo-4 N), RHOD(Rhod-2, Rhod-FF, Rhod-5 N), calcium green, calcium orange, calcium crimson and Oregon Green 488-BAPTA are used as probes to measure fluorescence.

Fluorescent pH probes that are useful tools for the measurement of pH changes within the cell are listed in the Table 12.2. They provide improved sensitivity and spatial sampling as compared to the application of microelectrodes. Changes in intracellular pH values play a role in the investigation of diverse physiological processes like muscle contraction, endocytosis, cell proliferation, apoptosis, and ion transport (Table 12.3).

Table 12.2 Different pH probes with their fluorescence excitation ratio

pH probe	pH range	Measurement mode
HPTS(pyranine)	7.0–8.0	Exc. Ratio 450/405 nm
Oregon Green dyes	4.2–5.7	Exc. Ratio 510/450 nm
SNARF indicators	6.0–8.0	Exc. Ratio 580/640 nm
Fluorescein and carboxyfluoresceins	6.0–7.2	Exc. Ratio 490/450 nm
LysoSensor Yellow/Blue DND-160	3.5–6.0	Exc. Ratio 450/510 nm
2′,7′-Bis-(2-Carboxyethyl)-5-(and-6-)-carboxyfluorescein	6.5–7.5	Exc. Ratio 490/440 nm

Table 12.3 Fluorescence methods and their Application with extrinsic fluorescence dyes for protein characterization [14]

Method	Information	Application with non-covalent extrinsic dyes
Steady state fluorescence	Spectral information (emission spectrum and fluorescence intensity)	Detection of protein structural changes by dye-protein interactions
Time-resolved fluorescence	Fluorescence lifetime	Detection of protein structural changes by dye-protein interactions
Anisotropy (steady-state and time-resolved)	Rotational motions	Study of rotational dynamics, Determination of size of dye-protein complexes
Fluorescence correlation spectroscopy(FCS)	Translational motions/diffusion	Determination of size of dye-protein complexes
Fluorescence microscopy	Visualization of particles	Determination of size and morphology of large aggregates, fibrils, etc.

Organic dyes are also used as fluor molecules which bind to molecules and fluoresce. There are five classes of organic dye: Xanthenes {Fluoresceins (pH sensitive and not photostable), Rhodamines (tendency to aggregate in aqueous solution), coumarins (low extinction coefficients), BOPIDYs (non-polar and insoluble in water) [4], Phthalocyanines (photostable but difficult to synthesize) and Cyanines (high extinction coefficients).

12.3 Fluorescence Based Study on the Conformation of Protein

Several useful reviews and monographs have summarized the spectral properties of proteins. Proteins contain three amino-acid residues that contribute to their ultraviolet fluorescence which are usually described by their three- or one letter abbreviations. These are tyrosine (Tyr, Y), tryptophan (Trp, W), and phenylalanine (Phe, F). Trp fluorescence is more sensitive to the environment as compared to Tyr and Phe due to its large redistribution of electron density in the asymmetric indole ring of the Trp residue after the excitation by the photons, while practically no redistribution occurs in Tyr and Phe symmetric rings. The dependence of intrinsic protein fluorescence on the unique location of tryptophan residues was elaborated by the study of the intrinsic fluorescence of model compounds and proteins in different structural states. The combined analysis of the characteristics of protein intrinsic fluorescence and protein three-dimensional structure with the peculiarities of the location of tryptophan residues in protein structure was performed for the first time on azurin [6]. Emission of proteins is dominated by tryptophan, which absorbs at the longest wavelength and displays the largest extinction coefficient. Energy absorbed by phenylalanine and tyrosine is often transferred to the tryptophan residues in the same protein. Phenylalanine displays the shortest absorption and emission wavelengths. Phenylalanine displays a structured emission with a maximum near 282 nm. The emission of tyrosine in water occurs at 303 nm and is relatively insensitive to solvent polarity. The emission maximum of tryptophan in water occurs near 350 nm and is highly dependent upon polarity and/or local environment. Indole is sensitive to both general solvent effects. Indole displays a substantial spectral shift upon forming a hydrogen bond to the imino nitrogen, which is a specific solvent effect. Additionally, indole can be quenched by several amino-acid side chains. As a result, the emission of each tryptophan residue in a protein depends on the details of its surrounding environment. Protein fluorescence is generally excited at the absorption maximum near 280 nm or at longer wavelengths. Consequently, phenylalanine is not excited in most experiments. Furthermore, the quantum yield of phenylalanine in proteins is small, typically near 0.03 so, emission from this residue is rarely observed for proteins. The absorption of proteins at 280 nm is due to both tyrosine and tryptophan residues. At 23 °C in neutral aqueous solution the quantum yields of tyrosine and tryptophan are near 0.14 and 0.13, respectively, the reported values being somewhat variable [15]. At wavelengths longer than 295 nm, the absorption is primarily due to tryptophan. Tryptophan fluorescence can be selectively excited at 295–305 nm. This is why many papers report the use of 295 nm excitation, which is used to avoid excitation of tyrosine. Tyrosine is often regarded as a rather simple fluorophore. However, under some circumstances tyrosine can also display complex spectral properties. It can undergo excited state ionization, resulting in the loss of the proton on the aromatic hydroxyl group. In the ground state the pKa of this hydroxyl group is about 10 while in the excited state the pKa decreases to about 4. In neutral solution the hydroxyl group can dissociate during the lifetime of the excited state, leading to quenching of the tyrosine fluorescence. Tyrosinate is weakly fluorescent at 350 nm, which can be confused with tryptophan fluorescence. Tyrosinate emission is observable in some proteins, but it appears that excited state ionization is not a major decay pathway for tyrosine in proteins [16]. Because of their spectral properties, resonance energy transfer can occur from phenylalanine to tyrosine to tryptophan. Energy transfer has been repeatedly observed in many proteins and is one reason for the minor contribution of phenylalanine and tyrosine to the emission of most proteins. Also, blue-shifted tryptophan residues can transfer the excitation to longer wavelength tryptophan residues. The anisotropies displayed by Tyr and Trp are sensitive to both overall rotational diffusion of proteins, and the extent of segmental motion during the excited-state lifetimes. Intrinsic fluorescence of Trp is a powerful tool for investigating the structure, conformation (because we can follow the local conformational changes around Trp), dynamics and folding-unfolding of proteins. This is due to high sensitivity of various parameters of the fluorescence of tryptophan residues (spectrum position, quantum yield, anisotropy, etc.) to their microenvironment and to the peculiarities of their location in the protein macromolecules [17].

In the study, the significance of selective excitation of surface Trp of protein by 305 nm in the place of 280 nm that is used to monitor the total tryptophan. A number of approaches have been used to develop a method for the selective monitoring of "surface" tryptophans from the "buried" tryptophans in a protein using selective excitation of surface tryptophan at 305 nm. It was also observed that the quantum yield obtained at 305 nm excitation is about 10 times lower than that obtained by using 280 nm. It is also noted that excitation of Trp using 280 nm showed fluorescence at 332 nm while excitation at 305 nm showed the fluorescence at 340 nm. Vashist et al. [18] used excitation at 305 nm to monitor the surface Trp of a bacterial outer membrane protein, OmpAb. They performed iodide quenching experiment using excitation at both 305 nm as well as 280 nm and revealed that there are more fluorescence quenching in the case of 305 excitation, which further confirm that fluorescence after excitation at 305 nm comes only from the surface Trp. Spectroscopy result was further validated by N-Bromosuccinimide (NBS) oxidation methods that are used to calculate the number of Trp residue in the protein. NBS oxidation of protein (OmpAb) also suggests that about half of the total Trps are present on the molecular surface of protein (OmpAb), which is available for oxidation by NBS. It was also confirmed that excitation at 305 nm leads to the fluorescence of the "surface tryptophan" of β-lactamase. Effect of pH and temperature on the conformation of protein is examined by selective excitation of surface tryptophan of protein using 305 nm light. This is important for studying the protein conformation by monitoring the surface tryptophans of the protein. The results of the comparison of the fluorescence of surface Trp and total Trp at different pH (Fig. 12.11a, b) as well as temperature (Fig. 12.11c, d) by selective excitation at 305 nm (plot 2) and 280 nm (plot 1) respectively, are shown in Fig. 12.11. Similar profile of the fluorescence emission intensity as a function of pH was observed when the tryptophans were excited by 280 nm (plot 1, Fig. 12.11a), and 305 nm (plot 2, Fig. 12.11a). The result shows that pH below 4, the Asp and Glu residue get protonated that causes such characteristic fluorescence changes as seen in the Fig. 12.11a. This result also suggests that the side chains of neighbouring Asp and Glu residues influence the same tryptophan residues during excitation at 280 nm and 305 nm. Bioinformatics data highlighted that modeled OXA-51 have shown to have four (positions 102, 112, 218 and 220) out of seven Trps on the surface of protein [19] that appear to emit fluorescence when excited at 305 nm. Therefore, it can be concluded that it is only surface Trp that is responsible for change in fluorescence

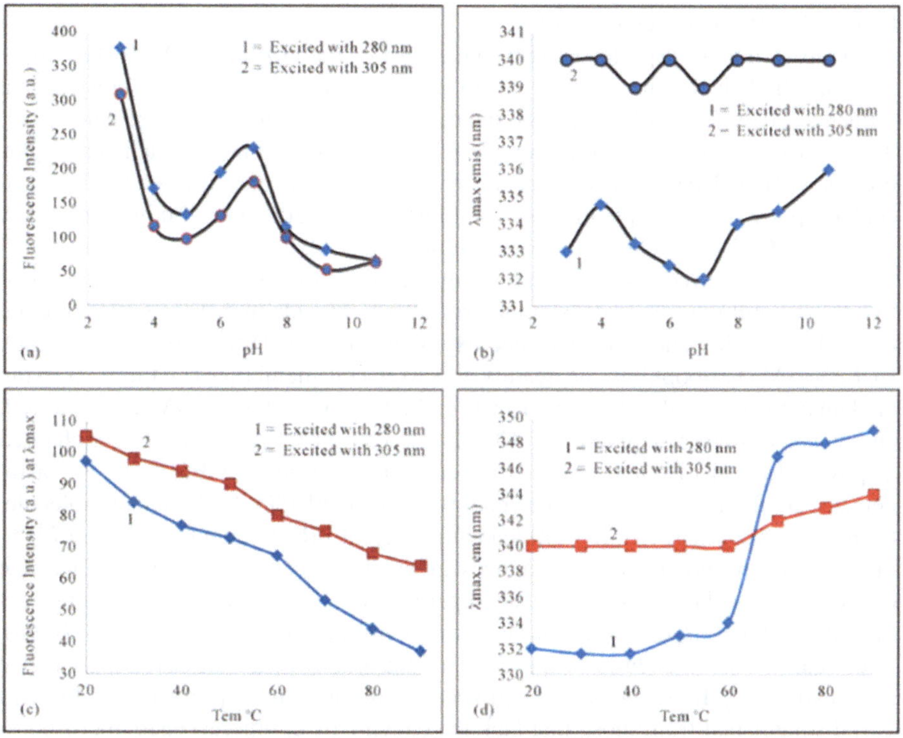

Fig. 12.11 Comparative display of the change in fluorescence intensity and fluorescence wavelength of the surface and total tryptophan by selective excitation at 205 nm and 280 nm respectively at different pH ((**a**) and (**b**)) and temperature ((**c**) and (**d**))

at different pH. Interestingly, when excited by 305 nm, Trp fluorescence emission maximum (340 nm) slightly change throughout the pH (plot 2 Fig. 12.11b) while when excited at 280 nm it showed a large shift (plot 1 Fig. 12.11b). This explains that some interior Trp move from core to the surface of protein that causes an increase during excitation at 280 nm. These results again support the fact that the fluorescence after excitation at 305 nm comes from surface Trps in the protein. Surface Trp residues were also monitored at different temperature after excitation at 305 nm and 280 nm (Fig. 12.11c, d). It is found that λ at 340 nm is marginally influenced with the increase in the temperature till 60 °C (plot 2, Fig. 12.11d) while influence of λ at 332 nm starts above 40 °C (plot 1, Fig. 12.11d). There is increase (3 nm) in the λ at 340 nm above 60 °C and decrease in the intensity of fluorescence above 60 °C (plot 2, Fig. 12.11c). This is due to the movement of interior tryptophan to the surface and their contribution in the fluorescence of surface Trp. Hence, it can be concluded that selective excitation of Trp at 305 nm may be emerged as important approach to explore the effect of surrounding environmental such as pH, ionic strength on the fluorescence of surface Trp which will significantly improve the study of bio-molecular structure and dynamics of the protein. Although limited information is available regarding selective excitation of surface Trp of protein using light with 305 nm wavelength. [20].

The β-lactamase reporter system makes use of the enzyme TEM-1β-lactamase (BLA), which lacks the N-terminal 23 amino acid periplasmic secretory signal sequence. BLA is a 29 kDa enzyme encoded by the ampicillin resistance gene (*amp*) that is active either as a monomer or when fused N or C terminally to a heterologous protein [12] and can cleave β-lactam-containing molecules like penicillins and cephalosporins with simple kinetics and high catalytic efficiency. Fluorogenic BLA substrates, CCF4, is relatively more soluble than CCF2 in aqueous solutions and is therefore preferred over CCF2 for some applications. CCF2/4 is composed of two fluorescent dyes, 7-hydroxycoumarin-3-carboxamide and fluorescein, bridged by cephalosporin. The esterified (acetoxymethylated; AM) form of this substrate, CCF2/4-AM, is lipophilic and readily traverses cell membranes without stressing or damaging live cells. As shown in Fig. 12.12, once CCF2/4-AM enters a cell, endogenous esterases convert it to CCF2/4. Because CCF2/4 is negatively charged at physicological pH, it becomes trapped inside the cell. CCF2/4 can be easily detected by fluorescence resonance energy transfer (FRET). In the intact molecule, excitation of coumarin at 408 nm leads to efficient FRET (quantum yield 0.74) to the fluorescein derivative and produces green fluorescence (λ530 nm). Cleavage of CCF2/4 by β lactamase, however, separates the two fluorophores, causing loss of FRET; excitation of coumarin now results in blue fluorescence ($\lambda = 460$ nm). Thus, based on the change in the color of the fluorescence emission signal, live cells expressing BLA can be unequivocally, simply, and rapidly (in as little as 1 h) distinguished from those that lack the enzyme by epifluorescence microscopy, fluorescent plate reader, or flow cytometry [21].

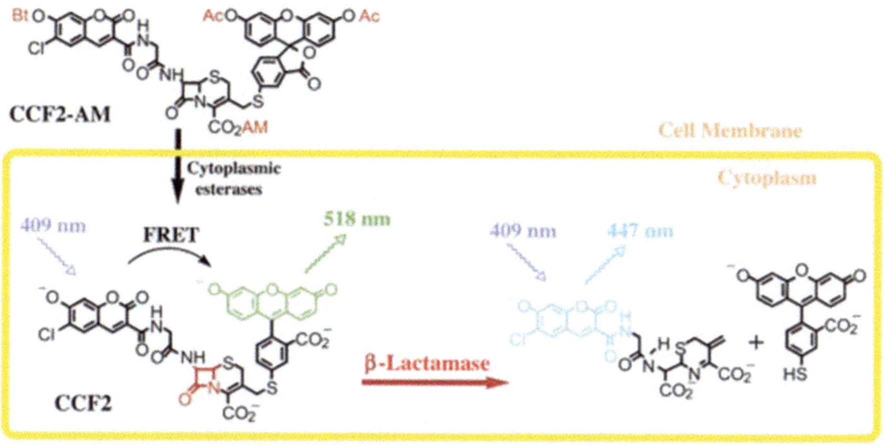

Fig. 12.12 Illustration of the β-lactamase reporter system. After an esterified (acetoxymethylated; AM) form of this substrate CCF2/4 (CCF2/4-AM) enters a cell, endogenous esterases convert it to CCF2/4, thereby trapping it inside the cell. Exciting CCF2/4 at 408 nm leads to efficient FRET from the coumarin moiety to the fluorescein derivative and produces green fluorescence detectable at 530 nm. After β-lactamase cleaves CCF2/4, the two fluorophores separate, causing loss of fluorescence resonance energy transfer (FRET); excitation at 408 nm now results in blue fluorescence detectable at 460 nm [21]

Fluorescence quantum yield of RNase A, at pH 7 and at 25 °C, was determined by comparison to a parent compound N-acetyl-tyrosinamide (NAYA), whose quantum yield was in turn determined using a standard solution of p-terphenyl in cyclohexane (0.77).

The absorption and emission spectra of RNaseA at pH 7.2 and 23 °C (Fig. 12.13) show maximum absorption at $\lambda_{max} = 278$ nm and maximum emission at $\lambda_{em} = 306$ nm. The absorption spectrum is 1 nm blue-shifted with respect to NAYA in the nonpolar solvent dioxane ($\lambda_{max} = 279$ nm) and 3 nm red shifted with respect to NAYA in water ($\lambda_{max} = 275$ nm) [23]. The maximum fluorescence emission of NAYA varies between $\lambda_{max} = 303$ nm in dioxane and $\lambda_{em} = 305$ nm in water; hence, RNase A shows a slight red shift in comparison to NAYA in water. The fluorescence quantum yield of RNaseA at 23 °C ($\Phi F = 0.013 \pm 0.001$), determined using NAYA in water as a standard, is in agreement with the value ($\Phi F = 0.014$) previously determined by Cowgill [24]. Compared to NAYA ($\Phi F = 0.049$), the quantum yield of RNaseA indicates that some of the tyrosine residues in RNaseA are quenched.

Another protein barnase is an extracellular protein that is often used as a model for protein folding. It is relatively a small protein: 110 amino acids, 124 KDa. The wild type protein contains three tryptophan residues, one of which is located close to histidine. pH dependent intensity of the tryptophan of wild type burnase exhibit excitation at 295 nm and emission at 340 nm. The emission intensity of wild type barnase, which contains all three tryptophan residues, increase nearly twofold as the pH is increased from 7 to 8.5. The increase in fluorescence is due to the presence of Trp94 and the histidine in the close proximity [25].

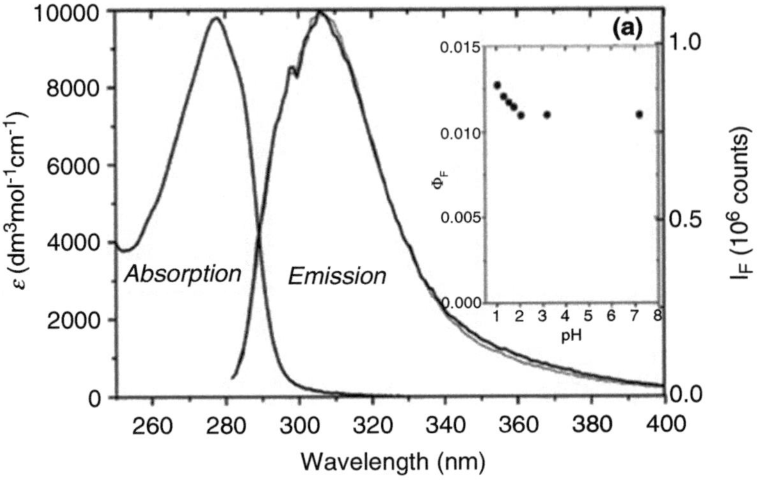

Fig. 12.13 Absorption (λ_{max} abs = 278 nm) and emission (λmax = 306 nm) spectra of RNase A in 10 mM sodium phosphate, pH 7.2 at 23 °C. The emission spectra of RNase A at pH 2.2 (*darker line*) at 23 °C show no change in the fluorescence intensity from pH 7 to pH 2.2. (*Inset*) Fluorescence quantum yield of RNase A in phosphate buffer at 23 °C from pH 7.2 to pH 1 [22]

12.4 Fluorescence Based Study on the Conformation of Nucleic Acid

Though nucleic acids have high absorbance at 260 nm, they do not emit fluorescence. The purine and pyrimidine bases present in nucleic acids absorb radiation but dissipate energy by various radiation less transitions. Therefore, to study the DNA/RNA structure, one needs to attach an extrinsic fluorophore to the nucleic acid. Some of these molecules either bind to DNA minor groove (Hoechst 33,258 for DNA and SYBR-GREEN II for RNA) or intercalate between base pairs of the helical form of nucleic acid (ethidium bromide, acridine orange). DNA/RNA quantification is routinely done by using different fluorescent probes like Hoechst 33,258 dye and ethidium bromide (for DNA), SYBR-Green II (for RNA). These dyes show a linear relationship of their fluorescence with subsequent increase in nucleic acid concentration. This method of nucleic acid quantification is very popular because it is rapid, highly sensitive accurate, and less expensive. Structural studies of DNA/RNA are also done by following the fluorescence of dye/ligand which is a flourophore and binds to DNA. For example, DNA-durg binding assay is followed for adraimycin binding to DNA (from the promoter region of c-met oncogene) [13].

Triplex-forming oligonucleotides (TFOs) are oligonucleotides that can associate with duplex DNA in the major groove through Hoogsteen hydrogen bonding between nucleobases. Currently, TFOs with natural nucleobases can bind only to duplex DNA with oligopurine or oligopyrimidine sequences, generating three types of triple helices: (i) a TC triplex, in which the third strand is parallel to the polypurine strand of the DNA duplex in the Hoogsteen conformation, forming TA*T and CG*C+ triplets; (ii) a GA triplex, in which the third strand is antiparallel to the polypurine strand in reverse Hoogsteen conformation, forming CG*G and TA*A triplets; and (iii) a GT triplex, in which the third strand is either antiparallel to the polypurine strand in reverse Hoogsteen conformation, forming CG*G and TA*T triplets, or parallel to the polypurine strand in Hoogsteen conformation, forming CG*G and TA*T triplets. The association between TFOs and duplex DNA is sequence specific, making TFOs good ligands for DNA targeting without the need for denaturing the double-stranded DNA. Recently, TFOs have been applied in several biomedical research studies such as transcription inhibition, sequence-specific cleavage, cross-linking, and chromosome mapping. TFOs have also been explored for their therapeutic use. In addition to gene targeting, TFOs have the potential to serve as sequence-specific probes for detecting DNA *in vitro*.

Modification of the backbones or the sugars, modification of the 3′ or 5′ termini, and insertion of intercalating nucleic acids in TFOs improve the affinity to their target duplex and may benefit the association with unfavourable sequences. *In vitro* selection with SELEX (systematic evolution of ligands by exponential enrichment) has been used to identify the most stable TFO from a library of oligonucleotides, but this method has yet to be applied to DNA duplexes of arbitrary sequences. Several researchers have used FRET and melting analysis to monitor the formation of triple helices. This approach uses a pair of fluorophores, a donor and an acceptor, to label the target duplex and its TFO, respectively. The formation of triple helix brings the two fluorophores close to each other, allowing FRET to occur. Darby and co-workers [26] reported a modified design that labels a fluorophore and a quencher on the oligonucleotide that can fold into duplex, triplex, and quadruplex structures. Different structures result in different fluorescence melting profiles that can be measured by a real-time polymerase chain reaction (RT-PCR) instrument such as a Roche LightCycler 480. Measuring the specific emission from FRET between the fluorophore pair on the TFO and the target duplex allows sensitive, realtime, and high-throughput detection of triplex formation. However, this format is not suitable for *in vitro* detection of DNA because in most situations the target DNA is unlabeled. Here we demonstrate a novel FRET design to detect the formation of triple helices without prelabeling the target duplex. This design uses a fluorophore-labeled TFO probe that undergoes FRET with a double-stranded DNA intercalating dye (the concept is depicted in Fig. 12.14). Thus, the association and dissociation of the TFO probe can be monitored. The main concept of using a TFO as a detection probe without prelabeling the target duplex is based on FRET between the fluorophore on the TFO probe and the fluorescent dye (e.g., SYBR green I) intercalated into the target duplex, as shown in Fig. 12.14. In this study, the TFO probes were labeled at their 3′ ends with a ROX fluorescent dye, which acted as boths a FRET acceptor and a 3′ blocker (to prevent chain elongation). The SYBR green I dye intercalated into target duplex acted as a FRET donor. Thus, formation of a triple helix could be detected by exciting SYBR green I and measuring the specific

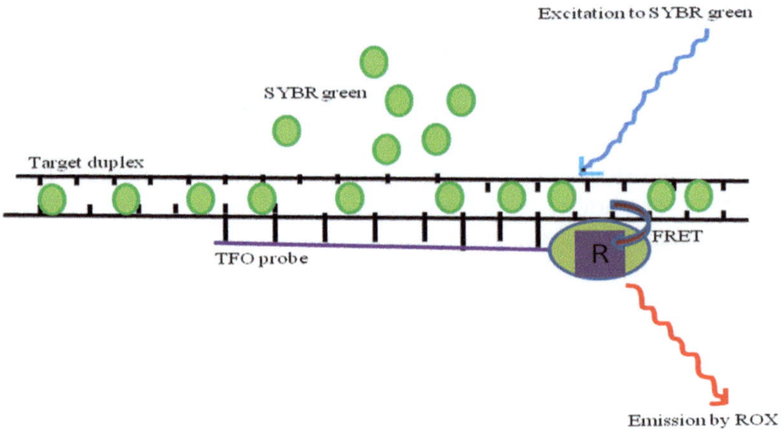

Fig. 12.14 Schematic diagram of the TFO probe is labeled with ROX(R), which undergoes FRET with SYBR green 1 intercalating into the target DNA duplex [27]

emission of ROX. We synthesized a serial of target duplexes with an oligopurine - oligopyrimidine sequence and two ROX-labeled TFO probes: one with a GA-rich sequence (GA–TFO) and the other with a TC-rich sequence (TC–TFO). The target duplexes were expected to form triple helices with the GA–TFO probe through reverse Hoogsteen hydrogen bonding or with the TC–TFO probe through Hoogsteen hydrogen bonding. Analysis of triplex status was performed on a LightCycler 480. Two sets of filter combinations on the machine were used in this study: excitation at 483 nm and detection at 533 nm (channel 533) for measuring SYBR green I fluorescence and excitation at 483 nm and detection at 610 nm (channel 610) for measuring ROX emission generated by FRET.

FRET not only is used to study G- and C- quadruplexes but is also used to analyse large divergence between the fluorescence properties of the folded and unfolded forms such as the study of duplex–quadruplex equilibria, conformational heterogeneity, quadruplex kinetics, and protein–DNA interactions. They can also be used as potassium sensors, aptamers, nanodevices or molecular beacons. Fluorescent dye-oligonucleotide conjugates have became the preferred method for detection of nucleic acids in molecular biology. Some common examples include automated fluorescent DNA sequencing, fluorescent genotyping, and quantitative target detection techniques (e.g., molecular beacons, etc.). Recent works have also reported that the application of fluorescent oligonucleotides have been extended to the field of ion-sensing by developing the oligonucleotide-based molecular sensor for potassium ion (potassium sensing oligonucleotide; PSO). The PSO is a FRET -based probe consisting of two fluorophores (tetramethylrhodamine, TAMRA and fluorescein, FAM) attached to the both termini of the 21-mer oligonucleotide with a human telomeric DNA sequence, d((GGGTTA) GGG). This oligonucleotide is able to form stable G-quadruplex (G-4) complexes with metal cations. The following observations were made for PSO systems with different metal cations:

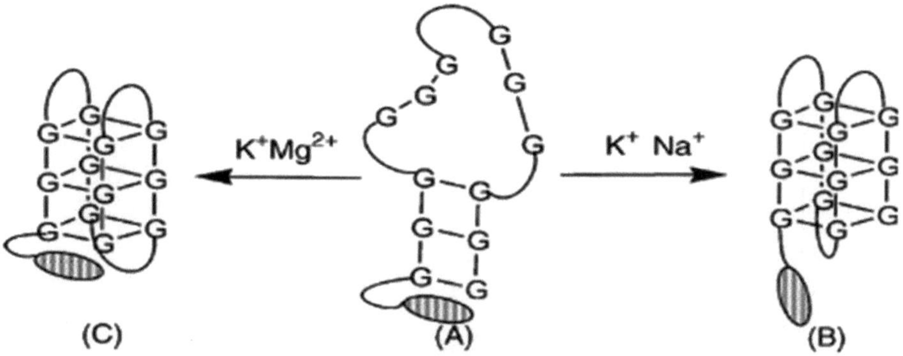

Fig. 12.15 Schematic representation of G-quadruplex structures of 21-TAMRA, showing possible interactions of TAMRA moiety with terminal guanine planes. A hairpin type structure (**a**) (without cation) enables guanine to quench TAMRA fluorescence Na (I) and partially K(I) ions form fluorescent intramolecular antiparallel "basket-type" G-quadraplexes (**b**) With a diagonal loop that prevents the dye from quenching interactions with terminals guanines. The chair type structures (**c**) (Mg^{+2} and partially K^{+}) allows TAMRA moiety to remain associated with the end of quadruplex and to be quenched [28]

(i) FRET signal (sensitized fluorescence of TAMRA) appeared only for K(I) and Na(I) complexes, (ii) much lower concentration of potassium ion was required to 3 generate FRET signal, (iii) the intensity of sensitized fluorescence of TAMRA finally attained was larger for Na^{+} than that for K^{+} ions and (iv) multivalent metal cations (Mg^{2+}) caused only quenching of the donor fluorescence (FAM) without sensitization of the TAMRA emission (Fig. 12.15). Variations in FRET efficiency for particular cation/PSO complexes are surprising and probably reflect differences in structures of quadruplexes as well as can result from specific effects of metal cations on spectral properties of the donor-acceptor pair. It is well known that spectral properties of fluorescent dyes are affected by their environment including solvent, pH, added salt and conjugation to nucleic acids. For example, fluorescent properties of fluorescein change upon conjugation and are seriously influenced by pH, contrary to TAMRA emission that is less sensitive to environmental effects resulting from conjugation. Additionally, fluorescence may be affected by the sequence and secondary structure of oligonucleotide. Depending on the peculiarity of the system, both fluorescence quenching and enhancement were reported. Therefore, the broad experimental data concerning FRET-labeled oligonucleotides are necessary in order to understand processes in these systems and to develop future FRET probes with predictable spectral performance. Such probes, based on the G-quadruplex structural motif, are potentially useful for sensing of specific proteins and cations. Systematic studies have been performed to assess the impact that complexation of metal cation by PSO has on the spectral properties of FRET donor (FAM) and acceptor (TAMRA). Three different fluorescent oligonucleotides were investigated: a dual-labeled PSO probe (Fig. 12.16) and two reference telomeric 21-mers, FAM-21 and 21-TAMRA, single-labeled at 5′- and 3′-end, respectively.

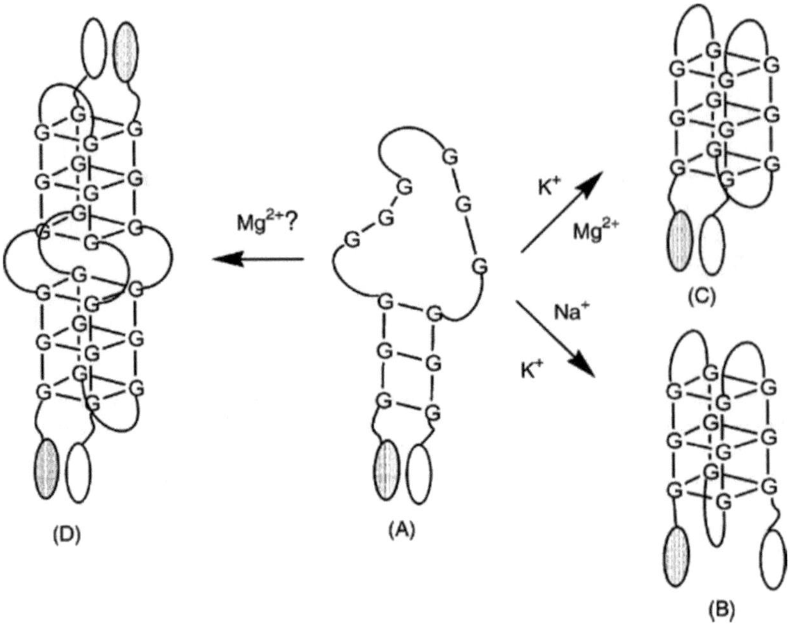

Fig. 12.16 Schematic representation of the formation of PSO quadruplexes in the presence of metal cations. The structures (**a**) (*hairpin-types*), (**c**) (*chair-types*) and (**d**) (*hairpin dimer*) are nonfluoresent or possess very low emission as result of direct stacking due to the short distance between the donor and the acceptor. The basket-type (**b**) structures shows efficient FRET, since both fluorophores are separated by a diagonal loop that prevents them from stacking interactions [28]

Steady state fluorescence and anisotropy measurements were carried out in the presence of selected cations (Et_4N^{++}, Li^+, Na^+, K^+, Mg^{2+}) circular dichroism (CD) spectra were recorded to distinguish between quadruplex structures and to follow the process of quadruplex formation.

Guanine-rich DNA and RNA sequences can form four-stranded helical structures called G-quadruplexes based on stacking of G-G-G-G tetrads. G-quadruplexes formed by natural DNA sequences in the telomeres and oncogenic promoters and by RNA sequences in the 5′ untranslated region (5′ UTR) of oncogenic transcripts have been established as attractive anticancer targets. On the other hand, engineered G-quadruplexes can have potential applications ranging from medicine to supramolecular chemistry and nanotechnology. G-quadruplex structures are highly polymorphic. They give rise to specific spectroscopic signatures in IR, UV-absorption thermal difference spectra (TDS), CD, Raman, and NMR spectroscopy. The intrinsic fluorescence yield of DNA is very low as it corresponds to quantum yields of the order of $Q \sim 10^{-4}$, and the detection of DNA usually relies on targeting DNA selectively with fluorescent dyes. This approach has also been used in search of selective quadruplex binders. In analogy to molecular beacons as fluorescent probes in DNA conformational analysis, G-quadruplex formation has been followed by pyrene excimer emission and FRET in the ensemble and at the single-molecule level. In both cases the fluorescent reporters were attached at the termini of DNA sequences. Recently, the detection of G-quadruplex formation using an internal guanine derivative as a fluorescent probe has been reported. However, all these strategies involving exogenous fluorophores might affect the G-quadruplex structure. In recent reports, it has been pointed out that DNA G-quadruplexes have higher intrinsic fluorescence quantum yields than their less-structured counterparts. This notion has been further substantiated by large amplitude and long lifetime components of excited states. The formation and structure of G-quadruplexes verified using NMR, UV absorption and CD spectroscopy reveals that the formation of G-quadruplexes as a function of cations and temperature can be followed by their fluorescence and fluorescence excitation spectra. These features are shown to discriminate, for example, between G-quadruplex and Z-DNA structures.

12.5 Fluorescence Based Study on Macromolecules Assembly (Membrane Assembly)

The discovery of amyloid fibrils as a major constituent of protein deposition in various tissues and correlation of their accumulation with various diseased states has led to intense study of the formation and structure of such fibrils, and their interaction with possible targets such as membranes. Two types of covalently attached fluorescent probes have been used in investigations of amyloidogenic proteins with membranes. The first type involves the attachment of a fluorophore, usually a derivative of naphthalene, fluorescein or rhodamine, to the protein, often via a specific Cys thiol residue, or less specific labeling of lysine residues. The covalently bound probes can then be used in a variety of types of fluorescent experiments, ranging from simple changes in the environment of the probe on fibrillation or interaction with a membrane to FRET studies [29]. The second type of covalent probe involves intracellular expression of a fusion protein containing the amyloidogenic polypeptide and a fluorescent tag, usually either GFP or a homolog or a tetra-Cys motif to bind a fluorescence arsenic compound e.g. FLASH. Membrane fluorescent probes are used for studying lateral diffusion, fluidity, positioning of lipids and proteins in the lipid bilayer, and surface potential of membranes. Lateral diffusion is usually studied by excimer-forming fluorophores such as pyrene. Excimer emission (due to dye–dye interaction) is red-shifted in comparison to emission of the monomeric dye. Fluorescein is another fluorophore, which can be used for studying lateral diffusion: the dimerization of fluorescein results in homo-FRET and quenching of fluorescence. The commonly employed probe 1,6-diphenyl-1,3,5-hexatriene (DPH) is localized in the lipid bilayer and serves as a sensor of the disordering of fatty acyl chains that occurs during the transition from gel to liquid crystalline phase. DPH also detects the perturbation of the fatty acyl chain packing upon insertion of proteins. Anthroylstearate is another

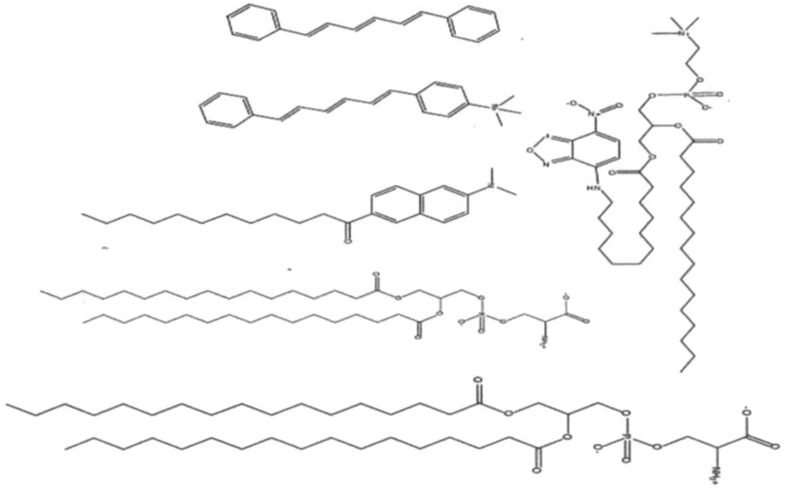

Fig. 12.17 Chemical structures and approximate location of the flurocent probes for commonly used fluorescent membrane probes: NBD-PC, DPH, laurdan and TMA-DPH in a model DPPS monolayer [30]

fluorescent probe detecting lipid dynamics at different depths into the lipid bilayer. The dansyl (dimethylaminonaphthlenesulfonyl) group and 2-N-(4-nitrobenzo-2-oxa-1,3-diazole (NBD) attached to the phospholipid head group are responsive to surface-solvent effects. NBD is sensitive to the environmental polarity: it is weakly fluorescent in water and strongly fluorescent in hydrophobic environments. Dansylated lipids and 2-(p-toluidinyl) naphthalene-6-sulfonate (TNS) have been used to study solvent relaxation processes in membranes. Figure 12.17 shows the structures of some commonly used fluorescent membrane probes.

Protein deposition diseases are characterized by accumulation of proteinaceous aggregates frequently composed of insoluble fibrils. All amyloid fibrils possess cross-beta structure and consist of one particular protein or peptide. A parallel arrangement of beta-sheets with β-strands perpendicular to the axis of the fibril appears to be the major structural feature of amyloid fibrils. However, the native structures of amyloidogenic proteins differ greatly and several amyloidogenic proteins are intrinsically disordered or contain highly flexible unfolded regions (e. g. A-beta, α-synuclein, the N-terminal domain of the prion protein, tau, the poly(Q) regions of huntingtin and the ataxins), whereas others are stably folded monomers, dimers (e. g. immunoglobulin light chains), tetramers (e. g. transthyretin), hexamers (e. g. insulin) or exist as a part of larger macromolecular complexes (e. g. beta-2 microglobulin of the major histocompatibility complex I, and apolipoprotein I and serum amyloid A, which are components of high density lipoproteins). Their folds are also different from each other. Insulin, lysozyme, apolipoproteins, serum amyloid A, the C-terminal of animal prions all contain several α-helices, whereas beta-2 microglobulin, immunoglobulin light chains, and transthyretin are folded into beta-sandwiches. Several amyloidogenic proteins are stabilized by intramolecular disulfide bonds. The nature of the structural elements guiding amyloid formation is poorly understood. Such structural analysis is important in order to understand the pathological function of amyloid and to find inhibitors that prevent the formation of toxic species or their interaction with targets such as cell membranes or the extracellular matrix. The discovery of transient oligomers of several amyloidogenic proteins poses a new challenge in order to understand their relation to amyloid fibrillogenesis and toxicity. The oligomeric species could be precursors of fibrils or off-pathway intermediates, and have been proposed to be cytotoxic. Time-resolved FRET measurements were used to study the structure and dynamics of α-synuclein in three different conditions: at physiological pH 7.4, acidic pH 4.4, and in the presence of SDS micelles. Tryptophan and 3-nitrotyrosine at six different locations in the protein were chosen as donor–acceptor pairs with a Förster critical distance of 26 Å. In acidic solutions, the C-terminal of α-synuclein becomes more compact, whereas in the presence of SDS micelles the N-terminal is more compact and the C-terminal is more extended. FRET was also used to analyze the conformational changes of α-synuclein during aggregation. The triple mutant Y125W, Y133F, Y136F of α-synuclein has one tyrosine–tryptophan donor–acceptor pair with a Förster critical distance of 15 Å. The calculated distance between Y39 and W125 in monomeric α-synuclein is 25 Å, which decreases during aggregation to 22 Å (early oligomeric species) and 19 Å (late oligomeric species). Alpha-synuclein binds preferentially to small unilamellar vesicles containing acidic phospholipids such as phosphatidate, phosphatidylserine, but not significantly to vesicles with a net neutral charge (e.g. phosphatidylcholine). Monounsaturated and polyunsaturated acyl groups in lipid vesicles increase the binding affinity of α-synuclein. The interaction of α-synuclein with membranes is complex and dependent on the composition of the membranes, their curvature, as well as the ratio of membrane lipids to protein. Far-UV CD and NMR data show that α-synuclein adopts α-helical conformation on binding to membranes. The N-terminal and central parts of the protein form a broken α-helix with a distorted twist at residues or two helices with a linker in the 39–44 region. A Thr residue is embedded into the hydrophobic core of the bilayer. It is possible that the hydroxyl group of Thr interacts with the carboxyl group of acyl chains through hydrogen bonding. A Gln residue occupies an interface position. In solution, the monomeric Y39W mutant of α-synuclein has a λ at 350 nm and, in the presence of lipid vesicles, the maximum is shifted to ~340 nm, indicating a change from hydrophilic to a significantly more hydrophobic environment. However, there is no corresponding change in λ max for the Y125W mutant. The previously studied data correlate with NMR results showing that only the N-terminal and central parts of α-synuclein interact with lipid vesicles, and are similar to those for single Trp mutants (F4W, Y39W, and F94W) in the presence of SDS micelles. If the binding to lipid vesicles and membranes is strong, the α-helix will be stable and the formation of fibrils or pre-fibrillar oligomers, will be slowed or prevented. Under *in vitro* conditions of a high ratio of protein to lipid, membranes can promote the self-association of α-synuclein and accelerate fibrillation. The mechanism of this aggregation is still unknown. However in the majority of the conditions, the interaction with lipid vesicles, especially synthetic acidic phospholipids such as phosphatididate and phosphatidylserine, leads to the inhibition of α-synuclein self-association and prevents aggregation. The addition of monomeric α-synuclein to synaptic vesicles or synthetic vesicles composed of brain sphingomyelin, DPPC, or POPC/POPG increases the anisotropy of DHP, a fluorescent membrane probe sensitive to the ordering of fatty acid acyl chains. α-Synuclein not only stabilizes the packing of the acyl chains but also increases the gel–liquid crystalline transition temperature. The stabilization of the outer region of the lipid bilayer has been also detected with the fluorescent dye Laurdan. Laurdan is a polarity-sensitive, amphiphilic fluorophore, positioned 5 Å beneath the lipid–water interface. An increase in the intensity of Laurdan fluorescence and a decrease in the generalized polarization correspond to penetration of water molecules into the lipid bilayer. In the presence of α-synuclein Laurdan fluorescence decreases, and the general polarization increases, indicating stabilization of the upper region of the bilayer, and decreased water penetration.

Prions are infectious protein particles causing several human and animal diseases including scrapie in sheep, bovine spongiform encephalopathy in cattle and Creutzfeldt–Jakob disease in humans. In these transmissible diseases, the cellular form of the prion protein (PrP scrapie isoform (PrPSc) is converted into the infectious). The normal function of PrP is unknown; however, the protein may be involved in copper and zinc metabolism, cell signaling, synaptic transmission, and cellular resistance to oxidative stress. PrPC is a GPI anchored membrane protein, abundantly expressed in neurons and other cell types. The protein has three structurally distinct regions; the N terminal part (residues 23–90) is unfolded and consist of four repeats (in humans) of PHGGGWGQ involved in copper binding, whereas the C-terminal part (121–231) consist of two motifs: a beta–helix–beta and a helix–turn–helix fold stabilized by a disulfide bond. The flexible middle part (91–120) is important for fibril formation and undergoes a transition to beta-sheet conformation during fibrillogenesis [30].

12.6 Other Fluorescence-Based Methods and Their Significacnes

12.6.1 Fluorescence Resonance Energy Transfer

The technique of fluorescence resonance energy transfer is exploited to determine the Försten's distance. A fluorophore has a characteristic λ_{max} in both its absorbance and emission spectra. FRET occurs when the emission spectra of one fluorophore (A) overlaps with the absorbance λ_{max} of a second fluorophore (B). If the fluorophores (extrinsic or intrinsic) have unique locations within the protein or complex, it is possible for emission light energy from A to be absorbed by B and to be emitted as part of B's emission spectrum. This phenomenon is called Fluorescence

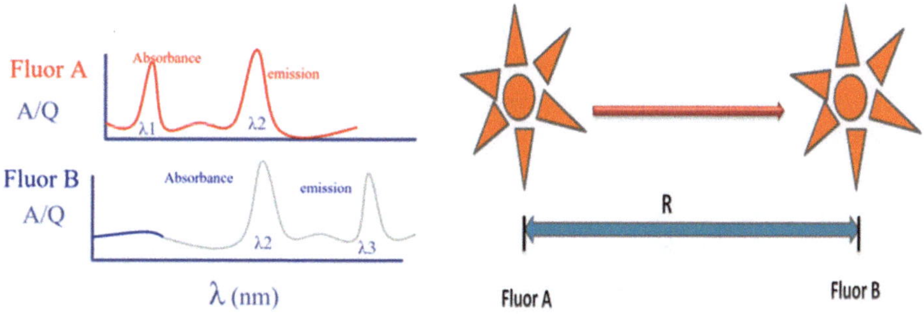

Fig. 12.18 Efficiency of the energy transfer (E) from donor to acceptor

Resonance Energy Transfer or FRET. It is dependent on the distance, R, between the two fluors. The chromophore, B need not necessarily be part of the molecules but, has to be in the close proximity (protein, membranes and macromolecular assemblies) of within distance in 50 Å apart. The importance of the energy-transfer phenomenon to biochemistry is that the efficiency of transfer is a function of the separation of the fluors and can therefore be used to measure molecular distance [31]. The efficiency of the energy transfer (E) from donor to acceptor (A to B) (Fig. 12.18) is defined by

$$E = \frac{\mathrm{Ro}^6}{(Ro^6 + R^6)}$$

R = distance between A and B, Ro is a constant calculated from absorption and emission spectra (Fig. 12.18).

E can be calculated from fluorescence intensity (F)

$$\mathrm{E = Fda/Fd}$$

Fda = F in the presence of acceptor, Fd = F in the absence of acceptor.

Once E is known, R can be calculated from the first equation if Ro is known

$$R = Ro\left[\frac{1-E}{E}\right]^{1/6}$$

Generally Ro is in the range of 10–50 Å and determined upto a distance of 80 Å. For example, in a polypeptide which is attached to a D and A on both N and C terminals, for examples Dansyl $(\mathrm{pro})_n$-NH-NHCONH-Naphyl. Determination of the distance between donor and acceptor in a peptide R, gives information about the conformation of the polypeptide. For proteins, dansyl chloride is used as environmental probe and ethidium bromide for nucleic acids.

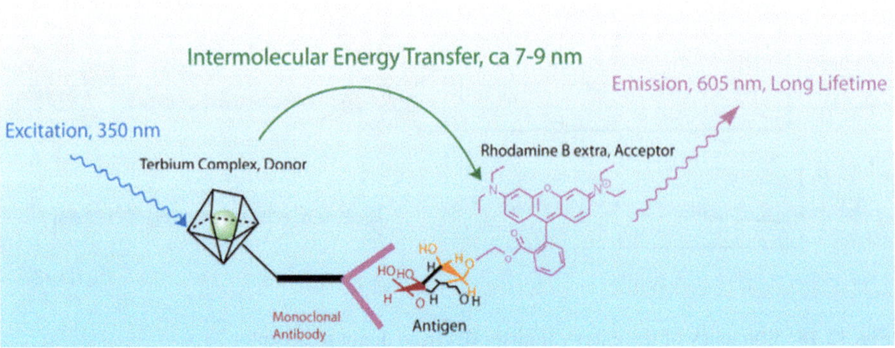

Fig. 12.19 Diagrammatic representation of HTRF principle

12.6.1.1 Uses of FRET

- Probe the topology of membrane proteins which can't be done by X-ray or NMR.
- Study the movement in contractile proteins e.g. actin & myosin.
- Determine changes upon receptor ligand binding, monitoring the formation of tetraplex & duplex DNA formation & mapping of active sites.
- Homogenous time resolved immunoassay (HTRF) in durg discovery and durg discovery (Fig. 12.19).

12.6.1.2 Limitations

- Limited range (10–80 Å) if R is > 80 Å. E is very small
- 2 fluorophores are required, a donor & an acceptor. Trp is a good donor.
- Assumptions are made in the calculation of R

12.6.1.3 Significance

- Quite fast & simple compared to 3-D structure determination.
- Tyr to Trp transfer explains why we don't see Tyr fluorescence in proteins.
- Spatial distribution and assembly of protein complexes
- Receptor/ligand interactions
- Probing interactions of single molecules
- Structural and conformational changes of nucleic acids
- Detection of nucleic acid hybridization
- Primer-extension assays for detecting mutations
- Automated DNA sequencing
- Distribution and transport of lipids

12.6.2 Congo Red Binding and Birefringence

Congo red staining is commonly used for the identification of amyloid in tissue samples. An alkaline solution of Congo red stains amyloid with an intense red colour, which leads to a shift in the wavelength of maximal absorption from 490 nm to 540 nm upon binding to amyloid samples. This characteristic helps in the spectroscopic detection of amyloid and measurement of fibril formation. Binding of Congo red to amyloid yields a unique blue– green birefringence under cross-polarized light, enabling the visual determination of amyloid formation. This phenomenon was first observed by the Belgian physician Paul Divry in 1927, while studying degeneration of the ageing brain. It is hypothesized that the Congo red dye intercalates between β-strands parallel to the peptide chains and perpendicular to the fibril direction. Binding also occurs through interaction between the dye's negatively charged sulfonate groups with the positive N-terminals of polypeptides, in such a way that the dye's axis is perpendicular to the length of the peptide and parallel to the direction of the fibril axis. High salt concentrations and alkaline pH are required for binding and birefringence [33].

12.6.3 Thioflavin T and Thioflavin S Fluorescence

Thioflavin T has both polar and non-polar functional groups, which form micelles in aqueous solutions with the non-polar dimethylaminophenyl groups in the core and the charged benzothiazole groups exposed to the solvent. Thioflavin T micelles bind to β-sheet structures; resulting in relatively intense fluorescence signals compared to unbound thioflavin T. Vassar and Culling (1959) introduced the use of this dye in 1959, for detecting amyloid tissue in the kidney. In the unbound state, excitation and emission occur at 342 nm and 430 nm, respectively, while when bound to amyloid samples; the excitation and emission of thioflavin T undergo a bathochromic shift to 444 nm and 482 nm, respectively. Thioflavin T has been used to characterize amyloid aggregates *in vitro* and *ex vivo*, whereas uncharged thioflavin derivatives have been used to detect amyloid *in vivo*. This dye can also be used as a staining agent for fluorescence microscopy and confocal microscopy of amyloid proteins [32].

12.6.4 Steady-State and Time-Resolved Spectroscopy

Steady-state measurements are important in the analysis of changes in the intensity, excitation and emission maxima, and polarization of fluorophores. However, steady-state spectroscopy is inadequate in the analysis of heterogeneous systems with multiple states of the fluorophore. In addition, intensity measurements are highly dependent on the fluorophore concentration and, in cases where small changes have to be detected; the experimental error can exceed the level of subtle changes. Time-resolved (lifetime) fluorescence spectroscopy allows estimating the lifetime of the fluorophore, which is not dependent on concentration thus providing more accurate data. Lifetime measurements are used in anisotropy, energy transfer, and quenching experiments [33]. The fluorescence decay analysis based on the discrete multiexponential or lifetime distribution model permits to estimate the number of different microenvironments of the fluorophore. The binding and stabilization of the fluorophore usually results in an increase in the lifetime, whereas quenching decreases the fluorescence lifetime [30].

12.6.4.1 Significance

- Identify or separate species by their fluorescence lifetime
- Monitor changes in the environment
- Study protein folding or signaling pathways
- Detect singlet oxygen for photodynamic therapy
- Study membrane rigidity or enzyme/substrate interactions

12.6.5 Fluorescence Correlation Spectroscopy

Fluorescence correlation spectroscopy (FCS) is based on correlation analysis of fluctuations of the fluorescence intensity. It offers insights into the photophysics that cause these characteristic fluorescence intensity fluctuations as well as diffusion behaviour and absolute concentrations of detected particles. Eventually, important biochemical parameters as the concentration and size or shape of the particle (molecule) or viscosity of the environment can be determined. Most experiments employ an autocorrelation function. The theory of FCS has been developed and applied to various fields of fluorescence spectroscopy and microscopy [34]. It is especially valuable in studies where a significant background affects the direct measurement of fluorescence changes. FCS also employs a cross-correlation function, which allows comparison of the changes in two fluorophores residing in either one or two molecules. The analysis of FCS correlation curves allows estimation of the diffusion coefficients and rates of fluorophore transitions [35]. FCS can be applied to study the self assembly of macromolecules such as fibril formation or the interaction of proteins with membranes [30].

12.6.5.1 Significance

- Measure diffusion rates that are typically expressed as the diffusion coefficient of the molecule or molecular complex.

- Detect and distinguish between single and multiple component diffusion.
- Investigation of lateral and rotational diffusion of fluorophores as well as conformational dynamics providing various informations like hydrodynamic radii and singlet- triplet dynamics.
- Provides a direct and independent, calibration-free measure of molecular concentration within a sample. Based on the concentration measurement and cross-correlation between two fluorophores (dual-color FCCS), data are used to detect molecular association and dissociation and to determine the stoichiometry of molecular complexes.
- Determination of kinetic rate constants, i.e. on and off kinetics of complex formation, as well as enzyme dynamics and intramolecular dynamics *in vitro* and in living cells.

12.6.6 Fluorescence Lifetime Correlation Spectroscopy (FLCS)

The fusion of Time-Correlated Single Photon Counting and Fluorescence Correlation Spectroscopy, called Fluorescence Lifetime Correlation Spectroscopy (FLCS), is a method that uses picosecond time-resolved fluorescence detection for separating different FCS-contributions. FLCS is of particular advantage when using spectrally inseperable fluorophores that differ in their lifetime for Fluorescence Cross-Correlation Spectroscopy (FCCS) because it offers elmination of spectral cross talk and background. It also offers a way around detector after pulsing artifacts. In FLCS, a separate autocorrelation function is calculated for each fluorophore component determined by its decay pattern, emitted, for example, by various species in the sample. The only assumption is that various emissions have different TCSPC (Time-Correlated Single Photon Counting) histograms (i.e., different fluorescence lifetimes), which is practically always the case. The core of the method is a statistical separation of different intensity contributions with similar lifetimes, performed on a single photon level.

12.6.7 Fluorescence Lifetime Imaging (FLIM)

Fluorescence Lifetime Imaging (FLIM) produces an image based on the differences in the excited state decay rate from a fluorescent sample. Thus, FLIM is a fluorescence imaging technique where the contrast is based on the lifetime of individual fluorophores rather than their emission spectra. The fluorescence lifetime is defined as the average time that a molecule remains in an excited state prior to returning to the ground state by emitting a photon. As the fluorescence lifetime does not depend on concentration, absorption by the sample, sample thickness, photobleaching and/or excitation intensity it is more robust than intensity based methods. At the same time, the fluorescence lifetime depends on different environmental parameters such as pH, ion or oxygen concentration, molecular binding or the proximity of energy acceptors making it the technique of choice for functional imaging of many kinds.

12.6.7.1 Significance

Local Environment Sensing

The fluorescence lifetime can change depending on the fluorophore environment (which means polarity, pH, temperature, ion concentration etc.) and is therefore used as a parameter for biological sensors. Specifically, fluorophores return to the ground state through radiative and non-radiative processes. Quenching of the excited state by external factors decreases the fluorescence lifetime. The resulting lifetime shortening provides information about the molecular environment of the fluorophore and even allows for quantitative distinction between subpopulations of quenched and unquenched fluorophores.

Detection of Molecular Interactions

A special case for the influence of the local environment on fluorescence lifetime is FRET, where the donor dye is quenched by the presence of an acceptor dye. Thus, a decrease of the fluorescence lifetime is indicative for FRET. In this way, the fluorescence lifetime serves as a sensor parameter for intra- and intermolecular interactions allowing for distance measurements in the nanometer range.

Detection of Conformational Changes

Applying an intramolecular labelling approach, the distance between the dye and the quencher or FRET acceptor can also vary along with different conformations of the labeled biomolecule. Hence intramolecular changes like folding or action of molecular motors are detectable.

Discrimination of Multiple Labels or Background Removal

With the advancement of fluorophores and microscopy, researchers now are able to use several fluorescent markers in parallel to assess different processes simultaneously. One challenge is that the employed fluorophores need to be distinguishable and have commonly used spectral characteristics. This limits the number of useful fluorescent markers. The analysis of the fluorescence lifetime can help to overcome these limitations. Additionally, the fluorescence lifetime enables to discriminate label fluorescence from the fluorescence background of the sample (e.g. cell or tissue) and thereby allows a higher detection efficiency and more accurate marker localization.

Tissue Characterization by Autofluorescence

The autofluorescence can be characteristic for a certain tissue and therefore be used e.g. for tumor detection. Typically, Two-Photon Excitation (TPE) is combined with Non-Descanned Detection (NDD) for deep tissue imaging, as these applications are generally more prevalent in tissues or organisms.

12.6.8 Fluorescence Microscopy Method

Fluorescence microscopy is an invaluable tool to study cellular processes *in vivo*. Recently, it has been extensively applied in single-molecule research including studies on protein–membrane interactions and macromolecular assembly [36]. New approaches include total internal reflection, confocal and two or multi-photon fluorescence microscopy.

12.6.8.1 Total Internal Reflection Fluorescent Microscopy (TIRFM)

Total internal reflection fluorescent microscopy (TIRFM) allows visualization of wide areas but relatively thin regions of the sample. It is based on the evanescent wave excitation of fluorophores located less than 200 nm from the glass–water interface. The evanescent wave is generated when the incident light is totally reflected from the glass water interface and its efficiency to penetrate into the sample decreases exponentially [30].

12.6.8.2 Time-Resolved Fluorescence Microscopy

Time-resolved fluorescence microscopy in the deep UV can be employed in microfluidic environments and enables label-free detection and identification of various aromatic analytes in chip electrophoresis.

12.6.8.3 Stimulated Emission Depletion Microscopy (STED)

Stimulated emission depletion microscopy (STED) is a fluorescence microscopy technique imaging below the optical diffraction limit that overcomes the diffraction limited resolution of confocal microscopes. The resolution enhancement is essentially based on switching off the fluorescence of dye molecules by stimulated emission using intense laser light in the outer regions of the diffraction limited excitation focus. This intense radiation causes almost all of the excited molecules to return to the ground state. Fluorescence from the remaining excited dye molecules in the center of the excitation focus is then detected and used to form the high resolution images. An even further resolution enhancement is possible by applying time gates to the collected data (gated STED or gSTED). As STED creates an effectively smaller observation volume, it can also be applied to other methods such as FCS.

Author Disclosure Statement The authors have declared that no competing interests exist.

References

1. Elmer P (2000) An introduction to Fluorescence Spectroscopy. PerkinElmer Ltd Post Office Lane Beaconsfield Buckinghamshire HP9 1QA
2. McGlynn SP (1966) Fluorescence and phosphorescence analysis. Principles and applications. J Am Chem Soc 88(23):5688–5688
3. Szmacinski H, Terpetschnig E, Lakowicz JR (1996) Synthesis and evaluation of Ru-complexes as anisotropy probes for protein hydrodynamics and immunoassays of high-molecular-weight antigens. Biophys Chem 62(1–3):109–120
4. Schade SZ, Jolley ME, Sarauer BJ, Simonson LG (1996) BODIPY-alpha-casein, a pH-independent protein substrate for protease assays using fluorescence polarization. Anal Biochem 243(1):1–7
5. Huang X (2003) Fluorescence polarization competition assay: the range of resolvable inhibitor potency is limited by the affinity of the fluorescent ligand. J Biomol Screen 8(1):34–38
6. Jameson DM, Seifried SE (1999) Quantification of protein-protein interactions using fluorescence polarization. Methods 19(2):222–233
7. Yu W, So PT, French T, Gratton E (1996) Fluorescence generalized polarization of cell membranes: a two-photon scanning microscopy approach. Biophys J 70(2):626–636
8. LeTilly V, Royer CA (1993) Fluorescence anisotropy assays implicate protein-protein interactions in regulating trp repressor DNA binding. Biochemistry 32(30):7753–7758
9. Kim J, Felts S, Llauger L, He H, Huezo H, Rosen N, Chiosis G (2004) Development of a fluorescence polarization assay for the molecular chaperone Hsp90. J Biomol Screen 9(5):375–381
10. Lu Z, Yin Z, James L, Syto R, Stafford JM, Koseoglu S, Mayhood T, Myers J, Windsor W, Kirschmeier P, Samatar AA, Malcolm B, Turek-Etienne TC, Kumar CC (2004) Development of a fluorescence polarization bead-based coupled assay to target different activity/conformation states of a protein kinase. J Biomol Screen 9(4):309–321
11. Zaman GJ, Garritsen A, de Boer T, van Boeckel CA (2003) Fluorescence assays for high-throughput screening of protein kinases. Comb Chem High Throughput Screen 6(4):313–320
12. David Freifelder WH (1982) Physical blochemistry, applications to biochemistry and molecular biology. W. H. Freeman and Company, San Francisco Second Edition
13. Rajeshwari MR (2013) An introduction to biophysics. Rastogi Publications, New Delhi
14. Hawe A, Sutter M, Jiskoot W (2008) Extrinsic fluorescent dyes as tools for protein characterization. Pharm Res 25(7):1487–1499

15. Ladokhin AS (2006) Fluorescence spectroscopy in peptide and protein analysis. In: Encyclopedia of analytical chemistry. Wiley, Chichester
16. Ladokhin AS, Jayasinghe S, White SH (2000) How to measure and analyze tryptophan fluorescence in membranes properly, and why bother? Anal Biochem 285(2):235–245
17. Alexander Ross JB, Laws WR, Rousslang KW, Wyssbrod HR (2002) Tyrosine fluorescence and phosphorescence from proteins and polypeptides. In: Lakowicz JR (ed) Topics in fluorescence spectroscopy. Springer US, Boston, pp 1–64
18. Vashist J, Tiwari V, Kapil A, Rajeswari MR (2010) Quantitative profiling and identification of outer membrane proteins of beta-lactam resistant strain of Acinetobacter baumannii. J Proteome Res 9(2):1121–1128
19. Tiwari V, Moganty RR (2014) Conformational stability of OXA-51 beta-lactamase explains its role in carbapenem resistance of Acinetobacter baumannii. J Biomol Struct Dyn 32(9):1406–1420
20. Tiwari V, Tiwari M (2015) Investigation of surface tryptophan of protein by selective excitation at 305 nm. J Biophys Chem 6:87–90
21. Qureshi SA (2007) Beta-lactamase: an ideal reporter system for monitoring gene expression in live eukaryotic cells. Biotechniques 42(1):91–96
22. Noronha M, Lima JC, Paci E, Santos H, Macanita AL (2007) Tracking local conformational changes of ribonuclease A using picosecond time-resolved fluorescence of the six tyrosine residues. Biophys J 92(12):4401–4414
23. Noronha M, Lima JC, Lamosa P, Santos H, Maycock C, Ventura R, Macxanita AL (2004) Intramolecular fluorescence quenching of tyrosine by the peptide carbonyl group revisited. J Phys Chem A 108:2155–2166
24. Cowgill RW (1964) Fluorescence and the structure of proteins. Iii. Effects of denaturation on fluorescence of insulin and ribonuclease. Arch Biochem Biophys 104:84–92
25. Loewenthal R, Sancho J, Fersht AR (1991) Fluorescence spectrum of barnase: contributions of three tryptophan residues and a histidine-related pH dependence. Biochemistry 30(27):6775–6779
26. Darby RA, Sollogoub M, McKeen C, Brown L, Risitano A, Brown N, Barton C, Brown T, Fox KR (2002) High throughput measurement of duplex, triplex and quadruplex melting curves using molecular beacons and a LightCycler. Nucleic Acids Res 30(9):e39
27. Chiou CC, Chen SW, Luo JD, Chien YT (2011) Monitoring triplex DNA formation with fluorescence resonance energy transfer between a fluorophore-labeled probe and intercalating dyes. Anal Biochem 416(1):1–7
28. Juskowiak B, Galezowska E, Zawadzka A, Gluszynska A, Takenaka S (2006) Fluorescence anisotropy and FRET studies of G-quadruplex formation in presence of different cations. Spectrochim Acta A Mol Biomol Spectrosc 64(4):835–843
29. Lakowicz JR (2006) Principles of fluorescence spectroscopy, 3rd edn. Springer US, New York
30. Munishkina LA, Fink AL (2007) Fluorescence as a method to reveal structures and membrane-interactions of amyloidogenic proteins. Biochim Biophys Acta 1768(8):1862–1885
31. Willy Wriggers, P. D. (n.d.)Introduction to spectroscopy and fluorescence. http://staff.bath.ac.uk/bssmdb/,http://biomachina.org/courses/structures/05.html
32. Tiwari V, Solanki V, Tiwari M (2015) In-vivo and in-vitro techniques used to investigate Alzheimer's disease. Frontiers in Life Science 8(4):332–347
33. Ed Lakowicz J (1992) Topics in fluorescence spectroscopy: biochemical applications, vol 3. Springer; Plenum Press, New York, pp. 1–63
34. Haustein E, Schwille P (2003) Ultrasensitive investigations of biological systems by fluorescence correlation spectroscopy. Methods 29(2):153–166
35. Bacia K, Schwille P (2003) A dynamic view of cellular processes by in vivo fluorescence auto- and cross-correlation spectroscopy. Methods 29(1):74–85
36. Michalet X, Weiss S, Jager M (2006) Single-molecule fluorescence studies of protein folding and conformational dynamics. Chem Rev 106(5):1785–1813

Index

C.D. Geddes (ed.), *Reviews in Fluorescence 2016*, Reviews in Fluorescence,
DOI 10.1007/978-3-319-48260-6

MIX
Papier aus verantwortungsvollen Quellen
Paper from responsible sources
FSC® C105338

If you have any concerns about our products,
you can contact us on
ProductSafety@springernature.com

In case Publisher is established outside the EU,
the EU authorized representative is:
Springer Nature Customer Service Center GmbH
Europaplatz 3, 69115 Heidelberg, Germany

Printed by Libri Plureos GmbH
in Hamburg, Germany